Descartes's Mathematical Thought

DESCARTES'S MATHEMATICAL THOUGHT

by

CHIKARA SASAKI

The University of Tokyo,
Japan

KLUWER ACADEMIC PUBLISHERS

DORDRECHT / BOSTON / LONDON

A C.I.P. Catalogue record for this book is available from the Library of Congress.

ISBN 1-4020-1746-4

Published by Kluwer Academic Publishers,
P.O. Box 17, 3300 AA Dordrecht, The Netherlands.

Sold and distributed in North, Central and South America
by Kluwer Academic Publishers,
101 Philip Drive, Norwell, MA 02061, U.S.A.

In all other countries, sold and distributed
by Kluwer Academic Publishers,
P.O. Box 322, 3300 AH Dordrecht, The Netherlands.

Printed on acid-free paper

Printed in the Netherlands.

To the Memory of

Prof. Thomas Samuel Kuhn

(July 18, 1922 – June 17, 1996)

My Teacher of the History and Philosophy of Science

Preface

The present book is a much revised and enlarged version of my dissertation, for which I received the degree of Doctor of Philosophy from Princeton University on January 21, 1989. The Japanese edition appeared in February of 2003 with the University of Tokyo Press.

I began to prepare for the dissertation in 1978 when I was a graduate student at Princeton. After having returned to Tokyo, I completed the first draft in the spring of 1987, the 350th anniversary of the publication of René Descartes's *Discours de la méthode*. Thus, I spent nine years on my Ph. D. dissertation. I publish it now after having kept it for much more than nine years, although I do not completely endorse the words of Horace beloved by Descartes: "If you write something, keep it back till the ninth year (*Si quid* [...] *scrispseris* [...], *nonumque prematur in annum* [...]. *De arte poetica, ll.* 386–388).

The Japanese version of this book is the most detailed, third part of a trilogy on "The Historical Theory of Modern Science." The first part is titled, *Kagakukakumei no Rekishi Kōzō* (The Historical Structure of Scientific Revolutions, published in Japanese in 2 volumes by Iwanami Shoten in 1985; 2nd paperback edition by Kodansha in 1995), and it treats the development of mathematical and scientific thought in Europe from the 17th to 20th centuries. The second part is titled, *Kindai Gakumon Rinen no Tanjō* (The Birth of Modern Scientific Ideals, published in Japanese by Iwanami Shoten in 1992 and awarded the Suntory Prize for Social Sciences and Humanities for the year 1993), and it treats scientific thought in 17th-century Europe. At the same time, the present book is the first volume of a series, "Collection: History of Mathematics," appearing with the University of Tokyo Press, for which I will serve as the series editor.

Few words seem required to justify a study about the formation of Descartes's mathematical thought and its philosophical significance, for historians of both mathematics and philosophy have almost unanimously assented to the importance of these topics. How Descartes arrived at his notions has been a conundrum. To reconstruct Descartes's early mathematical career, I have had not only to analyze his own mathematical writings up to the time of the publication of the *Géométrie* of 1637, which accompanied the *Discours de la méthode*, but also to examine the works of mathematicians who were his contemporaries, and whose works I suppose he studied. In particular, it has been my hope for many years to know whether or not he read a work (or works) of François Viète, the founder of modern algebraically analytic mathematics. To solve this problem seems to me to amount to revealing one of the 'secrets', as it were, of the birth of modern European mathematics. Part I is devoted to this key problem.

Part II considers Descartes's philosophy of mathematics. His theory of mathematics is contrasted in particular to that of Aristotle. I examine the philosophy of mathematics of Aristotle and Aristotelians, which Descartes discarded without pity. I treat this problem by examining both philologically and philosophically the evolution of the notion thought to correspond to Descartes's concept of 'mathesis universalis'. This procedure will help present Descartes's mathematical thought in its global historical perspective. It will be found that 'mathesis universalis' plays a crucial role in Parts I and II. In fact, it was the expression of the most fundamental mathematical discipline during the seventeenth century, and for Descartes a key notion which may have lead to his novel mathematics.

The present monograph may differ from writings by other historians of science. The author had been studying the classification theory of linear algebraic groups, following the late French mathematician Claude Chevalley and the Japanese mathematician Ichiro Satake and seeking a professional mathematical career, before shifting his field of study to the history of science. Thus, the present study is the product of an ex-practitioner of mathematics. In addition to this, the author is a product of non-Western culture. For the Japanese in general, European thought in the seventeenth century has been a great enigma. In contrast to Japan, which started to take a way of rapid modernization or Westernization in the late nineteenth century for an 'external' reason, in other words, 'by force', Western Europe took steps for modernization 'spontaneously', My trilogy on "The Historical Theory of Modern Science" as a whole offers a program to explore this question historically. It is my sincere hope to contribute to understanding a core part of modern Western thought equally with colleagues who have grown to maturity entirely within the cultural tradition of modern Europe.

The series "Collection: History of Mathematics," which consists of 5 volumes at the start, will publish research on the history of mathematics originating in Ph. D. dissertations on Isaac Newton, Gottfried Wilhelm Leibniz, and traditional Japanese mathematics which I have advised at the University of Tokyo since 1980, in addition to works of my own and by my colleague there during the years from 1994 to 1997, Professor Roshdi Rashed, an authority on the history of Arabic mathematics. At the present time at the University of Tokyo, I am studying the history of mathematics not only with excellent Japanese students, but also with young historians of mathematics overseas, mainly from China, and sometimes from Europe, in the Graduate School of Arts and Sciences. Moreover, I am regularly delivering lectures in the Graduate School of Mathematical Sciences on the historical development of mathematics to students who are expected to become professional mathematicians.

Mathematical culture, I believe, has existed and developed not simply by seeking original results in mathematics. The mountain of mathematical culture will be higher and more graceful when the historical and philosophical plains at the foot of the mountain become broader. The series "Collection: History of Mathematics" aims at heightening and broadening mathematical culture through elaborating its historical bases.

For drafting this monograph, I am indebted to many libraries in the United States of America, Europe, and Japan which have allowed me to use their facilities mainly through making photographic copies and microfilms, especially Columbia University Library, Harvard University Library, The Joseph Regenstein Library of the University of Chicago, The New York Public Library, Princeton University Library, The University of Michigan Library, The University of Southern California Library, Bibliothèque de la Sorbonne, Bibliothèque Nationale of Paris, Cambridge University Library, Glasgow University Library, Stadtbibliothek Ulm, Staat- und Stadtbibliothek Augusburg, Komaba Library and the Library of the Graduate School of Mathematical Sciences at the University of Tokyo, The Library of the Mathematical Institute of Tohoku University, and Tohoku University Library.

I would like to express my heartfelt thanks to the late Professor Koshiro Nakamura and to Professor Kokiti Hara, both of whom pioneered in Japan the study of mathematics of seventeenth-century Europe.

I am deeply grateful to Professor Michael S. Mahoney, my academic advisor at Princeton University, whose uncanny but warm eye has always been encouraging the present study. It was his brilliant Ph. D. dissertation, titled, "The Royal Road: The Development of Algebraic Analysis from 1550 to 1650, with Special Reference to the Work of Pierre de Fermat," written in 1967 under the guidance of the late Professor Thomas S. Kuhn, that contributed to my having chosen the history of mathematics as profession. No words can express my deep gratitude for his helpful criticisms of the draft of this book when it was in the form of a dissertation. Professor Charles C. Gillispie's exceptional kindness and encouragement bolstered me during my Princeton years. The late Professor Kuhn's penetrating scholarship was instrumental in redirecting my interests toward the history of science. I was honored to have spent pleasant times in Tokyo and Nikko with the three professors and their wives when they visited Japan in 1999, 1981, and 1986, respectively.

When I was studying the history of mathematics at Princeton for four years during the late 1970s, these three professors and Professor Gerald L. Geison, a leading historian of medicine known for his study of Louis Pasteur, constituted the faculty of the Program in History and Philosophy of Science (now the Program in History of Science). According to a remark of an observer, Princeton's program at that time was in a 'golden age'. Professor Kuhn passed away in the early summer of 1996 after his long-time struggle with cancer, and Professor Geison died a sudden and premature death in the summuer of 2001. Where have those great times gone, as the French poet François Villon wonders: "Où sont les neiges d'antan?" It was my resolution at the time of returning from Princeton to Tokyo that I would bring the academic ethos of Princeton's program to Japan. Reflecting on the situation of our discipline at the beginning of the 21st century, historians of mathematics in East Asia are proceeding to the forefront of the historical study of mathematics. It is not too much to say that Tokyo is one of the international centers in the study of history of mathematics. Over the past quarter century, the field of history of mathematics which underwent a complete change was Arabic mathematics, due mainly to Professor Rashed. My dream is to heighten the quality of research on the history of mathematics in China and Japan to the level that Professor Rashed has achieved for the history of Arabic mathematics.

A close colleague of my generation in the history of science, Professor Lewis Pyenson of the University of Louisiana, was very helpful in achieving the present English edition. I wish him to know how grateful I am. Mr. Charles Erkelens, editor of Kluwer Academic Publishers, and Ms. Jolanda Voogd, publishing assistant, provided great assistance. I appreciate the hospitality they showed when I visited Dordrecht in February of 2003.

It was Professor Kuhn who was encouraging me to publish my study on Descartes in book form and who introduced appropriate publishers immediately after it was accepted as a Ph. D. dissertation. An incomparable thinker who played a pivotal role in reformulating our notion of science, he made an enduring impression on me as a teacher of the history and philosophy of science. I dedicate the present book to his memory.

May 2, 2003, Yokohama

Chikara SASAKI

Contents

LIST OF ABBREVIATIONS

The following abbreviations are used in the notes of this book.

AM: Descartes, *Correspondance*, publiée par Charles Adam et Gérard Milhaud, 8 Vols. (Vols. 1–2, Paris: Félix Alcan, Vols. 3–8, Paris: Presses Universitaires de France, 1936–1963; Reprinted Edition, Nendeln/Liechtenstein: Kraus Reprint, 1970–1983).

AT: *Œuvres de Descartes*, éditées par Charles Adam et Paul Tannery, 11 Vols. (Paris: Cerf, 1897–1913; New Edition, Paris: Vrin, 1969–1975).

CM: *Correspondance du P. Marin Mersenne*, commencée par Mme Paul Tannery et éditée par Cornelis de Waard et Armand Beaulieu, 16 Vols. (Vols. 1, 2, and 4, Paris: PUF, Vols. 3 and 5–16, Paris: CNRS, 1945–1986).

DSB: *Dictionary of Scientific Biography*, edited by Charles Coulston Gillispie, 16 Vols. (New York: Charles Scribner's Sons, 1970–1980).

Olscamp: Descartes, *Discourse on Method, Optics, Geometry and Meteorology*, translated with Introduction by Paul J. Olscamp (Indianapolis: Bobbs-Merrill, 1965).

* * *

A NOTE ON THE QUOTATION AND TRANSLATION

In most cases, the original material is quoted in footnotes. Therefore, French words without *accents*, for example, are following the custom of the time of their composition.

As to the translation of the works of Descartes, for *Rules for the Direction of the Mind, Discourse on the Method, Principles of Philosophy, and Meditations of First Philosophy* (including Objections and Replies), and his letters, I have used with some alterations *The Philosophical Writings of Descartes*, translated by John Cottingham, Robert Stoothoff, Dugald Murdoch, and Anthony Kenny, 3 Vols. (Cambridge: Cambridge University Press, 1984–1991). This translation indicates the pagination of *AT*. Unless indicated, all other translations are mine.

INTRODUCTION

René Descartes and Modern European Mathematics

To debate the mathematical thought of René Descartes is to debate modern European mathematics as a whole. The central role Descartes played in forming the tradition of modern European mathematics can be well understood from the facts that the style of contemporary mathematics is basically algebraic and that the degree of mathematization of a science is taken to be a mark of the degree to which it has become a true science. Few contemporary thinkers would question this characterization of mathematics and science. Yet we must not forget that the Cartesian vision of the world is the product of particular circumstances, which figure prominently in the following pages.

Descartes is known not only as one of the founders of algebraically analytic mathematics but also as a thinker who attempted to establish a fundamentally new philosophy attaching particular importance to pure mathematics and to the mathematization of physics. Combining these two ideas we may say that what he accomplished, or, more appropriately, attempted to accomplish, was placing his reformed analytic mathematics into the pivotal role in a new scheme of learning which was to replace the one that stemmed from the Middle Ages. The goal of the following pages is to reexamine critically the main characteristics of the Cartesian scheme of learning, first by reconstructing the historical process of the formation of his mathematics which is best represented by the *Géométrie* (although it was originally published as a part of a book, we will regard it as a single treatise, since it became later considered to be an independent monograph), published in 1637 as one of the three essays of his *Discours de la méthode*. A second focus shall be examining both historically and philosophically Descartes's theory of mathematics, which appeared mainly in his *Regulae ad directionem ingenii*, drafted in his youth, and in his mature *Meditationes de prima philosophia*, first published in 1641. The two different objects of inquiry roughly correspond to Parts I and II of this book.

Before beginning to pursue the historical traits of Descartes's mathematical thinking, we shall present the problems we are going to resolve and the essential core of our solutions to them in advance, as an Ariadne's thread, which will prevent us from losing our way.

In Part I, "The Formation of Descartes's Mathematical Thought," we inquire about how Descartes formed his mature algebraically analytic mathematics. There are some well-written works which examine the *Géométrie* rather closely from a mathematical point of view. Such achievements may be represented, first of all, by Frans van Schooten's commentaries on two Latin editions published in 1649 and in 1659–61 and Claude Rabuel's comprehensive *Commentaires sur la Geometrie de M. Descartes* published in 1730. (As to the details of works referred to, see the bibliography at the end of this book.) A. Mouchot's *La Réforme cartésienne étendue aux diverses branches des mathématiques pures* of 1876 and John F. Scott's *Scientific Work of René Descartes* published in 1952 may also be said to belong to the same category. But unfortunately there have been very few works which are devoted to the formation of his mathematical thought. Gaston Milhaud's *Descartes savant* (1921) is, of course, an exceptional case. But even this excellent monograph already seems dated when compared with today's Cartesian scholarship. How Descartes created his mathematics is an enigma for us even though historians of mathematics and sometimes those of philosophy have almost unanimously admitted the importance of reconstructing that history. This is certainly due to the fact that his mathematics consists mainly of his own discoveries, rather than being simply derivative of the works of others. Even while fully recognizing this, historians have to note that he generally didn't pay sufficient respect to the works of the contemporaries by whom he is supposed to have been influenced. Descartes's mathematical work, however original it may be, is not a 'proles sine matre creata'.

Chapter 1 shows that Descartes was not an exception to the general rule that institutional settings play a crucial role in forming a thinker's ideas. He was educated at a celebrated Jesuit college during the Counter Reformation, exactly when mathematical studies began to be emphasized. There he seems to have learned not only classical Euclidean geometry but also higher mathematics, which had been fully restored during the Renaissance with mathematical disciplines begun in medieval Islamic civilization. The most important mathematician of the late sixteenth and early seventeenth centuries within the Jesuit Order was Christoph Clavius, who was commonly nicknamed 'the sixteenth century Euclid'. What Descartes was attracted to at the Jesuit college of La Flèche and later tried to reform was the mathematical thought incorporated into the educational program of the Society of Jesus. It is known that Clavius played the crucial role of drawing up its program of mathematical study. Clavius wrote many well-read mathematics textbooks which had been influenced by the surge in mathematical studies among some Renaissance humanists. Among these textbooks Descartes became familiar with at least two, either as textbooks of a regular course, as extracurricular reading, or as readings immediately after his graduation: Euclid's *Elements*, compiled and annotated by Clavius, of which the first edition was published in 1574 and the second enlarged and revised version in 1589, and his *Algebra*, published in 1608 (see Chapter 2, § 1). Such an education constituted the most fundamental background that made up both Descartes's basic intellectual and technical sustenance, and his main target when he began to resist and reform traditional mathematical thought.

With his encounter with Isaac Beeckman during the years 1618–1619 as a turning point, Descartes started to part from the old-fashioned mathematics which had been placed basically either in the scholastic intellectual milieu or in a situation where simply practical purposes were valued. On March 26, 1619 he wrote to Beeckman about the significant idea of constructing "an entirely new science" which would encompass the two branches of pure mathematics, arithmetic and geometry (see Chapter 3, § 1). This idea is supposed to have had a new stage with his having known a rather high mathematical culture of the German *Rechenmeister* in the late autumn of the same year. What could such a new discipline which he attempted to build in 1619 possibly be like? What is certain at least is that from this time till about the mid-1620s Descartes's mathematical endeavors centered on constructing his own algebra to serve as a tool for mathematical analysis, some exemplars of which were shown to Beeckman in the winter of 1628–1629 (Chapter 4, § 1). The manuscript of the monograph entitled *Algebra*, considered to have been completed by 1628, is now lost, but it is exactly this treatise on which the young Descartes concentrated his mathematical efforts up to the mid-1620s, and which functioned as the model for mathematics in his *Regulae ad directionem ingenii*, left incomplete and published posthumously in 1701.

Here we should keep in mind that Descartes confessed on March 31, 1638 to Mersenne, "You know that more than 15 years have already passed since I declared I would disregard geometry and never to dwell on solving any problem unless it is at the request of some friend."[1] If Descartes's statement is true (in fact it does not seem to be contradicted by other evidence), the principal part of his mathematical thought had crystallized before about 1623. From his autobiographical confession cited above, we cannot determine precisely when he wrote his *Algebra* of 1628, but we may safely consider him to have possessed its central idea before about 1623.

Of course, two important events in Descartes's mathematical career had yet to occur. First, toward the end of 1631 the Dutch mathematician Jacobus Golius challenged Descartes, who was already renowned as a talented mathematician, to solve a problem contained in Pappus's *Collection* (Chapter 5, § 1). After tackling the problem for about five or six weeks, Descartes succeeded in discovering a solution to it and reported the fruit of his study to the proposer. This experience surely enhanced Descartes's confidence that his mathematical

[1] "René Descartes, à (Santpoort), à Mersenne, à Paris. 31 mars 1638," *AT*, I, p. 95; *AM*, II, p. 222; *CM*, VII, p. 132: "Et vous scavez qu'il y a desja plus de 15 ans que je fais profession de negliger la Geometrie, et de ne m'arester jamais à la solution d'aucun problesme, si ce n'est à la priere de quelque ami [...]"

method was superior to that of the ancients. It is no accident, then, that he incorporated his solution of the Pappus problem into his final *Géométrie* as its essential core. One of the aims he had entertained since his youth had been accomplished with the composition of the *Géométrie*. He composed it in a rather short time, and seems to have obtained, as he himself admitted, some significant results and to have revised his previous symbolism in the midst of its composition. This constituted the second and most important event in his mathematical career after 1628.

What follows from the brief description above is that Descartes's program to reform mathematics, the end result being the *Géométrie* of 1637, had been essentially completed before the mid-1620s at the latest. Its preliminary outcome was the old *Algebra* of 1628. An important problem to be asked here is what position Cartesian mathematics occupies in the tradition of modern algebraically analytic mathematics. In order to answer this question we have to consider a conundrum in the history of mathematics. It seems to be well-established that it was François Viète who founded the analytic tradition of modern mathematics. His *In artem analyticem isagoge*, published in 1591, was a manifesto of that current. The conundrum to be resolved is whether or not our mathematician was influenced by Viète's mathematics. Descartes surely had contact with Viète's work, including *In artem analyticem isagoge*, shortly before May 3, 1632. But he publicly denied that Viète had influenced his basic mathematical thought when a charge of plagiarism was brought against him soon after the publication of the *Géométrie*. On the one hand, we cannot help admitting there are similarities between the mathematics of Viète and Descartes. On the other, there may be a way which leads to a solution without either supposing that Descartes told an untruth or that he was not influenced at all by Viète's mathematics.

Recent scholarship on the editorial problems concerning the *Regulae ad directionem ingenii* seems to offer a clue to a solution of this enigma. The key concept for grappling with the problem was 'mathesis universalis'. It is known that this concept appears only in the latter half of Rule IV of the *Regulae ad directionem ingenii* insofar as Descartes's corpus is concerned. In his *Constitution du texte des Regulae* published in 1964, Jean-Paul Weber has suggested that Descartes's idea of 'mathesis universalis' was essentially a mathematical concept borrowed from previous scholars, such as Adriaan van Roomen. Furthermore, Giovanni Crapulli has studied most exhaustively what the concept of 'mathesis universalis' meant in the sixteenth century in his historical study under the title *Mathesis universalis: Genesi di una idea nel XVI secolo*, which appeared in 1969. According to Crapulli, 'mathesis universalis' in the sixteenth century had several denotations, but most ordinarily indicated a common mathematical discipline which was considered prior to both arithmetic and geometry. The most celebrated promoter of that notion was again Van Roomen.

But one problem with these two historians of philosophy is that they have not sufficiently examined mathematical works by those who used the concept of 'mathesis universalis', for instance, Van Roomen. Van Roomen was a close friend of Clavius, and a sometime rival and later friend of Viète. He is known to have exchanged difficult mathematical problems with Viète beginning in 1593. Through some years of rivalry he was impressed by the talent of the French mathematician and began to accept his rival's style of mathematics, that is, symbolic algebra. The most telling evidence for this acceptance was his *Mathematicae analyseos triumpus*, published in 1609. Van Roomen was, therefore, a proponent of both 'mathesis universalis' and the analytic mathematics founded by Viète. Indeed Viète's algebraic analysis, although its terminology is basically geometrical, is an art applicable to and prior to both arithmetic and geometry. Thus a possible missing link connecting Descartes to Viète has been discovered. Of course, that Descartes knew Viète's analytic program indirectly through the mediator Van Roomen remains basically a conjecture. There is still the alternative that Descartes invented his algebraic tool totally independently of Viète's work, having simply studied Clavius's *Algebra*. But our conjecture seems to be verisimilar and worthy of examination in futher historical inquiries. From the fact that Descartes used 'mathesis universalis' in his *Regulae*, he may be supposed to have caught a hint, even though it may have been very slight, by coming in contact with some work of Van Roomen before the establishment of his own reformed algebra, perhaps before March 26, 1619.

To reconstruct Descartes's mathematical career is a difficult task. We possess his own biographical narration on the formation of his mathematical thought in Parts I and II of his *Discours de la méthode*. Whether autobiographical statements are a secure path to historical truth should be questioned, for self-justification at a later time is sometimes unavoidable. First of all, then, we have to trust solid evidence obtained from outsiders and institutional studies. Further it should be compared with statements by the person concerned which are not directly related to the question of the moment. Statements proposed by the historical person directly to justify the problem must be used by the historian only as the last resort. In this categorization, our conjecture is not, at least, one which relies simply on statements of the third genre. In this sense, our hypothesis must be favorably evaluated as a topic for future investigation, even though it remains at present a conjecture with a high probability.

In addition, Descartes was different from other contemporary mathematicians who, like his rival Pierre de Fermat, carried on in the algebraically analytic tradition. Fermat might be judged to have been superior to our mathematician in his technical competence in solving mathematical problems.[2] Descartes, on the other hand, no doubt surpassed Fermat in his ability to bring revolutionary change into the fundamental notions of mathematics as a whole. Fermat did

[2] For Fermat as an analytic mathematician, see Michael S. Mahoney, *The Mathematical Career of Pierre de Fermat 1601–1665* (Princeton, 1973, $_2$1994); Jean Itard, *Pierre de Fermat* (Basel, 1950).

not write his own *Géométrie*, having simply produced splendid mathematical memoirs. Descartes published a textbook *Géométrie*, which with its two Latin editions became the paradigm of his genuinely original mathematical thought. Therewith his mathematics acquired the crucial precondition for its institutional, as well as intellectual, acceptance. While the conceptual and technical components of mathematics cannot be separated essentially, and we will discuss the latter category when necessary, we have emphasized fundamental conceptual changes rather than technical ones because of the revolutionary character of Descartes's mathematical thought.

The *Géométrie* was composed as one of the three essays, products of his quite general "method of rightly conducting one's reason and seeking the truth in the sciences."[3] Descartes's mathematics was an application of his philosophical method. In other words, his mathematical thought becomes fully understandable when his philosophical framework is examined comprehensively. What Descartes accomplished was a revolution in the overall structure of learning in addition to the one within mathematics. In his conception of the structure of the sciences, he considered mathematics to play a pivotal role. He judged that he had radically reformed even metaphysics, or first philosophy, with mathematics, especially algebraic analysis, being the model for certainty. At least in *Regulae*, he claimed that the sciences should achieve a certainty equal to that of pure mathematics. Afterward as well, he never stopped seeking a certainty in some way or other.

Mathematics has enjoyed an extremely high status in our modern scheme of learning for certain historically ascertainable reasons. Historically that scheme was establised in seventeenth century Europe, due to the endeavors of our mathematician-philosopher in particular to its later intellectual and social acceptance. For this reason we must attempt to situate historically the position of mathematics within the scheme of learning. We fulfill this task in Part II "The Concept of 'Mathesis Universalis' in Historical Perspective."

*

The key notion of Part I was 'mathesis universalis'. The same concept will take a central role in Part II as well. As we have indicated above, at the beginning of the seventeenth century 'mathesis universalis' usually meant a mathematical discipline applicable to the two main branches of pure mathematics. A representative mathematician who used this concept was Van Roomen. In his *Apologia pro Archimede* (1597) and *Universae mathesis* [sic!] *idea* (1602), he redefined this notion into a kind of art of computation and again as "first mathematics" (*prima mathesis*) by analogy with the first philosophy of Aristotle. "First

[3]From the full title of the *Discours de la méthode* published by Ian Maire in Leiden on June 8, 1637: *DISCOURS DE LA METHODE Pour bien conduire sa raison, & chercher la verité dans les sciences. Plus LA DIOPTRIQVE. LES METEORES. ET LA GEOMETRIE. Qui sont des essais de cete METHODE.*

mathematics" then had a status among the mathematical disciplines similar to that which first philosophy (i. e. metaphysics) had among the various branches of philosophy (see Chaper 7, § 2, among others). In Rule IV of his *Regulae ad directionem ingenii*, Descartes is supposed to have followed Van Roomen's use of that concept. For Descartes, 'mathesis universalis' seems to have connoted some kind of algebra.

'Mathesis universalis' in the seventeenth century was thus placed in a position similar to set theory of Georg Cantor, or a totally formalized system of logic and arithmetic by David Hilbert in the late nineteenth and early twentieth centuries in the sense that both took on the role of 'first mathematics' over the whole domain of the mathematical disciplines. Thus 'mathesis universalis' can be reinterpreted, without committing a serious error, as a kind of foundation of mathematics; it was construed to be prior to any other discipline of mathematics.

We should come to understand that to inquire into the true meaning of Descartes's 'mathesis universalis' is of special importance for a sufficient understanding of both his philosophy of mathematics and his overall philosophical scheme. We undertake this inquiry in a roundabout way, that is to say, we discuss 'mathesis universalis' in its global historical perspective from Aristotle up to Descartes and Leibniz.

In his stimulating monograph *Universal Mathematics in Aristotelian-Thomistic Philosophy: The Hermeneutics of Aristotelian Texts Relative to Universal Mathematics* (1980), Charles B. Crowley has recently claimed that several passages in Aristotle's *Metaphysics* which are today typically interpreted as being related to 'universal mathematics' (almost synonymous with Descartes's 'mathesis universalis') should be understood as they were by Thomas Aquinas. In Thomas's exegesis, Crowley maintains, the Aristotelian passages were explained as referring to metaphysics, the unique 'universal science' rather than 'universal mathematics'.

We will reexamine Crowley's thesis in the same philological and philosophical manner as his, but more extensively. By analyzing the works of Aristotle and his Greek commentators, it is shown that the Aristotelian passages in question can be interpreted as referring to what is related to mathematics, 'mathematics as a whole' or 'universal mathematics'. We will show that while Crowley's Thomistic interpretation can be refuted, his further assertion that Aristotle did not grant a high epistemological status to a mathematical domain which may be defined as a kind of "universal mathematics" can be, in a sense, supported. Moreover, we pursue how Thomas's interpretation became possible philosophically and historically, examining appropriate medieval commentaries, for example, by Ibn Rushd (Latinized name: Averroës) and Albertus Magnus. We will find that medieval interpreters of Aristotle on a whole did not admit the possibility of 'universal mathematics' in its strict sense (Chapter 6).

Nevertheless, the neoPlatonist Proclus gave a certain favorable place to a mathematical domain common to arithmetic and geometry, although he still maintained, following Plato, that it is inferior to dialectic, the universal art of critical discourse. Influenced partly by Proclus, several scholars of the sixteenth century began to recognize the possibility of a certain common science which some called 'mathesis universalis'. In his *Apologia pro Archimede*, Van Roomen, for example, proposed "an idea of a kind of mathesis universalis, which we call first mathematics." However, influenced by Aristotelian epistemology, even he does not seem to have believed that 'mathesis universalis' could exist as an independent and autonomous discipline (Chapter 7).

The young Descartes, in his *Regulae*, granted 'mathesis universalis' an extremely high position, disdaining dialectic and insisting on the reform of philosophy, with 'mathesis universalis' being the paradigm. But this point of view does not appear to have been maintained by the mature Descartes after 1630. On closely examining tactics which the skeptics used to attack the supposedly genuine certainty of mathematics, he was compelled to admit that mathematical statements (even those which are called axioms) cannot be demonstrated in the most rigorous sense. For this reason, he made efforts to discover the first truth, irrefutable even by the most intransigent skeptic. His first truth was, as is well known, "Cogito, ergo sum." With this keystone, Descartes fought to reestablish the authority of mathematics and mathematical physics. The records of these efforts, although it is highly questionable whether they were finally successful, are nothing other than his *Discours de la méthode* (1637), *Meditationes de prima philosophia* (1641), and *Principia philosophiae* (1644). In these later philosophical writings, the term 'mathesis universalis' did not appear at all (Chapter 8, §1).

Consistently throughout history, 'mathesis universalis' was considered to be a mathematical discipline which treated more or less general and abstract quantities common to both discrete numbers and continuous magnitudes, that is, the two more concrete concepts of quantity. To give a high position to this notion rather than to arithmetic, geometry or any other more concrete discipline will be favorable for a Platonist but not always for an Aristotelian. To admit that 'mathesis universalis' is possible as a science which can be consistent and complete within itself is to think that mathematical activities can be totally autonomous and independent of other sciences and, above all, of this life-world. We question whether mathematics has such transcendent foundations. We contend that mathematical activities, although relatively autonomous, are carried out fruitfully only through interacting with other intellectual and social activities. That mathematics is fundamentally a historical phenomenon is our starting presupposition and at the same time our goal to be proved.

Descartes was, of course, an extremely ingenious mathematician and philosopher. He was not a naive thinker in any sense. The spectrum of activity which the historical Descartes has shown us through his mathematical and philosophical career is magnificent and fascinating. His mathematical thought is still vividly alive in the present day because of both its greatness with its limitations. Towards the end of her ***Descartes: Biographie***, today's most trustworthy biography of our philosopher published in 1995, Geneviève Rodis-Lewis has stated clearly, "Cependant, c'est en mathématiques que Descartes a ouvert les voies les plus profondes."[4] Certainly, Descartes's mathematical thought is still vividly alive even today because of its greatness, and, at the same time, is waiting for a sincere and drastic revision because of its limitations. Encouraged by Rodis-Lewis's words, with Ariadne's thread in hand, let us start our journey, asking the problems Descartes himself faced.

[4] Genevieve Rodis-Lewis, *Descartes: Biographie* (Paris, 1995), p. 312; An English translation, *Descartes: His Life and Thought*, tr. by Jane Marie Todd (Ithaca/London, 1998), p. 226.

Part I

The Formation of Descartes's Mathematical Thought

"Je commence en cela par ou Viète avoit fini."

—Descartes à Mersenne, 31 mars 1638.

CHAPTER 1

DESCARTES AND JESUIT MATHEMATICAL EDUCATION

§ 1. Descartes and the Jesuit College of La Flèche

Descartes studied at the Jesuit college of La Flèche, "one of the most famous schools in Europe,"[1] for about eight and half years probably from Easter day of 1607 to September 1615. Named after Henri IV, the Collège Henri IV de La Flèche was planned under the king's patronage in 1603, and actually began to receive students at the end of 1604. The school was built as a college for externs, one of a type designed principally for students not of the Jesuit Order. Both Henri and the Jesuits hoped the students there, chosen from the French elite, would be loyal both to the French monarchy and to the Catholic Church. Descartes, from a family which would qualify as *noblesse de robe*, was one of those students who were expected to "mount the stage of the theater of the world"[2] and play an essential role in French society.

[1] *AT*, VI, p. 5. G. Rodis-Lewis, *Descartes* (Introduction, n. 4), p. 25; English translation, p. 8. Adrien Baillet claimed that Descartes was at La Flèche between the years 1604 and 1612. Baillet, *La Vie des Monsieur Des-Cartes*, t. I (Paris, 1691), pp. 16–17. But this description was corrected by, for example, Charles Adam, *Vie et Œuvres de Descartes: Étude historique*, Supplément à l'Édition de Descartes (Paris, 1910), pp. 564–565; Karl Six, "Descartes im Jesuitkolleg von La Flèche," *Zeitschrift für katholische Theologie*, **38** (1914), pp. 498–508. Both claimed that Descartes was in La Flèche between the years 1606 and 1614. See also Descartes, *Discours de la méthode: Texte et commentaire*, par Étienne Gilson (Paris, $_5$1967), pp. 103–105. On the other hand, Antonella Romano has ascribed Descartes's student years at La Flèche to the period between 1605 and 1614. See her *La Contre-Réforme mathématique* (n. 13, below), p. 483. We have found Rodis-Lewis's discussion in her aforementioned book to be the most persuasive and will discuss this problem in detail in Chapter 2, § 5. Incidentally, the collège was transformed into a military school by Louis XV, and finally into today's "Le Prytanée national militaire" by Napoléon Banaparte.

[2] This expression is taken from the young Descartes's "Cogitationes privatae," *AT*, X, p. 213.

Our main goal in Part I is to show how Descartes created the algebraically analytic mathematics presented in his *Géométrie* of 1637 and formulated the general methodological rules stated in Part II of his *Discours de la méthode*. The two processes of formation were very closely interwoven. To begin with let us look at Descartes's autobiographical narrative where it concerns his philosophical method. He described in his *Discours de la méthode* how he formulated it:

> I continued practicing the method I had prescribed for myself. Besides taking care in general to conduct all my thought according to its rules, I set aside some hours now and again to apply it more particularly to mathematical problems. I also applied it to certain other problems which I could put into something like mathematical form by detaching them from all the principles of the other sciences, which I did not find sufficiently secure (as you will see I have done in many problems discussed later in this book [i. e. *Discours de la méthode* and the three accompanying essays]).[3]

Descartes formed his method of thinking from certain procedures of mathematics. And conversely he elaborated his mathematics according to this method. So now we ask: Why and how did mathematics come to play a central role in Descartes's thought? What kind of mathematics did he learn from others and how did he transform this older form into his own? According to the description in his *Discours de la méthode*, he received the first intimation of his mathematical thought in the college of La Flèche.

Let us look, then, at Descartes's own impression of the mathematical instruction at La Flèche. Descartes thought that at La Flèche "there must be learned men if they existed anywhere on earth."[4] There he learned "everything that the others were learning; moreover, not content with the subjects they taught" the students, he went through all the books that fell into his hands "concerning the subjects that are considered most abstruse and unusual."[5] He pursued mathematics among his subjects. In Part I of the *Discours de la méthode*, he stated:

> I knew [...] that mathematics contains some very subtle devices which serve as much to satisfy the curious as to further all the arts and lessen man's labours.[6]
>
> [...]
>
> Above all I delighted in mathematics, because of the certainty and self-evidence of its reasonings. I did not yet notice its real use; and since I thought it was of service only in the mechanical arts, I was

[3] *AT*, VI, p. 29.
[4] *AT*, VI, p. 5.
[5] *Ibid.*, pp. 14–15.
[6] *AT*, VI, p. 6.

> surprised that nothing more exalted had been built upon such firm and solid foundations.[7]

Here Descartes makes the very general statement that he was attracted to mathematics because of its usefulness and certainty.

In Part II of the *Discours de la méthode*, however, Descartes gives more detailed information about what he studied during his sojourn at La Flèche. There he tells us explicitly how he formed his method. Three disciplines were crucial to its formation: logic, analysis and algebra. He thought he "had to seek some other method comprising the advantages of these three subjects but free from their defects."[8] It is supposed to have been in a 'poèle', sometimes 'poêle' or 'poële' (stove-heated room) in Germany when he was twenty three years old that he made up his mind to formulate his own universal method out of the three disciplines. But the grounding of this endeavor is of earlier origin. Listen to Descartes's statement:

> When I was younger, I had done a little study of logic among the branches of philosophy, and among those of mathematics, the analysis of the geometers and algebra. These three arts and sciences, it seemed, ought to contribute something to my plan.[9]

According to Étienne Gilson,[10] the expression "when I was younger" (*étant plus jeune*) refers to the time when Descartes was at La Flèche, for Descartes is conjectured to have returned to his first mathematical studies only in the autumn of 1618 at the instigation of Isaac Beeckman. Also the wording, "among the branches of philosophy" (*entre les parties de la Philosophie*) and "among those of mathematics" (*entre les Mathematiques*), makes us recall the curriculum of the college and reconfirms Gilson's conjectue.

Descartes, however, was not necessarily satisfied with the three arts taught at the college. He expresses his dissatisfaction:

> But on further examination I observed with regard to logic that syllogisms and most of its other techniques are of less use for learning things than for explaining to others the things one already knows or even, as in the art of Lull, for speaking without judgement about matters of which one is ignorant.
> [...]
> As to the analysis of the ancients and the algebra of the moderns, they cover only highly abstract matters, which seem to have no use. Moreover the former is so closely tied to the examination of figures that it cannot exercise the intellect without greatly tiring

[7] *Op. cit.*, p. 7.
[8] *Op. cit.*, p. 18.
[9] *Op. cit.*, p. 17.
[10] Gilson, *Texte et commentaire* (n. 1), p. 181.

> the imagination; and the latter is so confined to certain rules and symbols that the end result is a confused and obscure art which encumbers the mind, rather than a science which cultivates it.[11]

We may suppose that Descartes felt this dissatisfaction deeply after the encounter with Beeckman, not during his stay at La Flèche, although a seed of this feeling might have been sown during his school days. Judging from Descartes's description, the analysis and the algebra which he learned at La Flèche seem to have been old-fashioned. The analysis was tied to geometrical figures and, therefore, must have been geometrical analysis. And the algebra was performed with certain rules and symbols (*chiffres*) and can be safely characterized as cossist algebra. Gilson thought that the sources of Descartes's reading at La Flèche came from Christoph Clavius's popular textbooks *Geometria practica* (1606) and *Algebra* (1608).[12] We will discuss this problem in Chapter 2.

So far we have examined the kind of mathematics Descartes studied at the Jesuit college of La Flèche, based on his accounts of the teaching given in the Part I and II of the *Discours de la méthode*. We should look at the whole structure of the curriculum of the college from another side, that is, from institutional point of view.

§ 2. The Curriculum at La Flèche

Descartes received the mathematical education at a Jesuit college under the Counter Reformation. As Antonella Romano's *La Contre-Réforme mathématique: Constitution et diffusion d'une culture mathématique jésuite à la Renaissance*,[13] the most comprehensive study of Jesuit mathematical education, has pointed out, a crucial historical background against which Descartes emerged was a mathematical culture of this period which the Society of Jesus created and diffused.

The Society of Jesus was an order founded by Ignatius de Loyola and officially admittted by the Roman Catholic Church in 1540 with the preparatory period between the years 1534 and 1538.[14] The curriculum at the Jesuit colleges of the early seventeenth century was essentially grounded upon the *Constitutions* of the Society of Jesus and the *Ratio atque institutio studiorum Societatis Iesu*, or concisely *Ratio studiorum* of 1599. The *Constitutions* originated in

[11] *AT*, VI, pp. 17–18.

[12] Gilson, *Op. cit.*, pp. 182–183.

[13] Antonella Romano, *La Contre-Réforme mathématique: Constitution et diffusion d'une culture mathématique jésuite à la Renaissance (1540–1640)* (Rome, 1999). Descares is discussed in "Le cas Descartes," pp. 482–491.

[14] Luce Giard, "Le Devoir d'intelligence, ou l'insertion des jésuites dans le monde du savoir," in *Les Jésuites à la Renaissance: Système éducatif et production du savoir*, sous la direcion de L. Giard (Paris, 1995), pp. XII–LXXIX. Cf. Philippe Lécrivain, *Pour une plus grande gloire de Dieu: Les Missions jésuites* (Paris, 1991).

the founder Ignatius de Loyola's educational program. The formulation of the *Ratio studiorum* of 1599 was based upon the Jesuits' educational activities dating back to the foundation of the first fully constituted classical college of the Society, opened in 1548 at Messina in Sicily. It may be said that in its formative period the Society's method of teaching as a whole followed that of the University of Paris, Ignatius's alma mater, namely the *Modus Parisiensis*.

For determining the educational method of Society of Jesus, the dates of its founder Ignatius's sojourn in Paris, from 1528 to 1535, are crucial. In Paris, Ignatius moved from the Collège de Montaigu to the Collège de Sainte-Barbe in 1529. At the Collège de Montaigu, the humanist movement was extremely weak, but at Sainte-Barbe it found fertile soil. As a historian of Jesuit education states, Ignatius "witnessed the final phase of the change from the old to the new education, which was in a sense the transition from Medievalism to the Renaissance."[15] At this college the humanistic curriculum made the classics the foundation of teaching. "Ignatius witnessed the change, kept it in mind, and afterwards made it the inevitable rule of his educational legislation."[16] But the scholastic tone of medievalism was still alive in the main branches of the curriculum.

The curriculum of the college at La Flèche was scholastic, but modified by the humanist reforms. The courses at La Flèche were divided into two parts. The lower part was called the humanistic curriculum; and the upper part was the philosophical curriculum. The humanistic curriculum formed the basis for the higher philosophical studies and needed five years to be completed. The first three years were devoted to learning Latin grammar and a little Greek grammar. In the fourth year class, the students learned humanities, the aim of which was to lay the foundation for eloquence. They read Cicero, the historians such as Caesar and Sallust, and the poets such as Virgil and Horace. In the fifth and last year of the humanistic curriculum, rhetoric was taught as training to perfect eloquence. Cicero was a central author for reading and writing. While throughout the Middle Ages oral discussion in Latin had been essential, here composition in Latin and Greek was emphasized.

It is more important for us, however, to understand the philosophical curriculum. The course on philosophy lasted for three years. The first year of philosophical studies consisted of logic and introduction to natural philosophy. The students mainly read Aristotle's works on logic. The teacher explained first the *Isagoge* of Porphyry, the *Categories*, the *De Interpretatione*, the first five chapters of the *Prior Analytics*, the eight books of the *Topics*, and the *Posterior Analytics*. The Jesuit commentators, Pedro da Fonseca and Francisco de Toredo, served as the guides to Aristotle's logic. During the period between 1606 and 1626, the second year was devoted to the study of natural philosophy

[15] P. Farrell, *The Jesuit Code of Liberal Education: The Development and Scope of the Ratio Studiorum* (Milwaukee, 1938), p. 31.
[16] *Ibid.*

and mathematics.[17] The students read the eight books of Aristotle's *Physics*, the four books of the *On the Heavens*, and the first book of the *On Generation*. As for the study of mathematics, they learned "the *Elements* of Euclid and more difficult mathematical problems,"[18] one hour a day. In the third year, the students spent their time on metaphysics and moral philosophy. The second book of the *On Generation*, the *On the Soul*, and the *Metaphysics* were explained. In moral philosophy the *Nicomachean Ethics* of Aristotle was read.

In the philosophical curriculum we are left with a strong impression that the tone was rigidly Aristotelian. The philosophy of the Society of Jesus was Aristotelianism as expounded by Thomas Aquinas, or more exactly, Aristotelianism as modified by the sixteenth-century Jesuit commentators such as Francisco Suárez, Fonseca, and Toredo. As a recent institutional study on the philosophy and sciences of the sixteenth century has reported, the curriculum of higher educational institutions was with few exceptions strictly Aristotelian.[19] We may call the philosophy of the Jesuit colleges of the sixteenth and early seventeenth centuries 'Renaissance Aristotelianism.' But the Aristotelianism of that period consisted of a great variety of philosophies agreeing only on the most fundamental issues, and thus may be said to have been not unitary but formed of "plusieurs aristotelismes."[20] We here attempt to examine the Jesuits' philosophy only as it concerns their approach to mathematical knowledge.

Generally the style of philosophy in the Jesuit colleges was conservative, but nevertheless critical. While the professors rarely made constructive contributions to their fields of study, they were sometimes very quick at introducing new thought.[21] For example, in 1611 students at La Flèche read a sonnet on "the discovery of some new planets or stars rounding about Jupiter made this

[17]Camille de Rochemonteix, *Un Collège de jésuites aux XVIIe et XVIIIe siècles: Le Collège Henri IV de La Flèche*, t. 4 (Le Mans, Leguicheux, 1889), p. 32.

[18]Farrell, *Op. cit.*, p. 343.

[19]Charles B. Schmitt, "Philosophy and Science in Sixteenth-Century Universities: Some Preliminary Comments," in J. E. Murdoch and E. D. Sylla, eds., *The Cultural Context of Medieval Learning* (Dordrecht, 1975), pp. 485–530, and reprinted as "V" in Ch. B. Schmitt's *Studies in Renaissance Philosophy and Science* (London, 1981).

[20]Charles H. Lohr, "Les jésuites et l'aristotelisme du XVIe siècle," in L. Giard, dir., *Op. cit.* (n. 14), pp. 79–91, on p. 85. Cf. John E. Murdoch, "From the Medieval to the Renaissance Aristotle," in John Henry and Sarah Hutton, eds., *New Perspectives on Renaissance Thought: Essays in the history of science, education and philosophy in memory of Charles B. Schmitt* (London, 1990), pp. 163–176; Ch. B. Schmitt, *A Critical Survey and Bibliography of Studies on Renaissance Aristotelianism 1958-1969* (Padua, 1971), pp. 16–17; *Idem*, "L'introduction de la philosophie platonicienne dans l'enseignement des universités à la Renaissance," in *Platon et Aristote à la Renaissance: XVIe Colloque International de Tours* (Paris, 1976), pp. 93–104; *Idem*, *Aristotle and the Renaissance* (Cambridge, Mass., 1983); and his various papers reprinted in his *The Aristotelian Tradition and Renaissance Universities* (London, 1984).

[21]This general characterization has been adopted from Bernhard Jansen, "Die Pflege der Philosophie im Jesuitenorden wärend des 17.–18. Jahrhunderts," *Philosophisches Jahrbuch der Görres-Gesellschaft*, **51** (1938), pp. 172–215, 345–66, and 435–456.

year by Galileo, the celebrated mathematician of the grand duke of Florence."[22] Needless to say, "the discovery by Galileo" here means a discovery of satellites of Jupiter by a telescope. This episode tells us that the teaching at La Flèche was not entirely dogmatic but to some degree open to new knowledge. This harmonizes with another story about Clavius in Rome accepting the new discovery by Galileo.[23] Of course, we must be very careful, for this does not necessarily mean that the entire Jesuit Order admitted Galileo's theory of planetary motions.

Descartes frankly expressed his dissatisfaction with the curriculum of his alma mater in his *Discours de la méthode*. At the same time, however, he wrote a letter in 1638, stating that "there is no place in the world where, in my judgment, philosophy is taught better than at La Flèche."[24] Descartes, as he himself admitted, acquired one of the best educations in the Europe of his time.

Since it was by means of certain procedures of mathematics that Descartes tried to break the firm unitary character of the scholastic scheme of learning, we should more closely observe Jesuit mathematical education as it appears in the *Ratio studiorum* of the Society of Jesus.

§ 3. Mathematical Studies in the *Ratio Studiorum*

It can be safely said that Descartes was educated fundamentally in accordance with the 1599 definitive edition of the *Ratio studiorum*. The *Ratio studiorum* was a practical handbook in educational method and college management. It devotes only a very brief space to a description of mathematical studies. Under the title "Rules for the Professor of Mathematics," we read:

> 1. *What Authors Are To Be Explained, and at Which Time and to Which Students.*—Let him explain in class to the students of physics for about three-quarters of an hour the *Elements* including in the explanations, after they have become somewhat familiar during two months, something of geography or of the sphere or other matters which students are glad to listen to, and this along with Euclid either on the same day or on alternate days. (The *Constitutions*, Part IV, Chapter 12. C.)
>
> 2. Problems.—And let him arrange that every month or every other month some one of the students before a large gathering of students of philosophy and theology has some famous mathematical problem to work out and afterwards, if it seems well, to defend his solution.

[22] C. de Rochemonteix, *Op. cit.* (n. 17), t. 1 (Le Mans, Leguicheux, 1889), p. 147. Galileo discovered three satellites around Jupiter in January 1610.

[23] Stillman Drake, *Galileo at Work: His Scientific Biography* (Chicago, 1978), p. 165.

[24] "Descartes à (Debeaune), 12 septembre 1638," *AT*, II, p. 378; *AM*, III, p. 75.

3. Repetition.—Once a month and generally on Saturday in place of the lecture, let the principal problems which have been explained during that month be publicly repeated.[25]

Both the subject matter of mathematics and the method of teaching are concretely described. Euclid was explained together with geography or astronomy. And the teaching method was designed to be rather effective from a practical point of view. In other words, the teacher gave students well-known mathematical problems to be solved, and some important problems were repeatedly studied.

The *Constitutions* help us understand the purpose of the study of mathematics. Part IV, Chapter 12 is concerned with "the Sciences Which Are To Be Taught in the Universities of the Society of Jesus."[26] The first section of this chapter declares that "the aim of the Society and of its pursuits is to help neighbors to a knowledge and love of God and to the salvation of their souls." For this purpose, the faculty of theology plays a central role. In other words, the professors of philosophy and mathematics lean heavily upon the faculty of theology. Every discipline taught in the Jesuit colleges serves this same religious aim. This is understandable since the raison d'être of the Society of Jesus was "Ad majorem Dei gloriam." In Section 3, we read a description concerning mathematics:

> So, since the arts and natural sciences (Explanation C) dispose the mind for theology, and serve to perfect its knowledge and application and of themselves help toward the same end, let them be treated with whatever diligence is proper and by learned professors, seeking in all things sincerely the honor and glory of God.
>
> (Explanation C) There will be treated logic, physics, metaphysics, moral science, and even mathematics in so far as it helps toward the attaining of our proposed end. To teach others to read and write could also be a work of charity, if the number of persons in the Society were sufficient for us to undertake all things, but on account of the lack of members we do not ordinarily teach them.

Mathematics is studied to enhance "the honor and glory of God." In this passage, the end of mathematical studies is not clearly stated. But for

[25] *Monumenta Paedagogica Societatis Iesu*, Nova editio penitus retractata, ed. Ladislaus Lukács, t. V, *Ratio atque Institutio Studiorum Societatis Iesu (1586, 1591, 1599)* (Rome, 1986), p. 402: A. Romano, *Op. cit.* (n. 13), p. 617; *Monumenta Germaniae Paedagogica*, ed. K. Kehrbach, Band V, *Ratio Studiorum et Institutiones Scholasticae Soc.* J. 2. t. II, *Ratio Studiorum ann. 1586. 1599. 1832*, ed. G. M. Pachtler (Berlin, 1887), p. 348. The translation is taken from Edward A. Fitzpatrick, ed., *St. Ignatius and the Ratio Studiorum* (New York/London, 1933), p. 175.

[26] *Monumenta Germaniae Paedagogica*, ed. K. Kehrbach, Band II, *Ratio Studiorum et Institutiones Scholasticae Soc. J. 1.* (Berlin, 1887), pp. 52–54. For the translation see Fitzpatrick, *Ibid.*, pp. 100–101.

the Jesuits of the sixteenth and early seventeenth centuries the religious aim had a very concrete import. We shall analyze this problem in the next section where we try to understand the motives for mathematical studies. Then we shall acquire a concrete sense of the religious aim and the reason why Euclid was read along with textbooks of geography and astronomy. But in order to understand better the unstated aspects of the teaching of mathematics, we should examine not only the *Ratio studiorum* of 1599 and the *Constitutions*, but also historical documents which contributed to the composition of the official version of the *Ratio studiorum*.

It is in the historical documents related to the first Jesuit college at Messina that we find the first program for mathematical education.[27] This program was composed as early as 1548, the year of the foundation of the college. The author was the first rector of the college, Jerome Nadal. From the following passage, we can ascertain which textbooks were used:

> In an extraordinary case, he [i. e. the professor] will lecture on the mathematics which will be judged to be the most appropriate by the rector at the right time; first several books of Euclid so long as they [i. e. the students] are familiar with demonstrations; then the *Practica arithmetica* of Oronce [Fine] and the *Sphera* of the same author, the *Astrolabium* of Stöffler and the *Theorica* of Peurbach.[28]

The *Arithmetica practica libris IV absoluta* (1535) of Oronce Fine is an elementary textbook of practical arithmetic. With the exception of Euclid, the other works mentioned above are all related to astronomy. Fine's *De Mundi sphaera, sive cosmographia* was published in 1542 in Paris. The *Elucidatio fabricae ususque astrolabii* of Johannes Stöffler, professor of the University of Tübingen, and the *Theoricae novae planetarum cum scholiis Rheinholdi* of Georg von Peurbach appeared in 1512 and 1542, respectively. According to the diary of a Jesuit father, Nadal himself taught Euclid as well as Greek and Hebrew in 1549.[29]

[27] For the following analysis of the *Ratio studiorum*, see Giuseppe Cosentino, "Le Matematiche nella *Ratio Studiorum* della Compagnia di Gesù," *Miscellanea Storica Ligure*, Anno II (n. s.), N. 2 (1970), pp. 171–213 (I was indebted to the late Professor Charles B. Schmitt, when I visited the Warburg and Courtauld Institutes at the University of London in December 1979 for having obtained a copy of this article. He passed away suddenly in 1986.); *Idem*, "L'Insegnamento delle matematiche nei collegi gesuitici nell' Italia settentrinale," *Physis*, **13** (1971), pp. 205–217; Romano, *Ibid.*

[28] *Monumenta Paedagogica Societatis Iesu*, ed. L. Lukács, t. I (1540–1556), (Rome, 1965), p. 26: "Praeleget extra ordinem mathematicen, quo tempore commodissimum esse ab ipso Rectore censebitur. Primum aliquot libros Euclidis, donec assuescant demonstrationibus. Deinde practicam arithmeticam Orontii et eiusdem spheram, astrolobium Stoflerini et theoricas Purbachii." I refer to the editor's footnotes for the information about the authors and the works cited in the passage.

[29] *Ibid.*, p. 515: "Exerpta eorum quae monumenta paedagogia illustrant in Chronico Patris Polanco."

Nadal wrote almost the same as the above in another document under the title *De Studii generalis dispositione et ordine.* Nadal seems to have intended to send this treatise to Ignatius in order to acquire influence over the other Jesuit colleges including the Collegio Romano, which was founded in 1551. But only one thing was added to the earlier document. Nadal suggested studying the *De Triangulis omnimodis libri V* (1533) of Regiomontanus (Johannes Müller) or the *De Triangulis* of Jordanus de Nemore.[30] This suggestion was probably made with the intention of providing astronomy with a mathematical background.

Nadal's experimental program at Messina and Ignatius's comprehensive section on mathematical education in the *Constitutions* were integrated when the Collegio Romano opened in 1551. As seen above, it was the *Modus Parisiensis*, the method of teaching at the University of Paris, that Ignatius adopted as the exemplar of scholastico-humanistic education in the Jesuit colleges. After the establishment of the college in Rome, the *Modus Romanus* was to serve as the model system for the world of higher education of the Society of Jesus. In the beginning period of the Collegio Romano, it seems that we can detect almost nothing notable about mathematical education there. We find a paragraph on mathematics in the *De Artium liberalium studiis* (1565–1570) among the documents of James Ledesma's *De Ratione et ordine studiorum Collegii Romani*. But it does not provide us with any important information other than that "mathematical disciplines should not be neglected" and that "at least the knowlege of the sphere should be possessed."[31] The latter statement was for the teacher of the college. What was important about the opening of the Collegio Romano was that the Society of Jesus began to possess colleges which had a full course of liberal arts beyond the study of grammar and rhetoric. In fact, in 1553 the Collegio Romano had its first professor of mathematics, Balthassar Torres.[32]

But, when the composition of the *Ratio studiorum* governing the entire Jesuit educational system was planned, the teaching of mathematics became full-fledged and systematic. The most important scholar who was directing the composition of a draft on mathematical studies was almost certainly the second professor of mathematics at the Collegio Romano, Christoph Clavius (see Figure 1.1). Clavius was the most influencial mathematician not simply of the Collegio Romano but of the entire Society of Jesus, after he became professor of mathematics at the college in 1563.[33] Sixtus V, who was Pope in the years

[30] *Ibid.*, p. 149.

[31] *Monumenta Paedagogica Societatis Iesu*, ed. L. Lukács, t. II (1557–1572), (Rome, 1974), p. 256: "Mathematicae disciplinae praeteriri non debent." "Sed spherae saltem cognitio habenda est [...]."

[32] Riccardo G. Villoslado, *Storia del Collegio Romano dal suo inizio (1551) alla soppressione della Compagnia di Gesu (1773)* (Rome, 1954), p. 335, and Ugo Baldi, *Legem impone subactis: Studi su filosofia e scienza dei Gesuiti in Italia, 1540–1632* (Rome, 1992), p. 568; *Idem*, *Saggi sulla cultura della Compania di Gesù (secoli XVI–XVIII)* (Padova, 2000), p. 55.

[33] Ugo Baldi, *Legem impone subactis*(n. 32) pp. 568, 572 & 573.

of the late 1580s when Clavius was widely and highly reputed, is said to have stated, "Had the Jesuit order produced nothing else than this Clavius, on this account alone should it be praised."[34] And later, Pierre Gassendi characterized Clavius to be "the greatest mathematician among men of the Society of Jesus" (*hominum Societatis Iesu γεωμετρικώτατος*).[35]

Before the final edition of the 1599 *Ratio studiorum* was determined, two previous versions of it were made in 1586 and 1591. It was without doubt Clavius who played the crucial role in the formation of the mathematical part of the *Ratio studiorum* of 1586, 1591, and 1599.

Figure 1.1 Christoph Clavius (1538–1612)

The *Ratio studiorum* of 1586 was written in the process of preparing the official version and was in no sense definitive. The *editio princeps* of the *Ratio studiorum* was composed after examining correspondence from various provinces of the Jesuit Order, the statutes of noted universities, and the fourth part of the *Constitutions* related to education. In 1586 it was sent to the provincials of the order for further consultations.[36]

The section under the title "De Mathematicis" of the *Ratio studiorum* of 1586 begins with a quotation from the *Constitutions* which we have already examined. The *Constitutions* required that mathematical studies be encouraged. But they had not so far sufficiently taken root. The *Ratio studiorum* describes the unsatisfactory situation of mathematical education within the Jesuit Order. Furthermore, it adduces reasons why the content of the *Constitutions* must be reaffirmed. It emphasizes the usefulness of mathematics:

> They [i. e. mathematical disciplines] offer and explain to poets the risings and settings of stars; to historians the shapes and distances of places; to students of the *Analytics* examples of solid demonstra-

[34] Quoted from David Eugene Smith, *History of Mathematics*, Vol. 1: *General Survey of the History of Elementary Mathematics* (New York, 1923), p. 334.

[35] P. Gassendi, "Viri illustris Nicolai Claudii Fabricii de Peresc Senatoris Aquisextiensis vita," in *Opera omnia*, t. V (Lyon, 1658; Stuttgart/Bad Cannstatt, 1964), p. 265.

[36] Farrell, *Op. cit.* (n. 15), p. 225.

> tions; to statesmen the distinctly admirable arts for conducting well in domestic and military services; to physicists the forms and differences of celestial changes, lights, colors, transparent bodies, and sounds; to metaphysicians the number of the spheres and angels (*intelligentiae*); to theologians the distinguished parts of the divine maker; to the court of justice and ecclesiastical community accurate computations of times.[37]

For these purposes, mathematical studies should be encouraged and teachers of mathematics must be fostered at the Jesuit colleges.

The next two papagraphs of the *Ratio studiorum* mention the case of the Collegio Romano. "In order to cure so great a paucity and deficiency, we need two professors of mathematics in the Collegio Romano. One of them should arrange a short course of daily lectures running one and half years on mathematical subjects to be audited by us and externs."[38] He should teach Euclid's *Elements* to the auditors of logic who are prepared for "the *Posterior Analytics* which could scarcely be understood without mathematical examples."[39] The first several books of the *Elements* must be studied as a necessary preparation for geography and astronomy. The *Ratio studiorum* also mentions a mathematical compendium which Clavius produced for use as a textbook in the college. According to Giuseppe Cosentino, a historian of Jesuit mathematical education, this is probably Clavius's *Epitome arithmeticae practicae*, published in Rome in 1583.[40]

"Another professor," states the *Ratio studiorum*, "who could only be Father Clavius, should be installed." He "should set forth the fuller teaching of mathematics every three years and explain privately to eight or perhaps ten

---

[37] *Monumenta Paedagogica Societatis Iesu*, t. V (n. 25), p. 109: A. Romano, *Ibid.*, p. 614; Monumenta Germaniae Paedagogica, Bd. V (n. 25), p. 141: "Illae namque suppeditant atque exponunt poetis ortus occasusque syderum, historicis locorum facies atque intervalla; analyticis solidarum exempla demonstrationum; politicis artes plane admirabiles rerum bene gerendarum domi militiaeque; physicis coelestium conversionum, lucis, colorum, diaphanorum, sonorum formas et discrimina; metaphysicis sphaerarum atque intelligentiarum numerum; theologis praecipuas divini opificii partes; iuri et consuetudini ecclesiasticae accuratas temporum supputationes."

[38] *Monumenta Paedagogica Societatis Iesu*, t. V, p. 109: A. Romano, *Ibid.*, p. 614; *Monumenta Germaniae Paedagogica*, Bd. V, p. 142: "Ut tantae paucitati ac penuriae medeamur, duobus in Romano Collegio Mathematicis Professoribus indigemus. Quorum unus sesquianno quotidianis lectionibus breve curriculum mathematicarum rerum conficiat a nostris et ab externis audiendum [...]."

[39] *Ibid.*: "[...] Posteriora Analytica, quae sine mathematicis exemplis vix possunt intelligi [...]."

[40] Cosentino, "Le Matematiche nella *Ratio Studiorum*" (n. 27), p. 209. Cosentino refers to the 1585 edition. But the first edition appeared in 1583. Cf. C. Sommervogel, ed., *Bibliothèque de la Compagnie de Jesus*, t. II (Bruxelles/Paris, 1891), col. 1215. The Chinese translation of this monograph *Tongwen suanzhi* was made by Ricci and Li Zhizao, and published in 1614. It is now being discussed among Japanese historians of mathematics whether or not this was imported and had an influence on mathematics of Edo Japan.

of our students who would be at least of reasonable talent, be not ignorant of mathematics, and audit philosophy, who should be assembled from various provinces, one from each if possible."[41] The position seems to have been planned as having a higher status than the former professor. It is noteworthy that the *Ratio studiorum* tells us clearly that up until then only Clavius was eligible for this position. The Order required this kind of mathematician to train students who would become future teachers of mathematics at the Jesuit colleges.

The *Ratio studiorum* of 1591 proposed that, in addition to the version of 1586, students of the second and third year of the philosophical course should listen to public lectures on mathematics every weekday, if possible, with these lectures being followed the curriculum determined by Clavius.[42]

From 1581 to 1594 in the process of preparing for the versions of *Ratio studiorum*, furthermore, Clavius composed several important documents in order to promote mathematical education in the Socety of Jesus. These documents are certainly characterized as preparatory drafts, but are quite important as they, in fact, supplement the understanding of the *Ratio studiorum*. Among them, the treatise entitiled "A Method of Promoting Mathematical Disciplines in the Schools of the Society," which is supposed to have been drafted in 1582, is worthy of attention, first of all. In this treatise Clavius develops his ideas on the teaching of mathematics in two ways. One concerns the quality of teachers, the other the close relationship between mathematics and natural philosophy. Clavius introduces his convictions about mathematical education with a discussion of the teachers of the Jesuit colleges.

> First a teacher must be chosen with uncommon knowledge and authority; for if either of these is absent, as experience shows, students seem unable to be attracted to the mathematical disciplines. [...] [I]t will easily happen that they [i. e. the students] will be convinced that philosophy and the mathematical sciences are connected, as they are—especially because up to now students seem almost to have despised these sciences, for the sole reason that they think these sciences are considered not valuable, even useless. [...]
> It also seems necessary that the instructor should have a certain inclination and desire for lecturing on these sciences, and should not be taken up with many other concerns; otherwise he will scarecely be able to help his students. Now in order that the society may always have capable professors of these sciences, some people should

[41] *Monumenta Paedagogica Societatis Iesu*, t. V, p. 110: A. Romano, *Ibid.*, p. 615; *Monumenta Germaniae Paedagogica*, Bd. V, p. 143: "Professor alter, qui modo P. Clavius esse posset, constituatur, rerum mathematicarum pleniorem doctrinam conferat in triennium, explicetque privatim nostris octo circiter aut decem, qui mediocri saltem sint ingenio, nec a mathematicis alieno, et philosophiam audierint, qui ex varijs essent convocandi provinciis, unus ex qualibet, si fieri posset."

[42] *Monumenta Paedagogica Societatis Iesu*, t. V (n. 25), p. 234: A. Romano, *Ibid.*, p. 615.

> be selected who are fit and capable for carrying out this task, so that they may be instructed in various mathematical subjects in a private school. [...] I omit the fact that natural philosophy without the mathematical disciplines is lame and incomplete.[43]

This part is clearly related to the passage mentioned above on the two professors at the Collegio Romano. High quality is required here for the teachers of mathematics. The author is demanding that teachers be duly occupied with the teaching of mathematical sciences.

The next passage asserts that mathematics is necessary for the study of natural philosophy:

> Second, then, it is necessary that the students should understand that these sciences [i. e. the mathematical sciences] are useful and necessary for rightly understanding the rest of philosophy, and that they are at the same time an embellishment to all other arts, so that perfect knowledge might be acquired; indeed, these sciences and natural philosophy have so close an affinity with one another that unless they give each other mutual aid they can in no way preserve their own status. For this to happen, it will be necessary first that physics students should study mathematical disciplines at the same time—a habit that has always been retained in the society's schools thus far. For if these sciences were taught at a different time, philosophy students would think, and understandably so, that they were in no way necessary to physics. And so very few would want

[43]"Modus quo disciplinae mathematicae in scholis Societatis possent promoveri": *Monumenta Paedagogica Societatis Iesu*, t. VII, *Collectanea de Ratio Studiorum Societatis Iesu (1588-1616)* (Rome, 1992), pp. 115–116; *Monumenta Paedagogica Societatis Jesu Quae primam Ratiomem Studiorum anno 1586 editam praecessere* (Madrid, 1901), pp. 471–472: "Primum deligendus erit magister eruditione atque authoritate non vulgari. Alterutra enim si absit, disciputi, ut experientia docet, non videntur ad disciplinas enim si absit, discipuli, ut experientia docet, non videntur ad disciplinas mathematicas allici posse. [...] [P]ersuadeant sibi coniunctas esse, ut vere sunt, philosophiam scientiasque mathematicas; praesertim, quia hactenus discipuli contempsisse videntur fere has scientias hac una adducti ratione, quod putent eas non haberi in pretio, imo inutiles esse [...]. Necessarium etiam videntur, ut praeceptor habeat inclinationem quandam et propensionem ad has scientias praelegendas, et non sit multis aliis occupationibus distentus; alias vix discipulos iuvare poterit. Ut autem Societas semper habere possit idoneos harum scientiarum professores, eligi deberent aliquot ad hoc munus obeundum apti et [12] idonei, qui in privata academia instituentur in variis rebus mathematicis [...]. Omitto philosophiam naturalem sine disciplinis mathematicis mancam esse et imperfectam [...]." A discussion on these sentences appears in Robert Schwickerath, *Jesuit Education: Its History and Principles* (St. Louis, 1903), pp. 133–135. Cf. Edward C. Phillips, "The Proposals of Father Clavius, S. J., for Improving of the Teaching of Mathematics," *Bulletin of the American Association of the Jesuit Scientists, Eastern Section*, Vol. 18, No. 4 (1941), pp. 203–204. The translation is taken from Roger Ariew, John Cottingham, and Tom Sorell, eds., *Descartes' Meditations: Background Source Materials* (Cambridge, 1998), pp. 25–26.

> to understand them [...].[44]

Then, the author goes on to demonstrate concretely the necessity of mathematics for the study of natural philosophy. For instance, mathematics is essential for understanding the following subjects:

> the number and motion of celestial orbs; the multitude of angels (*intelligentiae*); the effects of the stars, which depend on various conjunctions, oppositions, and other distances between them; the division of continuous quantities into infinities; of the tides, the winds, comets, the rainbow, the halo, and other meteorological matters; the proportion of motions, qualities, actions, passions, reactions, and so on concerning which the calculators (*calculatores*) wrote much.[45]

Here Clavius, as contemporary scholars usually did, regarded a comet as a meteorological phenomenon rather than as a heavenly body. "The calculators" in the quoted passage refer to those scholars who applied their specific theory of proportion to the theory of motion, such as Richard Swineshead, a fourteenth century philosopher of Merton College, Oxford, who composed *Liber calculationum* and was nicknamed "the Calculator."[46] Clavius insists that he can present many more examples besides those above:

> I omit countless examples in Aristotle, Plato, and their more illustrious commentators, which can by no means be understood without some knowledge of the mathematical sciences. Indeed, because of their ignorance of these, some professors of philosphy have very often committed many errors, and those most grave; and what is worse

[44] *Ibid.*, p. 116; p. 472: "Secundo ergo loco, necesse est, ut discipuli intelligant, has scientias esse utiles et necessarias ad reliquam philosophiam recte intelligendam, et simul magno eas ornamento esse omnibus aliis artibus, ut perfectam eruditionem quis acquirat. Immo vero tantam inter se habere affinitatem hasce scientias et philosophiam naturalem, ut nisi se mutuo iuvent, tueri dignitatem suam nullo modo possint. Quod ut fiat, necessarium erit primo, ut auditores physices audiant simul disciplinas mathematicas. Qui mos hactenus in scholis Scietatis semper fuit. Nam si alio tempore praelegentur hae scientiae, existimarent philosophiae auditores, neque immerito, eas nullo modo esse necessarias ad physicam; atque adeo pausissimi eas intelligere vellent [...]." Cf. Phillips, p. 204; R. Ariew *et al.*, p. 26.

[45] *Ibid.*: "[...] de numero et motu orbium caelestium, de multitudine intelligentiarum, de effectibus astrorum, qui pendent ex variis coniunctionibus, oppositionibus et reliquis distantiis inter sese, de divisione quantitatis continuae in infinitum, de fluxu et refluxu maris, de ventis, de cometis, iride, halone et aliis rebus meteorologicis, de proportione motuum, qualitatum, actionum, passionum et reactionum etc., de quibus multa scribunt calculatores." Cf. Phillips, pp. 204–205; R. Ariew *et al.*, p. 26.

[46] See, for example, John E. Murdoch and Edith E. Sylla, "The Science of Motion," in David Lindberg, ed., *Science in the Middle Ages* (Chicago, 1978), pp. 206–264, esp. pp. 223-231; E. D. Sylla, "The Oxford Calculators," in Norman Kretzmann, Anthony Kelley, and Jan Pinborg, eds., *The Cambridge History of Later Medieval Philosophy* (Cambridge, 1982), pp. 540–563; William A. Wallace, "The Calculatores in the Sixteenth Century," in his *Prelude to Galileo* (Dordrecht, 1981), pp. 78–90.

> they have even published them; it would not be difficult to produce examples.[47]

Clavius does not view the existing situation optimistically. The common professors even "teach that the mathematical sciences are not sciences, do not have demonstrations, and abstract from being (*ens*) and the good (*bonum*)."[48] For that reason in particular, Clavius states the following:

> Moreover, Schoolmen will be greatly encouraged to study mathematics, if once a month all the philosphers (*philosophi*) are assembled in one place, where a student would read a short appreciation of the mathematical disciplines, and then with one or two others, would explain a problem from geometry or astronomy, as would be pleasant to hear and be of use for the humanities; such problems can be found in abundance. Or let him explain a mathematical text from Aristotle or Plato, in whose works such texts are numerous. Or even let him propose new and original demonstrations of some of the propositions of Euclid. Let praise be given there to those who best solve the proposed problem [...].[49]

Clavius's proposal is rather concrete. Noteworthy in the above passage is that Plato is placed together with Aristotle. We might be able to characterize the philosophy of Clavius only as scholastic Aristotelianism. But we can also detect a considerable and conscious Platonic element in his philosophy. As we will witness in the next chapter again, Clavius's philosophy was Aristolelianism containing a certain notable Platonic tendency, influenced by the Neoplatonist Proclus, and may be compared to the philosophy of Jacopo Mazzoni, Galileo's colleague at the University of Pisa, who is known for his attempt to make

[47] *Monumenta Paedagogica Societatis Iesu*, t. VII (n. 42), p. 116; *Monumenta Paedagogica Societatis Jesu* (n. 42), pp. 472–473: "Omitto infinita exempla in Aristotele, Platone et eorum interpretibus illustrioribus, quae nulla ratione intelligi possunt sine mediocri scientiarum mathematicarum cognitione. Immo propter earum ignorationem nonnulli philosophiae professores saepissime multos errores, eosque gravissimos, commiserunt, et (quod peius est) scriptis etiam mandarunt; quorum aliquos in medium proferre non esset difficile." Cf. Phillips, p. 205; R. Ariew *et al.*, pp. 26–27.

[48] *Ibid.*, p. 116; p. 473: "[...] docent scientias mathematicas non esse scientias, non habere demonstrationes, abstrahere ab ente et bono etc." Cf. Phillips, p. 205; Ariew *et al.*, p. 27. For "the common professor" whom Clavius is supposed to have referred to, see Chapter 7, Chapter 7, § 2. One of them was clearly B. Pereira.

[49] *Ibid.*, p. 117; pp. 473–474: "Praeterea, ad haec studia maxime incitabuntur scholastici, si singulis mensibus omnes philosophi in unum aliquem locum conuenirent, ubi unus disciplulorum habeat brevem commendationem disciplinarum mathematicarum; deinde cum uno aut altero explicet problema aliquod geometricum vel astronomicum, quod et iucumdum esset auditoribus et utile rebus humanis; qualia problemata plurima reperiri poterunt; vel declaret locum aliquem mathematicum ex Aristotele vel Platone, qualia loca apud ipsos non pauca sunt; vel etiam afferat novas demonstrationes quarumdam propositionum Euclidis a se excogitatas. Ubi laudari possent ii, qui melius problema propositum solvissent [...]." Cf. Phillips, p. 205; Ariew *et al.*, p. 27.

a syncretism of Aristotle and Plato.[50] In any case, Clavius was a scholastic who placed mathematics in a very high status and criticized any dogmatic epistemology that claimed mathematics is not a proper science.

What is going to be referred to among those proposals by Clavius on mathematical education is a draft entitled "On Teaching Mathematics" which is conjectured to have been composed toward the year 1593. This treatise is designed to give a concrete program of mathematical studies "which were being almost neglected."[51] Here Clavius sets out a rather ambitious plan for the study of mathematics. Probably it was written not for ordinary students in the colleges of the order, but for more advanced students within it. There he suggests mathematical texts to be learned in addition to the first six books of Euclid's *Elements*:

> They [i. e. the students] can begin studying at the seventh book [of Euclid] and go through to the twelfth inclusively; then they add the elements of *Spherics* of Theodosius, and some of the known works of Apollonius. This could be done easily in one year, if they attend two lectures a day, which is what ought to be done.[52]

Further Clavius recommends that theology students have lessons "in the theory of planets, in gnomonics, and on the astrolabe, something from Archimedes and algebra."[53] It is known that studies of astronomy had been part of the university curriculum since the fourteenth century. Clavius adds to them "something from Archimedes and algebra." With the rise of Renaissance humanism, Archimedes not only became an object of admiration as a mathematician of rare talent but also acquired the reputation of an ingenious engineer.[54] Clavius recommended theology students to study Archimedes and algebra, because they were considered to be necessary for teaching later generations within the Order and mission activities. In a memorandum drafted by Clavius toward the year 1581 he recommends concerning the astrolabe that its sturcture and use should be explained.[55] The same note suggests that as to Archimedes some writings of his should be studied with the method of discovering mean proportionals of two line segments, the problem of doubling the cube, and the approximate quadra-

[50] See F. Purnell, Jr., "Jacopo Mazzoni and Galileo," *Physis*, **14** (1972), pp. 272–294.

[51] "De re mathematica instructio": *Monumenta Paedagogica Societatis Iesu*, t. VII (n. 42), p. 117; *Monumenta Paedagogica Societatis Jesu* (n. 42), p. 474: "[...] mathematicae studia, quae pene iam negligebantur [...]." Cf. Phillips, p. 206.

[52] *Ibid.*, p. 118; p. 476: "[...] poterunt a septimo initium audiendi facere usque ad duodecimum inclusive; tum vero addere Theodosii sphaerica elementa, et aliqua ex cognitis Apollonii. Quod satis commode uno anno fieri posset [...]." Cf. Phillips, p. 207.

[53] *Ibid.*: "[...] exercentur [...] theoricas planetarum, gnomonicen, astrolabium, aliquid ex Archimede et ex algebra [...] audire [...]."

[54] W. R. Laird, "'Archimedes among the Humanists," *Isis*, **82** (1991), pp. 629–638.

[55] "Ordo servandus in addiscendis disciplinis mathematicis": *Monumenta Paedagogica Societatis Iesu*, t. VII (n. 42), p. 112.

ture of the circle.[56] As textbooks of algebra it refers to those by Michael Stifel, Johannes Scheubel, and Jacques Peletiers.[57] A textbook on algebra by Clavius himself was published in 1608 in Rome much later than these.

Now we know that the *Ratio studiorum* of 1599 reflects the intentions of some influential Jesuits to reform the teaching of mathematics. We cannot necessarily think that Clavius's proposals for promoting mathematical education were totally incorporated in it, for some conservative philosophers did not accept them, as Clavius himself admitted. In any case, for all intents and purposes, the curriculum of mathematics was contained basically within the scholastic Aristotelian framework. Mathematics was studied to make it easy for students to understand, for example, Aristotle's works and other sciences such as geography and astronomy. But, at the same time, it is undeniable that the Renaissance idea of raising the status of mathematics within the scheme of learning as a whole was introduced into the attempts to teach mathematics. So to the program of mathematical education we may apply our general conception on the overall curriculum of the Jesuit colleges which we have obtained in Section 2.

During the late sixteenth and early seventeenth centuries, the Jesuit program of mathematical education was rapidly diffused "du centre romain à la périphérie française,"[58] to use a phrase of Romano's monograph. Descartes was almost on the frontline of this wave. He studied mathematics at the Jesuit college of La Flèche no doubt on the basis of the 1599 *Ratio studiorum*. The chair of mathematics was founded there in 1608, and the most probable candidate for Descartes's mathematics teacher was Jean François. He came to La Flèche in 1612, and taught mathematics there until 1617. That is to say, Jean François is supposed to have been a *maître de mathématique* at La Flèche when Descartes was studying there. This will be discussed in details in Chapter 2, § 5. This position was established in answer to the following request of the General of the Society, Claudio Aquaviva:

> If there are not ways to create the chair, it will be necessary to take care to make the resources which will permit bringing it into existence someday.[59]

§ 4. Motives for the Teaching of Mathematics in the Jesuit Colleges

Before shifting our theme of discussion from the institutional background of the teaching of mathematics at La Flèche to the content of the mathematics which

[56] *Ibid.*, p. 113.

[57] *Ibid.*, p. 111.

[58] A. Romano, *Ibid.*, p. 181: "Du centre romain à la périphérie française (Seconde moitié du XVIe siècle)".

[59] Quoted from François de Dainville, *L'Éducation des jésuites, XVIe–XVIIIe siècles* (Paris, 1978), p. 326.

Descartes is supposed to have studied, in this section we discuss the reasons why the Jesuits introduced mathematics into their curriculum. Trying to undertand their motives for mathematical education will help us comprehend the whole scheme of learning of the Society of Jesus more deeply.

It has been well established among historians that there was a revolution in education in the second half of the sixteenth and the first half of the seventeenth centuries.[60] The facilities for higher education expanded very rapidly during that period. In France of the sixteenth century both Catholics and Protestants made efforts to form or adopt new pedagogical theories. Among Catholics of the Counter Reformation, Jesuits' attempts were remarkable.[61] The first college of the Society of Jesus in France opened at Billom in 1556. In the second half of the sixteenth century, Jesuit schools were widespread. From small beginnings the number of the schools increased and the Jesuits acquired renown in the field of learning. Adding up the number of colleges belonging to the Society, by 1581 there were over 150 for secular youth, while in Ignatius's lifetime there had been only 35.[62] One of the reasons that the Jesuits planned the *Ratio studiorum* was to provide their colleges with common standard regulations and methods of teaching. A historian of the *Ratio studiorum* tells of an episode in Paris: "When the College of Clermont was opened in Paris, so many students flocked to it from the university that that venerable institution saw its classes in the Humanities and arts decimated and its prestige seriously threatened."[63] The college of La Flèche was one of those schools that met the demands of French society.

Our understanding of this general background leads us to maintain that the Jesuits had to make up deficiencies in the old scholastic curriculum. Certainly in the medieval university the mathematical disciplines were taught as branches of the quadrivium.[64] But, generally speaking, only elementary arithmetic and the principal parts of Euclid's *Elements* were taught, and even they were studied in a very specific way: They were studied basically not for their own sake but for the sake of the other branches of learning.[65] The Jesuits

[60] As for England, see Lawrence Stone, "The Educational Revolution in England, 1560–1640," *Past and Present*, No. 28 (July 1964), pp. 41–60; and about the Continent, see Henry Kamen, *The Iron Century* (London, 1971).

[61] See, for example, Émile Durkheim, *L'Évolution pédagogique en France* (Paris, 1938; $_2$1969), Deuxième partie, Chapitres 5–7, pp. 261–303; Eugenio Garin, *L'Éducation de l'homme moderne: La Pédagogique de la Renaissance (1400–1600)*, tr. de l'italien par J. Humbert (Paris, 1968), "Les Contre-Reforme et les jésuites," pp. 183–189; Roger Chartier, Dominique Julia et Marie-Madeleine Compère, *L'Éducation en France du XVIe au XVIIIe siècles* (Paris, 1976), Ch. V, "Naissance du Collège," pp. 147–173. In relation with the Counter Reformation, see A. D. Wright, *The Counter-Reformation: Catholic Europe and the Non-Christian World* (London, 1982), Ch. 3, "Scholasticism and Science," pp. 84–120.

[62] Farrell, *Op. cit.* (n. 15), pp. 219–220.

[63] *Ibid.*, p. 220.

[64] For characteristic aspects of the medieval quadrivium, see articles in David L. Wagner, ed., *The Seven Liberal Arts in the Middle Ages* (Bloomington, 1983).

[65] For a general characterization of mathematics in the Middle Ages, see Michael S. Ma-

emphasized and fully institutionalized humanistic learning and arts in addition to scholastic disciplines first of all in order to compete with other educational institutions, above all, those of the Protestants. It was necessary for them to educate their students to be of good erudition and reputation in learning. This was necessary for their purpose of achieving the tasks of the Counter Reformation.

Why, then, did the Jesuits introduce mathematics into their curriculum? We could answer this question in various ways. As we have seen in the 1586 *Ratio studiorum*, mathematics was appreciated as a useful subject for other sciences. Mathematics can serve astronomers, physicists, even historians and theologians as a tool for their studies. This had been recognized in the Middle Ages and emphasized particularly during the Renaissance. In France, especially after the establishment of the royal chair of mathematics by François I in 1532, whose first occupant was Oronce Fine, and by the proselytization of such pedagogues as Peter Ramus, the usefulness of mathematics had beeen gradually recognized.[66] A similar trend could be witnessed in England as well, and England of the same period became a field of "the mathematicians' apprencticeship,"[67] so to speak. Such a general current of the Renaissance began to spread among middle and higher educational institutions. So for secular youth of the Jesuit colleges mathematics was a necessary discipline because of its utility.

We note further a more concrete and clearer motive for the teaching of mathematics. As we see in the 1599 *Ratio studiorum*, mathematics was taught in liaison with geography and astronomy. This fact suggests that the Jesuits thought that mathematics, geography, and astronomy should be studied for a special aim, that is, as part of missionary training. This was true especially for the students of theology. It is no accident that some volunteers studied privately under the guidance of appropriate tutors to make secure their familiarity with mathematics and astronomy.

Missionary activities were extremely crucial for the Jesuits not simply in Europe but also overseas. Here we provide some examples from missionary experiences in East Asian countries, especially in Japan and China, as to why missionary works demanded mathematical training in the Jesuit colleges. These two countries were particularly important, since they already had traditional cultures of their own no less sophisticated than European culture.

Francisco Xavier, Ignatius de Loyola's close colleague even before the

honey's contributions in David Lindberg, *Op. cit.* (n. 45), Ch. 5, "Mathematics," pp. 145–178, and "Mathematics," in *Dictionary of the Middle Ages*, **8** (1987), pp. 205–222.

[66]See F. de Dainville, "L'Enseignement des mathématiques," XVIIe siècles: *Bulletin de la Société d'Étude du XVIIe siècle*, no. 30 (1956), pp. 62–68; R. Hooykaas, *Humanisme, science, et réforme: Pierre de la Ramée* (Leiden, 1958); Jean-Claude Margolin, "L'Enseignement des mathématiques en France (1540–1570): Charles de Bovelles, Fine, Peletier, Ramus," in Peter Scharratt, ed., *French Renaissance Studies* (Edinburgh, 1976), pp. 109–155.

[67]See Mordechai Feingold, *The Mathematicians' Apprenticeship: Science, Universities and Society in England, 1560–1640* (Cambridge, 1984).

founding of the Society of Jesus, attempted a missionary work in Japan from 1549 to 1551. Reflecting upon his missionary experience in Japan, he wrote the following in his letter on January 29, 1552 from Cochin, port town in Southern India, to the Jesuits in Europe:

> They [the Japanese people] are so full of curiosity, so quick to put questions, and have such a strong thirst of knowledge, that their questions are endless. Moreover, they are telling to others what we have answered to their questions. They were ignorant of the roundness of the earth and of the orbit of the sun. They asked about these things and about others, for example, about comets, thunder, rain, snow, and other similar matters. As we answered these questions, they were greatly satisfied and regarded us as learned men. This kind of thing was was very helpful in order that they believe in our words.[68]

Such a report that the Japanese, full of curiosity, have a thirst of knowledge about astronomy and meteorology, in a sense, more than about the Christian doctrine inevitably had a strong impact on the Jesuits.

Xavier wrote to Ignatius himself, too, from Goa on April 9, 1552:

> Furthermore, as the Japanese ask many things, it is also necessary to have knowledge for answering them. It would be good if the fathers were good artists; and they would not be wasting their time if they become well informed on sophistries, in order to be able to catch contradictions while discussing with the Japanese. And they should also know something on the Sphere, as the Japanese wish very much to know about celestial motions, solar elipse, waxing and waning of the moon, they want to know how the water of the rain, snow, hail, thunder, lightning, comets, and other natural things are generated. Their success in explaining them is useful for them to acquire goodwill from people. I report to you this kind of thing about the people in Japan so that you may understand well what kinds of virtue are necessary for the fathers who are about to go to Japan.[69]

---

[68] *Epistolae S. Francisci Xavierii*, t. II (Rome, 1945), p. 265: "São tam curiosos e emportunos em preguntar, tão desejosos de saber, que numqua acabão de pregumtar e de falar aos outras as coussas que lhes respomdemos aas suas pregumtas. Nom sabião eles ho mundo ser redomdo, nem sabião ho cursso do sol; pergumtando eles por estas coussas e per outras como das cometas, relampagos, chuva, e neve, e outras semelhamtes, a que nós respomdendo e declaramdo-lhas, ficavão muito contemtes e satisffeytos, temdo-nos por homens doctos, o que ajudou não pouquo pera darem credito a nossas palavras."

[69] *Ibid.*, p. 373: "También es necesario que tiengan letras para responder a las muchas preguntas que hazen los gipones. Sería bueno que fuesem buenos artistas; y no perdirían nadie [!] que fuesem ssofistas para en las disputas tomar los gipones un contradición; que supiesen alguna cosa de la esphera, porque huegan en grande manera los gipones en saber los movimien-

What is stated is similar to the previous contention. But, this letter is conjectured to have been influential to a considerable degree, since it was from Xavier, "your smallest child in the most remote place," to the general of the Society of Jesus.

The annual report "Things in Japan in 1605" which the Society of Jesus in Japan sent to the headquarters in Rome wrote that "princes" in Japan became interested not only in "the matters of the God" but in mathematics. "And some," wrote that report, "visited the priests' houses":

> driven by the curiosity of listening to new things, especially mathematics, astronomy, other natural secrets the priests talked about. Hearing them, they were greatly astonished, realized the ignorance of their Buddhist priests, and poured ridicule upon lies and inconsistent stories they told about these same subjects. And they have a sharp talent and admired the proofs and the clarity with which the priests explained things, they concluded that the priests are telling the truth about these natural things, making them discover what they did not either know or understand. Consequently they deduce that they are also telling the truth in what concerns God and the salvation of the soul. Through this method, thus, many people are caught by the net of Christ.[70]

The headquarters of the Society of Jesus must have been very much impressed by such an official report telling that leaders in Japan showed special interests in Western mathematics and astronomy.

Carlo Spinola (see Figure 1.2), an Italian Jesuit who arrived in Japan in the summer of 1602, was a theology student who had studied mathematics under Clavius in Rome in 1587. He taught Western mathematics to Japanese people while being engaged in missionary works in "Miyako," namely Kyoto, from

---

tos del cielo, lo[s] eclipsis del sol, mengoar y creser la luna; como se engendra el agua de la lluvia, la nieve y piedra, trovanes, relánpagos, cometas y otras cosas ansi naturales. Mucho aprovecha lla declaración destas cosas para ganar la voluntad al pueblo. Esta informatión de la gente de Gipón me pareció ser cosa conveniente escrivir a vuestra santa Charidad, para que esté al cabo de las virtudes que han de tener los Padres que allá an de yr."

[70] *Relação Anual das coisas que fizeram os Padres da Companhia de Jesus nas suas Missiões, Nos Anos de 1600 a 1609*, Pelo P. Fernão Guerreiro, Tomo Secunda (Coimbra, 1931), p. 280: "Imitamno também nisto alguns senhores grandes da côrte, e de outros reinos quando veem ao Meaco, que é ordinariamente cada ano; os quais muitas vezes vão a casa dos padres; uns por desejo de ouvir as coisas de Deus, outros levados por curiosidade de ouvir coisas novas, e principalmente de matemática, astrologia e mais segredos naturais que os padres lhes declaram, de que ficam por extremo mara vilhados, e conhecendo a ignorância dos seus bonzos, rindo-se das patranhas e disparates que sôbre estas mesmas coisas lhes diziam. E como são de agudo engenho e caem bem pelas demonstrações e clareza com que os padres lhes explicam, inferem bem daqui que, pois os padres pestas coisas naturais lhes falam tanta verdade, descobrindo-lhes o que até agora não sabiam nem entendiam, não poderão deixar de também lha falar no que lhes pregam de Deus e da salvação; e assim por êste meio ficam muitos na rêde de Cristo."

1604 to October 1611. He wrote the following in a letter from Miyako to João Alvarez, assistant of the Portuguese Province of the Society of Jesus in Rome on December 3, 1606:

> Mathematics is quite useful to engage princes in a friendly atmosphere. They are very pleased with this kind of science. Having heard my facility with it, the Court and Mr. General invited me. What is necessary most is to acquire respect from the Japanese. It was good for me that I came to Japan after having studied mathematics. Those people who will come here will acquire respect if they have knowledge of mathematics. One regrettable thing is that I have not brought books. I do not remember things which may arouse curiosity, because I have lost books I had brought from Italy together with notebooks compiled over three years in Milan. These must astonish the Japanese. Thus, I ask you to send to me notes and new books which you don't need and which the fathers within or outside the Society of Jesus wrote. I don't mind that they are written in Italian. I need, for example, detailed books on arithmetic— for I have brought to Japan small books by Father Clavius on the spheres, horology, and the astrolabe—, and general treatises on the measurement of the earth, various kinds of machines, architecture, perspective, and pictures. Here we will obtain great fruits, if there are books about those problems rather than others on theology.[71]

"Mr. General" in the cited passage is the second shogun Tokugawa Hidetada, a son of the founder of the Tokugawa government Tokugawa Ieyasu. Here an impressive thing is reported. For missionary activities among the Japanese, the knowlege about mathematics and natural sciences such as astronomy is said to be more useful than books on theology.

In his letter from Nagasaki on May 25, 1612, too, Spinola wrote that

[71]Archivum Romanum Societatis Iesu, Japonica–Sinica 36, f. 145v: "La Matematica mi serve molto per insinuarmi nella famigliarità di questi Toni principali, I quali si dilettano assai di cotale scienza; e per questa eziandio il Dayri e il Xougunsama hanno di me notizia, e mi hanno fatta chiamare. Anzi per essere da Giapponesi stimato è la cosa più necessaria, sicchè atteso che sapea la Matematica, è etato meglio che venissi nel Giappone; e coloro che sono per venirci se la sapranno saranno stimati. Una cosa mi duole, ed è, che abbiamo libri; e siccome ho smarriti quei portati d'Italia, insieme cogli scritti di quanto aveva letto in 3 anni a Milano, non mi ricordo più di molte curiosità, le quali senza dubbio farebbero stupire questi Giapponesi. Il perchè la prego per carità di voler mandarmi alcuni acritti e libri moderni o sieno composti dai nostril padri, o dagli esterni, di quei che sopravvanzano nei Collegi di costi, ancorchè sieno nella ligua italiana, per esempio qualche Aritmetica copiosa (chè la piccolo del P. Clavio, colla Sfera, de Hologiis, e de Astrolabio ho qui), qualche trattato di misurare i campi ecc., di machine diverse, di architettura, prospettiva, pittura, ed altri. Qualunque libro di siffalte materie ora farà qui più frutto, che non altri libri di Teologia." I am indebted to Prof. Miyazaki Kentarō of Nagasaki Junshin Catholic University for providing a photocopy of this letter which has not been published before.

the Japanese were interested in astoronomy and meteorology.[72] He observed a lunar eclipse in Nagasaki in the autumn of 1612, and estimated, collating with an observation datum of Macao, that Nagasaki is located at a longitude of 15 degrees east of Macao, thus, in latitude 32° 44′ 35″ N and in longitude 127° 31′ 55″ E. These data are known to have been the standard showing the location of Japan for about 200 years.[73] He is said to have baptized more than

Al P. Giovanni Alvares d. C. d. G. Assist.te di Portogallo

Pax Christi

(a)

(c)

(b)

Figure 1.2 (a) The Beginning Part of Spinola's Letter to Alvarez on December 3, 1606 (Archivum Romanum Societatis Iesu, Jan.-Sin. 36, f. 145r); (b) The Cited Part of Spinola's Letter (f. 145v); (c) Spinola (1564–1622) as a Martyr on September 10, 1622 (ARSI, Jan.-Sin. 36, f. 136)

[72] Archivum Romanum Societatis Iesu, Jap.-Sin. 36, f. 156.

[73] See OBARA Satoru, "*Kirisitan Jidaino Kagaku Shisō: Pedro Gomez cho Tenkyū-ron no Kenkyū*" (Scientific Thought of the Christian Era: A Study of Pedro Gomez's *De Sphaera*," *Kirisitan Bunka Kenkyūkai* (Research Society of Christian Culture), ed., *Kirisitan Kenkyū* (Studies in Christianity), Vol. 10 (Tokyo, 1976; first published in 1965), p. 167. Obara's paper is supplemented by the original Latin text of Gomez: "De Sphaera, Textus Ms. editus a Satoru Augustino Obara S. J.," *Ibid.*, pp. 1–78, and its Japanese translation. Generally speaking, the scientific thought of the Christian era in Japan has not suffciently explored yet. It should be attemped with that in Europe and China together.

two thousand Japanese people. Unfortunately, however, he was arrested by the central goverment's policy of banning the Christian religion and burned to death at Nagasaki on September 10, 1622.[74] He has been remembered as a martyr of Christianity. As the basic character of Japanese mathematics since the seventeenth century called *wasan* is thought to have originated from China and the supposed influence of his mathematical education in Japan has remained to be vague, it is certain that he encouraged mathematical studies among the Japanese in the early seventeenth century. In any case, it is memorable that Spinola was one of the first important Jesuits who transmitted Occidental mathematics into early modern Japan.

The eager Christian mission under the leadership of the Society of Jesus lasted less than one hundred years since the arrival of Xavier in 1549, and the Tokugawa goverment issued a decree proscribing the Christian faith in the 1630s along with establishing a centralized feudal regime. In the backgound of this policy there existed ambitions of territorial expansion of Spain and Portugal, which were then maritime empires and supported missionary activities, in addition to Japan's narrow-minded chauvinistic parochialism. In fact, in 1599 when missionary works became apparently placed in a critical situation, a Jesuit sent a letter asking the General of the Society of Jesus Aquaviva asking for secular forces to defend Catholic lords who were emerging in Western Japan.[75] At any rate, the systematic reintroduction of Western mathematics had to await for the institutionalization in 1855 of the *Nagasaki Kaigun Denshūjo* (Nagasaki Naval Training Institute) after the arrival of Perry from the United States of America in 1853.[76]

More interesting stories can be heard from the Jesuits' missionary works in China. Important for us is an episode about Matteo Ricci (see Figure 1.3). Ricci is known above all as the first translator of Euclid's *Elements* into Chinese. He was born in Italy in 1552 and first learned jurisprudence in Rome, but, after having been admitted by the Jesuit order in 1571, studied philosophy

---

[74]On Spinola's biography, see Fabio Ambrosio Spinola, *Vita del P. Carlo Spinola della Campagnia di Giesù morto per la Santa Fede nel Giappone* (Bologna, 1706).

[75]See, for example, Takase Kōichirō, *Kiristan Jidai no Kenkyū* (Studies in the Christian Era) (Tokyo: Iwanami Shoten, 1977). Essential source materials are *Iezusukai to Nippon* (The Society of Jesus and Japan), 2 Vols., edited and translated by Takase Kōichirō (Tokyo: Iwanami Shoten, 1981-88). On the Jesuit missionary woks in Japan, see J. K. Moran, *The Japanese and the Jesuits: Alessandro Valignano in sixteenth-century Japan* (London/New York, 1993). Around the year 1612, Japan had exactly 127 members in the Society of Jesus, and the precise total number of Christians are not known. But, it is estimated that there were about 370,000 Christians in 1614. This seems to be extremely impressive.

[76]Sasaki Chikara, "The Adoption of Western Mathemtics in Meiji Japan, 1853–1903," Sasaki Chikara *et al.*, eds., *The Intersection of History and Mathematics* (Basel/Boston/Berlin, 1994), pp. 165–186, on p. 173; *Idem*, "The Emergence of the Japanese Mathematical Community in the Modern Western Style, 1855–1945," Chapter 12 of Karen Hunger Parshall and Adrian C. Rice, eds, *Mathematics Unbound: The Evolution of an International Mathematical Research Community, 1800–1945* (Providence/London, 2002), pp. 229–252, on p. 231.

and theology at the Collegio Romano. It was Clavius who taught mathematics to him there. Ricci entered China in 1582, went to its capital Beijing in 1601, there translated the first six books of Euclid's *Elements* orally to his Chinese colleague Xu Guangqi, based on the Clavius first edition of 1574, and published in 1607 under the title *Jihe Yuanben* (literally, Source of Quantity).[77] He died in Beijing in 1610. On his death he was given a graveyard there by the Chinese Emperor. Ricci's fate was quite different from Spinola, who attempted missionary actitivties in Japan and finally died a martyr.

Figure 1.3 Matteo Ricci (1552–1610) with a Confucianist robe

Ricci's diary entitled *Storia dell'Introduzione del Cristianesmo in Cina* describes in a quite interesting style what he experienced in China, succeeding the work Xavier had left behind for him. He attracted many Chinese people through his noble-minded personality. But his erudition in mathematics, bestowed on him by Clavius in Rome, played an important role no less than his noble character. No sooner had he entered China and revealed his knowledge of mathematical disciplines containing astronomy and of geography than he acquired "a reputation that he was the greatest mathematician in the world" (*fama che era il maggiore matematico di tutto il mondo*).[78] Then he had a Chinese disciple in mathematics named Chiutaisu

[77] See the following comprehensive study: Peter Engelfriet, *Euclid in China: The Genesis of the First Translation of Euclid's* Elements *in 1607 & Its Reception up to 1723* (Leiden, 1998); Jean-Claude Martzloff, "Clavius traduit en chinois," in *Les Jésuites à la Renaissance*, sous la direcion de L. Giard (n. 14), pp. 308–322; Li Yan and Du Shiran, *Chinese Mathematics: A Concise History*, translated by John N. Crossley and Anthony W.-C. Lun (Oxford, 1987), pp. 190–201; J.-C. Martzloff, *A History of Chinese Mathematics*, translated by Stephen S. Wilson (Berlin/Heidelberg/New York, 1997), pp. 21–22. On Xu Guangqi's life and works, we have now an excellent collection of papers: *Statecraft & Intellectual Renewal in Late Ming China: The Cross-Cultural Synthesis of Xu Guangqi (1562-1633)*, edited by Catherine Jami, Peter Engelfriet, & Gregory Blue (Leiden/Boston/Köln, 2001). Among others, see Chapters 9 and 10 of this volume: Keizo Hashimoto and Catherine Jami, "From the *Elements* to Calendar Reform: Xu Quangqi's Shaping of Mathematics and Astronomy," pp. 263–278, and Peter Engelfriet and Siu Man-Keung, "Xu Guangqi's Attempt to Integrate Western and Chinese Mathematics," pp. 279–310.

[78] Matteo Ricci, *Storia dell'introduzione del cristianesimo in Cina: Fonti Ricciane*, ed. Pasquale M. D'Elia, Vol. I (Rome, 1942), p. 212. The translation is mine. The English translation certainly exists, but it is not trustworthy. *China in the Sixteenth Century: The Journals of Matthew Ricci: 1583–1610*, translated by Louis J. Gallagher (New York, 1953),

(Qu Taisu, in pinyin in today's mainland China) from the city Suceo (Suzhou). At first the disciple knew simply the method of calculation by the Chinese abacus, but learned through Ricci the Western art of calculation in written form, and further obtained the knowledge of Clavius's *Sphaera*, of the first book of Euclid's *Elements*, of various designs of sundials, and of the art of measuring height and length.[79] Thus Ricci's erudition was acknowledged so that the Chinese no longer called "our country 'barbarous'" (*nostra terra* 《*barbara*》),[80] although before that time they had thought that any territories othen than theirs were "barbarous."

In *Storia dell'Introduzione del Cristianesmo in Cina*, we find a further episode telling that Ricci succeeded in his mathematical instruction in Nanjing in the 1590s. There appeared a Chinese scholar named Ciamiammue (Zhang Yangmo), who came to Nanjing attracted by the fame of the erudition of Ricci from the Hanlin Academy (*Hanlinyuan*) in Beijing. According to Ricci's report, knowing that "our intention was to propagate the sacred doctrine of Christ by exterminating the false thought of idol worship", Ciam said to Father Ricci that "You don't need to refute the doctrine of idol worship. What you have to do is only to instruct mathematics, for the Chinese will naturally realize the falsity of the doctrine of idol worship if they become to understand the truth concerning heavenly and earthly matters."[81] We see that a young Chinese man appreciated mathematical sciences highly because of their certainty as well as their usefulness.

Based upon such experiences, Ricci translated Euclid's *Elements* into Chinese with "il Dottor Paolo," namely Xu Guangqi and published it in Beijing. As to why he translated it, he wrote as follows: "Everybody complains that although mathematics is esteemed in China, it lacks a foundation (*fundamento*). When we want to teach the other things scientifically (*scienticamente*), we do nothing without this book. Especially because demonsrations in this book are extremely clear (*chiaro*)."[82] Besides the Chinese translation of the first six books of Clavius's edition of Euclid's *Elements*, Ricci translated Volume 3 of Clavius's *Geometria practica* into Chinese with the assistance of Xu, and published it in 1607 under the title *Celiang fayi* (Explanation of Practical Geometrical Methods). And with "il Dottor Leo," namely Li Zhizao as a scribe, he translated *Astrolabium* (originally published in 1593) into Chinese and published as *Hungai tongxian tushuo* (Illustrated Explanation of the Current Model of the Astrolabe) in 1607, *De figuris perimetris*, a treatise contained in the 1585 version of the *In Sphaeram Ioannis de Sacro Bosco commentarium* as *Yuan rong jiao yi* (The Measurement of Comparisons between Areas and Volumes of That Which Is Contained 'in the Round') posthumously in 1614, and *Epitome*

p. 169. Idem, *Storia: Fonti Ricciane*, Vol. II (Rome, 1949), p. 44. See, Gallagher, p. 322.

[79] Ricci, *Storia*, Vol. 1 (n. 78), pp. 297–298; Gallagher, p. 231.

[80] Ricci, *Storia*, Vol. II (n. 78), p. 51; Gallagher, p. 326.

[81] Ricci, *Storia*, Vol. II , p. 54; Gallagher, p. 328.

[82] Ricci, *Storia*, Vol. II , p. 356; Gallagher, p. 476.

arithmeticae practicae as *Tongwen suanzhi* (Rules of Arithmetic Common to Cultures) in 1614. All the translations were published in Beijing.[83] Ricci published maps of the world from 1584 on, which charmed Chinese and Japanese.[84] Thus, he transmitted the knowledge of mathematics, astronomy, and geography in Renaissance Europe to China.

During the period of total seclusion in Japan, the Chinese translations by Ricci, or "Li Madou's books" were supressed most carefully. The telling evidence is that "32 Books of the European Li Madou and Others and Books to Propagate the False Creed [Christianity]" were listed clearly in the Index of Prohibited Books in 1630.[85] But, it is conjectured that some books on mathematics and astronomy were imported and gave a certain influence on Japanese scholars, certainly before the Index of Prohibited Books was issued[86] and specially after the eighth Shogun Tokugawa Yoshimune relaxed partially the banning of books in 1720.

It is known that Ricci obtained various books of Clavius in China for making these Chinese translations. Ricci wrote a letter to Clavius in Rome from Nanjing on December 25, 1597. That was a letter of appreciation for having received *Astrolabium* from Clavius.[87] He sent two copies of its Chinese translation *Hungai tongxian tushuo*, one to General Aquaviva of the Society of Jesus and the other to Clavius.[88] As can be imagined, Ricci's library was filled with Clavius's books on mathematics.

After Ricci moved his base of missionary works to Beijing, Chinese emperors, in their encounters with Jesuit missionaries, seem to have been interested not in Western religion, i. e. the doctrine of Christianity, but in Western mathematics and astronomy. The Chinese court needed the mathematical sciences in order to reform their calendar, which was of no use. Chinese astronomy was not based on geometrical models and had not been accurate. Only the Western calendar could offer accurate data for astronomical observations. The Roman Catholic church had just accomplished a calendar reform under Pope Gregory XIII, predecessor of Sixtus V, with the aid of a commission of mathematicians including Clavius as a leader. The Gregorian Calendar was promulgated in

---

[83] Martzloff, *A History of Chinese Mathematics* (n. 69), pp. 375, 383, & 385; *Opere storiche del P. Matteo Ricci*, ed. del Comitato per le Onaranze nationali, Vol. II (Rome, 1913), pp. 543–548.

[84] AYUSAWA Nobotarō's studies are known on the maps made by Ricci and the distribution of them to China and Japan.

[85] On the policy of prohibition of imported books, see ITAZAWA Takeo, *Nichiran Bunka Kōshōshi no Kenkyū* (Studies in the Cultural Interaction between the Netherlands and Japan) (Tokyo: Yoshikawa Kōbunkan, 1959), pp. 440–463.

[86] For example, a collection of works by Matteo Ricci under the title *Tianxue chuhan* (A Fundamental Collection of Heavenly Studies), edited by Li Zhizao (*c.* 1629–30) was bought by the Tokugawa family in Nagoya in 1632.

[87] *Opere storiche del P. Matteo Ricci*, Vol. II, pp. 241–243; Christoph Clavius, *Corrispondenza*, Edizione critica a cura di Ugo Baldini e Pier Daniele Napolitani, Volume IV (1597–1601) (Pisa, 1992), Lettera # 140, pp. 29–31.

[88] Ricci, *Storia*, Vol. II , p. 174.

1582.[89] It would be relatively east for those Western scholars who had already experienced, although indirectly, calendar correction to undertake consulting work for Chinese administrators. Thus Ricci became the first Jesuit scholar who served the Chinese imperial court as an astronomer. He wrote to the Jesuit João Alvarez—the same person to whom Spinola addressed the previous letter we referred to above—in Rome on May 12, 1606, that Western mathematics intersted people in Beijing and asked him to convey this message to "my teacher Father Clavius." He then required to "send us an astronomer."

> Let one or two astronomers be sent directly to China, even to Peking [Beijing], for they would be little use in another place. Let them take care to bring with them the necessary books, without relying on Goa, or other places: in fact in the colleges in India, these [mathematical] books do not exist, and if there were any, they would not send them to any other house.[90]

Ricci was trusted at the Chinese court because of his knowledge of the mathematical sciences, not because of his enthusiasm for Christianity.

Another example which we should mention in relation to China is the eminent mathematician of the seventeenth century, Gregoire de St. Vincent. His talent in mathematics was recognized by his teacher Clavius. In about 1614, one missionary in China requested the Jesuit Order to send celebrated mathematician-astronomers to China. This request was favorably received by Rome. Gregoire was one of those who volunteered to go to China, but for some unknown reason, he did not go there.[91]

Certainly we know the example of Gregoire de St. Vincent, who did not go to China. But, after Ricci, not a few Jesuits went to China for missionary activities and teaching the Western doctrine of astronomy. The German Jesuit Johann Adam Schall von Bell (Chinese name, Thang Jo-Wand, or Tang Ruowang) served the Chinese court as astronomer and became the director of the Bureau of Astonomy there.[92] Thanks to such contributions of Jesuit astronomers, Western astronomy succeeded in coping with traditional Chinese and Islamic calendars through the efforts of the Belgian Jesuit Ferdinand Ver-

---

[89] Ugo Baldini, "Christoph Clavius and the Scientific Scene in Rome," in George V. Coyne, M. A. Hoskin, and O. Petersen. eds., *Gregorian Reform of the Calendar: Proceedings of the Vatican Conference to Commemorate its 400th Anniversary, 1582–1982* (Vatican City, 1983), pp. 137–169. Cf. J. K. O'Connell, "Calendar Reform," *New Catholic Encyclopedia*, **2** (New York, 1967), pp. 1065–1066; D. R. Campbell, "Gregory VIII, Pope," *Ibid.*, **4** (1967), pp. 779–781.

[90] *Opere storiche del P. Matteo Ricci*, Vol. II (n. 83), p. 285. Cf. Henri Bernard, *Matteo Ricci's Scientific Contribution to China*, translated by Edward C. Werner (Peiping, 1935), p. 66.

[91] Pasquale M. D'Elia, *Galileo in China* (Cambridge, Mass., 1960), p. 23.

[92] Cf. L. Pfister, *Notes biographiques et bibliographiques sur les jésuites de l'ancienne mission de Chine*, t. I (Changhai, 1932); F. de Dainville, *La Geographie des humanistes* (Paris, 1940), pp. 450–452.

biest (Chinese name, Nan Huai-Jen, or Nan Huairen) in 1668, soon after the death of Adam Schall von Bell in 1666 and its excellence was acknowledged by the Chinse court.[93] Nevertheless, the Jesuit astronomers' contribution to the Chinese court "was not an unmixed blessing," and the Chinese esteemed simply as "new" what Western missionaries insisted as "Western," as Joseph Needham rightly pointed out.[94]

These episodes tell us how important the mathematical sciences were for the Jesuits around 1600, although we don't know exactly to what extent a demand derived from the experiences of their missionary activities influenced the *Ratio studiorum* of 1599. It was necessary that mathematics be institutionalized for such religious purposes.

We might contend that there is one more motive which is not directly connected with the above utilitarian reasons. As we have noted in the previous section, the most influential mathematician in the drafting of the *Ratio studiorum* was Clavius. It was not only for its usefulness but also for its specific epistemological status among the sciences that Clavius advocated the study of mathematics. As will be examined more closely in Chapter 2, §2, he was an admirer of his friend Francesco Barozzi, the translator of Proclus's *Commentary on the First Book of Euclid's Elements*. Moreover, in his "Prolegomena to the Mathematical Disciplines," the introduction to his editions of Euclid's *Elements* (First published in 1574), Clavius declared his sympathy with Proclus. Proclus combined a Platonic belief in mathematics with the Aristotelian theory of scientific demonstration, rejecting Aristotle's epistemology. Barozzi is known for his rebuttal of Alessandro Piccolomini's standpoint which defended Aristotle's epistemological barrier between mathematics and physics. Clavius seems to have shared Barozzi's position in a series of debates about the certainty of mathematics which were argued by Piccolomini, Barozzi, Pietro Catena, Benedictus Pereira, and others in the second half of the sixteenth and the early seventeenth centuries.

As was stated previously, by the middle of the sixteenth century Platonic ideas were in general known in major universities although the dominant docrines there were still strongly Aristotelian. Clavius sometimes invokes Plato and Proclus to defend the study of mathematics. Of course, mathematical studies could be encouraged within the framework of Aristotelian epistemology. In fact, Clavius's metaphysical foundation was a highly eclectic version of Renaissance Aristotlelianism. When Clavius insisted that mathematics was an essential clue to understanding the *Posterior Analytics*, he was pointing beyond ordinary Aristotelians such as Piccolomini. With such a philosophical basis, as Charles B. Schmitt points out, "Clavius emphasized the superiority of mathe-

[93]P. Adrien Greslon, *Histoire de la Chine sous la donmination des Tartares, Ou l'on verra les choses les plus remarquables qui sont arrivées dans ce grand Empire, depuis l'année 1651, qu'ils ont achevé de le conquerir, jusqu'en 1669* (Paris, 1671).

[94]Joseph Needham, *Science and Civilisation in China,* Vol. 3*: Mathematics and the Sciences of the Heavens and the Earth* (Cambridge, 1970), pp. 447 & 449.

matics over dialectic in terms of its reasoning power, in some ways reminding us of Galilei."[95] Clavius clearly intended to go beyond ordinary Aristotelians in praising the certainty of mathematics. It must be admitted at the same time that it was in later scholars such as Galileo and Descartes that even such an eclectic version of Aristotelianism was rejected.

In any case, we witness a rather impressive scene in the history of mathematical education after the middle of the sixteenth century. Full-time teachers of mathematics were created in the Jesuit colleges. For example, in the German Provinces, teaching positions of mathematics were established at Prag in 1556; at Cologne in 1559; at Vienna in 1560; at Mainz in 1566; at Ingolstadt in 1580; at Graz in 1582; at Olmütz and Würzburg in 1590; at Dillingen in 1594; at Louvain in 1597; at Trier in 1599; at Molsheim and Paderborn in 1612; at Douai in 1617; at Freiburg-im-Breisgau in 1620; at Liège in 1625; and at Münster in 1626.[96] In France, mathematical teachers began to appear at Pont-à-Mousson in 1592; at Avignion in 1600; at Tournon in 1604; at Lyon in 1605; at La Flèche in 1608; at Dole in 1615; at Toulouse in 1619; at Paris and Reims in 1620; at Dieppe in 1633; at Poitiers in 1635; and at Aix in 1637.[97] These examples are taken only from the positions established before 1638.

Therefore, it would not be surprising that the Society of Jesus produced many excellent mathematicians in the seventeenth and eighteenth centuries such as Clavius, Ricci, Paul Guldin, Gregoire de Saint Vincent, Joannes Ciermans, Francesco Grimaldi, Jacques de Billy, Athanasius Kircher, and Girolamo Saccheri.[98] We could not even begin to mention all the mathematicians who were educated in the Jesuit colleges but did not belong to the Order. Here it is enough to note that the very first Jesuit college at Messina had already produced the influential Greek mathematician Francesco Maurolico.[99] There is much evidence that the Jesuits of the early modern period encouraged mathematical studies to a considerable degree.[100]

A certain part of the Roman Catholic Church clearly played a crucial role for the rising of modern science. As to this, William B. Ashworth writes: "There is one order that stands out from all others as the scientific order without rival in

---

[95] C. B. Schmitt, *Aristotle and the Renaissance* (n. 20), p. 105.

[96] K. A. F. Fischer, "Jesuiten-Mathematiker in dem deutschen Assistenz bis 1773," *Archivum Historicum Societatis Iesu*, **47** (1978), pp. 159–224.

[97] F. de Dainville, *L'Éducation des jésuites* (n. 59), pp. 323–354; Romano, *Ibid.*, p. 366.

[98] J. MacDonnell, "Jesuit Mathematicians before the Suppression," *Archivum Historicum Societatis Iesu*, **45** (1976), pp. 139–147.

[99] Mario Scaduto, "Il matematico Francesco Maurolico ed i gesuiti," *Archivum Historicum Societatis Iesu*, **18** (1949), pp. 126–141.

[100] In addition to Romano, *Op. cit.* (n. 13), see, for example, Farrell, *Ibid.*, pp. 370–373; R. Schwickerath, *Op. cit.* (n. 43), pp. 155–157; A. Schimberg, "L'enseignement des sciences et des mathématiques dans les collèges de la Compagnie de Jesus," in *L'Education morale dans les collèges de la Compagnie de Jésus en France sous l'Ancien Régime* (Paris, 1913), pp. 513–533.

seventeenth-century Catholicism, and that of course is the Society of Jesus."[101] Further, he points out in relation to other scientific societies: "One might well argue that the Society of Jesus, rather than the Accademia del Cimento or the Royal Society, was the first true scientific society."[102] One cannot talk sufficiently about the history of science in Europe or even in the world in the 16th and 17th centuries without studying the sciences of the Society of Jesus.[103] The study of the history of science of the Society of Jesus in future should be carried out with the perspective of the world including East Asia.

During the period in which the Society of Jesus began to regard mathematical education as important and to increase rapidly the number of full-time teachers of mathematics, Descartes could receive superior mathematical education.[104] He was standing exactly at the forefront of waves of Jesuit missionary works which would enhance the spirituality of the elite of France.[105]

Now we have to admit that Descartes entered "the stage of the theater of the world" at an auspicuous time. At the Jesuit college of La Flèche, he "delighted in mathematics." Nevertheless importantly, he "was surprised that nothing more exalted had been built upon such firm and solid foundations." To build something more exalted was Descartes's own historical task. Despite their emphasis on the teaching of mathematics, the Jesuits' overall scheme of learning was scholastic-Aristotelian. Even Clavius's program of improvement did not necessarily break through this framework. To reject that scheme itself and reform the mathematics that they were teaching was to be our mathematician-philosopher's work. In all events, it was at the Jesuit college of a small town in France that Descartes was provided with a background which he would later attempt to overcome by creating an entirely new mathematical thought.

Before starting to examine Descartes's own work, in the next chapter we shall turn to an investigation of the content of mathematics which he is considered to have learned at La Flèche.

---

[101] William B. Ashmorth, Jr., "Catholicism and Early Modern Science," in David C. Lindberg and Ronald L. Numbers, eds., *God and Nature: Historical Essays on the Encounter between Christianity and Science* (Berkeley/Los Angeles/London: 1986), p. 154.

[102] *Ibid.*, p. 155.

[103] In "Preface" of Mordechai Feingold, ed., *The New Science and Jesuit Science: Seventeenth Century Perspective* (Dordrecht/Boston/London, 2003) at p. vii, the editor has encouraged the history of science of the Society of Jesus, citing George Sarton's words: "[O]ne cannot study the history of mathematics in the 16th and 17th centuries without coming across Jesuits at every corner" (G. Sarton, "An Appeal for the Republication in Book Form of Father Bosmans' Studies on Belgian Mathematics in the Sixteenth and Seventeenth Centuries," *Isis*, **40** (1949), pp. 3–6, at p. 3). This collection contains useful essays on Christoph Grienberger (Clavius's successor), Giuseppe Biancani, and Honoré Fabri, and others.

[104] Steven J. Harris, "Les Chaires de mathématiques," in *Les Jésuites à la Renaissance*, sous la direcion de L. Giard (n. 14), pp. 239–261, esp. on p. 246.

[105] A. Lynn Martin, *The Jesuit Mind: The Mentality of an Elite in Early Modern France* (Ithaca/London, 1988).

Chapter 2

The Mathematical Thought of Christoph Clavius

§ 1. Descartes and Clavius

In Chapter 1 we took a general view of the institutional side of Descartes's mathematical background at the college of La Flèche. We found that in the Jesuit colleges of the late sixteenth and early seventeenth centuries the study of mathematics was generally encouraged, and that there the thought of Christoph Clavius, professor of mathematics at the Collegio Romano, was influential. However, the content of mathematical education in the Jesuit colleges, particularly in the college of La Flèche, and how it provoked the young Descartes still remain to be examined. Through examining these questions, we will better understand the continuities and discontinuities between the mathematical education which Descartes received and the mathematical thought which he began to develop in 1618.

The most influential mathematician in the Jesuit colleges during the years Descartes studied at La Flèche was Clavius. This is ascertained by Descartes's own admission as well as by the institutional evidence surveyed in Chapter 1. As is sometimes pointed out, Descartes talked very little about the sources of his mathematical thought. He almost always insisted upon the novelty of his own ideas rather than acknowledging his debt to other mathematicians. But there was at least one occasion when he revealed his mathematical background. He disclosed his studies of other mathematicians in 1646 when he met the English mathematician John Pell, who was then a public professor of mathematics at Amsterdam.

Descartes's pronouncement about his reading other mathematicians' works can be found in a letter from John Pell to Charles Cavendish dated March 2/12, 1646. (March 2 indicates the date of the Britich calendar of that time, while March 12 being the date of the Gregorian calendar.) Pell describes his meeting with Descartes in Amsterdam, giving details of the conversation between the two. It is worth quoting *in toto*:

> Last Thirsday Des Cartes came into our Auditory and heard me reade, though when I had done he excused it, saying y^{t} if his guide had known my chamber so well as my publike houre and place, he would rather have come thither to me: he went with me to my lodging, where we had long discourse of Mathematicall matters, though I sought not so much to speake myselfe as to give him occasion to speake.[1]

Pell seems to have caught a very rare opportunity to trace back Descartes's mathematical career. The English mathematician first reports on Descartes's evaluation of the Greek mathemticians Euclid, Apollonius, Diophantus, and Archimedes:

> He sayes he hath penned very few demonstrations in his life (understand after ye style of ye old Grecians which he affects not) THAT he never had an Euclide of his owne but in 4 dayes, 30 yeares agoe, (He is now 50 yeares old, wanting 3 months). He posted over it and so restored it to him y^{t} had lent it him. He then thought all ye theorems very manifest, but cares not to remember any more of y^{m} y^{n} 47. I de triangulo rectangulo [Book I, Proposition 47 on the rectangular triangle, i.e. the Pythagorean theorem] and 4. VI de triangulis similibus [Book VI, Proosition 4 on similar triangles]. He hath a highe opinion of Euclid and Apollonius for writing so largely y^{t} which he conceives may be put into so little roome. He suspects Diophantus might be excellent in the books wch are loste, but in most of his questions which he hath wrought, he finds they might have been solved with lesse adoe than Diophantus makes. Of all ye Ancients he magnifies none but Archimedes, who he says, in his bookes de Sphaero & Cylindro [*On the Sphere and the Cylinder*] and a piece or two more, shows himself fuisse bonum Algebraicum & habuisse vere-magnum ingenium [to have been a good algebraist and to have had a truly great talent].[2]

From the passage cited above we know that Descartes told Pell that he did not possess his own copy of Euclid's *Elements* and and that he praised both Euclid and Apollonius to some extent. But the French mathematician's praise of the two Greek mathematicians sounds ironical and does not appear

[1]This letter was first published in Helen Hervey, "Hobbes and Descartes in the Light of some Unpublished Letters of the Correspondence between Sir Charles Cavendish and Dr. John Pell," *Osiris*, **10** (1952), pp. 62–90. It was also reprinted in *AT*, IV, pp. 729–732. The part quoted above is on p. 78 and on pp. 729–730, respectively.

[2]Hervey, p. 78; *AT*, IV, p. 730.

to have been without reservation. It is truly impressive that he thought there existed algebra (we should read it as 'analysis') in the works of Archimedes and that he was fascinated by the talent of the ancient mathematician. Judging from Descartes's opinion of Diophantus and Archimedes together with his short comment on Euclid and Apollonius, Greek mathematical works altogether should be simplified, following his new mathematical method.

Pell also obtained some information as to how Descartes evaluated several modern mathematicians, i. e. Viète, Fermat, Roberval and Golius, but he did not write about this to Cavendish. Then follows the most important statement about the sources of Descartes's algebra: "He says he had no other instructor for Algebra than ye reading of Clavy Algebra above 30 yeares agoe."[3] Descartes, according to Pell, here refers to Clavius's *Algebra*, published in Rome in 1608. If Descartes's statement is trustworthy, he read before 1616 Clavius's monograph on algebra. It is conjectured that he graduated from the college of La Flèche in September 1615 and it is certain that he took a bachelor degree of law from the University of Poitiers in 1616. Thus, it may be reasonably safe to state that he studied Clavius's *Algebra* after the academic year 1613–1614 when it was determined at La Flèche that he should study mathematics as a second-year student of the philosophical curriculum, and before 1616. Moreover, from his statement that he read it more than 30 years ago, namely almost at the same time as Euclid's *Elements*, we may suppose that its reading was when he was going to graduate from the college at La Flèche. Of course, we should not trust autobiographical testimony without due consideration. Keeping this in mind, we may be permitted to rely on Descartes's claim that he read Clavius's *Algebra* before 1616, although it would be doubtful that Clavius's was the *only* textbook of algebra for Descartes.

There is other evidence which may reconfirm Descartes's careful reading of some of the works of Clavius. This case might be taken as contradictory to Descartes's assertion to Pell about his reading of Euclid. The passages cited above could give us the impression that Descartes looked at Euclid's *Elements* only superficially. The following description dispels such an impression if we grant that Descartes's narrative was basically trustworthy. On November 13, 1629 Descartes wrote to Mersenne from Amsterdam on the curve called quadratrix. Here again Clavius was referred to. In this letter Descartes answered a mathematical question proposed by Mersenne, mentioning the curve

> which is called 'quadratrix', because it is used for squaring the circle and even for dividing the angle into any equal parts just as this one, and for many other uses that you can see in the *Elements* of Euclid annotated by Clavius.[4]

[3] Hervey, p. 78; *AT*, IV, pp. 730–731.

[4] "Descartes, à (Amsterdam), à (Mersenne, à Paris). (13 novembre 1629)," *AT*, I, pp. 70–71; *AM*, I, pp. 83–84; *CM*, II, p. 315.

The work which Descartes touched on was *Euclidis Elementorum libri XV* with extensive annotations by Clavius, first published in Rome in 1574, and since then widely read, having gone through several impressions which included a considerable number of additions and emendations. The part referred was a rather long scholium by Clavius to Book VI of Euclid's *Elements*, which the commentator hit upon while pursuing Book IV of Pappus of Alexandria's *Collection*, and attached to the second edition, which appeared in Rome in 1589, for the first time. Clavius's argument seems to have influenced even the mature Descartes and will be discussed thoroughly in Section 3.

It seems clear that Clavius's mathematical thought had a considerable impact on Descartes in his formative period. The mathematics of the Society of Jesus in the early seventeenth century may be characterized as "L'Architecture mathématique bâtie par Clavius,"[5] this characterization can be ascertained by testimonies of Descartes himself.

Christoph Clavius was born on March 25, 1538 in the German town of Bamberg. He was admitted to join the Jesuits by Ignatius himself in 1555 and, soon, entered the University of Coimbra, Portugal. There he surely studied philosophy under Pedro da Fonseca, but it is conjectured that he learned mathematics under Pedro Nuñes Salaciense, although we don't have certain documentary evidence to support this conjecture. After returning to Rome, he began to teach mathematics at the Collegio Romano in 1563. From this year on until his death on February 6, 1612, he was the most influencial mathematician among the Jesuits. He visited Messina in 1574 to bring many writings of Francesco Maurolico back with him. He is also known as the principal promoter of the Gregorian calendar reform (1577–82) and a mentor of the young Galileo.[6] Incidentally, in China he became familiar with the Chinese name of "*Ding xiansheng*" (丁先生) which originated from his Latin name 'Clavius'.[7] The Latin

[5] *Les Jésuites à la Renaissance*, dir. par L. Giard (Ch. 1, n. 14), p. 237.

[6] For Calvius's works, see Calros Sommervogel, *Bibliothèque de la Companie de Jésus: Bibliographie*, t. II, Nouvelle édition (Bruxelles/Paris, 1891), cols. 1212–1224. As for Clavius's career and works, see Edward C. Philips, "The Correspondence of Father Christopher Clavius S. I. Preserved in the Archives of the Pont. Gregorian University," *Archivum Historicum Societatis Iesu*, **8** (1939), pp. 193–222; J. B. Easton, "Clavius, Christopher," *New Catholic Encyclopedia*, **3** (New York, 1967), pp. 923–924; H. L. L. Busard, "Clavius, Christopher," *DSB*, III, pp. 311–312; Charles Naux, "Le père Christophore Clavius (1537–1612): Sa vie et son œuvre," *Revue des questions scientifiques*, **154** (1983), Pt. 1 "L'Homme et son temps," Pt. 2 "Clavius astronome," and Pt. 3 "Clavius mathématicien," pp. 55–67, 181–193, & 325–347; Eberhard Knobloch, "Sur la vie et l'œuvre de Christophore Clavius (1538–1612)," *Revue d'histoire des sciences*, **41** (1988), pp. 331–356; *Idem*, "Christoph Clavius: Ein Namen und Schriftenverzeichnis zu seinen *Opera mathematica*," *Bolletino di storia delle scienze matematiche*, **10** (1990/2), pp. 135–189; *Idem*, "L'Œuvre de Clavius et ses sources scientifiques," in L. Giard, dir., *Op. cit.* (n. 5), pp. 263–283; *Idem*, "Vorwort" zu Clavius, *In Sphaeram Ioannis de Sacro Bosco Commentarius* (Hiedesheim/Zürich/New York, 1999); James M. Lattis, *Between Copernicus and Galileo: Christoph Clavius and the Collapse of Ptolemaic Cosmology* (Chicago/London, 1994), pp. 1–29.

[7] See Li Yang and Du Shiran, *Chinese Mathematics* (Ch. 1, n. 77), p. 193.

word 'clavus' means 'nail' (釘, *ding*), and because of the same pronounciation it seems to have become his Chinese name 'ding' (丁).

Clavius wrote numbers of books on mathematics and astronomy, in addition to his version and commentary of his Greek predecessor Euclid's *Elements*. Here we will consider his editions of Euclid's *Elements* which we have already referred to, *Epitome arithmeticae practicae* (1583), *Geometria practica* (1604), and *Algebra* (1608). These writings are all contained in his *Opera mathematica*, in 5 volumes, published in Mainz in 1611–12. According to Eberhard Knobloch, who examined Clavius's writings in details, they are classified into three categories, (1) manuals, (2) commentaries and editions, (3) polemical writings.[8] In his contemporary Viète's evaluation, Clavius was not necessarily a creative mathematician.[9] Clavius was a representative mathematician in the late sixteenth century belonging to the generation between Federico Commandino, who translated many Greek mathematical treatises into Latin and was called "restorer of mathematics" (*restraurator mathematicarum*), and Viète, who accomplished a radical reform in mathematics. The character of his mathematics is expressed by a word 'erudite'. As an astronomer, he was located in the generation between Tycho Brahe, who came soon after Tycho Brahe, and Galileo. But, as his big volume on astronomy *In Sphaeram Iohannis de Sacro Bosco commnetarius*, first published in 1570, shows, he had a character of the Middle Ages. *De Sphaera* was a short treatise by the British scholastic Johannes de Sacrobosco, namely, John of Holywood ('Holy Wood' corresponds to 'Sacro Bosco' in Italian), and was widely used in medieval universities as an astonomy textbook. It became customary that later scholars transmited their knowledge on geocentric astronomy originating from Claudius Ptolemaius (Ptolemy) in the form of commentaries on this textbook. Clavius was the last heir of this tradition.[10] In other words, he was an extremely great eclectic. Nevertheless, such an intellectual charater was also a general feature of the institutional matrix of the Jesuits. Despite this character, as a mathematician Clavius radiated an enormous influence upon Europe, even more the world, in the seventeenth century on the institutional basis of the Society of Jesus, and played a crucial role in producing original scholars of later generations.

Needless to say, even the examination of Clavius's essential writings is a painstaking work. Our discussion will naturally be confined to the problem of the extent that Descartes is supposed to have been influenced by Clavius and how he responded to this authoritative mathematician of the Jesuit Order. For this purpose we will concentrate our arguments on three important aspects of Clavius's mathematical thought, that is, his philosophy of mathematics, analysis in his geometrical treatises, and the main characteristics of his algebra.

---

[8]Knobloch, "L'Œuvre de Clavius et ses sources scientifiques," *Op. cit.* (n. 6), pp. 264–265.

[9]François Viète, *Opera mathematica*, recognita Francisci à Schooten (Leiden, 1946; Hildesheim/New York, 1970), "Francisci Vietæ Vita," p. [VIII].

[10]Lattis, *Between Copernicus and Galileo* (n. 6).

§ 2. The Philosophy of Mathematics of Clavius

As we noted briefly in Chapter 1, in the Jesuit colleges the teaching of mathematics was encouraged not only for usefulness but also for an intellectual reason. We witness this clearly in Clavius's philosophy of mathematics, which played an incomparably influential role in the Jesuits' program of mathematical education. It is in the "Prolegomena" to his edition of Euclid's *Elements*, first published in 1574, that Clavius's attempt at promoting his discipline can be seen most comprehensively. In this "Prolegomena" we observe again, but much more distinctly, Clavius's epistemological evaluation of the mathematical disciplines, which we have already examined in the documents related to the *Ratio studiorum*.

The "Prolegomena" is divided into nine sections, the titles of which are: (1) The Mathematical Disciplines, Why They Are So Called, (2) The Division of the Mathematical Disciplines, (3) Discoverers of the Mathematical Disciplines, (4) The Nobility and Preeminence of the Mathematical Sciences, (5) Various Uses of the Mathematical Disciplines, (6) The Recommendation for Euclid and for Geometry, (7) The Division of Geometry and of Euclid's *Elements*, (8) What Is a Problem, a Theorem, a Proposition and a Lemma According to Mathematicians, (9) What Are the Principles of Mathematicians.[11]

The most remarkable characteristic of Clavius's "Prolegomena" seems to be that the author was deeply influenced by the Neoplatonist Proclus's philosophy of mathematics developed in the latter's *Commentary of the First Book of Euclid's Elements*. How Clavius came to know about Proclus's commentary on Greek mathematics provides a clue for elucidating his own standpoint vis-à-vis the philosophy of mathematics. In Clavius's day, two printed editions of Proclus's commentary were available. One was the *editio princeps* of the Greek text attached to the well-known Simon Grynaeus version of Euclid's *Elements*, which appeared in Basel in 1533.[12] The other was a Latin edition translated from Greek manuscripts which were much better than the Basel version of 1533, by the Venetian mathematician Francesco Barozzi and published in Padua in 1560.[13] It was almost beyond doubt from the 1560 Barozzi edition that Clavius

[11] As to Clavius's edition of *Euclidis Elementa*, we refer to Clavius's *Opera mathematica*, t. I (Maintz, 1611). The original titles in Latin are as follows: "Mathematicæ Disciplinæ Cvr Sic Dictæ Sint; Disciplinarvm Mathematicarvm Divisio; Inventores Mathematicarum Disciplinarum; Nobilitas Atque Præstantia Scientiarum Mathematicarvm; Vtilitates Variæ Mathematicarvm Disciplinarvm; Euclidis Atqve Geometriæ Commendatio; Divisio Geometriæ Et Elementorvm Euclidis; Qvid Problema, Qvid Theorema, Qvid Propositio, Et Qvid Lemma apud Mathematicos; Qvænam Sint Principia apvd Mathematicos." Some parts of the "Prolegomena" have been translated into French as "Clavius: Extraits des Prolégomènes aux Disciplines mathématiques, par Michelle Beyssade," in *Le Discours et sa méthode*, sous la direction de M. Grimaldi et J.-L. Marion (Paris, 1987), pp. 207–211.

[12] *Commentariorum Procli editio prima quae Simonis Grynaei opera addita est Euclidis Elementis graece editis* (Basel, 1533).

[13] *Procli Diadochi Lycii in primum Euclidis Elementorum commentariorum libri IV a Fran-*

obtained almost all his knowledge of the Proclus commentary, particularly on the history of Greek mathematics and on the division of the mathematical disciplines. This is evident not only from Clavius's explicit reference to Barozzi's translation in the "Prolegomena",[14] but also from the two mathematicians' close friendship which can be testified to by the correspondence between them. For example, soon afer Barozzi published *Cosmographia* in Venice in 1585, Clavius wrote to Barozzi from Rome on January 29, 1586, thanking the Venetian for sending him a copy: "I admired your Proclus in 1560 and now I admire your fluent and erudite *Cosmographia*."[15]

In the philosophy of mathematics Clavius was a thinker who can be located "entre Proclus et Descartes."[16] As a preliminary step to understanding properly the historical significance of Clavius's theory of mathematics, we should consider the philosophical controversies in which his Venetian friend Barozzi was deeply involved. Indeed in his "Prolegomena", echoes resound of these historically significant debates in the mathematical thought of the sixteenth century. The debates were initiated at the University of Padua on the 'Quaestio de certitudine mathematicarum' between the followers of Alessandro Piccolomini and those of Francesco Barozzi. The episode which gave birth to the disputes was the publication of Piccolomini's *Commentarius de certitudine mathematicarum disciplinarium* in Rome in 1547, which was attached to the same author's paraphrase of the pseudo-Aristotelian *Questions of Mechanics*.[17]

---

cisco Barocio Patritio Veneto Editi (Padua, 1560).

[14]Clavius, *Op. cit.*, p. 3: "Quam quidem diuisionem, quoniam eleganter, copioseque docet, ad quænam sese extendant Mathematicae disciplinae, ferme ad verbum ex Proclo iuxta interpretationem Francisci Barocij Patricij Veneti excerptam hic subijcere statui."

[15]Christoph Clavius, *Correspondenza*, a cura di Ugo Baldini e Pier Daniele Napolitani, t. II (1570–1592) (Pisa, 1992), Lettera #26, p. 65; also cited in Paul Lawrence Rose, "A Venetian Patron and Mathematician of the Sixteenth Century: Francesco Barozzi (1537–1604)," *Studi venetiani*, N. S. **1** (1977), p. 136. The letter is preserved in the Bibliotheque Nationale of Paris.

[16]Echhard Kessler, "Clavius entre Proclus et Descartes," in L. Giard, dir., *Op. cit.* (n. 5), pp. 285–308.

[17]Our analysis is indebted to Giulio Cesare Giacobbe, "Il *Commentarium de certitudine mathematicarum disciplinarum* de Alessandro Piccolomini," *Physis*, **14** (1972), pp. 162–193. See also Paul Lawrence Rose, *The Italian Renaissance of Mathematics* (Geneva, 1975), pp. 285–286. Piccolomini's *Commentarium* and his paraphrase of the *Questions of Mechanics* are placed in the Renaissance context of the revival of interests in mechanics in general. See P. L. Rose and S. Drake, "The Pseudo-Aristotelian *Questions of Mechanics* in Renaissance Culture," *Studies in the Renaissance*, **18** (1974), pp. 65–104. Rufus Suter, "The Scientific Work of Allesandro [*sic*!] Piccolomini," *Isis*, **60** (1969), pp. 210–222, may be useful for learning about Piccolomini's general contributions to science, but touches very little on our present problem. In the following discussion on the debates as a whole, I refer to Paolo Galluzzi, "Il 《Platonismo》 del tardo Cinquecento e la filisofia di Galileo," in *Ricerche sulla curtura dell'Italia moderna*, a cura di Paola Zambelli (Bari, 1973), pp. 37–79; A. C. Crombie, "Mathematics and Platonism in the Sixteenth-Century Italian Universities and in Jesuit Educational Policy," ΠΡΙΣΜΑΤΑ: *Naturwissenschaftsgeschichtliche Schriften, Festschrift für Willy Hartner*, hrsg. von Y. Maeyama und W. G. Saltzer (Wiesbaden, 1977), pp. 63–94; Adriano Carugo and Alistair C. Crombie, "The Jesuits and Galileo's Ideas of Science and of Nature," *Annali*

In his interpretation of the *Questions of Mechanics*, Piccolomini showed a certain understanding of the application of mathematics to some physical problems through the medium of the so-called mixed mathematics (roughly corresponding to 'applied mathematics' in today's terminology) such as mechanics. In the *Commentarius*, however, he began to caution his readers not to think that the mathematical tool was useful in any field of physics. The status of mathematics, in his opinion, is not so high as that of Aristotelian logic.

Piccolomini starts his consideration on the certainty of mathematics by citing Averroës's, namely Ibn Rushd's, commentary on a well-known passage in Book II (α) of Aristotle's *Metaphysics*: "Averroës asserts that mathematical demonstrations are in the first order of certainty."[18] But the Paduan Aristotelian promptly adds to this assertion, "Mathematical certainty should not be demanded in all cases."[19] Piccolomini's point in the treatise was, to be brief, that mathematical certainty is only narrowly applicable, whereas Aristotelian logic or dialectic is quite universal and can be extended to the entire domain of being. In his own terminology, logic can deal with a "most powerful demonstration" (*demonstratio potissima*) which mathematics in general cannot.

Here we should provide a supplementrary explanation to Piccolomini's discussion. Aristotle's theory of demonstration was developed in Book 1, Chapter 13 of the *Posterior Analytics* and so forth. According to it, the knowledge obtained from demonstration or inference is classified into two kinds: 'knowledge of fact' (τοῦ ὅτι) and 'knowlege of reason' (τοῦ διότι). There concepts were translated into the Latin words 'quia' and 'propter quid', respectively. Further, Averroës, in his commentaries on the aforementioned chapter of Aristotle's *Posterior Analytics* and at the beginning of his commentaries of the *Physics*, paid a special attention to the third category of demonstration integrating both which had been called "unqualified demonstration" (ἀπόδειξις ἁπλῶς), and this concept became later called 'demonstratio absoluta', 'simpliciter', or 'potissima' in Latin.[20] Piccolomini's discussion presupposed such preliminary knowledge.

---

dell'Istituto e Museo di storia della scienza di Firenze, **8** (1983), Fasc. 2, pp. 3–68; Paul L. Rose, "《Certitudo mathematicarum》 from Leonardo to Galileo," *Leonardo nella scenza e nella tecnica: Atti del Simposio internazionale di storia della scienza, Firenze-Vinci 23–26 giugno 1969* (Firenze, 1975), pp. 43–49.

[18] *Alexandri Piccolominei in Mechanicas Quæstiones Aristotelis, paraphrasis* [...] *eivsdem Commentarivm de certitvdine mathematicarum disciplinarum* (Roma, 1547), f. LXXIv: "Mathematicas demonstrationes in primo esse ordine certitudinis [...] testatur Auer." Cf. *Aristotelis Metaphysicorum libri XIIII cum Averrois commentariis* (Venezia, 1562), f. 35v.

[19] Piccolomini, *Ibid.*: "Certitudo mathematica non in omnibus expetenda." It is on Aristotle's following passage that both Averroës and Piccolomini make comments. *Aristotelis opera*, ed. Immanuel Bekker, II (Berlin, 1831), 995a14–17: "The accuracy [or certainty] of mathematics is not to be demanded in all cases, but only in the case of things which have no matter. Therefore its method is not that of physics, for presumably all nature has matter." The translation is taken with some modifications from Jonathan Barnes, ed., *The Complete Works of Aristotle: The Revised Oxford Translation*, Vol. 2 (Princeton, 1984). This version, as usual, indicates pages and lines of the Bekker edition.

[20] E. g. *Aristotelis Opera cum Averrois commentariis*, Vol. I, Part. 2a: *Posteriorum Res-*

According to him, a comprehensive demonstration which provides "facts and reason at the same time" (*simul & quia et propter quid*) is called "most powerful" (*potissima*).[21] Namely, a logical 'potissima' demonstration can present both consequences of events and their reason. Moreover, in this jucture, it is necessary to be understood that logic or dialectic, in Aristotle's understanding, can argue synthetically various reasons or 'cause' ($\alpha\grave{\iota}\tau\acute{\iota}\alpha$ = causa) such as formal cause, material cause, efficient cause, and final cause. On the other hand, according to Piccolomini, "it should be understood that the mathematician cannot demonstrate efficient and final causes out of the four genera of causes."[22] Therefore, mathematical demonstrations are not most powerful, for quantity, the category with which the mathematical disciplines are concerned, provides only an ontological characterization of the lowest level among all accidents. Thus, even if mathematics may be certain, such a characteristic is by virtue of its abstraction from concrete beings. And this implies that mathematical demonstrations are incomparable with dialectical demonstrations and that they are far from 'most powerful'.

Piccolomini's argument was challenged by the Venetian nobleman Barozzi (see Figure 2.1) in his *Opusculum, in quo una Oratio, & duae Quaestiones: altera de certitudine, et altera de medietate mathematicarum continentur*, published in Padua in 1560.[23] The first "oration" is a lecture delivered at the University of Padua in 1559. The latter part on "two questions" is nothing other than the ducument which announced the beginning of the long debates on the 'Quaestio de certitudine mathematicarum', which today are very well-known. In the treatise the author Barozzi does not mention explicitly Piccolomini's name, only indicating his opponent, for instance, as "a very bright man who is recently renowned" (*Clarissimus Recentior*).[24] But contemporary readers do not seem to have had any difficulty in identifying to whom the author was referring by that appelation. In the part discussing the first problem, i. e. 'Quaestio de certitudine mathematicarum', Barozzi, opposing Piccolomini's reasoning, emphasizes the preeminence of mathematical demonstrations. Just as Piccolomini does, he quotes the Aristotelian passage "mathematical certainty should

---

olutiorum libri duo (Venezia, 1562), ff. 374r–376v; Vol. IV: *De Physico Libri octo* (Venezia, 1562), f. 4r. Cf. Nicholas Jardine, "Epistemology of the Sciences, " in *The Cambridge History of Renaissance Philosophy*, edited by Charles B. Schmitt and Quentin Skinner (Cambridge, 1988), pp. 685–711, esp. pp. 693–697.

[21] Picolomini, *Ibid.*, f. CXXXIv.

[22] Piccolonimi, *Ibid.*, f. CIIv: "Primum igitur tenendum est ex quatuor generibus causarum, per causam efficientem, & finalem Mathematicum demonstrare non posse."

[23] See G. C. Giacobbe, "Francesco Barozzi e la *Quaestio de certitudine mathematicarum*," *Physis*, **14** (1972), pp. 357–374; P. L. Rose, *The Italian Renaissance of Mathematics* (n. 17), p. 286; *Idem, Op. cit.* (n. 15), pp. 119–178. As for the context of the publication of Barozzi's book, see Manfredo Tafuri, *Venice and the Renaissance*, translated by Jessica Levine (Cambridge, Mass./London, 1989), pp. 133–135.

[24] Barozzi, *Opsculum, etc.* (Padua, 1560), f. 12v.

not be looked for in everything, but in those which do not have matter."[25] Nevertheless, he parts with Piccolomini, adducing wide ranging documentary evidence to prove that mathematical demonstrations are 'potissimae' in their own way. In his judgment, they are not inferior to logical demonstrations, and for this reason he invokes Proclus as an authority. For Barozzi, certainty is the most important criterion for knowledge. It may be said, maintains Barozzi, that divine science (*divina scientia*, i. e. theology) is the most certain not necessarily because of the power of its demonstrations but because of its subject's immutability. To the human mind, however, the certainty of mathematics should be evaluated as that of the first order.[26] Thus, the author's insistence on the preeminence of mathematical demonstrations may be reduced to an ontological problem.

Figure 2.1 A Portrait of Francesco Barozzi (1537–1604) at His Age of 22

In the treatise entitled "Quaestio de medietate mathematicarum" discussing the second problem, Barozzi tries again to justify his claim in the previous treatise by an epistemology which is fundamentally eclectic but influenced most by Proclus. As can be imagined from the title itself, it discusses the intermediateness of mathematics in the hierarchical structure of the sciences. Barozzi argues that Plato and Aristotle differ as to which is the more respectful science, mathematics or natural science. "Plato and his school prefer mathemtics to natural science. On the other hand, Aristotle and all Peripatetics have said that natural science is prior to mathematics, and even they have granted it an intermediate position between natural and divine science."[27] Why does this difference occur, Barozzi asks, between Platonists and Aristotelians? He tries to answer this question in such a way as to reconcile the views of Plato and Aristotle and reaches an ingenious conclusion. Barozzi's concluding remarks are concerned with the rank of theology, mathematics, and natural science:

> In the order for us, natural science is first, mathematics second, and divine science third; on the other hand, in the order of nature, divine science is first, mathematics second, and natural science last, so that

[25] *Ibid.*, f. 7r: "[C]ertitudinem verò Mathematicam non oportet in cunctis quærere, sed in ijs, quæ non habent materiam."

[26] *Ibid.*, f. 32r–32v.

[27] *Ibid.*, f. 37r: "Plato .n. eiusqúe sectatores mathematicam naturali præposuere. Arist. autem, omnesqúe peripatetici naturalem mathematica priorem esse dixerunt, ipsiqúe medium inter naturalem, & diuinam locum tribuerunt."

> mathematics is in truth in the middle between natural and divine science. It should be learned thoroughly, however, for the sake of convenience and in the order of reading before all the other arts and sciences.[28]

The opposition between Plato and Aristotle is, according to Barozzi, not intrinsic but originates in the difference in their points of view, that is, either for us or in nature. What is important for him in any case is that the study of mathematics should be encouraged.

Piccolomini and Barozzi should not be strictly classified as Aristotelian or Platonist. Both cite widely from Aristotle, Plato, Proclus and other authorities. What is surely different between the two is where they place emphasis. Piccolomini insists, on the one hand, that mathematics cannot treat things which contain matter and Barozzi, being consciously inclined to ignore Piccolomini's point, emphasizes, on the other hand, that mathematics has an incomparable certain demonstrative power and that it is closer to theology ontologically. What separates them might be said to be, in a sense, something like a gestalt switch.

The problem Piccolomini and Barozzi tackled attracted the interests of several contemporary scholars as well as those of the following generations. The professor of mathematics at the University of Padua, Pietro Catena, followed Piccolomini's lead by making a detailed comparison between Euclid's proofs and Aristotle's syllogisms.[29] The Jesuit philosopher Benedictus (or Benito) Pereira, Clavius's colleague at the Collegio Romano, also fundamentally supported Piccolomini's position in his book of 1576, developing the argument that mathematics is not properly a science which on a whole can treat causes.[30]

It was against such a problematic milieu that Clavius composed the "Prolegomena" to his edition of Euclid's *Elements*. Indeed, his "Prolegomena" can be rightly situated in the history of mathematics according to this background knowledge about the debates on the 'Quaestio de certitudine mathematicarum'. Clavius clearly sided with his friend Barozzi's following.

In the section of the "Prolegomena" under the title "The Nobility and Preeminence of the Mathematical Sciences," Clavius proclaims his attitude toward the problems of the certainty and of the intermediateness of mathematics:

---

[28] *Ibid.*, f. 40r: "[N]aturalem esse primam ordine, quoad nos, Mathematicam secundam, Diuinam tertiam: ordine verò naturæ diuinam esse primam, secundam mathematicam, postremam naturalem, ita vt Mathematica re vera media semper sit inter naturalem, & diuinam, quæ tamen commoditatis gratia, & lectionis ordine ante omnes artes, atque scientias perdiscenda est."

[29] See G. C. Giacobbe, "La Rifresione metamatematica di Pietro Catena," *Physis*, **15** (1973), pp. 178–196; P. L. Rose, "Professors of Mathematics at Padua University 1521–1588," *Physis*, **17** (1975), pp. 300–304.

[30] See G. C. Giacobbe, "Un gesuita progressista nella 《Quaestio de certitudine mathematicarum》 rinascimentale: Benito Pereyra," *Physis*, **19** (1977), pp. 51–86; P. L. Rose, "Pereira [or Pererius], Benedictus," *DSB*, X, pp. 512–513.

> Since the mathematical disciplines treat things which are considered without any sensible matter (although they are actually impressed in matter), it is clear that they occupy the intermediate place between metaphysics and natural science. If we consider the subject of these, as Proclus rightly proves, that of metaphysics is truly separated from all matter in reality and in theory; On the other hand, the subject of physics is connected to sensible matter in reality and in theory. Then, as the subject of the mathematical disciplines is considered beyond all matter (although matter is really found in it), it holds clear that this is in the middle between the other two. If the nobility and preeminence of a science is truly judged by virtue of the certainty of the demonstrations it uses, undoubtedly the mathematical disciplines will occupy the first place among all.[31]

Here Clavius repeats Barozzi's assertion. Mathematics, according to him, is placed between metaphysics and natural science, but with regard to the certainty of demonstrations, it is more preeminent than any other discipline.

Then Clavius argues as to why mathematics must be greatly respected. From his discussions we come to understand what kind of intellectual milieu he responded to chiefly. We read the following passage in the part which follows immediately after the sentences quoted above:

> For the mathematical disciplines demonstrate and justify by the most solid reasons everything they may raise for disputation, so that they beget knowledge in, and completely drive out all doubts from, the mind of the student. This is what we can hardly attribute to other sciences, because in them most of the time the intellect remains hesitating and dubious about the truth value of the conclusions, so manifold are the opinions and so conflicting the judgments. The many sects of the Peripatetics are enough to prove it (leaving aside other philosophers). All issued from Aristotle, as the various branches of a common trunk, they disagree so completely with each other, and sometimes with Aristotle himself, who is their source, that it is quite impossible to know what Aristotle was really after, or whether his philosophy was primarily concerned with words (*nomina*) or with things (*res*). Such is the reason why, among his

[31] Clavius, *Opera mathematica*, t. 1 (n. 11), p. 5: "Qvoniam disciplinæ Mathematicæ de rebus agunt, quæ absque vlla materia sensibili considerantur, quamuis reipsa materiæ sint immersæ; perspicuum est; eas medium inter Metaphysicam, & naturalem scientiam obtinere locum, si subiectum earum consideremus, vt recte à Proclo probatur, Metaphysices etenim subiectum ab omni est materia seiunctum, & re, & ratione: Physices vero subiectum & re, & retione materiæ sensibili est coniunctum: Vnde cum subiectum Mathematicarum disciplinarum extra omnem materiam consideretur, quamuis re ipsa in ea reperiatur, liquido constat hoc medium esse inter alia duo. Si vero nobilitas, atque præstantia scientiæ ex certitudine demonstrationum, quibus vtitur, sit iudicanda; haud dudie Mathematicæ disciplinæ cæteras omnes principem habebunt locum."

> interpreters, some will follow the Greeks, some others will favor the Latins, or the Arabs, or the Nominalists, or the so-called Realists (and yet all boast that they are Peripatetics). I suppose that everyone sees how far all that is from mathematical demonstrations. For the theorems of Euclid, as well as those of the other mathematicians, are just as purely true today, as safe in their demonstrations, as they already were in schools many centuries ago. In addition to this, Plato stated in *Philebus* or the dialogue about the supreme good: the science superior in accuracy and truth is more worthy and excellent [*Philebus*, 57D]. Since, therefore, the mathematical disciplines are so exclusively dedicated to the love and cultivation of truth, that nothing is received there of what is false, nor even of that which is mere probable, and that nothing is admitted that is not confirmed and corroborated by the most certain demonstrations, there is no doubt that the first place among sciences should be conceded to mathematics.[32]

Clavius praises the certainty of mathematics in the highest possible terms, asserting that there are no disagreements in mathematics, whereas there are many schools and inumerable conflicts even in the one and same sect in other branches of the sciences, especially in philosophy.

Clavius's emphasis on the importance of certainty in science seems to have been motivated at least partly by the specific background against which he reacted. He was struggling to find the means of pacifying conflicts among philosophers and to refute, among others, skeptic thinkers who were very pervasive in the intellectual scene of the late sixteenth and early seventeenth centuries.

---

[32] *Ibid.*: "Demonstrant enim omnia, de quibus suscipiunt disputationem, firmissimis rationibus, confirmantque, ita vt vere scientiam in auditoris animo gignant, omnemque prorsus dubitationem tollant: Id quod alijs scientijs vix tribuere possumus, cum in eis sæpenumero intellectus multitudine opinionum, ac sententiarum varietate in veritate conclusionum iudicanda suspensus hæreat, atque incertus. Huius rei fidem aperte faciunt tot Peripateticorum sectæ, (vt alios interim Philosophos silentio inuoluam) quæ ab Aristotele, veluti rami è trunco aliquo, exortæ, adeo & inter se, & nonnumquam a fonte ipso Aristotele dissident, vt prorsus ignores, quidnam sibi velit Aristoteles, num de nominibus, an de rebus potius disputationem instituat. Hinc fit, vt pars interpretes Græcos, pars Latinos, alij Arabes, alij Nominales, alij denique Reales, quos vocant, (qui omnes tamen Peripateticos se esse gloriantur) tanquam ductores sequantur. Quod quam longe à Mathematicis demonstrationibus absit, neminem latere existimo. Theoremata enim Euclidis, cæterorumque Mathematicorum, eandem hodie, quam ante tot annos, in scholis retinent veritatis puritatem, rerum certitudinem, demonstrationum robur, ac firmitatem. Huc accedit id, quod Plato ait in Philebo, seu dialogo, qui de summo bono inscribitur; Eam scientiam esse digniorem, præstatioremque, quæ magis synceritatis, veritatisque est amans. Cum igitur disciplinæ Mathematicæ veritatem adeo expetant, adament, excolantque, vt non solum nihil, quod sit falsum, verum etiam nihil, quod tantum probabile existat, nihil denique admittant, quod certissimis demonstrationibus non confirment, corroborentque, dubium esse non potest, quin eis primus locus inter alias scientias omnes sit concedendus." Cf. Étienne Gilson, *The Unity of Philosophical Experience* (New York, 1937), pp. 130–131.

It is in the section on "The Recommendation of Euclid and of Geometry" that Clavius suggests the existence of such an intellectual background. There he puts it:

> Unless one was revived by the recognition and knowledge of arithmetic, geometry and dialectic (these arts had been established by forefathers), he would almost arrive at the undecidedness of the Pyrrhonists (they were philosophers who decided nothing but doubted everything).[33]

For Clavius the mathematical disciplines together with dialectic were an effective tool for confronting the skeptics' predilection to hesitate in judgment and deny any certain human knowledge, without distinction between the scientific or the theological.

Pyrrhonism was, of course, one of the most consistent brands of ancient skepticism, of which founder was Pyrrho. It is known today that in 1575 a letter questioning the authority of mathematical demonstrations was addressed to Clavius by the renowned skeptic Francisco Sanches,[34] who published the influential work of skepticism, *Quod nihil scitur*, in Lyons in 1581. Clavius seems to have known to some extent how the skeptics attacked human knowledge. He is also reported to have reacted against a skeptical attitude in astronomy, which took an instrumentalist view concerning the status of astronomical hypotheses, defending a kind of realist philosophy together with Johannes Kepler.[35] As for the passage cited above, our attention should be drawn to the fact that Clavius referred to dialectic along with the mathematical disciplines as tools for combatting the Pyrrhonists. With respect to this, he remained within the framework of the scholastic scheme of learning.

In the section of the "Prolegomena," "Various Uses of the Mathematical Disciplines," we can detect a strong Platonist tone. In composing this section, Clavius appears to have been influenced at least partly by Proclus's chapter

---

[33]Clavius, *Op. cit.*, p. 7: "Ita vt ad Pyrrhoniorum fere (etant Pyrrhonii Philosophi, qui nihil decernebant, sed de omnibus dubitabant) hæsitantiam deuenturus fuerit, nisi Arithmeticæ, Geometriæ, Dialecticæque (quibus artibus ab auis & patre fuerat institutus) esset cognitione scientiaque reuocatus."

[34]As to Sanches's letter to Clavius, see J. Iriarte-Ag., "Francisco Sánchez el Escéptico disfrazado del Carneades en discussión epistolar con Cristóbal Clavio: Un autógrafo inédito y una revalorización de su doctrina," *Gregorianum: Commentarii de re theologica et philosophica*, **21** (1940), pp. 413–451. Sanches's assertion may be summarized by the following sentence: "In mathematics there are many doubts not only in its principles but also in its development." (Multa [...] sunt in Mathematicis dubia, non solum in principiis, sed etiam in progressu.), *Ibid.*, p. 427; Clavius, *Corrispondenza*, Vol. VI, Lettera # 336, p. 187. See also Thomas F. Glick, "Sanchez (or Sanches), Francisco," *DSB*, XII, pp. 97–98.

[35]See Nicholas Jardine, "The Forging of Modern Realism: Clavius and Kepler against the Sceptics," *Studies in History and Philosophy of Science*, **10** (1979), pp. 141–171. Jardine is examining the third edition of Clavius's astronomical work *In Sphaeram Ioannis de Sacro Bosco commentarius*, published in Rome in 1581.

which the translator Barozzi titled "On the Use of Mathematical Sciences."[36] There the Jesuit mathematician quoted widely as evidence of the promotion of mathematical studies not only Proclus's and Plato's works, such as the *Philebus*, the *Epinomis*, the *Timaeus*, and Book 7 of the *Republic*, but also those of the Christian Church Fathers, St. Augustine, St. Jerome, St. Gregory of Nazianzus, and St. Basil. Clavius clearly thought that mathematics was of use for purifying and heightening the Christian spirit in addition to rendering humble service in the worldly everyday life. So the Plato and Proclus which Clavius mentioned were Christianized rather than simply pagan. Invoking Proclus's and Plato's works, Clavius states the following:

> The entrance to metaphysics does not open to anyone, as Proclus elegantly makes clear, without going through the mathematical disciplines. For, if by any means we try to exalt the power and eye of our intellect from sensible things which the physicist (*physicus*) considers to things which are separated and distinguished from all sensible matters that the metaphysician contemplates, we shall become blind just as it happens to one who is freed from a certain darkened cave where one has lain hidden for a long time into the clearest light of the sun. Accordingly before the intellect ascends from physical things, with which matters, subjected to senses, are connected, to metaphysical things, which are very separated from the same, it is necessary for the intellect, in order not to be filled with the clearness of these things, first to be accustomed to such less abstract things as are considered by mathematicians so that it can comprehend them more easily.
>
> Thus the divine Plato suitably confirms that the mathematical disciplines raise up the soul and stimulate the power of the mind to the contemplation of divine things. The blessed Augustine in Book 2, Chapter 16 of *De Doctrina Christiana*, explains to us the most beautifully how much of an advantage these disciplines give for the precise understanding of sacred letters, demonstrating that many things which are placed figuratively and mystically in the Scripture are not to be understood by many people, if ignorant of numbers.[37]

---

[36] *Procli in primum Euclidis Elementorum commentariorum libri IV a Francisco Barocio* (Padova, 1560), "Cap. VIII. De vtilitate Mathematicæ scientiæ," pp. 11–14. Cf. Proclus, *A Commentary on the First Book of Euclid's Elements*, translated by Glenn R. Morrow (Princeton, 1970; 1992), pp. 17–22.

[37] Clavius, *Op. cit.*, p. 5: "Neque enim ad Metaphysicam, vt eleganter ostendit Proclus, vlli patet aditus, nisi per Mathematicas disciplinas. Nam si à rebus sensibilibus, quas Physicus considerat, ad res ab omni materia sensibili secretas, seiunctasque, quas contemplatur Metaphysicus, vires, aciemque nostri intellectus attollere absque vllo medio tentemus, nosmetipsos excæcabimus, non secus, ac ei contingit, qui è carcere aliquo tenebricoso, in quo diu latuit, in lucem Solis clarissimam emittitur. Quam ob rem, antequam à rebus physicis, quæ materiæ sensibus obnoxiæ sunt coniunctæ, ad res Methaphysicas [*sic*!], quæ sunt ab eadem

The well-known story of the allegory of a darkened cave Clavius refers to in the first half part of the quoted passage is, of course, taken from Book 7 of Plato's *Republic* and adduced once by Proclus in the introductory part of the aforementioned chapter "On the Use of Mathematical Sciences." Clavius here invokes this story along with Aurelius Augustinus's *De Doctrina Christiana* for the encouragement of the study of mathematics.

Thus Clavius's philosophy of mathematics can be judged to have been extremely syncretic and similar to Barozzi's in the sense that both relied above all on Proclus as the authority for promoting mathematical knowledge. Together with Federico Commandino, who accepted Proclus's theory of mathematics to a considerable degree in the "Prolegomena" to his Latin translation of Euclid's *Elements* published in 1572,[38] Clavius was confronted with the Piccolomini-Catena-Pereira tendency in the debates on the 'Quaestio de certitudine mathematicarum' Clavius's allusion in a document related to the *Ratio studiorum* to those who insisted that mathematics is not properly a science was clearly connected to this confrontation (see p. 28). Those philosophers whom Clavius criticized belonged to the Piccolomini tradition.

Clavius might be also said to have been among of the first to praise the certainty of mathematics in order to refute merely probable statements and, more fundamentally, skeptic philosophies. In the scholastic-Aristotelian tradition, to acknowledge mathematical certainty was not unusual. As we have seen above, Aristotle himself and even Piccolomini accepted its certainty. And Thomas Aquinas certainly followed Aristotle's lead while adducing the Aristotelian passage in the *Metaphysics* which both Piccolomini and Barozzi referred to, and Ptolemy's statement in the *Almagest* that among the three theoretical sciences, physics, mathematics, and theology, only mathematics can provide incomparable certainty.[39] We will cover Thomas's argument thematically in detail when we examine his theory of mathematics in Chapter 6, § 3. Here we only note the fact that Barozzi also cited Ptolemy's assertion to confirm his own position toward the end of his treatise "Quaestio de certitudine mathematicarum."[40]

---

maxime auulsæ, intellectus ascendat, necesse est, ne harum claritate offundatur, prius eum assuefieri rebus minus abstractis, quales à Mathematicis considerantur, vt facilius illas possit comprehendere. Quocirca recte Diuinus Plato Mathematicas disciplinas erigere animum, & ad diuinarum rerum contemplationem exacuere mentis aciem affirmat. Quantum vero emolumenti hæ disciplinæ ad sacras litteras recte percipiendas, interpretandasque conferant, multis pulcherrime nobis exponit B. Augustinus lib. 2. cap. 16. de Doctrina Christiana, demonstrans, numerorum inscitia multa non intelligi à multis, quæ translate, ac mysticè posita sunt in Scripturis."

[38]For a discussion of Commandino's point of view and of his relationship to Clavius, see Frederick A. Homann, "Christopher Clavius and the Renaissance of Euclidean Geometry," *Archivum Historicum Societatis Iesu*, **52** (1983), pp. 233–246.

[39]Claudius Ptolemaeus, *Syntaxis mathematica*, ed. J. L. Heiberg, *Opera omnia*, I (Leipzig, 1898), pp. 4–8; *Ptolemy's Almagest*, translated and annotated by G. J. Toomer (New York/Berlin/Heidelberg/Tokyo, 1984), pp. 35–37.

[40]Barozzi, *Opusculum*, f. 33r–33v.

In comparison with the orthodox tradition of scholasticism of which Thomas Aquinas was a representative, Clavius's point of view had two new elements. One was a rather strong leaning toward Plato's epistemology. But, although he emphasized the importance of mathematics, we cannot say that he went beyond the basic Thomistic scheme, the general philosophical framework of the Collegio Romano. The difference between orthodox Aristotelians and Clavius can be seen as "not really a difference of theory but a difference of emphasis within the same general theory."[41] The other was his insistence that mathematics could be used to react against the threat of skepticism. It is known that during the Counter Reformation the Catholic Church as a whole and the Society of Jesus in particular made efforts to fight against the threat of infiltration by skeptical philosophies.[42] We do not know exactly what kind of philosophers Clavius was referring to as the "Pyrrhonists" in his "Prolegomena". But it is reported that at the University of Coimbra, Clavius's alma mater, skeptical philosophies were known and commented on, of course, with great antipathy.[43] Moreover, as we have already pointed out above, he received a letter on skepticism from Sanches. What is significant is that Clavius appealed to the certainty of mathematics together with the effectiveness of dialectic to reconcile confrontations among philosophers and to bypass the skeptics' attack on human knowledge, although at present we do not bring into question whether his attempt was successful or not (see Chapter 8, § 1).

By maintaining the significance of the Neoplatonic philosophy of mathematics advocated by Proclus, Clavius surely brought about a new and fresh intellectual atmosphere, especially in the colleges of the Jesuit Order. After his mathematical thought infiltrated there through his mathematics textbooks and teaching at the Collegio Romano in addition to the *Ratio studiorum*, which reflected his program of mathematical education, not a few scholars came to stress the importance of mathematical studies. For example, his disciple at the Collegio Romano, Giuseppe Biancani (Blancanus), in his *Aristotelis loca mathematica* published in 1615, ranged extensively over the loci in which Aristotle argued on mathematics, and developed Barozzi's and Clavius's doctrine concerning the philosophy of mathematics.[44] He not only contributed to the de-

[41]Roger Ariew, "Christopher Clavius and the Classification of Sciences," *Synthèse*, **83** (1990), pp. 293–300, at p. 299.

[42]See A. Carugo and A. C. Crombie, "The Jesuits and Galileo's Ideas of Science and of Nature" (n. 30), pp. 4–5.

[43]Charles B. Schmitt, *Gianfrancesco Pico della Mirandola (1469–1533) and His Critique of Aristotle* (The Hague, 1967), pp. 172–174.

[44]Iosephus Blancanus, *Aristotelis loca mathematica: Ex vniuersis ipsius Operibus collecta, & explicata* (Bologna, 1615). A part of this book discussing "the nature of mathemtical sciences" has been translated into English in Paolo Mancosu, *Philosophy of Mathematics and Mathematical Practice in the Seventeenth Century* (New York/Tokyo, 1996), "Appendix," translted by Gyula Klima, pp. 178–212. See G. C. Giacobbe, "Epigoni nel Seicento della ⟪Quaestio de certitudine mathematicarum⟫ : Giuseppe Biancani," *Physis*, **18** (1976), pp. 4–40; William A. Wallace, *Galileo and His Sources: The Heritage of the Collegio Romano*

bates through his discussion on the nature of mathematics but also later enjoyed a considerable reputation as a mathematician within the Jesuit Order.

By the time the Savilian professor of geometry John Wallis published *Mathesis universalis* in 1657 in which the 'Quaestio de certitudine mathematicarum' was touched upon, it seems to have become evident that the viewpoint of Barozzi's side was predominant.[45] An echo of Pereira's discussions can be heard even in Isaac Barrow's Lucasian series *Lectiones mathematicae*, delivered from 1664 to 1666 at the University of Cambridge, on the philosophy of mathematics and the theory of proportion.[46] These long debates on 'Quaestio de certitudine mathematicarum' clearly contributed to arousing an intellectual trend in which natural philosphers viewed nature experimentally and mathematically.[47]

So in the end we ask: Did Descartes read Clavius's "Prolegomena" and was he affected by his philosophy of mathematics? We surely know, as seen in Section 1, that he glanced at at least a part of Clavius's edition of Euclid, but we have no firm evidence that he did read carefully the Jesuit mathematician's influential "Prolegomena." What we can affirm at least is that both Clavius and Descartes developed their intellectual activities in milieux in which skeptical thinking was widespread as an effect of the Counter Reformation movement and that Descartes was educated in an institution in which Clavius was influential in the mathematical subjects. Clavius's Germany, Coimbra, Rome, and Descartes's France were all shaken by both intellectual and religious confrontations. The two thinkers also struggled to overcome such an intellectual and social situation. Romano, in her monograph on the mathematical thought and institution of the Society of Jesus, has reported that there existed a book to which a Jesuit infulenced by Clavius referred with a title stating that the math-

in Galileo's Science (Princeton, 1984), pp. 141–148. William A. Wallace, Galileo and His Sources: The Heritage of the Collegio Romano in Galileo's Science (Princeton, 1984), pp. 141–148.

[45]See Paolo Mancosu, "Aristotelian Logic and Euclidean Mathematics: Seventeenth-Century Developments of the Quaestio de certitudine mathematicarum," *Studies in History and Philosophy of Science*, **23** (1992), pp. 241–265. Especially on Wallis's *Mathesis universalis*, see pp. 251–258.

[46]Isaac Barrow, *Lectiones mathematicae* (London, 1683), in *The Mathematical Works of I. Barrow*, ed. William Whewell (Cambridge, 1860), Lect. V & Lect. VI, pp. 76–103, in which Pereira is referred to. This work was translated into English by John Kirkby as *The Usefulness of Mathematical Learning* (London, 1734). Lectures V and VI of this translation (pp. 66–100) have the following titles, respectively: "Containing Answers to the Objections Which Are Usually Brought against Mathematical Demonstration" and "Of the Causality of Mathematical Demonstration." According to Mancosu, Barrow's target of critique was Pierre Gassendi. See, Mancosu, *Op. cit.* (n. 45), pp. 258–265. An extended discussion can be seen in P. Mancosu, *Philosophy of Mathematics and Mathematical Practice in the Seventeenth Century* (n. 44), pp. 19–24.

[47]See Peter Dear, "Jesuit Mathematical Science and the Reconstruction of Experience in the Early Seventeenth Century," *Studies in History and Philosophy of Science*, **18** (1987), pp. 133–175. See also the same author's *Discipline & Experience: The Mathematical Way in the Scientific Revolution* (Chicago/London, 1995), Ch. 2: "Experience and Jesuit Mathemical Science: The Practical Importance of Methodology," pp. 32–62.

ematical sciences can meet three scientific criteria of dignity (*dignitas*), utility (*utilitas*), and certainty (*certitudo*).[48] Clavius and Descartes clearly shared the same opinion that these three scientific criteria were crucial. In these circumstances, just as the young Galileo learned much about the contemporary scholastic-Aristotelian theory of knowledge through the Jesuit scholars of the Collegio Romano,[49] the young Descartes seems to have absorbed an invaluable intellectual apparatus which was to be of use in the future. Before looking at Descartes's own thought, we now outline the main characteristics of Clavius's mathematics.

§ 3. Pappus in the Works of Clavius

The three constituents which played an essential role in forming Descartes's universal method were, as was discussed in Chapter 1, § 1, the logic of the schools, the geometrical analysis of the ancients, and the algebra of the moderns. Our problem in this section is how Clavius in his works approached ancient geometrical analysis, especially Pappus's *Collection*, Book VII of which is known as the 'Domain of Analysis' (τόπος ἀναλυόμενος, literally, 'analyzed place'). We tackle this problem principally in order to compare concisely the mathematical thought of Descartes with that of Clavius. In the following examination we will find that the Jesuit mathematician generally ignored the analytic method in mathematics, simply concentrating the synthetic side.

It is known that Renaissance mathematicians obtained information about the meaning of analysis in ancient geometry mainly from the interpolations by a scholiast to Propositions 1–5 of Book XIII of Euclid's *Elements* and the beginning part of Book VII of Pappus's *Collection*.[50] At the outset let us examine how Clavius dealt with (or ignored) the two important topoi on analysis, to which he had access.

As to the scholia to Propositions 1–5 of Euclid XIII, the 1533 Grynaeus

---

[48]A. Romano, *La Contre-Réforme mathématique* (Ch. 1, n. 13), p. 149. The work refered to is A. Possevino's *Bibliotheca selecta de ratione studiorum*, 2 Vols. (Venezia, 1603).

[49]The bibliography related to this problem is extensive. Among others, see C. C. Crombie, "Sources of Galileo's Early Natural Philosophy," in M. L. Righini Bonelli and W. R. Shea, eds., *Reason, Experiment, and Mysticism in the Scientific Revolution* (New York, 1975), pp. 157–175; A. Carugo and A. C. Crombie, *Op. cit.* (n. 30); William A. Wallace, *Galileo's Early Notebooks: The Physical Questions* (Notre Dame, 1977); *Idem, Prelude to Galileo: Essays on Medieval and Sixteenth-Century Sources of Galileo's Thought* (Dordrecht/Boston, 1981); *Idem, Galileo and His Sources* (n. 43); *Idem, Galileo, the Jesuits and the Medieval Aristotle* (Aldershot/Brookfield, 1991).

[50]See Marco Panza, "Classical Sources of the Concepts of Analysis and Synthesis," in Michael Otte and Marco Panza, eds., *Analysis and Synthesis in Mathematics: History and Philosophy* (Dordrecht/Boston/London, 1997), pp. 365–414, esp. pp. 383–389; Richard Robinson, "Analysis in Greek Geometry," *Mind*, **45** (1936), pp. 464–473 and the notable historical analysis, Michael S. Mahoney, "Another Look at Greek Geometrical Analysis," *Archive for History of Exact Sciences*, **5** (1968), pp. 318–348.

Greek edition contained these interpolations[51] and Commandino's Latin translation of 1572 also provided the same scholia on analysis and synthesis.[52] In the Preface of his edition of Euclid, Clavius complimented Commandino on his restoration of the Greek geometer's brilliance,[53] indicating that he clearly read the Commandino version. But in contrast to the previous two versions, Clavius's way of presenting Proposition XIII, 1 (i. e. Proposition 1 of Book XIII) exhibits a distinguishing characteristic. In Proposition 1 we read: "If a straight line be cut in extreme and mean ratio, the square on the greater segment added to the half of the whole can be five times the square of that which is described by the half of the whole."[54] There follows a proof which can be seen in ordinary editions of Euclid's *Elements*. Then we observe something interesting.

Clavius proposes another synthetic demonstration without adducing its analysis. The new paragraph begins with the following words: "Otherwise: From the scholium of Proposition 4 of Book 2, the square of the straight line AB is four times the square of the straight line AD, because the line AB is taken double the line AD."[55] The essential line of Clavius's inference is as follows (see Figure 2.2): $AB^2 = AB \cdot AC + AB \cdot BC$, again $AB \cdot AC = 2AD \cdot AC$. On the other hand, $AB \cdot BC = AC^2$ from the definition of extreme and mean ratio. Therefore, $AC^2 + 2AD \cdot AC = 4AD^2$. Then, $AC^2 + AD^2 + 2AD \cdot AC = 5AD^2$. The left side is equal to CD^2. Thus we have the conclusion $CD^2 = 5AD^2$.

Figure 2.2

While Clavius's terminology is completely geometrical and our notation are algebraic, it does not distort the author's thought process. We witness similar descriptions in Propositions 2–5 as well. Thus we are permitted to say that Clavius was clearly aware of the interpolations to Propositions 1–5 on analysis and synthesis, but that he consciously ignored the analytic parts, offering simply synthetic demonstrations beginning with the word "otherwise" (*aliter*), which, together with its analytic counterparts, were usually placed in the scholia.

Is it accidental that Clavius did not incorporate analytic methods in his presentation of geometrical theorems? The case observed in Propositions 1–5 does not seem to be exceptional in the works of Clavius. To my knowledge, there are few places in which he presents analytic procedures explicitly.

---

[51] Εὐκλείδου Στοιχείων βιβλ. ΙΕ̄. *ἐκ τῶν Θεῶνος συνουσιῶν. εἰς τοῦ αὐτοῦ τὸ πρῶτον, ἐξηγημάτων* Πρόκλου *βιβλ.* δ̄. ed. Simon Grynaeus (Basel, 1533), pp. 239–243. Cf. *Euclidis Elementa*, IV, post I. L. Heiberg, ed. E. S. Stamatis (Leipzig, 1973), pp. 198–204.

[52] *Euclidis Elementorum libri XV cum scholis Antiquis et commentariis Federici Commandini Vrbinatis* (Pesaro, 1572), ff. 229r–232v.

[53] Clavius, *Opera mathematica*, t. I, p. 10.

[54] Clavius, *Ibid.*, p. 335: "Si recta linea secundum extremam & mediam rationem secetur; maius segmentum assumens dimidiam totius, quintuplum potest eius, quod à dimidia totius describitur, quadrati."

[55] *Ibid.*: "Aliter. Ex scholio propos. 4. lib. 2. quadratum rectæ AB, quadruplum est quadrati rectæ AD, quòd linea AB, dupla lineæ AD, ponatur."

Can we affirm that in general Clavius ignored geometrical analysis? Insofar as it is concerned with the first type which Pappus called 'theorematic' (or theoretical) as contrasted with 'problematic',[56] we have a vaguely suggestive clue as to the answer to this question. The analyses used in Propositions XIII, 1–5 are typical examples of theorematic analysis. Clavius seems to have understood mathematical analysis in the same way as scholars of the sixteenth century as having a close connection with logical analysis which was revealed in Aristotle's *Prior* and *Posterior Analytics*. In ancient times generally, analysis in any field was defined as "a process of resolution, of a whole into its parts, of a compound into its elements, of the complex into the simple."[57] Aristotle was well aware of both geometrical and non-geometrical analyses and did not make a clear distinction between the two. Many scholastics and humanists of the sixteenth century fundamentally appear to have followed Aristotle's lead when they tried to restore the classical methods of analysis.[58]

The clue which we were about to mention is that Clavius surely 'analyzed' the very first proposition of Euclid I, which requires the construction of an equilateral triangle on a given finite line. More precisely, he analyzed not in Pappus's manner but in Aristotle's manner, that is to say, he rewrote it into syllogistic form in several ways. For example, the first three lines of Clavius's rewritten form read: "Every triangle having three equal sides is equilateral. The triangle ABC has three equal sides. Therefore, the triangle ABC is equilateral."[59] After having decomposed (or resolved) the given problem into three syllogisms, including the previous one, Clavius notes, "All other propositions not only of Euclid but also of the other mathematicians can be analyzed (*resolvi*) in quite the same way. Nevertheless, mathematicians neglect such an analysis (*resolutio*) in their demonstrations, for they do demonstrate that which is proposed more briefly and more easily without it, as can be made clear from a superior demonstration."[60] Clavius did not further attempt to analyze logically

[56]For the two types of analysis, Mahoney, *Op. cit.* (n. 50), pp. 329–330. Cf. Pappus of Alexandria, *Book 7 of the Collection*, Part 1: *Introduction, Text, and Translation*, ed. Alexander Jones (New York/Berlin/Heidelberg/Tokyo, 1986), pp. 82–83.

[57]This definition is taken from Norman Gulley, "Greek Geometrical Analysis," *Phronesis*, **3** (1958), p. 4. On the relation between geometrical and non-geometrical analyses, see Robinson, *Op. cit.* (n. 50), pp. 466–467, footnote 3, where Robinson lists eight passages in ancient works; Benedict Einarson, "On Certain Mathematical Terms in Aristotle's Logic," *American Journal of Philology*, **57** (1936), pp. 33–54 & 152–172, esp. pp. 36–39; J. Hintikka and U. Remes, *The Method of Analysis* (Dordrecht, 1974), pp. 31–32.

[58]See Angelo Crescini, *Le Origini del metodo analitico: Il Cinquecento* (Udine, 1965), which consists of the following three parts, Part I on the analytic method of logic, Part II on that in natural science, and Part III on that in mathematics.

[59]Clavius, *Opera mathematica*, t. I, p.28: "Omne triangulum habens tria latera æqualia, est æquilaterum. Triangulum ABC, tria habet æqua latera. Triangulum igitur ABC, est æquilaterum."

[60]*Ibid.*: "Non aliter resolui poterunt omnes aliæ propositiones, non solùm Euclidis, verùm etiam cæterorum Mathematicorum. Negligunt tamen Mathematici resolutionem istam in suis demonstrationibus, eò quòd breuiùs, ac faciliùs sine ea demonstrent id, quod proponitur, vt

any other proposition of Euclid. We do not know exactly the reason why he discontinued logical analysis, but we speculate that he thought such an attempt to be useless or that only one example is enough in order to show that mathematical propositions have their place in the formal structure of the Aristotelian analytics.

But, this kind of attitude toward analysis as we have seen in Clavius was not an exception. For instance, we have a similar marked example in *Analyseis geometricae sex librorum Euclidis* published in 1566 by Cunradus Dasypodius with his teacher Christian Herlinus. In this book, all the propositions in the first six books of the *Elements* are rewritten into logical syllogisms.[61] Such an attempt seems to have been efficient to compare mathematical demonstarations with logical ones not only for students familiar with the scholastic curriculumn regarding logic as a center of the arts and sciences but for such scholars who considered demonstrations in mathematics being inferior to those in logic as Piccolomini.

We should not extend too hastily Clavius's attitude toward logical analysis to theorematic geometrical analysis. Be that as it may, what can be ascertained is that he did not pay any special attention to geometrical analysis, simply presenting synthetic demonstrations.

A similar attitude may be expected on the part of Clavius toward Pappus's *Collection* as well. During the late 1580s, Clavius at times communicated with Barozzi on Commandino's Latin translation of the *Collection* made between the years 1574–75, and on Barozzi's project to revise the Commandino manuscript.[62] Commandino's text was published posthumously as *Pappi Alexandrini Mathematicae collectiones* in Pesaro in 1588.[63] Generally speaking, Pappus appears to have attracted Clavius's mathematical interest, but it is doubtful whether Book VII of the *Collection*, the most important ancient literature on the method of analysis, impressed the Jesuit mathematician.

Let us examine how Clavius reacted to Pappus's *Collection*. He referred to it here and there in his *Geometria practica*. Clavius's *Geometria practica* was a work following the tradition of practical geometry which began in the twelfth century with Hugh of St. Victor's *Practica geometriae*.[64] Its topics consisted,

---

perspicuum esse potest ex superiori demonstratione." Clavius's attempt is discussed in Neal W. Gilbert, *Renaissance Concepts of Method* (New York, 1960), p. 90.

[61]Cunradus Dasypodius, *Analyseis Geometricæ sex librorum Euclidis* (Strasbourg, 1566). On Book I, Proposition 1, see, f. IIr.

[62]See P. L. Rose, "A Venetian Patron etc.," (n. 15), pp. 128–135; A. P. Treweek, "Pappus of Alexandria: The Manuscript Tradition of the *Collectio Mathematica*," *Scriptorium*, **11** (1957), pp. 230–231; Jones, *Op. cit.* (n. 56), p. 63.

[63]As to Commandino's translation, see P. L. Rose, *The Italian Renaissance of Mathematics* (n. 17), Chapter 9: "The Urbino School: I Commandino *Restaurator Mathematicarum*," pp. 185–221.

[64]For the tradition of practical geometry in the Middle Ages, see the introductory part of Stephan K. Victor, *Practical Geometry in the Middle Ages: Artis cuiuslibet consummatio and Pratike de Geometrie* (Philadelphia, 1979). See also, *Practical Geometry [Practica Ge-*

first of all, of the measurement of plane and solid figures both theoretically and by using mathematical instruments. But the final Book VIII is devoted to techniques for mathematical problem-solving. At the beginning of that book Clavius states that he took many examples from Pappus's *Collection*:

> In order to complete this *Geometria practica* of ours, we conclude it with many various theorems and problems of geometry, sometimes collected from other geometers, sometimes by my own efforts. And we have an illustrious example in Pappus of Alexandria, who has written eight books in all on the *Mathematical Collections*.[65]

We will examine only one typical problem out of 47 propositions (14 theorems and 33 problems). Problem 16 (Proposition 25) requires the reader: "To Divide the Given Rectilinear Angle into Three Equal Parts." Clavius preliminarily states the nature of the problem and proceeds to his method of solution:

> This problem has greatly stimulated geometers of the former time, and has not been solved geometrically by anybody up to the present. Pappus of Alexandria among others tried to solve it by describing a hyperbola. We will solve the same in this way by the Conchoidal line, which in Book 6, Proposition 15 of this work we have described following Nicomedes.

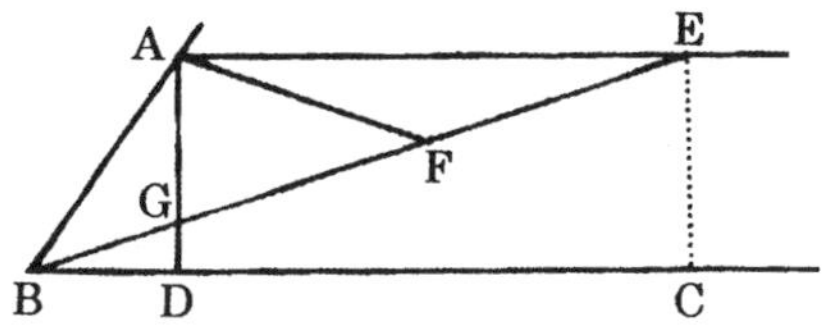

[Figure 2.3]

> Let ABC be the given acute angle: Now let AD be dropped perpendicular from any point A to BC and DC be taken double of AB; And let the Conchoidal line CE with the pole B and the interval DC, cutting the straight line AE drawn parallel to BC in E and let straight line BE be drawn. I say that the angle CBE is the third part [one third] of the given angle ABC.[66]

ometriae], Attributed to Hugh St. Victor, translated from the Latin with an Introduction, Notes and Appendice by Frederick A. Homann (Milwaukee, 1991).

[65]Clavius, *Geometria practica*, in *Opera mathematica*, t. II (Maintz, 1611), p. 195 (paginated as p. 174 by mistake): "Vt extremam manum Geometriæ huic nostræ practicæ imponamus, concludemus eam variis nonnullis Theorematibus, atque problematibus Geometricis, tum collectis ex Geometris aliis, tum proprio, vt aiunt, Marte excogitatis, ac demonstratis. Qua in re exemplum illustre habemus in Pappo Alexandrino, qui octo totos libros conscripsit de Mathematicis collectionibus."

[66]*Ibid.*, p. 208: "PROBL. 16. PROP. 25. Angvlvm datum rectilineum in tres æquales partes partiri. Problema hoc veteres Geometras diu, multumque, exagitauit, neque ab vllo ad hanc vsque diem Geometrice est solutum. Pappus Alexandrinus inter alios illud soluere conatus est per descriptionem hyperboles. Nos idem absoluemus per lineam Conchoideos, quam lib. 6. propos. 15. huius ex Nicomede descripsimus, hoc modo. Sit datus angulus acutus ABC: Demissa autem ex quouis puncto A, ad BC, perpendiculari AD, sumatur ipsius AB, dupla DC; Et polo B, interuallo autem DC, describatur linea Conchoideos CE, secans

Clavius, then, turns to a demonstration, which is not very difficult. According to the way of describing the Conchoid, DC = GE = 2AB (see Figure 2.3). Let F be the midpoint of GE. Then AF = GF = AB. Since the triangle EAG is a right angle, the angle AFG is double the angle FEA (= the angle FAE). As the triangle ABF is isosceles, the angle AFB equals the angle ABF. Therefore, the angle ABF is double the angle FBC. It follows from this that the trisection of the angle ABC has been accomplished.

The point of Clavius's solution is reducing the trisection of an angle to the description of the curve named the Conchoid of Nicomedes. Clavius has already shown how to draw this curve in Proposition 15 of Book VI of his *Geometria practica.*[67] Noteworthy in Clavius's manner of presenting his solution is that he does not state explicitly how the problem is analyzed, directly showing the construction of the problem and its synthetic proof.

In antiquity the trisection of any angle was usually treated as a typical problem based on a certain analytical technique called 'νεῦσις' (inclination), one of the most primitive methods of problematic analysis. In fact, Pappus presented his solution in Book IV of the *Collection* as an application of the technique 'neusis'.[68] Pappus's solution was, as Clavius states, by means of a hyperbola.

The use of the Conchoid of Nicomedes for his solution may have been suggested to Clavius by Pappus's statement that that curved line can be applied to the trisection problem.[69] In contrast to Pappus's placing an emphasis upon the methods of analysis and synthesis, Clavius's *deus ex machina* manner, so to speak, inevitably gives an impression that in general he accorded special attention neither to the method of problematic analysis nor to that of theorematic analysis.

This impression is deepened when we note that in other places as well Clavius makes no explicit mention of the method of analysis which was a subject of Book VII of Pappus's *Collection.* In his Geometria practica on the whole,

rectam AE, ipsi BC, ductam parallelam in E, ducaturque recta BE. Dico angulum CBE, esse tertiam partem dati anguli ABC; [...]."

[67]Clavius, *Ibid.*, pp. 162–163, under the title "Modvd Nicomedis in Libro de Lineis choidibvs." Thr manner of generationg the Conchoids of Nicomedes is also seen in Thomas L. Heath, *A History of Greek Mathematics*, Vol. 1 (Oxford, 1921), pp. 238–239. In the figure, the fixed point C is named the pole (πόλος) and the constant length PD the distance (διάστημα). In Clavius's terminology, they are called "polus" and "intervallum", respectively. In polar coordinates, the curve is denoted by the equation $r = a + b\sec\theta$, where the two constants a = HL and b = DE, and r = DH, $\theta = \triangle$CDH.

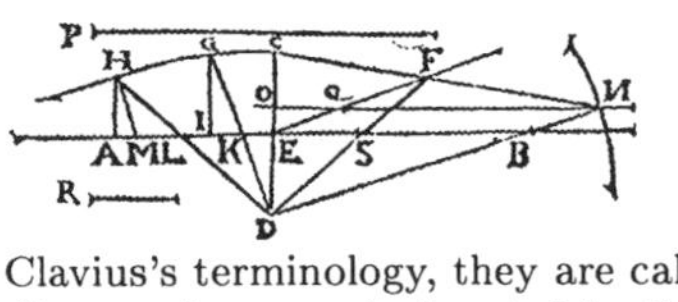

[68]Pappus, *Collectiones*, ed. F. Commandino (Pesaro, 1588), ff. 61r–67r; *Collectio quae supersunt*, ed. F. Hultsch, Vol. 1 (Berlin, 1876), pp. 270–285; *La Collection mathématique*, tr. Paul Ver Eecke, t. 1 (Paris/Bruges, 1933), pp. 206–220. Cf. Heath, *Op. cit.* (n. 67), pp. 235–244; Vol. 2 (Oxford, 1921), pp. 189–192. The epitome of the technique 'neusis' a problematic analysis, see Mahoney, *Op. cit.* (n. 50), pp. 332–333.

[69]Commandino, f. 56r; Hultsch, pp. 246–247; Ver Eecke, p. 187.

Clavius can be said to have borrowed many examples from Pappus's *Collection* without touching on the analytical procedure.

Yet, it is certainly undeniable that Clavius was attracted to Pappus's *Collection*, apart from the method of analysis incorporated into it. For example, his long scholium on the quadratrix in Book VI of his second edition of the *Elements* originated in his reading of Book IV of Pappus's *Collection.* The curved line quadratrix or 'τετραγωνίζουσα' was so named because it was considered to be useful for the quadrature, or squaring, of the circle, one of the three famous problems in Greek mathematics. This is none other than the curve in relation to which Descartes referred to Clavius's edition of Euclid in 1629. We will take a brief look at this example since the case of the quadratrix sheds light on how Clavius approached Pappus and then how Descartes read Clavius.

Pappus treated the curve in Book IV of his *Collection.* The generation of the quadratrix was interpreted roughly as follows: Let us take a square ABCD and describe a circular arc BED around the center A (see Figure 2.4). Rotate the straight line AB, the point A being the center, so that the point B moves uniformly along the arc BED. At the same time, move the straight line BC down always uniformly and parallel to the straight line AD. The two lines AB and BC will always intersect at a point (say F) and at the final stage will coincide with the line AB. Thus, a curved line BFG will be generated as the locus of intersecting points by the movements of the two straight lines. This curve will serve for finding a square equivalent in area to a given circle.[70]

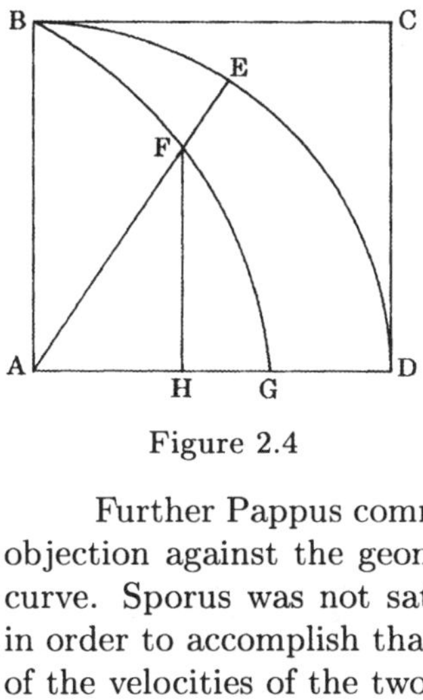

Figure 2.4

Further Pappus commented on this construction citing Sporus of Nicaea's objection against the geometrical legitimacy of this method of generating the curve. Sporus was not satisfied with this line, according to Pappus, because, in order to accomplish that construction, one has to know in advance the ratio of the velocities of the two movements (one along a straight line and the other along a circumference) which is equal to the ratios of the straight line AB to the circumference BED. In Sporus's judgment, the construction of the quadratrix presupposes knowledge of this ratio, and therefore, that of the value π and therefore commits a *petitio principii* (i. e. in order to draw the curve serving to square the circle, it is necessary to know p beforehand). Pappus considered

[70]Commandino, f. 57r; Hultsch, pp. 250–253; Ver Eecke, pp. 191–192. Cf. Heath, *Op. cit.*, Vol. I, pp. 226–229. In modern terminology, the quadratrix can be explained as follows: By its construction, we get $\angle$ BAD : $\angle$ EAD = arc BED : arc EB = AB : FH. If we pose AF = ρ, $\angle$ EAD = φ, AB = R, we will obtain the formula $\frac{\varphi}{\frac{1}{2}\pi} = \frac{\rho \sin\varphi}{R}$, i. e. $\rho = \frac{2\varphi}{\pi} \cdot \frac{R}{\sin\varphi}$. Letting $\varphi \to 0$, we will have AG $= \frac{2R}{\pi}$, or $\pi = \frac{2R}{\mathrm{AG}}$. This prima facie gives the value π, and the square equivalent to a given circle.

Sporus's objection to be valid.[71]

In contrast, Clavius was not persuaded by Sporus's argument or Pappus's agreement. The two Greek mathematicians had expelled the quadratrix from the territory of ordinary geometry, which restricted the means of construction to the rule and the compass. But that curve is, in Clavius's opinion, totally legitimate and geometrically acceptable, like the conic sections by Apollonius of Perga. His way of generating the quadratrix was by means of plotting arbitrarily many points rather than depending upon the two movements.[72] This manner of describing curves may be called 'pointwise construction'. Clavius fully believed in the legitimacy of such pointwise construction of curves and further claimed that the quadratrix "can be called geometrical in a certain way just as the descriptions of the conic sections which are also carried out through points (*per puncta*) are called geometrical as is explained by Apollonius."[73]

It would be rather easy to point out the defects in Clavius's argument from the viewpoint of today's mathematics, for in the pointwise construction of Clavius denumerably infinite points can be plotted, but the entire continuous curve is not necessarily describable.

What is important for us, however, is that Clavius's manner of constructing curves seems to have left a certain mark on Descartes. As was mentioned in Section 1, it was in a letter of November 13, 1629 from Descartes to Mersenne

[71] Commandino, f. 57r–57v; Hultsch, pp. 252–257; Ver Eecke, pp. 193–194. Heath reviews Sporus's argument in *Op. cit.*, Vol. I, pp. 229–230. Sporus appears in the corrupted Commandino Latin version of 1588 as "spero" (in small letters) instead of "Sporo."

[72] Clavius, *Euclidis Elementa*, in *Opera mathematica*, t. I, p. 296: "We will describe the same quadratrix geometrically in the following manner: Let the arc DB be divided into arbitrarily many eqaul parts, both of the sides AD, BC into as many equal parts as the arc DB. The division will be easiest if the arc DB and both of the sides AD, BC be bisected first, then again both of the halves be bisected, once more again each of the singular parts be bisected and in succession as arbitrarily many as one pleases. However, the more the number of divisions will have been carried out, the more accurately the quadratrix line will be described. We have cut the arc DB as well as the two sides AD, BC into eight equal parts in order that confusion should be avoided." (Quare nos Geometricè eandem lineam Quadratricem describemus hoc modo: Arcus DB, in quotuis æquales diuidatur, & latus vtrumque AD, BC, in totidem æquales partes. Facillima diuisio erit, si & arcus DB, & vtrumque latus AD, BC, secetur primùm bifariam, deinde vtraque semissis iterum bifariam, & singulæ rursum partes iterum bifariam, atque desemissis iterum bifariam, & singulæ rursum partes iterum bifariam, atque deinceps, quantum libuerit. Quò autem plures extiterint diuisiones, eò accuratius Quadratrix linea describetur. Nos ad confusionem vitandam secuimus tam arcum DB, quàm duo lateræ AD, BC, in octo partes aequales.) Incidentally, Clavius also discusses the quadratrix in his *Geometria practica*, in *Opera mathematica*, t. II, pp. 189–192.

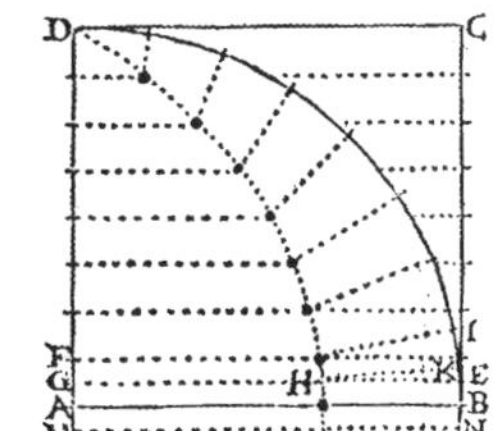

[73] *Ibid.*, p. 297: "Hæc igitur est descriptio lineae Quadratricis, quae Geometrica appellari potest, quemadmodum & conicarum sectionum descriptiones, quae per puncta etiam fiunt, vt ab Apollonio traditur, Geometrica dicuntur." Clavius's 'criticism' of Pappus is discussed in Lüder Gäbe, *Descartes' Selbstkritik: Untersuchungen zur Philosophie des jungen Descartes* (Hamburg, 1972), pp. 120–123.

that Descartes referred to Clavius's edition of Euclid. There Descartes set forth his view of Clavius's way of describing curves. He pointed out that various uses of the quadratrix could be seen "in the *Elements* of Euclid annotated by Clavius," but that this curve had not been traditionally accepted in geometry. He gave the reason why such curves as the spiral and the quadratrix were rejected by geometers:

> For although one can find out infinite points through which the spiral and the quadratrix pass, yet one cannot always find out geometrically any points which are necessary for the results desired from both of the curves; and one can only draw them completely by intersecting two movements which do not depend on each other [...].[74]

It is clear from this passage that Descartes read Clavius's scholium to Book VI of Euclid's *Elements* carefully and that he had a valid critical attitude toward Clavius's manner of constructing the quadratrix. In Descartes's judgment, some points which are necessary for drawing the entire curve cannot always be plotted when using Clavius's method. Accordingly the quadratrix cannot be accepted as a valid geometrical construction, contrary to Clavius's opinion.

Nevertheless, Descartes did not necessarily respond to Clavius's pointwise construction totally in the negative. In Book II of his *Géométrie* of 1637, Descartes proposed a criterion for distinguishing curves which can be accepted as geometrically valid. According to that criterion, algebraic curves can be allowed to enter the territory of geometry, and only for them is the manner of pointwise construction permissible.[75] Descartes's essential point is that the pointwise construction of curves can be permitted only for curves expressed by algebraic equations (i. e. algebraic curves in later terminology), and that it is prohibited for mechanical curves which are not representable in algebraic equations with finite terms. He listed the spiral, the quadratrix, and similar curves as examples of the mechanical lines.[76]

It is an algebraic criterion that constitutes Descartes's touchstone for distinguishing the geometrical from the mechanical curve. It seems to have been the lack of such a clear criterion that confused Clavius in his classifying curves. The latter insisted that there was no barrier between Apollonius's conic sections and the quadratrix because both are geometrical in the sense that they can be constructed by plotting points. But his bold attempt (it was too bold indeed) to extend the category of geometrical curves to those generated by pointwise construction appears to have a certain connection with Descartes's effort in the *Géométrie*, althought there still remains the slight possibility that Descartes learned his manner of constructing curves from mathematicians other

[74] *Ibid.* (n. 4): *AT*, I, p. 71; *AM*, I, p. 84; *CM*, II, pp. 315–316.

[75] *AT*, VI, pp. 411–412; Olscamp, p. 206. Cf. *AT*, VI, p. 386; Olscamp, pp. 188–189.

[76] *AT*, VI, p. 390; Olscamp, p. 191.

than Clavius or devised it totally independently. The younger methematician judged that pointwise construction was legitimate only for algebraic curves. Of course, one might still justly question whether even the pointwise construction of algebraic curves is fully legitimate, for not every algebraic equation is solvable even by means of radicals.

To sum up, Clavius does not seem to have been sympathetic to geometrical analysis, even though he read Pappus's *Collection*. His syle of mathematics was deductivist on the whole. Even in his *Geometria practica*, which inherited the medieval tradition of practical geometry and borrowed not a few problems from Pappus's *Collection*, he did not mention analytical methods explicitly. Accordingly, Descartes is supposed to have acquired his knowledge of geometrical analysis of the ancients either directly form classical authors such as Apollonius, Archimedes (both were referred to in the conversation with John Pell in 1646), and Pappus or indirectly from contemporary authors who were involved in restoring lost ancient works on geometrical analysis. Mathematicians who belonged to the latter category were, for example, Marino Ghetaldi and Willebrord Snel, who would be mentioned in a letter from Descartes to Mersenne toward the end of 1637 (see Chapter 5, § 3).

If we are allowed to divide mathematicians into two groups, deductivists and heurists, Euclid, who compiled and transmitted the body of mathematical knowledge already obtained in the axiomatic style, surely belongs to the former group and Pappus, who emphasized mathematical tools for inventing and solving problems, may be classified in the latter category. According to this categorization, Clavius was no doubt a Euclid, nay more there seems to have been a legitimate reason why he was called 'the Euclid of the sixteenth century'. Further, just as Euclid had been a mathematical hero in the scholastic-Aristotelian tradition of the Middle Ages, Clavius can be said to have been the most important mathematician in the entire educational system of the Society of Jesus. For this reason again, he was not a Pappus but a Euclid in the strict sense. Insofar as geometrical analysis is concerned, Descartes had to go far beyond Clavius.

§ 4. Diophantus in the Works of Clavius

The algebra of the moderns was an essential component which Descartes employed in forming his method. According to his testimony, he "had no other instructor for Algebra than ye reading of Clavy Algebra" (see p. 47). Clavius's *Algebra* was published in Rome in 1608. This monograph shows us a different facet from that which 'the Euclid of the sixteenth century' displayed in his attitude toward geometrical analysis. It is almost certain that Descartes learned from it various tools for solving mathematical problems. In the following reconstruction of Clavius's *Algebra*, we will bring into focus particularly how the author evaluates the discipline of algebra, its symbolism, and how he manipulates it.

Clavius begins his *Algebra* by bestowing lavish praise on the discipline of algebra. At the very beginning part of the "Preface on the Preeminence of Algebra," he states: "The art which our people have called by the accepted term 'algebra' is posterior to nothing in its highest dignity and in the laudations of its widest achievements."[77]

Then Clavius defines the subject of algebra as equations composed of known and unknown quantities. His definition of algebra is impressive in that he regards this art as restricted neither to numbers (*numerus*, the subject of arithmetic) nor to magnitudes (*magnitudo*, the subject of geometry). In a certain manner algebra goes across and beyond various mathematical disciplines. Clavius states the range of algebra:

> Consequently how widely does the object of algebra extend? It regards neither any kind of numbers nor the variety of any magnitude as irrelevant to itself in order that it may not only disclose all hiding-places of numbers but also obtain a finite magnitude of each mass, a measure of sounds, moments of weights,and certain boundaries of measurement. Nor is there any question subject to arithmetic that algebra does not recognize as its own and solve. Algebra deals with so many, so various, so unintelligible, and so difficult general parts of mathematics.[78]

This statement looks very novel particularly when it is compared with the standard standpoint of classical Greek mathematics. Clavius considers algebra to be a discipline traversing both arithmetic and geometry. In ancient Greece, the latter two branches of pure mathematics were regarded as basically mutually independent genera.

In Chapter 1 under the title "On the Invention and Name of Algebra," Clavius calls his art by various names such as the 'Ars maior', the 'Regula census et rei', the 'Regula radicis et quadrati', the 'Regula cosae' (or 'della cosa', and the 'Algebra et almuchabala'.[79] The 'Algebra et almuchabala' had been the traditional Arabic name since Al-Khwārizmī originated it, probably from two operations for transforming equations into canonical forms. Clavius translates 'algebra' and 'almuchabala' into Latin words, 'restauratio' and 'oppositio', re-

---

[77]Clavius, *Algebra*, in *Opera mathematica*, t. II (Maintz, 1612), p. 3. "Prooemium de Algebræ Præstantia": "Et dignitate summa, & amplissimarum laudum præconijs nulli postrema ea ars est, quam recepto vocabulo Algebram nostrates appellarunt."

[78]*Ibid.*: "Quam igitur scopus iste late vagatur, qui nec genus vllum numerorum, nec vllius magnitudinis diuersitatem, à se alienam putat? vt non numerorum modo latebras omnes detegat, sed vniuscuiusque etiam molis finitam magnitudinem, suorum[a] metrum, ponderum momenta, mensurarumque certos terminos assequatur; neque vlla Arithmeticæ quæstio subijciatur, quam non veluti suam agnoscat Algebra, atque expediat. Tam multas, tam varias, tam obscuras, tam difficiles Mathematicæ partes vna algebra pertractat vniuersas." ([a] "sonorum" instead of "suorum" in *Algebra* [Roma, 1608], p. 2. We have adopted the reading of the 1608 edition.)

[79]*Ibid.*, p. 4.

spectively. The 'Regula census et rei' or 'Regula cosae' is also of Arabic origin. The Latin words 'res' and 'census' are derived from the Arabic 'shay'' (thing) and 'mal' (wealth), respectively. Both the two Arabic words were borrowed from commercial parlance and grew into technical terms for the unknown quantities in the equation: 'shay'' was usally used to denote the unknown quantity of the first degree, and 'mal' that of the second degree.[80] The Italian word 'cosa' is synonymous with 'shay'' or 'res'. The words 'radix' and 'quadratus' denote the unknown quantity of the first degree and that of the second degree, repsectively, so that the 'Regula census et rei' and the 'Regula radicis et quadrati' have the same meaning. The *Ars maior* (greater art) meant a higher art, and was used in contrast with the 'Ars minor,' i. e. vulgar arithmetic.

As is already known from the names which Clavius gives to his discipline, his *Algebra* basically belonged to the tradition of cossist algebra which flourished especially in Italy and Germany during the late Middle Ages and the Renaissance. As the sources of his textbook, he explicitly mentions Diophantus of Alexandria, Jordanus de Nemore, Girolamo Cardano, Niccolò Tartaglia, Rafael Bombelli, and Pedro Nuñes Salasience.

Pedro Nuñes is known to have taught Clavius at the University of Coimbra.[81] Clavius is supposed to have consulted his *Libro de álgebra en arithmética y geometría*, published in Antwerp in 1567. Further Clavius refers to Jordanus as the author of the *Libri de numeris*,[82] which is clearly identical with *De numeris datis*. Jordanus's treatise was listed as a textbook of algebra at the University of Vienna around the mid-fifteenth century.[83] Clavius appears to have followed the textbook tradition of algebra of the Renaissance. *De numeris datis* is sometimes overrated as a promoter of algebraic analysis and, then, as a precursor of Viète,[84] since Jordanus attempted to use a literal symbolism and to introduce some (geometrical) analytical techniques which may be seen in Euclid's *Data* (*Δεδομένα*, in Greek) into arithmetical problems.

The names of Cardano, Tartaglia, and Bombelli appear where Clavius mentions the cubic equation in Chapter XII "On the Extraction of Roots which the Rule of Algebra Is Concerned with." Discussing the cubic equation of the

---

[80]One of the best accounts for the terms of algebra is seen in W. Hartner, "Al-DJABR WA'L-MUKĀBALA," *The Encyclopaedia of Islam*, New Edition, Vol. II (Leiden/London, 1965), pp. 360–362. See also Roshdi Rashed, "Algebra," Ch. 11 of R. Rashed, ed., *Encyclopedia of the History of Arabic Science*, Vol. 2 (London/New York, 1996), pp. 349–375, esp. at p. 370, n. 2.

[81]J. M. López de Azcona, "Nuñes Salaciense, Pedro," *DSB*, X, p. 160.

[82]Clavius, *Ibid.*, p. 78.

[83]See P. L. Rose, *The Italian Renaissance of Mathematics* (n. 17), p. 93; JoAnn S. Morse, "The Reception of Diophantus' *Arithmetic* in the Renaissance" (Princeton University Ph. D. Dissertation, 1981), p. 58.

[84]Barnabas Bernard Hughes's Introduction to, Jordanus de Nemore, *De numeris datis: A Critical Edition and Translation* (Berkeley/Los Angeles/London, 1981), pp. 5–6. A more critical historical evaluation can be seen in Michael S. Mahoney, "Mathematics," in David Lindberg, ed., *Science in the Middle Ages* (Chicago, 1978), pp. 160–161.

types of $x^3 = 4x^2 + 16$ and $x^3 = 10x + 24$, to use modern terminology, Clavius touches upon the history of the solution of cubic equations:

> The art by which roots of this sort should be surely discovered has not been invented yet, although Cardano and Niccolò Tartaglia found the value of a single root in certain individual cases. On the other hand, Rafael Bombelli believes that he has found how roots should be discovered from certain equations of that sort and some others. François Viète is also said to have shown a general rule for extracting roots of that sort, for it has not hitherto been permitted [for me] to see it, and Bombelli's methods are exceedingly obscure.[85]

Clavius was informed of the existing knowledge on the solution of the cubic equation through Cardano's *Ars magna* (1545) and Bombelli's *Algebra* (1572). Cardano and Tartaglia gave only one real root of the cubic equation whereas Bombelli tried to provide the complete solution to it by introducing imaginary quantities. Clavius also has certain information on Viète's method probably developed in the latter's *De Numerosa potestatum ad exegesin resolutione*. Viète's treatise on the numerical solution of the cubic and biquadratic equations was published in 1600 under the editorship of Marino Ghetaldi, a disciple of both Clavius and Viète.[86] His mention of Viète could have originated from another source which remained unpublished as of 1608. For example, *De Æquationum recognatione et emendatione tractatus duo*, published posthumously in 1615 by Alexander Anderson, also treats the cubic equation in a general way.[87] In any case, it would seem that Clavius learned Viète's attempt at second hand. Yet Clavius neither states distinctly why the technique of solving the cubic equation remains incomplete nor proposes his own method of solution.

In order to evaluate Clavius's *Algebra* in a historically just manner, his reference to Diophantus is crucial. He makes mention of the Alexandrian mathematician while citing Regiomontanus's oration on the ninth-century Muslim astronomer al-Farghānī delivered in 1464 at the University of Padua.[88] This lecture was historically significant because Regiomontanus tried to raise the

---

[85]Clavius, *Algebra*, in *Opera mathematica*, t. II, p. 26: "[N]ondum est inuenta ars, qua huiusmodi radices certò eruantur; quamuis Cardanus, & Nicolaus Tartalea in quibusdam exemplis singularibus inienerint æstimationem vnius radicis. Raphael autem Bombellus ex quibusdam etiam æquationibus eiusmodi, & alijs nonnullis putat se inuenisse, quo pacto eruendae sint radices. Franciscus quoque Vieta dicitur demonstrasse regulam generalem pro eiusmodi radicibus extrahendis: quam quia videre hactenus non licuit, & rationes Bombelli obscurae valde sunt."

[86]François Viète, *Opera mathematica*, ed. F. van Schooten (Leiden, 1646), pp. 162–228. For the English translation, see Viète, *The Analytic Art*, tr. T, Richard Witmer (Kent, 1983), pp. 311–370. Cf. Frédéric Ritter, "François Viète: Inventeur de l'algèbre moderne 1540–1603, Essai sur sa vie et son œuvre," *Revue occidentale philosophique, sociale et politique*, second série, **10** (1895), pp. 383–389.

[87]Viète, *Opera mathematica*, pp. 84–161; Witmer's translation, pp. 158–310. Cf. Ritter, pp. 389–401.

[88]Clavius, *Algebra* (n. 77), p. 4.

status of algebra to the level of the classical mathematical disciplines such as arithmetic and geometry. Until the time of Regiomontanus algebra had tended to be seen as a 'merchants' art'. But the situation began to change with his discovery of a manuscript of Diophantus's *Arithmetic*. Regiomontanus stated in the oration quoted above: "But no one up to now has translated from Greek into Latin Diophantus's thirteen most subtle books, in which the flower of entire arithmetic lies hidden, namely the *ars rei et census*, which today they call by the Arabic name *Algebra*."[89] The orator consciously tried to connect the Arabic art of algebra to the newly discovered classical *Arithmetic* of Diophantus. This is quite interesting because Regiomontanus's attempt was the same as Qusṭā ibn Lūqā of the nineth century entitled his Arabic translation of Diophantus's *Arithmetic*, *Ṣināʿat al-jabr* (Art of algebra).[90] It was Bombelli who followed Regiomontanus's lead. Regiomontanus did not write a textbook on algebra. Bombelli, on the other hand, took a step forward and wrote a treatise of algebra, borrowing freely from the problems in Diophantus's *Arithmetic*.[91] Indeed, he was the first to make known the work of Diophantus in modern Europe.

Clavius's treatise is, first of all, reckoned as one of the textbooks of cossist algebra. Nevertheless, it did not necessarily belong to the tradition of cossist algebra in the narrow sense, that is to say, that of the practical art which responded mainly to the needs of commercial marketplaces. Bombelli's book succeeded in introducing a reformed symbolism, which was in marked contrast with the main trend of cossist algebra, and in contributing the solution of the cubic and biquadratic equations. Clavius's book was more conservative than Bombelli's in these aspects. But his *Algebra* certainly had characteristics which were similar to the textbook of Bombelli in the sense that both regarded algebra as a theoretical discipline and utilized many problems of Diophantus's *Arithmetic* in order to provide exemplars of their method of problem-solving. Problems 1–5 of Book I of Diophantus's *Arithmetic*, to give only a few examples, correspond to Problems 1 (which is essentially the same as Problem 2), 9, 10, 11, and 12 in the same order in Clavius's Chapter 29 "Various Problems (*Aenigmata*) of Abstract Numbers Explained by Algebra in Which an

[89] "Oratio Iohannis de Monteregio, Habita Patauij in praelectione Alfargani: Hac oratione compendiose declarantur scientiae Mathematicae, & utilitates earum," in Joannis Regiomontanus, *Opera collectanea*, hrsg. von Felix Schmeidler (Osnabruck, 1972), p. 46: "Diofanti autem tredecim libros subtilissimos nemo usque hac ex Graecis Latinos fecit, in quibus flos ipse totius arithmeticae latet, ars uidelicet rei & census, quam hodie uocant Algebram Arabico nomine."

[90] See R. Rashed, *Op. cit.* (n. 80), p. 356, and Diophante, *Les Arithmétiques*, T. III: *Livre IV*, Texte établi et traduit par. R. Rashed (Paris, 1984), Introduction, p. XI.

[91] For Bombelli's algebra, see J. E. Hofmann, "Bombelli's Algebra—eine genialische Einzelleistung und ihre Entwirkung uf Leibniz," *Studia Leibnitiana*, **5** (1972), pp. 196–252. Hofmann compares the problems of Bombelli's *Algebra* with those of Diophantus's *Arithmetic* and of Viète's *Zetetica* (1593), showing a table of concordance on p. 215, footnote 46. See also Karin Reich, "Diophant, Cardano, Bombelli, Viète—Ein Vergleich ihrer Aufgaben," *Rechenpfennige: Aufsätze zur Wissenschaftsgeschichte* (München, 1968), pp. 131–150.

Equation between Only Two Numbers Appears, Which Is in Fact Said to Be a Simple Equation" with only a minor modification.[92] Thus Clavius may be said to deserve a chapter (even though very brief) in a book on the reception of Diophantus's *Arithmetic* in the early modern period.[93] Clavius's textbook reflected the humanists' efforts to place the art of algebra on an equal footing with classical theoretical mathematics.

Let us examine some of problems which Clavius proposed and solved. To do this, we have to explain Clavius's cossic symbolism. His notations were, as is often remarked, basically the same as those of the *Arithmetica integra* (1544) of Michael Stifel,[94] although the Jesuit mathematician made no mention of the well-known sixteenth century cossist of his native country. Clavius, following Stifel, introduces cossic symbols representing the geometrical progression of an unknown quantity as follows:[95]

0.	1.	2.	3.	4.	5.	6.	7.	8.	9.	10.	11.	&c.
N.	𝔵.	𝔷.	𝔠𝔢.	𝔷𝔷.	ß.	𝔷𝔠𝔢.	Bß.	𝔷𝔷𝔷.	C𝔠𝔢.	𝔷ß.	Cß.	&c.
1.	2.	4.	8.	16.	32.	64.	128.	256.	512.	1024.	2048.	&c.

The numbers of the first row (0. 1. 2. 3. ...) are exponents (*exponentes or denominationes*) of the unknown. N of the second row stands for a constant (*numerus simplex et absolutus*); 𝔵(R), root (*radix*) or thing (*res* or *cosa*), that is, an unknown quantity; 𝔷(Z), wealth (*zensus*) or square (*quadratus*) of the unknown; 𝔠𝔢(C), cube (*cubus*); 𝔷𝔷(ZZ), wealth-wealth (*zenzizensus*) or square-square (*quadratiquadratus*), i. e. the biquadratic of the unknown; ß, surdisolid (*surdisolidus*), i. e. the fifth power of the unknown, and so on. The numbers of the third row represent powers of 2, i. e. 2^n.

In Stifel's and Clavius's symbolism, any powers of an unknown quantity can be expressed by introducing notation only for prime exponents. The following symbols which use the alphabet after B correspond to prime powers:

---

[92] Clavius, *Ibid.*, pp. 79–83. The numbering of the problems of Diophantus's *Arithmetic* is that of *Diophanti Alexandrini Opera omnia cum Graecis commentariis*, ed. Paul Tannery, 2 vols. (Leipzig, 1893–1895).

[93] For a discussion of the Renaissance recovery of Diophantus, we are very much indebted to JoAnn S. Morse's Ph. D. dissertation at Princeton (1981), quoted in n. 83. This thesis allocates its five chapters (Chapter 2–6; Chapter 1 is the Introduction) to Regiomontanus, Bombelli, Viète, Xylander (Wilhelm Holtzman), Bachet de Méziriac.

[94] For Stifel's symbolism, see, among others, Florian Cajori, *A History of Mathematical Notations*, Vol. 1: *Notations in Elementary Mathematics* (La Salle, Ill., 1928), pp. 139–147; Kurt Vogel, "Stifel, Michael," *DSB*, XIII, pp. 58–62; Johannes Tropfke, *Geschichte der Elementarmathematik*, 4. Auflage Bd. 4: *Arithmetik und Algebra*, Vollständig neu bearbeitet von Kurt Vogel, Karin Reich und Hermut Gericke (Berlin/New York, 1980); J. E. Hofmann, "Michael Stifel 1487?–1567," *Sudhoffs Archiv*, supp., no. 9 (1968). For the influence of Stifel on Clavius, see Charles Hutton, *Tracts on Mathematics and Philosophical Subjects*, Vol. II, Tract 33 (1812): "The History of Algebra," pp. 237–241 & 257; P. Treutlein, "Die dutsche Coss," *Abhandlungen zur Geschichte der Mathematik*, Zweites Heft (Leipzig, 1879), pp. 97–98.

[95] Clavius, *Ibid.*, pp. 5–7.

2.	3.	5.	7.	11.	13.	17.	19.	23.	29.	31.	37.	41.	&c.
𝔷.	𝔠𝔢.	ß.	Bß.	Cß.	Dß.	Eß.	Fß.	Gß.	Hß.	Iß.	Kß.	Lß.	&c.

Thus, for instance, 𝔷Bß stands for the 14 ($= 2 \times 7$)-th power.

This symbolism may appear very complicated, but it is exactly the one which Descartes equipped himself with and attempted to simplify after 1619, as will see in Chaper 3.

Now we can present the way in which Clavius solves Problem 1 of Chapter 29, which is concordant with Problem I, 1 of Diophantus's *Arithmetic*, Problem I, 1 of Jordanus de Nemore's *De numeris datis*, and Problem III, 1 of Bombelli's *Algebra*.[96] The problem and its solution read as follows:

> [Problem] 1. To divide any given number into two parts exceeding mutually by any given excess.
>
> [Solution] Let 100 be a number to be divided into two parts and 40 be their excess or difference. Let the lesser part be put 1𝔵, accordingly the larger 100 − 1𝔵. By taking away 1𝔵 from 100 − 1𝔵, there remains 100 − 2𝔵 as you see here. And the equation between 100 − 2𝔵 and 40 will be discovered. Indeed when subtracting the lesser part from the larger, 40 must remain. By adding 100 − 2𝔵 to both sides, there will be the equation between 100 and 40 + 2𝔵. And by taking away 40 from both, there will be the equation between 60 and 2𝔵. Therefore, if from the prescription of the rule of algebra 60 is divided by 2, 1𝔵 will be discovered to be 30. And this is the lesser part. Therefore, the larger will be 70, truly 100 − 30. The larger part will also be obtained if the proposed difference 40 is added to the lesser 30. Therefore, you see the larger part exceeds the lesser 30 by the number 40 as is proposed.[97]
>
> 100 − 1𝔵
+ 1𝔵
―――――
100 − 2𝔵

There follows another solution in which the larger part instead of the lesser is set to 1𝔵. In this case, the lesser will be 100 − 1𝔵, and the difference between the two will be 1𝔵−(100 − 1𝔵), i. e. −100 + 2𝔵. This must be 40. Then 2𝔵

[96] Diophantus, ed. P. Tannery (n. 92), Vol. 1, pp. 16–17; Jordanus de Nemore, ed. B. B. Hughes (n. 84), pp. 57 & 127; Bombelli, *L'Algebra* (Bologna, 1576), p. 416.

[97] Clavius, *Ibid.*, pp. 79–80: "1 Datum numerum quemcunque diuidere in duas partes se mutuo superantes in dato excessu quocunque. Sit numerus 100. diuidendus in duas partes, quarum excessus, vel differentia sit 40. Ponatur minor pars 1𝔵, ac proinde maior 100−1𝔵. Detracta 1𝔵 ex 100−1𝔵 remanent 100−2𝔵, vt his vides. Eritque inuenta æquatio inter 100−2𝔵 & 40. quippe cum dempta minore parte ex maiore, relinqui debeant 40. Additis vtrintque 2𝔵. erit æquatio inter 100. & 40 + 2𝔵. Et vtrintque ablatis 40. æquatio erit inter 60. & 2𝔵. Si ergo ex praescripto regulæ Algebræ diuidantur 60. per 2. inuenietur 1𝔵. esse 30. atque hæc est minor pars. Ergo maior erit 70. nimirum 100 − 30. quæ maior pars etiam habebitur, si ad minorem 30. adiiciatur differentia proposita 40. Vides ergo maiorem partem 70. superare minorem 30. numero 40. vt propositum est."

will be 140. We have 1𝔛 equal to 70. Then the lesser will be 100 − 70, i. e. 30. We have 1𝔛 equal to 70. Then the lesser will be 100 − 70, i. e. 30.

As can be noticed in the above example, Clavius solves the problem by proposing the concrete numbers 100 and 40 as the given number and excess. In this respect, he is following his precursors Diophantus, Jordanus, Cardano, Stifel, Bombelli, and so forth. Indeed in the examples of Clavius, all numbers except the unknowns expressed by cossic symbols, that is to say, coefficients and constants, are always represented by concrete numbers. This is one of the most significant differences between the algebraists from Diophantus through Bombelli and the modern analysts who followed François Viète. Viète in his *Zetetica* solved the same Diophantine problem expressed entirely in geometrical (instead of arithmetical) terms by substituing B and D quite generally for the difference and the total side.[98] However, we should not make the hasty conclusion that Clavius's method is not at all general in comparison with Viète's. The exemplar numbers taken for the given number and the given difference are essentially arbitrary. Then we may characterize Clavius's method as 'quasi-general', as it were, although it can be no match for the generality of Viète's totally analytical method.

Clavius treats equations having two terms in Chapter 29. In Chapter 30, he tackles equations having three terms. The method of this chapter does not differ essentially from that of the previous one. A new aspect emerges in Chapter 31 "Various Problems of Numbers Connected to Material Things Shown with Several Examples Pertaining to the Second Root." In the previous chapters, Clavius used cossic symbols for powers of only one unknown number. Here again, following Stifel, he introduces a new symbolism for more than one unknown. Stifel and Clavius represent the other unknown quantities by A, B, C, etc.[99] Accordingly if we rewrite 𝔛, A, B, etc. of the Stifel-Clavius symbolism as x, y, z, etc., then ʒ, Aʒ, Bʒ, etc. will denote x^2, y^2, z^2, etc.

To illustrate how Clavius uses this notation, let us observe Problem 58.

> [Problem] 58. Two friends have two numbers of golden coins. By subtracting their sum from the sum of the squares made from each, there remains 78. Now, when it is added to the number produced by their multiplication, the result makes 39. It is asked what these numbers are.[100]

To tackle this problem, Clavius alludes to the Lemma to Euclid X, 54[101]

---

[98] Viète, *Opera mathematica* (n. 86), p. 42; Witmer's translation, pp. 83–84.

[99] For Stifel's symbolism for more than one unknown, see F. Cajori, *Op. cit.* (n. 94), pp. 139–142. Clavius's interpretation can be seen in Clavius, *Ibid.*, pp. 36–38 & 161–162.

[100] Clavius, *Ibid.*, p. 163 : "58 Duo socij habent duos numeros aureorum, quorum summa à summa quadratorum ex ipsis procreatorum subtracta, relinquit 78. Addita vero ad numerum ex eorum multiplicatione productum facit 39. Quæritur, qui sint isti numeri."

[101] Lemma to Proposition in Clavius's edition: *Euclidis Elementa*, in *Opera mathematica*, t. I, p. 441. Cf. Heath, *The Thirteenth Books of Euclid's Elements*, Vol. 3 (New York, 1956),

and to its arithmetical version. The latter version may be written in a form in which AB begets a mean proportional between A^2 and B^2, with A, B being two numbers. This rewriting is not totally anachronistic, for Clavius in its demonstration uses such symbols as A, B, and "A *in* B" (A times B) to denote two numbers and their product.[102] His symbolism makes us recall that of Euclid's arithmetical books or Jordanus de Nemore's *De numeris datis*, in which the medieval mathematician certainly represented numbers by symbols, but neither operated on symbols nor with them.

Now we have acquired the tools sufficient for solving Problem 58. In the following we trace Clavius's essential line of reasoning. All the notation except the equal sign '=' is the original author's own. Let the lesser number be 1𝓍 (equal to the line AB in Figure 2-5) the larger 1A (= BC) and their sum 1B (= AC). The square of the straightline will be 1𝔷. From the first condition of the problem, we get (1𝔷 + 1A𝔷)−1B = 78. Therefore, 1A𝔷= 78 + 1B−1𝔷. From the second condition, we have 1B + 1𝓍 · 1A =39 (1𝔷 · 1A is our notation standing for the complement AD). Hence, 1𝓍 · 1A = 39−1B. From the figure, we get 78 + 1B + 2(39−1B) = 1B𝔷, which is identical with the quadratic equation 156−1B = 1B𝔷. To make a square, we rewrite this equation as $\frac{625}{4}$ =the square of ($\frac{1}{2}$ + 1B). Clavius chooses only the positive root. Consequently, 1B (= 1𝓍 + 1A) = 12. Substituting 12 for 1B, we have 39−1B = 27 and 78 + 1B−1𝔷 = 90−1𝔷. Applying the lemma mentioned above concerning a mean proportional number to the numbers which appear in the figure, we have the continous proportion (90−1𝔷): 27 = 27 :1𝔷. From this proportion, we have the equation 729 = 90𝔷−1𝔷𝔷. This equation is biquadratic in its form, but may be reinterpreted as a quadratic equation of an unknown 1𝔷. To make a square, we write the equation as 1296 = the square of (1𝔷−45). Again Clavius takes the positive root alone. Hence, 1𝔷−45 = 36, i. e. 1𝔷 = 81. Then taking the positive root again, we finally get 1𝓍 = 9. Hence, 1A = 1B−1𝓍 = 3. Finally he pays attention to the fact that 1𝓍 and 1A are mutually exchangeable. So, if 1𝓍 is actually the lesser number, it will be equal to 3, and 1A will be 9.

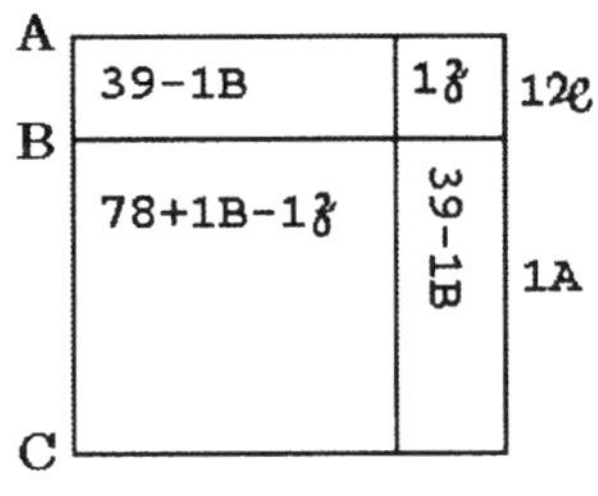

Figure 2.5

In short, Clavius resolves the problem by setting up the simultaneous equations 1𝔷+ 1A𝔷−(1𝓍 + 1A) = 78 and 1𝓍 + 1A + 1𝓍 · 1A = 39, with 1𝓍 and 1A being two unknown quantities. His solution might be said to be awkward, but we should recognize that he manipulates rather sophisticated symbolism.

Clavius adds a significant scholium to this problem.

---

pp. 115–116. Proposition 55 in Clavius's edition corresponds to Prop. 54 in Heath's.
[102] Clavius, *Algebra*, in *Opera mathematica*, t. II, p. 137.

> From this problem, it is easily understood that one who wants to solve questions by algebra should be best exercised in the science of geometry as we have stated in Chapter 1. For, this problem will not almost or by no means be solved by one ignorant of geometry as it is evident.[103]

According to Clavius, working with algebra requires the knowledge of the classical mathematical disciplines, arithmetic and geometry.

Clavius's *Algebra*, thus, may be placed historically at the intersecting point of the tradition of cossist algebra and Diophantus's *Arithmetic*. Having adopted the algebraic symbolism of Michael Stifel, Clavius could represent by symbols more than one unknown quantity. Besides, he sometimes denoted certain numbers by signs, as Jordanus's *De numeris datis* had done in the thirteenth century. On the other hand, his *Algebra* was basically freed from the practical tradition of algebra. Clavius reevaluated algebra as a respectable discipline, invoking the ancient authority of Diophantus and such Renaissance forerunners as Regiomontanus and Bombelli.

Nevertheless Clavius's treatise was a typical product of the sixteenth century textbook tradition of algebra even though it appeared at the beginning of the seventeenth century. It surely provided rather sophisticated techniques for solving both arithmetical and geometrical problems, but it is an open question whether or not it is a work leading straight to the idea of algebraic analysis, which made its full-scale appearance with Viète's *In artem analyticem isagoge* of 1591. In this sense, it did not face forward to the seventeenth century.

Clavius was not completely ignorant of the works of Viète. He certainly touched upon some of them. As was stated above, he mentioned Viète's way of solving the cubic equation. Moreover, as to the well-known debates between Joseph Justus Scaliger and his critics on the measurement of the circle, Clavius distinctly supported Scaliger's opponents Viète and the Belgian mathematician Adriaan van Roomen. In his *Refutatio cyclometriae Iosephi Scaligeri*, Clavius showed clear signs that he had read the works of the two mathematicians.[104] Van Roomen's work, which Clavius referred to, was no doubt the trilogy including the *Apologia pro Archimede ad clarissimum virum Josephum Scaligerum*, published in 1597, in which he intended to defend Archimedes's attempt at measuring the circle. Van Roomen evidently sent that work directly to Clav-

---

[103]Clavius, *Ibid.*, p. 163: "Ex hoc ænigmate facile intelligitur, eum qui quæstiones per Algebram soluere vult, debere optime esse exercitatum in Giometriæ scientia, vt cap. 1. diximus. Hoc enim ænigma ab eo, qui Geometricam ignorat, vix, aut nullo modo soluetur, vt patet."

[104]The names of Viète and Van Roomen appear twice respectively on pp. 2 and 9, in Clavius, *Refutatio Cyclometricae Iosephi Scaligeri*, in *Opera mathematica*, t. V (Maintz, 1612). Sommervogel states that he does not know whether this work was published separately besides the *Opera mathematica* version. See Sommervogel, *Bibliothèque de la Compagnie de Jésus: Bibliographie*, t. II (Bruxelles/Paris, 1891), col. 1218.

ius.[105] Clavius did mention Viète's name but did not state explicitly which of his books he perused. It is unquestionable, however, from Clavius's citation of a mathematical result of Viète that the work of Viète was the *Munimen adversus nova cyclometrica*, published in 1594.[106] Both Van Roomen's *Apologia pro Archimede* and Viète's *Munimen adversus nova cyclometrica* were composed to criticize Scaliger's dilettantish venture in attacking the famous problem of the squaring the circle in the latter's *Cyclometrica elementa duo*, published in 1594.

Additionally, Viète was known as a mathematician critical of the new Gregorian calendar. Clavius and Viète were rivals in this matter. In 1602, Clavius even exchanged letters with Viète on the mathematics related to the calendar reform.[107] But in both the *Munimen adversus nova cyclometrica* and the works related to calendar calculations, no traces of the new symbolic algebra appear.

What is significant for us is that Clavius had opportunities for becoming aware of Van Roomen's conception of 'mathesis universalis' as a discipline for unifying arithmetic and geometry, which was developed in the aforementioned *Apologia pro Archimede*, and Viète's symbolic algebra or algebraic analysis through the intermediary Van Roomen. Van Roomen wrote to Clavius on September 17, 1597, touching upon Viète's doctrine of plane and spherical trigonometry given in the French mathematician's *Variorum de rebus mathematicis responsorum libri VIII*, published in 1593:

> Has Your Reverend seen Viète's doctrine of triangles? Certainly it is excellent, but I do not comprehend to what extent further it may be able to approach perfection. The doctrine is easy. And indeed it originated in common [algebra] and thereafter was found out by equations, strictly speaking algebraic equations. I call the algebraic ones 'figurative' (*figuratus*) or (as Viète says) 'specious' (*speciosus*).[108]

---

[105]See Van Roomen's letter to Clavius dated on February 10, 1597 in Clavius, *Corrispondenza*, Vol. IV (1597–1601) (n. 15), Lettera # 141, p. 31.

[106]Viète, *Opera mathematica* (n. 86), pp. 436–446. For the dispute between Scaliger and Viète, see F. Ritter, *Op. cit.* (n. 86), pp. 259–262; J. E. Hofmann, "Vorwort: François Viète (1540–1603)—Leben, Wirken, Bedeutung," in a facsimile reprint edition of Viète, *Opera mathematica* (Hildesheim, 1970), pp. XXI–XXII. The proposition which Clavius mentioned on p. 9 of his *Refutatio* is clearly Proposition II in Viète's booklet *Munimen* (*Opera mathematica*, pp. 438–439).

[107]The topic of the letters was restricted to the calendar reform. Viète's letter to Clavius on 9 May 1602 and Clavius's reply to Viète on 8 July 1602 are preserved at the Archives of the Gregorian University in Rome, and are partly printed in Paul P. Bockstaele, ed., "The Correspondence of Adriaan van Roomen," *LIAS: Sources and Ducuments Relating to the Early Modern History of Ideas*, Volumen III/Separatum (Amsterdam, 1976), pp. 79–80, n. 4. For the related documents of the rivalry between the two mathematicians, see *Ibid.*, p. 70, n. 5, & p. 72, n. 10. Cf. Clavius, *Corrispondenza*, t. V (1602–1605) (n. 15), Lettere # 189–190. & 192–193. pp. 15–19 & 21–28.

[108]"Van Roomen to Clavius in Rome, Würzburg, 17 September 1597," Bockstaele, *Op. cit.*

In his *Varia* responsa, actually here and there we can observe some marks where Viète employed his symbolic algebra.[109] Indeed, Viète's algebra was one of the topics which appeared several times in the correspondence between Van Roomen and Clavius.[110]

It becomes clear that Clavius had certain information about Viète's *logistice speciosa.* But in spite of his having some knowledge of Van Roomen's '*mathesis universalis*' and Viète's algebra, the mathematician of the Collegio Romano continued within the tradition of cossist-Diophantine algebra, which Viète called *logistice numerosa* in contrast to his own reformed *logistice speciosa.*[111] Clavius's *Algebra*, nevertheless, held a thread, however thin it may have been, with which it may be connected to the notion of 'mathesis universalis' or with the new symbolic algebra, by its definition as a discipline of problem-solving techniques applicable to both numbers and magnitudes.

The textbook of algebra by Clavius might not have conveyed the author's original achievements, but it surely contained the essential achievement of Renaissance algebra. It is exactly this work that is considered to have constituted the most important mathematical background of the young Descartes. Starting from this monograph, we conjecture that Descartes traced back, as it were, to the mathematical thought of Van Roomen and then perhaps of Viète relatively easily. As may be said about his other mathematical works, it was by virtue of its educational value that Clavius's *Algebra* was read widely. Descartes was one of the readers who could have access to this fruit of sixteenth century algebra at the institution endowed with educational mathematics and who would finally bring about its efflorescence.

(n. 104), p. 46; Clavius, *Corrispondenza*, t. IV (n. 15), Lettera #138, p. 26: "An Reverentia vestra vidit Triangulorum doctrinam Vietae? Certe ea pulchra est, neque ulterius quid perfectioni accedere, video. Doctrina facilis est, atque ex vulgari orta postea per Aequationes dumtaxat Algebraicas inventa, dico Algebraicas figuratas aut (ut Vieta inquit) speciosas."

[109] E. g. Viète, *Opera mathematica*, pp. 372 & 397.

[110] One may add another important corresondent who connects Viète to Clavius, i. e. Marino Ghetaldi, who was a disciple of Clavius in Rome and of Viète in Paris. Several letters of Ghetaldi to Clavius from 1602 to 1608 are preserved at the Archives of the Pont. Gregorian Univeristy in Rome. See E. C. Philips, *Op. cit.* (n. 9), pp. 212–216; *Corrispondenza*, Vol. V, Lettere # 196, 200, 205, 209, & 220; Vol. VI, # 280, 283, & 287. For the relationship between Ghetaldi and Viète, see Karin Reich, "Quelque remarques sur Marinus Ghetaldus et François Viète," Actes du symposium internationale »*La Géométire et l'Algèbre au début du XVIIe siècle*«, *À l'occasion du quatrième centenaire de la naissance de Marin Ghetaldić, Dubrovnik, 29. IX - 3. X* 1968, ed. Z. Dadić *et al.* (Zagreb, 1969), pp. 171–174.

[111] Viète, *In artem analyticem isagoge* (Tours, 1591), f. 5r; *Idem*, *Opera mathematica*, p. 4; Witmer's translation, p. 17.

§ 5. Descartes's Mathematical Background before the Encounter with Beeckman

In the preceding sections we have studied the main characteristics of Clavius's mathematics which were considered to have been influential in the formative period of Descartes's mathematical thought. In this section we will try to secure a general conception as to what mathematical knowledge Descartes possessed at his departure from the college of La Flèche, probably in the fall of 1615, or at the latest just prior to his encounter with Beeckman in Breda in October of 1618, by examining some works of a mathematics teacher at La Flèche and considering Descartes's autobiographical statements of his later days.

At the college of La Flèche, was Descartes directed to study the works of Clavius of his own accord? This was possible since the general intellectual atmosphere there would have been sufficient to lead him to read some advanced books of mathematics. But it seems more natural to suppose that it was through the suggestion of a certain senior scholar of the college that he came to recognize the value of studying Clavius. The likeliest scholar would be Descartes's mathematics teacher.

The name of the *maître des mathématiques* at La Flèche began to appear in 1606, and the first teacher of mathematics there was Père Jacques Guernisac, who occupied that position until 1610. For the academic year 1610–1611, Maître Nicolas Laplace occupied the same position when he was a second year student of theology. We have no information about the next academic year. Subsequently Maître Jean François taught mathematics while studying theology from 1612 through 1616. As a Père he instructed the same discipline from the academic year 1616–1617, then for the cycle of the next three years of the philosophical curriculum, logic, physics, and metaphysics including arithmetic, geometry, music, and astronomy. In 1620 he moved to Paris to teach mathematics at the college of Clermont.[112]

Who was Descartes's teacher of mathematics? It is certain from Descartes's frequent statements that he was at La Flèche for eight or nine years.[113] There have been essentially three weighty theories as to exactly when he was there. First, Adrien Baillet in his biography of Descartes has him entering the college of La Flèche in 1604 and leaving in 1612. Second, according to some other biographers, on the other hand, Descartes was at the Jesuit college

[112] A. Romano, *Ibid.* (Ch. 1, n. 13), pp. 537–538. Cf. François Dainville, *L'Éducation du jésuites (XVIe-XVIIe siècles)* (Paris, 1978), p. 338; H. Beylard, "François (Jean)," *Dictionaire de biographie française*, t. 14 (Paris, 1979), col. 1041.

[113] "Descartes au P. [Hayneuve]. [Leyde, 22 juillet 1640]," *AT*, III, p. 100; *AM*, IV, p. 92: "[...] olim per nouem ferè annos, in vno ex vestris Collegijs fui institutus [...]." "Descartes au P. [Grandamy?]. [Leyde, 2 mai 1644?]," *AT*, IV, p. 122; *AM*, VI, p. 148: "[...] à la Flèche, ou j'ay demeuré huit ou neuf ans de suite en ma ieunesse." "Descartes au P. [Bourdin]. [Egmond, 9 février 1645]," *AT*, IV, p. 161; *AM*, VI, p. 188: "[I]l a eu la direction huit ans durant, pendant que i'estois à la Flèche [...]."

between 1606 and 1614. And third, in the recent persuasive theory by Rodis-Lewis which we introduced at the beginning of Chapter 1, Descartes was at the college of La Flèche from 1607 to 1615.[114] If Baillet were right, Descartes's mathematics teacher must have been a certain Nicolas Laplace. The following anecdote seems to confirm that the latter conjecture was the case.

On publishing his *Principia philosophiae* in 1644, Descartes asked Father Bourdin to forward its exemplars to several Jesuit fathers. Among those whom Descartes hoped to present his book of philosophy, we find the name "le R. P. F., mon ancien Maistre."[115] Charles Adam, one of the editors of the *Œuvres de Descartes*, at first could not identify who Descartes's former teacher "Father F." was, simply because of having presupposed, following Baillet, that Descartes left La Flèche in 1612 before Father François arrived there.[116] Even so, he does not seem to have considered any candidate for "le R. P. F." other than "le Révérend Père François." Once we adopt the different conjecture that Descartes left La Flèche in 1614 or 1615, the problem will be solved. Besides, Jean François surely looked over Descartes's *Principia philosophiae* and expressed his will to give his best wishes to Descartes even while criticizing the latter's laws of motion in his letter to Mersenne on September 28, 1647.[117]

Accordingly we entertain the conjecture that the person "le R. P. F." to whom Descartes wanted to forward his book was "the Reverend Father François, my former teacher." So we may proceed in our discussion assuming that Descartes's teacher of mathematics was Jean François. Even if this conjecture turns out to be untrue, some knowledge of François's thought will deepen our understanding concerning Descartes's educational circumstances.[118] From this, we will learn something about the college where Descartes was studying. In any case, we know almost nothing about Nicolas Laplace, François's predecessor, and our knowledge of François is virtually a little more than that he was a teacher of mathematics at the college of La Flèche around the period when Descartes studied there.

[114]About the authors supporting these conjectures, see Chapter 1, note 1.

[115]"Descartes au P. [Bourdin]. [Paris, octobre 1644]," *AT*, IV, p. 144; *AM*, VI, p. 167.

[116]Charles Adam's embarrassment can be witnessed in his note on *AT*, IV, p. 144.

[117]"Le Père Jean François de Nevers, à Mersenne, à Paris. 28 septembre 1647," *CM*, XV, pp. 450–451.

[118]In fact, H. Beylard's recent article in *Dictionnaire de biographie française* denies that François was a teacher of Descartes on the ground that the latter left La Flèche in April 1612: "En oct. 1619 [*sic!*] il [François] commença l'étude de la théologie au collège de La Flèche tout en y enseignant les mathématiques. Quoi qu'en aient écrit les PP. de Dainville et Delattre, it ne fut pas le maître de Descartes, qui avait quitté La Flèche en avr. 1612, avant son arrivée." *Op. cit.* (n. 112), col. 1041. "En oct. 1619" must be corrected to "En oct. 1612." Beylard's conjecture seems to be based on Jacques Feller's assertion that Descartes came to La Flèche in 1606 and left in 1612. Feller, "Descartes (René)," *Dictionnaire de biographie française*, **10** (Paris, 1965), cols. 1240–1244. We cannot side with Feller and then Beylard. See also P. Costabel's description: Pierre Costabel, "L'Initiation mathématique de Descartes," *Archives de philosophie*, **46** (1983), pp. 637–646, esp. pp. 637–638. As I have already stated above, I support G. Rodis-Lewis's assertion in her *Descartes: Biographie* (Ch. 1, n. 1).

Jean François was born in Saint-Claude in 1582 and was educated at the Jesuit college of Dole before being admitted to the Jesuit Order in 1605. For a year from 1611, he was "professeur en classes inférieurs" at Pont-à-Mousson. It was already pointed out that from 1612 to 1616 he taught mathematics at La Flèche while studying theology. For one year from 1617, he taught logic at La Flèche, for another year from 1618, he taught physics, and for the last year at La Flèche between 1619 and 1620, he taught metaphysics. After leaving La Flèche, he continued to occupy important teaching positions at several Jesuit colleges. He seems to have been a rather outstanding school teacher although he was not necessarily an original researcher of the first rank. In his later years, he was the vice-rector or rector at the colleges of Alençon, Amiens, and Nevers. He died at the beginning of 1668.[119]

We may gather what kind of teacher and scholar François was from his publications listed in Carlos Sommervogel's *Bibliothèque de la Companie de Jésus* and in Romano's monograph. The books written by François are with few exceptions related to mathematics in some sense or other, especially to practical mathematical sciences such as cosmology, geography, hydrography, chronometry, and astrology.[120]

A close examination of his book titled *L'arithmétique et la géométrie pratique* strengthens the impression that François was particularly interested in the practical facet of mathematics. This work consisted of the two monographs: one is *L'arithmétique ou L'art de compter toute sorte de nombre avec la plume et les jettons*, and the other is *La géométrie pratique ou L'art de mesures tant de loing que près toute sorte de lignes, de surfaces & de corps.*[121]

The first treatise apparently belongs to the category of practical arithmetic, which closely resembles Clavius's widely used *Epitome arithmeticae practicae*, first published in Rome in 1593. François's book is divided into four chapters. Chapter 1 "On the Principles of Arithmetic" explains integral numbers. Chapter 2 "On the Arithmetical Operations, Particularly and Firstly on Integral Numbers" concerns addition, subtraction, multiplication, and division of numbers. Chapter 3 "On Fractions or Broken Numbers (*nombres rompus*)" teaches how to use rational numbers. Finally, Chapter 4 "On Proportions between Quantities and on the Rule of Three" discusses the theory of arithmetical proportion and various practical rules of arithmetic. These rules of arithmetic are well-known techniques for solving practical arithmetical problems.[122]

[119]For François's biographical data, A. Romano, *Ibid.* (Ch. 1, n. 13), pp. 576–577.

[120]Carlos Sommevogel, *Bibliographie*, t. III, Nouvelle édition (Bruxelles/Paris, 1892), cols. 938–939. To list in chronological order, these books are *La science de la géographie* (1652), *La science des eaux* (1653), *La chronologie* (1655), *De la sphère* (1655), *Traité des influences célestes* (1655) and *L'art des fondaine* (1665), all of which were published in Rennes. See also A. Romano, *Ibid.*, pp. 632–633.

[121]The two monographs were first published separately in 1653. We refer to the one-volume fourth edition published in Paris in 1681, which is now at the Bibliothèque Nationale de Paris (cote #, V. 8756).

[122]The arithmetical rules that François elucidated are, e. g. the well-known rule of three

What is significant is that the structure of François's treatise is similar to Clavius's *Epitome arithmeticae practicae* and, in particular, all of François's arithmetical rules are contained in Clavius's as well.[123] François makes no mention of Clavius, and these rules were all very common in the tradition of practical arithmetic of the sixteenth and seventeenth centuries. But judging from the circumstantial evidence that the most influential mathematician in the Jesuit Order in those days was Clavius, we might not committ a serious error in conjecturing that François studied Clavius's *Epitome*.

The second treatise on practical geometry is a manual mainly on the measurement of plane and solid geometrical figures, placing special emphasis on land-surveying and its instruments. The content is very elementary and the author seems to intend to provide primitive knowledge which was transmitted in the tradition of practical geometry of the Middle Ages and Renaissance. As in the case of arithmetic, a model for François's textbook of practical geometry might have been the beginning part of Clavius's *Geometria practica*.

François did not completely lack interest in the theoretical aspect of the mathematical disciplines. He published in 1655 a philosophical monograph under the title *Traité de la quantité considérée absolument et en elle-mesme [...] relativement et en ses rapports, matériallement et en ses plus nobles sujets, pour servir d'introduction aux sciences et arts mathématiques et aux disputes philosophiques de la quantité*, which Gilson sometimes refers to in his commentaries on Descartes's *Discours de la méthode*.[124] As can be inferred from the title itself, the purpose of this book is to argue all aspects of quantity, both theoretical and practical, in order to present "the key of the mathematical sciences and arts which join the evidence of demonstrations with the certainty of operations and the enjoyment of speculative sciences with the uses of practices."[125]

As can be imagined from François's definition of quantity, "quantity is the

(*la regle de trois*), sometimes called the golden rule because of its importance, the inverse rule of three (*la regle de trois de la renversee*), the composed rule of three (*la regle de trois composée*), the rule of partnership (*la regle de société, ou de compagnie*), the rule of alligation (*la regle de l'alliage, et du mélange*), the rule of false position (*la regle de fausse position*) and the rule of double false position (*la regle de fausse position double*). These rules of applied arithmetic are all explained in detail in Lambert L. Jackson, *The Educational Significance of Sixteenth Century Arithmetic* (New York, 1906), pp. 127–170.

[123]The titles of Chapters 17–23 of Clavius's *Epitome arithmeticæ practicæ* are "17 Regula trium; quae alio nomine regula aurea, siue regula proportionum dici solet. 18 Regula trium euersa. 19 Regula trium composita. 20 Regula Societatum. 21 Regula alligationis. 22 Regula falsi simplicis positionis. 23 Regula falsi duplicis positionis." *Opera mathematica*, t. II, pp. 35–65. The historical position of Clavius's arithmetical book is discussed in P. Treutlein, "Das Rechen in 16. Jahrhundert," *Abhandlungen zur Geschichte der Mathematik*, Erstes Heft (Leipzig, 1877), pp. 1–100, esp. pp. 18–19. As for its global social and intellectual background, see Frank J. Swetz, *Capitalism and Arithmetic: The New Math of the 15th Century* (La Salle, Illinois, 1987).

[124]The Bibliothèque Nationale of Paris possesses two copies (The numbers of cotes are V. 6284 and Rés. p. V. 242). Gilson mentions this book in his annotations, *Op. cit.* (Ch. 1, n. 1), pp. 126–127.

[125]Jean François, *Traité de la quantité* (Rennes, 1655), "Avis au Lecteur."

common object of our senses,"[126] the philosophical authority for François was Aristotle. After speculation on abstract quantities which are the objects of study of arithmetic and geometry, material quantities become a topic of discussion. The argument there extends even to the systems of the world by Ptolemy, Copernicus, and Tycho Brahe. Among them, François admits the superiority of Tycho's system. He follows the predominant opinion of the Society of Jesus concerning cosmological problems.[127] It is also worthy to note that the concept of infinity is a subject of the discourse.

Further evidence corroborated that François was faithful to the orthodox doctrines of the Jesuit Order. This is obtained from his aforementioned letter to Mersenne dated on September 28, 1647. In this correspondence, while alluding to Descartes's *Principia philosophiae*, François comments on the laws of motion in that book. He writes to Mersenne:

> Please give great regards to Monsieur Descartes. I find all these laws of motion, except two, very doubtful and according to Father Fabri false.[128]

The letter tells us that François at least glanced over Descartes's *Principia philosophiae* and sided with his Jesuit colleague Honoré Fabri's criticism of Descartes.[129] Here again we witness a facet of Father François, who was gen-

---

[126] *Ibid.*, p. 5.

[127] *Ibid.*, p. 65: "Ie conclus, disant, Que l'Hypothese de Tycho à du moins égalité en l'Astronomie, l'auantage & le plus de probabilité en Physique, la certitude en Théologie, ce que ie déclare plus amplement en vn traité exprès & que l'on trouuera éclaircy dans plusieurs Autheurs." Incidently, referring to the same passage, Gilson asserts erroneously that "En 1655, le P. J. François tenait encore pour Ptolémée contre Copernic [...]." Gilson, *Texte et commentaire*, p. 127. Tycho enjoyed a great reputation among the Jesuit scholars in the mid-seventeenth century. See C. Doris Hellman, "Was Tycho Brahe as Influential as He Thought?" *The British Journal of the History of Science*, **1** (1963), pp. 295–324, esp. pp. 305, 313 & 315–317. According to Hellman, "Tycho's system as a stopgap for those not satisfied with the Ptolemaic and still unprepared for the Copernican appealed particularly to the Jesuits and to other Catholic orders and gained a high reputation among them."(p. 313.) For Clavius's position on cosmology, see the detailed analysis of William H. Donahue, "The Solid Planetary Spheres in Post-Copernican Natural Philosophy," in Robert S. Westman, ed., *The Copernican Achievement* (Berkeley/Los Angeles/London, 1975), pp. 244–275. Although Clavius's final position appears to have been Ptolemy's, he fully appreciated the Tychoic system of the world.

[128] *Op. cit.* (n. 117), p. 451: "Je lui prie de dire mille recommandations a Monsieur Descartes. Je trouue toutes ces [*sic!*] regles du mouuement, a la reserue de deux, bien douteuses et selon le P. Fabri, fausses."

[129] Fabri was a celebrated Jesuit scientist who wrote several books on mathematics and natural philosophy. François almost undoubtedly learned Fabri's criticism of Descartes from the latter's *Tractatus physicus de motu locali*, published in Lyons in 1646. This treatise on the theory of local motion was a fundamentally Aristotelian response to new mechanistic theories, mainly Galileo's and partly Descartes's. Many statements of Fabri's book are known to agree with those of the proponents of the newly proposed mechanical world view and at the same time, its fundamental philosophical concepts being Aristotelian, many of which contradict the latters'. So Fabri's position, as far as the treatise of 1646 is concerned, may be characterized

erally interested in contemporary intellectual matters and faithful to dominant doctrines within the Jesuit Order.

According to Descartes's later retrospection (see p. 14), he thought there must be learned men at the college of La Flèche, if they existed anywhere on earth. Although he was a young scholastic student when he arrived at La Flèche, François, at least, may have been counted as a candidate for such a learned man. Even though his works we examined briefly were all published in the 1650s, many years after François was at La Flèche, it must be admitted at least that he was not a mediocre teacher of mathematics. Whether it was the case or not that François was Descartes's actual mathematics teacher, because mathematics teachers are supposed to have been relatively well-trained, as we saw in Chapter 1, Descartes was in circumstances favorable to being led to the mathematical disciplines with a certain enthusiasm.

The well-ordered *Ratio studiorum*, well-written textbooks of mathematics, and good teachers with a firm sense of mission—perhaps these were not the only causes which spurred the young Descartes on to the study of mathematics, but there can be no doubt that he was stimulated by some of such institutional elements to become engaged on his own reading of advanced textbooks of mathematics. Notwithstanding, according to some attestations of his later days, it is also true that he was not always satisfied with the content of the mathematical disciplines which he studied during his college years.

Let us see, then, how the later Descartes evaluated the mathematical books which he read at La Flèche and why he felt dissatisfaction with them. Through this procedure, we may ascertain what the main characteristics of his mathematical knowledge before 1618 were. In Chapter 1, § 1, we have already examined how he thought of the mathematical studies of his school days in his

---

as Aristotelianism modified by some knowledge of Galileo and Descartes. For an analysis of several propositions of Fabri compared with Galileo's, see Stillman Drake, "Impetus Theory and Quanta of Speed before and after Galileo," *Physis*, **16** (1974), pp. 47–65, esp. pp. 54–65; *Idem*, "Free Fall from Albert of Saxony to Honoré de Fabri," *Studies in History and Philosphy of Science*, **5** (1975), pp. 347–366, esp. pp. 361–366. Descartes himself paid heed to Fabri's criticism and tried to refute it. On Part IV: "The Earth," Article 23 "How all the parts of the earth are driven downwards by the celestial matter, and so become heavy" of his *Principia philosophiae*, Descartes writes to Mersenne on 26 April 1647, "[Father Fabri] says that bodies called heavy cannot be attracted by the Earth, or pushed towards it by some subtle matter (which is against me), from which he concludes that they themselves must possess a quality that makes them descend [...]. He says that the subtle matter is light, according to the opinion of those who have invented it (that is to say myself), and that, in consequence, bodies in obscure cellars should not have as much gravity as if they were exposed to the sun; but our experiments show the contrary. From which it can be seen that he has really read what I have written, but that he has understood it very badly; for I have never said that subtle matter was light, nor that it was gravity, but that it has several different actions, one of which excites in us the sensation of light, and the other makes heavy bodies descend towards the Earth. And these two actions nowise prevent each other, as I have suffficiently proved." "Descartes, à Egmond, à Mersenne, à Paris. 26 avril 1647," *AT*, IV, p. 637; *AM*, VII, pp. 282–283; *CM*, XV, p. 210. Cf. René Dugas, *Mechanics in the Seventeenth Century*, tr. Freda Jacquot (Neuchatel, Switzerland, 1958), p. 190, note 3.

Discours de la méthode. There he stated that at the college he had acquired knowledge of logic, geometrical analysis, and algebra, the three components which were to crystallize as the unified general method. Here, we should read an autobiographical narration in Rule IV of his *Reguale ad directionem ingenii*.

Descartes reminisces about his first experience of studying pure mathematics:

> When I first appied my mind to the mathematical disciplines, I at once read most of the customary lore which mathematical writers pass on to us. I paid special attention to arithmetic and geometry, for these were said to be the simplest and, as it were, to lead into the rest. But in neither subject did I come across writers who fully satisfied me. I read much about numbers which I found to be true once I had gone over the calculations for myself. The writers displayed many geometrical truths before my very eyes, as it were, and derived them by means of logical arguments. But they did not seem to make it sufficiently clear to my mind why these things should be so and how they were discovered. [...] For there is really nothing more futile than so busying ourselves with bare numbers and imaginary figures that we seem to rest content in the knowledge of such trifles. And there is nothing more futile than devoting our energies to those superficial proofs which are discovered more through chance than method and which have more to do with our eyes and imagination than our intellect; for the outcome of this is that, in a way, we get out of the habit of using our reason. At the same time there is nothing more complicated than using such a method of proof to resolve new problems which are beset with numerical disorder.[130]

Among Descartes's expressions, there might be some exaggeration. But he reveals a truth about the arithmetic and geometry he learned at the school. The characteristic of geometry pointed out above is supposed to apply, especially to Clavius's edition of Euclid. Clavius's Euclid taught the reader almost nothing about how its propositions had been discovered, neglecting the method of discovering and solving mathematical truths, that is, analysis. We do not know what Descartes meant precisely by arithmetic. We only know that the textbook (or textbooks) of arithmetic he read was filled with "bare numbers" and "numerical disorder."

Of course, these negative aspects were only one side of what Descartes saw in the mathematics of his school days. In reality he studied, to say retrospectively, a kind of mixture of "vulgar mathematics" and "true mathematics." He goes on with the description of his speculation on the mathematical disciplines:

> Later on I wondered why the founders of philosophy would admit no

[130] *AT*, X, pp. 374–375.

> one to the pursuit of wisdom who was unversed in mathematics—as if they thought that this discipline was the easiest and the most indispensable of all for cultivating and preparing the mind to grasp other more important sciences. I came to suspect that they were familiar with a kind of mathematics quite different from the one which prevails today; not that I thought they had a perfect knowledge of it, for their wild exaltations and thanksgivings for trivial discoveries clearly show how rudimentary their knowledge must have been. [...] But I am convinced that certain primary seeds of truth naturally implanted in human minds thrived vigorously in that unsophisticated and innocent age—seeds which have been stifled in us through our constantly reading and hearing all sorts of errors. So the same light of the mind which enabled them to see (albeit without knowing why) that virtue is preferable to pleasure, the good preferable to the useful, also enabled them to grasp true ideas in philosophy and mathematics, although they were not yet able fully to master such sciences. Indeed, one can even see some traces of this true mathematics, I think, in Pappus and Diophantus who, though not of that earliest antiquity, lived many centuries before our time. But I have come to think that these writers themselves, with a kind of pernicious cunning, later suppressed this mathematics as, notoriously, many inventors are known to have done where their own discoveries were concerned. They may have feared that their method, just because it was so easy and simple, would be depreciated if it were divulged; so to gain our admiration, they may have shown us, as the fruits of their method, some barren truths proved by clever arguments, instead of teaching us the method itself, whch might have dispelled our admiration.[131]

We cannot determine precisely what "later on" in Descartes's statement means. We speculate this was the time when he had his first intimation about "true mathematics" (*vera Mathesis*), and therefore the period when he knew the science of algebra or more probably than after he met Beeckman and began to envisage the idea of "an entirely new science" which would be the kernel doctrine for both arithmetic and geometry. (For "entirely new science," see below Chapter 3, § 1.) A mathematical discipine, as Descartes calls "true mathematics" in the *Regulae ad directionem ingenii*, is a normative science for his general method. According to him, vestiges of this discipline can be found in the works of Pappus and Diophantus, which are thought to show not only valid mathematical results but also how they were discovered. That is to say, true mathematics is a science which can instruct us the method of analysis.

In the wake of the passage quoted above, Descartes tells us that even

[131] *Ibid.*, pp. 375–377.

before him there were restorers of true mathematics in his own day:

> In the present age some very gifted men have tried to revive this method, for the method seems to me none other than the art which goes by the outlandish name of 'algebra' (*ars illa, quam barbaro nomine Algebra vocant*)—or at least it would be if algebra were divested of the multiplicity of numbers and incomprehensible figures which overwhelm it and instead possessed that abundance of clarity and simplicity which I believe the true mathematics ought to have. It was these thoughts which made me turn from the particular studies of arithmetic and geometry to general investigation of mathematics.[132]

What Descartes states here is very significant indeed. He discovered a beginning to what he thought to be true mathematics in his contemporaries' algebra. It is true that the normative disciplines which served to form the universal method were logic, analysis, and algebra, as Descartes indicates in Part II of his *Discours de la méthode*. And in Rule IV, prior to the passage on his mathematical career we have just cited, he lists geometrical analysis and algebra in parallel as two components which figured in the creation of his method:

> For the human mind has within it a sort of spark of the divine, in which the first seeds of useful ways of thinking are sown, seeds which, however neglected and stifled by studies which impede them, often bear fruit of their own accord. This is our experience in the simplest of sciences, arithmetic and geometry: we are well aware that the geometers of antiquity employed a sort of analysis which they went on to apply to the solution of every problem, though they begrudged revealing it to posterity. At the present time a sort of arithmetic called 'algebra' is flourishing, and this is achieving for numbers what the ancients did for figures. These two disciplines are simply the spontaneous fruits which have sprung from the innate principles of this method.[133]

We suppose that these two or three disciplines must have played equally important roles as the principles of his method. But what Descartes suggests in his statement about his career is that among these components the most crucial discipline by which his mind had been nourished, even before he identified consciously what true mathematics should be, was algebra. The algebra which he had studied previously was still full of a "multiplicity of numbers and incomprehensible figures." If Descartes's testimonies in his *Regulae* and in his conversation with John Pell in 1646 are trustworthy, the textbook of algebra with which he had become familiar must be, among others, Clavius's *Algebra*

[132] *Ibid.*, p. 377.
[133] *Ibid.*, p. 373.

of 1608. It was this treatise that helped the young French noble later to hit upon the true method in mathematics and in the end in all sciences. Clavius's textbook contained one element which characterized traditional manuals of cossist algebra. But it surely excerpted not a few mathematical problems from Diophantus's *Arithmetic*, and importantly illustrated the nature of algebra as consisting in problem-solving techniques applicable to both arithmetic and geometry.

However, we should not forget that all of these attestations were derived from Descartes's retrospection in his later days, in the case of the *Regulae*, of the 1620s. It seems unlikely that he recognized very clearly what the mathematics he studied in the college lacked even if he received some intimation of this recognition. In 1618, Descartes resolved to travel to the Netherlands in order to seek true knowledge "in the great book of the world," leaving aside "the study of letters."[134] He did not yet expect that this departure would mark the separation from the mathematical world of 'the Euclid of the sixteenth century'.

[134] Descartes, *Discours de la méthode, AT*, VI, p. 9.

Chapter 3

The First Attempt at Reforming Mathematics

§ 1. "An Entirely New Science": The Idea for the Unification of Arithmetic and Geometry

The inspiration which led Descartes to his reform of traditional mathematical thought took place soon after his encounter in the autumn of 1618 with Isaac Beeckman, a Dutch scholar his senior by eight years. The significance of this meeting for him is shown clearly in his letter of April 23, 1619 to Beeckman: "You are truly the only one who awoke [me] from sloth, recalled erudition which had almost passed away from memory, and bettered my mind which was drifting away from serious occupations."[1]

Isaac Beeckman, having been cultivated in the spirit of Dutch Calvinism, especially Ramism, was an intellectual product of the newly born Dutch Republic at the dawn of the Golden century. As a student at Leiden from 1607 to 1610, he studied under the representative Dutch Ramists Rudolph Snel van Royen and his son Willebrord. The list of mathematical books which Rudolph Snel recommended that Beeckman read epitomizes the kind of intellectual milieu in which the young Dutch student was educated. According to the list, the students were encouraged to be familiar with the following authors: Ramus, Euclid, and Hero in geometry; Ramus, Boethius, and Euclid in arithmetic; Ramus, Clavius, and others in practical arithmetic; Ptolemy and Copernicus in astronomy; Hero, Commandino, and Pappus in mechanics; and several authors in astrology, optics, music, and other 'mixed' mathematics.[2] We do not know

[1] "Descartes, à Bréda, à Isaac Beeckman, à Middlebourg. 23 avril 1619," *AT*, X, pp. 162–163; *Journal tenu par Isaac Beeckman*, éd. C. De Waard, t. IV (Supplément), (La Haye, 1953), p. 62: "Tu enim reverâ solus es qui desidiosum excitasti, jam è memoriâ penè elapsam eruditionem revocasti et à serijs occupationibus aberrans ingenium ad meliora reduxisti."

[2] *Journal*, t. IV, pp. 17–19.

precisely to what extent Beeckman fulfilled this task assigned to him by Snel. That he accomplished it to a considerable degree we can judge to some extent from his *Journal*, covering the years 1604 to 1634, discovered by Cornelis de Waard in 1905 at the Provincial Library of Zeeland. In this *Journal*, he referred, for example, to some works of Euclid and Ramus. It is noteworthy that he quoted Clavius's *Geometria practica* and *Algebra*,[3] and wrote algebraic equations in cossic symbols.[4] At the beginning of the seventeenth century Clavius was admired in the Netherlands, especially by Simon Stevin and the Snels.[5]

Judging from the topics he was concerned with in his *Journal*, Beeckman's interests seem to have centered not primarily in pure mathematics but predominantly in practical mathematics and experimental natural philosophy. Such an intellectual tendency can generally be observed in the Dutch Ramist movement around 1600.[6]

Beeckman is also known to have basically accepted the Copernican hypothesis and to have been one of the first to formulate the principle of inertia, although it is undeniable that his scientific achievements were fundamentally unsystematic and fragmentary.[7] Descartes was not the only one who was inspired by Beeckman. In 1629, Beeckaman's atomistic doctrine prompted Pierre Gassendi, for example, to study seriously the ancient atomists Democritus and Epicurus. Gassendi called the Dutch scholar "le meilleur philosophe que j'aye encore rencontré."[8]

We should not make the hasty judgment that Beeckman unilaterally influenced Descartes. The interaction between the two seems to have been, in fact, a kind of collaboration. It is certain that the senior Dutch scientist enlightened Descartes about some important matters. This enlightenment, however, played the role of not necessarily implanting a seed of an entirely new learning but of reorienting or providing a fresh stimulus for the young French noble's already existing talent for mathematics and natural philosophy which had been lying

---

[3]Consult the "Table des ouvrages cites" and "Index général" attached to the end of Volume IV of Beeckman's *Journal*.

[4]For example, Beeckman writes "1ʒ aequatur 4174𝔵 + 276128" for today's $x^2 = 4174x + 276128$, in *Journal*, t. I (La Haye, 1939), p. 6.

[5]See Dirk J. Struik, *The Land of Stevin and Huygens* (Dordrecht/Boston/London, 1981), p. 77.

[6]See Walter J. Ong, *Ramus, Method, and the Decay of Dialogue: From the Art of Discourses to the Art of Reason* (Cambridge, Mass., 1958), p. 305; P. Dibon, "L'influence de Ramus aux universités néerlandaises du 17e siècle," *Proceedings of the XIth International Congress of Philosophy*, Vol. XIII (Amsterdam, 1953), pp. 307–311.

[7]As to the main characteristics of Beeckman's natural philosophy, see R. Hooykaas, "Science and Religion in the Seventeenth Century: Isaac Beeckman (1588–1637)," *Free University Quarterly*, **1** (1951), pp. 169–183; *Idem*, "Beeckman, Isaac," *DSB*, I, pp. 566–568; John A. Schuster, "Descartes and the Scientific Revolution, 1618–1634: An Interpretation" (Princeton University Ph. D. Dissertation, 1977), Chapter 1, B. "Beeckman and Descartes," pp. 53–71.

[8]"Pierre Gassendi, à Bruxelles, à Nicolas-Claude Fabri de Peresc, à Aix. 21 juillet 1629," *CM*, II, p. 246. For Beeckman's role in Gassendi's intellectual career, see Howard Jones, *Pierre Gassendi 1592–1655: An Intellectual Biography* (Nieuwkoop, 1981), p. 27.

dormant at the outset of their meeting.

We concentrate our discussion on the mathematical thoughts of these two young scholars. Beeckman's very first reference to Descartes in his *Journal* appears in the entry of November 11, 1618. With the margin heading "Descartes Has Proved That an Angle Is in Fact Nothing," Beeckman writes as follows:

> Yesterday, which was 10 November [1618], in Breda a Frenchman from Poitou made an effort to prove that an angle is in fact nothing, arguing the following:
>
> An angle is the combination of two lines at one point such as *ab* and *cb* at the point *b*. But if you cut the angle *abc* by the line *de*, you will divide the point *b* into two parts, so that one half would be attached to *ab* and the other half to *bc*. That is in contradiction to the definition of a point, according to which [its] part is nothing.
>
> But he has taken the point as a real magnitude, while the point is nothing other than the extremes of the lines *ab* and *cb*. And as no point constitutes a whole, so that a thousand points can be in the same place. Therefore, the line *de* passes through a point, in fact, *b*, but not cutting it, constitutes a whole, because a line is not extended [in width]. Therefore any point in the line *de* has the same position as the point *b*. A point in *fg* is also the same. Therefore the lines *fg*, *de*, cutting the angle, do not diminish the lines *ab* and *cb* in size as it happens when we cut something with a saw, but only separate one from another.[9]

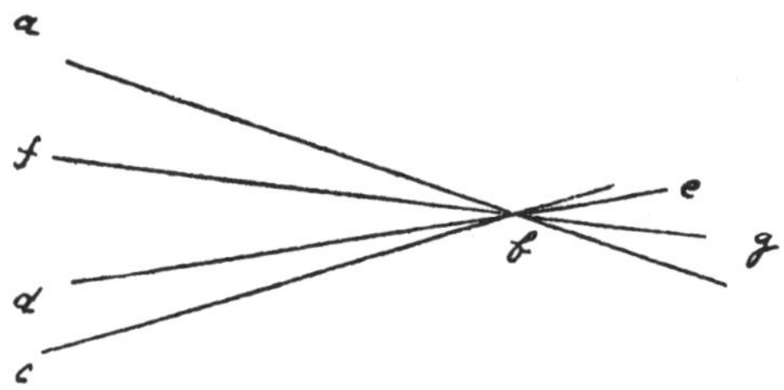

[Figure 3.1]

According to Beeckman's description, Descartes and he discussed a kind of mathematico-philosophical problem. The Frenchman proposed an problem concerning the definition of a plane angle. He defined the angle *abc* as consisting

[9] *AT*, X, pp.46–47; *Journal*, t. I, p. 237: "{Angulum nullum esse malè probavit Des Cartes.} Nitebatur heri, qui erat 10 Nov., Bredæ Gallus Picto probare nullum esse angulum reverâ, hoc argumento: Angulus est duarum linearum concursus in uno puncto, ut *ab* et *cb* in puncto *b*. At si seces angulum *abc* per lineam *de*, divides punctum *b* in duas partes, ita ut ejus dimidium *ab* adjungatur, alterum dimidium *bc*. Quod est contra puncti definitionem, cui pars nulla. At ille punctum sumpsit pro reali magnitudine, cùm punctus nihili aliud sit quàm extremitas lineæ *ab* et *cb*. Nec totum complet punctus, ita ut mille puncti possent esse eodem loco. Linea igitur *de* transit per punctum quidem *b*, sed id non secat, verùm totum complet, cùm linea non sit lata. Quare punctum aliquod in lineâ *de* eodem in loco est quo punctum *b*. Tale etiam punctum est *fg*. Non igitur lineæ *fg*, *de*, secantes angulum, minuunt lineas *ab* et *cb*, ut fit cùm serrâ quid secamus, sed solummodo separant unam *ab* aliâ." A margin headline is shown with the parentheses { }.

of the two lines *ab* and *bc*, and the point *b*. Then, he cut it by the line *de*. Through this procedure, contended Descartes, the point is divided into two parts. This contradicts the definition of a point: "A point is that a part of which is nothing."[10] So he concluded that an angle itself is nothing.

Beeckman demonstrated his logical talent by detecting the sophism in Descartes's inference. He argued that a point is not a substantial magnitude and that the Frenchman erred in claiming that a point can be divided into two parts. But did Descartes seriously believe that his argument was legitimate? It seems to us, at least, that he simply enjoyed devising a mathematico-philosophical puzzle and examining his senior Dutchman's ability to discover the sophism contained within. As the editor of Beeckman's *Journal* has suggested, similar mathematico-philosophical questions had been discussed since antiquity.[11] In 1618, Descartes was also interested in such mathematico-philosophical problems.

Sometime later, Descartes presented another mathematical problem to Beeckman. This problem is recorded in Beeckman's *Journal* in an entry dated between November 23 and December 26, 1618. There Beeckman begins: "René Descartes has proposed a problem to me: To give a square equal to the root of another square."[12] Beeckman tried to solve Descartes's question by finding a mean proportional between the given square and the unit. The terms of his argument were fundamentally geometrical but he illustrated his solution 'quasi-generally', providing arithmetical examples. One of these was stated as follows: "As 9 is to 1 so *ab* to *e*; but *cd* is a mean proportional between *ab* and *e*, therefore it is a side of the second square."[13] Clearly he took 9 as a root of the given square and found a mean proportional between it and the unit 1 (see Figure 3.2). This procedure essentially amounts to the extraction of the square root (or side) of the second square (the square on the side *cd*), the latter being equal to the square root of the first square on side *ab*. So the required square is the one con-

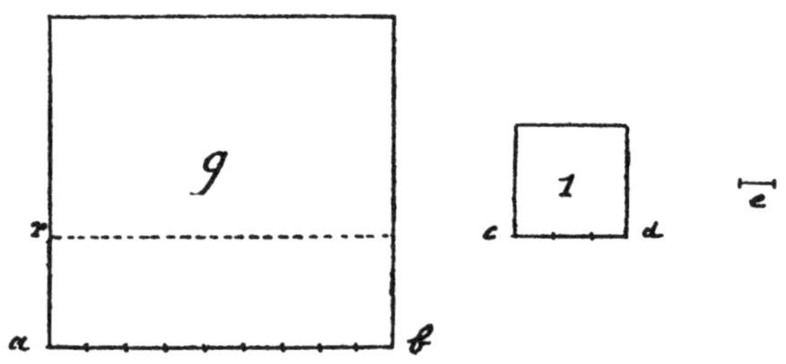

Figure 3.2

[10]Clavius, *Euclidis Elementa, Opera mathematica*, t. I, p. 13: "Pvnctvm est, cuius pars nulla est."

[11]As a contemporary book related to the above-mentioned problem, the editor C. de Waard refers to Henri de Monantheuil, *De Puncto primo geometriae principio Liber* (Lugd. Bat., 1600). *Journal*, t. I, p. 237, n. 4).

[12]*AT*, X, p. 54; *Journal*, t. I, p. 255: "Renatus Descartes mihi proposuit problema: *Dare quadratum æquale radici alterius quadrati.*"

[13]*AT*, X, p. 55; *Journal*, t. I, p. 255: "Ut se habet 9 ad 1 sic *ab* ad *e*; sed *cd* est medium proportionale inter *ab* et *e*, ergo est latus secundi quadrati."

structed on the square on side *cd*. Strikingly, Beeckman did not rely upon any algebraic technique which might have reduced the problem to one formulated in a certain algebraic equation.

Except for the two fragments examined above, almost all the topics of the discussions of 1618 recorded in Beeckman's *Journal* were about the mathematical theory of music and the mathematical formulation of the law of free fall. As to the first topic, Descartes worked up his own ideas into a monograph titled *Compendium musicae*, which he presented to Beeckman at the beginning of 1619. The two young scholars exchanged views on the second subject. Their views were analyzed in detail both historically and philosophically by Alexandre Koyré in his monumental *Études galiléennes* of 1939.[14]

Through this intellectual interaction, Beeckman and Descartes could affirm their common idea that mathematics is an extremely valuable tool in the study of natural philosophy. In his *Journal*, Beeckman writes a significant note under the margin heading "Physico-Mathematicians Are Extremely Few":

> The man from Poitou has much frequented the Jesuits, men of erudition and learning. He told me, however, that he had never met a man, apart from myself, who was in the habit of studying in the way in which I am pleased to do, joining, in precise fashion, physics and mathematics. And for my part, I have never spoken to anyone possessed of so much learning.[15]

Beeckman's memorandum suggests not only that he and Descartes were two of the very few scholars conscious of the importance of applying mathematics to physical problems, but also that before meeting Beeckman Descartes's system of knowledge was basically that which he had received from the Jesuits.

Almost nothing significantly new occurred in Descartes's mathematical career between his encounter with Beeckman and this point in time besides his having become interested in the above-mentioned mathematico-philosophical problem and in the use of mathematics in physics. But in 1619 Descartes began to confess that his senior friend truly awoke his theretofore slumbering intellect and stimulated him into expressing his own program for reorganizing the entire discipline of mathematics. His ambitious attempt appears in his correspondence

---

[14] A. Koyré, *Études galiléennes*, (Paris, 1939; actually published in 1940), pp. 99–128 (II-25–54). Koyré assigned the erroneous mathematical formulation of the law of free fall to Descartes's one-sidedness as a pure mathematician. A more historically sound analysis may be seen in J. A. Schuster, *Op. cit.* (n. 7), pp. 72–93. Further, on the development of Descartes's natural philosophy from 1618 on, see Vincent Jullien et André Charrak, *Ce qui dit Descartes touchant la chute des graves: 1618 à 1646, étude d'un indicateur de la philosophie naturelle cartésienne* (Villeneuve d'Ascq, 2002).

[15] *AT*, X, p. 52; *Journal*, t. I, p. 244: "{Physico-mathematici paucissimi.} Hic Picto cum multis Jesuitis alijsque studiosis virisque doctis versatus est. Dicit tamen se nunquam neminem reperisse, praeter me, qui hoc modo, quo ego gaudeo, studendi utatur accuratèque cum Mathematicâ Physicam jungat. Neque etiam ego, praeter illum, nemini locutus sum hujusmodi studij."

of 1619 with Beeckman.

It was in a farewell letter dated March 26, 1619 from Breda to the elder 'physico-mathematician' that Descartes revealed some of the concrete content of his reform project for the first time. At the beginning of the letter, he describes what he studied and obtained as a result of his research while residing in Holland: "Indeed in a short time I have discovered four conspicuous and entirely new demonstrations with the help of my compasses."[16] Then he explains what his four discoveries were: "The first is the very famous problem of dividing an angle into any number of equal parts." His explanation which follows reads:

> The other three pertain to cubic equations: of which the first kind between an absolute number, roots and cubes; the second between an absolute number, squares and cubes; the third, finally, between an absolute number, roots, squares and cubes. I have found three demonstrations corresponding to these, each of which should be extented to various terms according to the variation of the signs + and −; I have not yet discussed all, but in my judgment I will apply easily what I have found in some to the rest. And by this art four times as many and much more difficult questions will be able to be solved than by the common algebra; indeed I count thirteen different kinds of cubic equations, whereas there are only three [kinds] of common equations, namely between 1𝔷 and o℞+ *oN*, or o℞−*oN*, or, finally, *oN*−o℞. The other is that which I now look for on the extraction of roots composed together of various denominations; if I discover it, as I hope, I will clearly interpret that science in an orderly manner, provided that I could conquer my natural idleness and that the fate would grant my life freedom.[17]

Descartes's 'solution' to the first problem of dividing an angle into any number of equal parts was recorded in his manuscript *Cogitationes privatae*, which will be examined in the next section. He treated it by means of a compass, a mechanical instrument. The second part of the quoted paragraph is concerned

[16]"Descartes, à Breda, à Isaac Beeckman, à Middlebourg. 26 mars 1619," *AT*, X, p. 154; *Journal*, t. IV, p. 58: "Quatuor enim à tam brevi tempore insignes et planè novas demonstrationes adinveni, meorum circinorum adjumento."

[17]*AT*, X, pp. 154–156; *Journal*, t. IV, pp. 58–59: "Prima est celeberrima de dividendo angulo in æquales partes quotlibet. Tres aliæ pertinent ad æquationes cubicas, quarum primum genus est inter numerum absolutum, radices et cubos, alterum inter numerum absolutum, quadrata et cubos, tertium denique inter numerum absolutum, radices, quadrata et cubos. Pro quibus tres demonstrationes repperi, quarum unaquæque ad varia membra est extendenda propter varietatem signorum + et −; quæ omnia nondum discussi, sed facilè, meo judicio, quod in unis repperi, ad alia applicabo. Atque hac arte quadruplo plures quaestiones et longè difficiliores solvi poterunt quàm communi Algebrâ; tredecim enim diversa genera æquationum cubicarum numero, qualia tantùm sunt tria æquationum communium, nempe inter 1𝔷 et o℞+*oN*, vel o℞−*oN*, vel denique *oN*−o℞. Aliud est quod jam quæro de radicibus simul ex pluribus varijs nominibus compositis extrahendis, quod, si reperero, ut spero, scientiam illam planè digeram in ordinem, si desidiam innatam possim vincere et fata liberam vitam indulgeant."

with the solution of cubic equations. It is notable that Descartes wrote "common equations," i. e. quadratic equations, in cossic symbols identical with those of Clavius's. According to Descartes's classification, these three may be rewritten as follows: (1) $\pm a \pm bx = x^3$, (2) $\pm a \pm bx^2 = x^3$, (3) $\pm a \pm bx \pm cx^2 = x^3$, where a, b, c are positive known quantities, with the three cases $-a - bx = x^3, -a - bx = x^3, -a - bx - cx^2 = x^3$ being excluded, for every root is, in principle, considered to be positive. Similarly, he classified quadratic equations into three kinds: (1) $x^2 = ax + b$, (2) $x^2 = ax - b$, (3) $x^2 = b - ax$, excluding $x^2 = -ax - b$ as impossible.[18] In his way of sorting quadratic equations, there are $13 (= 16 - 3)$ different types of cubic equations, and $3 (= 4 - 1)$ types of quadratic equations. We still cannot understand well, however, why Descartes wrote, "by this art four times as many [...] questions will be able to be solved [...] by the common algebra." But if "the common algebra" means the algebra for handling quadratic equations, "four times" may have been derived from the two numbers of cubic and quadratic equations, $16 (= 4 \times 4)$ and 4. The three demonstrations which he made mention of were also discussed in detail in the *Cogitationes privatae*, and will be examined in Section 2 again.

Descartes takes up one more problem on "the extraction of roots composed of various denominations." This has been commonly interpreted to be the extraction of roots of quantitites in the form $a + \sqrt{b} + \sqrt{c} + \cdots$.[19] He might have acquired a hint for this problem, perhaps, from Chapter 28 "On the Extraction of Roots from Binomials and Apotomes, Wherewith Incidentally on the Other Irrational Lines on Which Euclid Discusses in Book 10" in Clavius's *Algebra*.[20]

Even from his statement quoted above, we know that as early as the spring of 1619 Descartes began trying to extricate himself at least in part from the confines of Clavius's *Algebra*. In his *Algebra* Clavius certainly mentioned, as we saw in Chapter 2, § 4, the attempts by Cardano, Tartaglia, Bombelli, and Viète to solve cubic equations, but did not present his own method for solving them. Descartes, on the other hand, attempted to solve cubic equations as well as to divide an angle into any number of equal parts by devising his own rather sophisticated compasses.

In the next paragraph of the same letter, a much more general and important statement appears concerning Descartes's project to create a new mathematical discipine. His program as described in this letter was uncommonly ambitious, so that it may be not too much to say that it governed his entire mathematical career. Beeckman, after receiving the letter, wrote in the margin of the top of the paragraph: "A General Art to Solve All Questions Selected." Descartes proclaimed what his "general art" was like:

---

[18] Cf. Gustav Eneström's remarks in *AT*, X, p. 155, nn. a. and d.

[19] Cf. G. Eneström's note a. in *AT*, X, p. 156.

[20] Clavius, *Algebra*, Cap. 28, "De Extractione Radicvm ex Binomiis, et Apotomis. Vbiobiter de alijs lineis Irrationalibus, de quibus Euclides in lib. 10. disputat." *Opera mathematica*, t. II, pp. 72–79.

> And certainly, to tell you openly what I am planning, I want to expound, not the *Ars brevis* of Lull, but an entirely new science, by which all problems that can be posed, concerning any kind of quantity, continuous or discrete, can be generally solved. Yet each problem will be solved according to its own nature, as, for example, in arithmetic some questions are resolved by rational numbers, others only by surd [irrational] numbers, and others finally can be imagined but not solved. So also I hope to show for continuous quantities that some problems can be solved by straight lines and circles alone; others only by other curved lines, which, however, result from a single motion and can therefore be drawn with new types of compasses, which are no less exact and geometrical, I think, than the common ones used to draw circles; and finally others that can be solved by curved lines generated by diverse motions not subordinated to one another, which curves are certainly only imaginary such as the rather well-known quadratrix. I cannot imagine anything that could not be solved by such lines at least, though I hope to show which questions can be solved in this or that way and not any other, so that almost nothing will remain to be found in geometry. It is, of course, an infinite task, not for one man only, incredibly ambitious; but I have seen some light through the dark chaos of this science, by the help of which I think all the thickest darkness can be dispelled.[21]

Descartes's program of "an entirely new science"(*scientia penitus nova*) seems to have been connected with an algebraic discipline. We should recall here that Clavius intended his *Algebra* to be applied to any kind of quantity, no matter what it may be, number or magnitude (see our discussion in Chapter 2, § 4, p. 73). Descartes called number "discrete quantity" (the subject of arithmetic) and magnitude "continuous quantity" (the subject of geometry). The passage

[21] *AT*, X, pp. 156–158; *Journal*, t. IV, pp. 50–60: "[Beeckman: {Ars generalis ad omnes quaestiones solvendas quæsita.}] Et certè, ut tibi nudè aperiam quid moliar, non Lullij *Artem brevem*, sed scientiam penitus novam tradere cupio, quâ generaliter solvi possint quæstiones omnes quæ in quolibet genere quantitatis, tàm continuæ quàm discretæ, possunt proponi. Sed unaquæque juxta suam naturam: ut enim in Arithmeticâ quædam quæstiones numeris rationalibus absolvuntur, aliæ tantùm numeris surdis, aliæ denique imaginari quidem possunt, sed non solvi, ita me demonstraturum spero, in quantitate continuâ, quædam problemata absolvi posse cum solis lineis rectis vel circularibus; alia solvi non posse nisi cum alijs lineis curvis, sed quæ ex unico motu oriuntur, ideòque per novos circinos duci possunt, quos non minus certos existimo et geometricos quàm communis quo ducuntur circuli; alia denique solvi non posse nisi per lineas curvas ex diversis motibus sibi invicem non subordinatis generatas, quæ certè imaginariæ tantùm sunt: talis est linea quadratrix, satis vulgata. Et nihil imaginari posse existimo, quod saltem per tales lineas solvi non possit, sed spero fore ut demonstrem quales quæstiones solvi queant hoc vel illo modo et non altero, adeò ut penè nihil in Geometriâ supersit inveniendum. Infinitum quidem opus est, nec unius. Incredibile quàm ambitiosum, sed nescio quid luminis per obscurum hujus scientiæ chaos aspexi, cujus auxilio densissimas quasque tenebras discuti posse existimo."

related to arithmetic can be easily understood. But it is not so easy to understand the author's clear emphasis on the geometrical part. As is well known, geometrical problems which can be solved by means of the ruler and compass alone were granted a privileged status in classical Greece.[22] The third genre of curved lines is the one which will be called "mechanical" (*Mechanique*) by Descartes in Book Two of his *Géométrie* of 1637. These curves are exemplified by the quadratrix, and will be excluded from theoretical geometry in the same treatise. The most crucial category of curved lines results from a simple motion and can be drawn by means of newly devised compasses.

These significant compasses were first explained minutely in the *Cogitationes privatae*. Here we only mention that they were invented to construct the n-th power x^n for an arbitrarily given quantity x and, therefore, had a close relation to Descartes's plan to reform algebra. In the algebraic treatment of mathematical problems, geometrical progressions of one or more unknown quantity play an extremely important role. As was seen in Clavius's *Algebra*, they were designated without exception with cossic symbols such as 𝔵, 𝔷, and 𝔠𝔢. Descartes seems to have thought that with the invention of these compasses one could easily construct the quantities which appear in algebraic equations. This insight of his affirms our introductory statement of the previous paragraph that Descartes's project of "an entirely new science" was related to algebra.

But an important question arises here: what inspired the young French mathematician to build "an entirely new science"? He believed that his reformed algebra, through the mediation of these mathematical instruments, would make the unification of arithmetic and geometry possible. There is no doubt that Descartes thought he had succeeded in inventing compasses to construct x^n. We cannot overstate the importance of various types of compasses in the history of mathematics in the late sixteenth and early seventeenth centuries.

Let us take only one famous example. Around the turn of the century Galileo Galilei made several geometrical compasses or sectors to simplify arithmetical and geometrical calculations. As a manual for handling these compasses, he published a booklet titled *Le Operatione del compasso geometrico et militare* in Padua in 1606. According to the manual, Galileo's compasses are capable of, for example, calculations usually carried out by the arithmetical rule of three, computations of compound interest, square root extraction, and the discovery of the mean proportional. Galileo's aim in inventing the mathematical compasses was stated in the preface entitled "To the Discreet Readers." Through his experience of teaching mathematics at the University of Padua, he states, Galileo had come to the following conclusion:

> [E]ven in our age very few can patiently travel the steep and thorny paths along which one first must pass before acquiring the precious fruits of this science; frightened by the long rough road, and not

[22] See Heath, *A History of Greek Mathematics*, Vol. 1, pp. 175–176.

> seeing or being able to imagine how those dark and unfamiliar paths can lead them to the desired goal, they falter halfway there and abandon the undertaking.[23]

So he undertook to devise his compasses in order that noble gentlemen could overcome "the length and difficulty of ordinary roads." He thereby tried to open "the royal road":

> I fell to trying to open this truly royal road (*via veramente regia*)—for with the aid of my Compass I do that in a few days, teaching everything derived from geometry and arithmetic for civil and military use that is ordinarily received only by very long studies.[24]

Improved versions of Galileo's model spread very rapidly all over Europe through the hands of the German 'Rechenmeister' Johannes Faulhaber in Ulm, for example.[25] Galileo's was one of the most widely disseminated models, but of course was not necessary unique.

Different types of geometrical compasses were also invented and sold by many instrument makers of those days. For example, Benjamin Bramer devised a rather subtle type of sector and wrote about it in his monograph on mathematical instruments published in Marburg in 1615. In that book Bramer cited Galileo.[26] Further, Descartes mentioned Benjamin Bramer in his *Cogitationes privatae*. We will discuss this topic again in the next section.

Needless to say, Descartes's compasses were not identical with Galileo's sector, which was usually called 'compasso di proportione' in Italian.[27] But it is also undoubtedly true that the young French mathematician was influenced by some mathematical practitioners who greatly respected mathematical compasses like those of Galileo. In the spring of 1619, Descartes intended to accomplish the unification of arithmetic and geometry with the aid of mathematical compasses, not necessarily in the purely written form of algebra. There

[23] *Opere di Galileo Galilei*, ed. Antonio Favaro (Firenze, $_2$1932), t. II, p. 369; Translated, with an Introduction by Stillman Drake (Washington, D. C., 1978), p. 41.

[24] *Ibid.*

[25] Cf. S. Drake's introduction in *Op. cit.*, esp. pp. 26–29.

[26] Benjamin Bramer, *Beschreibung und Underricht, wie allerley Theylungen zu den Mathematischen Instrumenten zu verfertigen: Neben dem Gebrauch eines newen Proportional Instruments* (Marburg, 1615), pp. 5 & 91. For Bramer's sector, see Ivo Schneider, "Der Proportionzirkel, ein universelles Analogrechnerinstrument der Vergangenheit," *Deutsches Museum Abhandlungen und Berichte*, **38** (1970), Heft 2, pp. 1–96, esp. pp. 58–62. I am indebted to Prof. MIURA Nobuo in Kobe University for my understanding of how to construct and use Bramer's sector.

[27] In Britain the sector has been distinguished from the proportional compass. "The European names for the sector are (German) Kreissektor; (Italian) compasso di proporzione; (French) compas de proportion. The proportional compass is known as (German) Proportionalzirkel; (Italian) compasso di reduzione; (French) compas de réduction." Gerard L'E. Turner, *Antique Scientific Instruments* (Poole/Dorset, 1980), p. 58. See also Maurice Daumas, *Scientific Instruments of the Seventeenth and Eighteenth Centuries*, translated and edited by Mary Holbrook (London, 1972), pp. 22–25.

was both a similarity and a difference betwen Galileo and Descartes. With his compass Galileo sought a mechanical approximation to practical mathematical problems. And he was concerned with problems, as he himself stated in the above-cited preface, derived from both arithmetic and geometry which were amenable to systematic treatment only through the science of algebra. Surely Descartes's attempt resembled Galileo's in the sense that both of these mathematicians relied on compasses for the unification of these two fundamental mathematical discipines. He seems to have had no more a sense of incompatibility in using mathematical instruments than Clavius in his *Geometria practica.*[28] An important difference between Galileo and Descartes was that the latter tried to provide his compasses with theoretical foundations through exact algebraic considerations.

Another motivation which spurred Descartes to reform the mathematical disciplines appears to have come from the general history of ideas outside the scope of mathematics itself. In the letter of March 26, 1619 he referred to the *Ars brevis* of Ramon Lull. But there he simply stated that he was to create "an entirely new science" instead of something similar to the Lullian art and expressed no further opinions about Lullianism.

In his letter to Beeckman of April 29, 1619, however, Descartes wrote much about Lull's art. According to his description in that letter, three days ago "I [Descartes] found a learned man in an inn of Dordrecht with whom I talked about Lull's *Ars parva*."[29] The book of Lull cited here is no doubt identical with the *Ars brevis*, which Descartes mentioned in the previous letter of March 26. The man in the letter, Descartes tells us, was an old rather talkative man. He boasted of being able to use the Lullian art successfully to such an extent that he could spend an hour about any subject and, if someone encourages him, 20 hours. But the old man's knowledge seemed to Descartes to have been not deep but superficial. Descartes asked him whether that art provided dialectical tools from which arguments could be drawn. The advocate of Lullism answered the question by stating that neither Lull nor Agrippa delivered in their books certain keys necessary to open the secrets of the art. Descartes complained about the old man's ignorance. But he still wanted to know whether the Lullian art contained something ingenious, so he asked Beeckman in Middelburg if anyone possessed a copy of Agrippa's commentary on Lull's *Ars brevis*.

Beeckman's reply can be read in his letter to Descartes of May 6, 1919. There Beeckman first apologized to the Frenchman for lacking the time to examine sufficiently the art of Lull. But he communicated his conclusion that in

---

[28]Clavius mentioned some mathematical instruments, for example, the ordinary compass and the quadrant in his *Geometria practica*, in *Opera mathematica*, t. II, Lib. I, Caput I, "Instrvmenti partivm constructio, atque vsus," pp. 5–14. Cf. I. Schneider, *Op. cit.* (n. 26), p. 46.

[29]"Descartes, à Amsterdam, à Isaac Beeckman, à Middlebourg," *AT*, X, p.164; *Journal*, t. IV, p. 63: "Repperi nudius tertius eruditum virum in diversorio Dordracensi, cum quo de Lullij *Arte parvâ* sum loquutus."

Agrippa's *In artem brevem Raymundi Lullij commentaria* there were no such keys as the old man and Descartes might desire to know. According to his opinion, he wrote to Descartes, the essential point of the Lullian art is to divide successively all that exists into elementary entities and to combine them with letters of the alphabet which are located in geometrical figures, such as various concentric circles. Combining memories with the "symbols of letters" (*characteres litterarum*) by imagination, one may be able to talk for almost infinite hours.[30]

Between the 2nd and 14th of May, 1619, almost at the same time that he composed his letter, Beeckman recorded a memorandum in the *Journal* with the margin subtitle "Lull's art compared with Logic." By logic he understood among others the logic of Peter Ramus. According to him, "particular sciences stand in the place of the Lullian art, but Lull's art evidently cannot stand in the place of Logic."[31] Beeckman does not seem to have been as sympathetic toward Lull's art as he was toward the Ramist logic.

As was observed in the correspondence between Descartes and Beeckman and in Beeckman's note in his *Journal*, the two scholars were generally antipathetic to the Lullian art. A similar stance in Descartes is also seen in Part Two of his *Discours de la méthode*. There he regards the art of Lull as the technique "for speaking without judgment about matters of which one is ignorant."[32] What we witness in his letters to Beeckman is that as early as 1619 Descartes had unhesitatingly rejected Renaissance naturalism, of which one of the representative thinkers was Henry Cornerius Agrippa, a commentator of Lull's *Ars brevis*.[33] Such an antipathetic attitude toward Renaissance naturalism had been fostered in the Jesuit Order. But, at the same time with his relationship with Beeckman providing momentum, Descartes seems to have begun to part from the revisionist Scholasticism of the Jesuits and to form his own way of thinking. In any event, in his letter to Beeckman he showed an interest in knowing what the keys of the Lullian art were. His hope had no fruits.

---

[30] "Isaac Beeckman, à Middlebourg, à Descartes, à Copenhague. 6 mai 1619," *AT*, X, p. 168; *Journal*, t. IV, p. 65. We are aided in understanding Beeckman's description of Lull's *Ars brevis* by Frances A. Yates, "The Art of Ramon Lull: An Approach to It through Lull's Theory of Elements," *Journal of the Warburg and Courtauld Institutes*, **17** (1954), pp. 115–173, esp. pp. 116–117; Reprinted in *Eadem, Lull and Bruno: Collected Essays*, Vol. I (London, 1982), pp. 9–77, esp. pp. 10–11. Cf. Heinrich Cornelius Agrippa von Nettesheim, *In artem brevem Raymundi Lullij commentaria*, in *Operum pars posterior* (Lyon, 1600; Hildesheim/New York, 1970).

[31] *AT*, X, p. 65; *Journal*, t. I, p. 295: "{Ars Lullij cum Logicâ collata.} Particulares scientiae igitur sunt vice artis Lullianae, ars verò Lullij non potest planè esse vice Logicae." The editor of Beeckman's *Journal* remarks that Beeckman was taught the method of Ramus while studying at Leiden and that his teacher Rudolph Snel published the *Commentaria in Dialecticam Petri Rami* (Herborn, 1587).

[32] *AT*, VI, p. 17

[33] Agrippa, *Op. cit.* (n. 30). For Agrippa's thought in general, see Richard H. Popkin, Introduction to Agrippa, *Opera*, I (Hildesheim/New York, 1970); Charles G. Nauert, Jr., *Agrippa and the Crisis of Renaissance Thought, Illinois Studies in the Social Sciences* (Urbana, 1965).

How should we understand Descartes's interest in the Lullian art, notwithstanding that his final judgment on it was negative? Two things seem important about it. First, through his contact with Lull's art he may possibly have noticed the importance of symbolism for the reform of the mathematical disciplines, especially algebra. We have already seen that before 1618 he had become familiar with Clavius's *Algebra*, in which the author used cossic symbols, such as those of Michael Stifel. It is noteworthy that Stifel himself was known to have attached particular importance to algebraic symbolism. It does not appear a mere accident that Stifel was preoccupied for a time with a 'word calculus' (*Wortrechnung*), by means of which he tried to interpret the Bible in a cabalistic way.[34] Here attention should be drawn to the fact that both Lull's art and Ramus's logic, which are supposed to have been topics of discussion between Beeckman and Descartes, aimed at reorganizing the sciences by stressing mnemonic aspects.[35] Descartes will fully acknowledge the importance of symbolism in mathematics in his *Regulae ad directionem ingenii*.

Second, the Lullian art may have been significant to the French mathematician as a pansophic symbol for the unification of the mathematical sciences. The Ramean program to reorganize the system of learning through an effective pedagogical method, of which Descartes had a fair possibility of coming to know through Beeckman, might have played a similar role. Paolo Rossi describes how Lullism was accepted among sixteenth-century intellectuals as follows:

> Agrippa [...] emphasized in the preface to his commentary two of the fundamental characteristics of the Lullian art as it appears to the Renaissance. Firstly, it is seen as a general and universal science which, dealing with absolutely certain principles and infallible proofs, makes it possible to establish an absolute criterion of truth; secondly, because it is conceived as the science of sciences, it offers a key to the exact and rational ordering of all knowledge, whose various aspects are all comprised in, and verified by, it.[36]

Rossi thought that the art of Lull interested Descartes because of its character as the common bond of all sciences.[37]

---

[34] Kurt Vogel, "Stifel, Michael," *DSB*, XIII, p. 58.

[35] See Paolo Rossi, *Clavis universalis: Arti della memoria e logica combi-natoria da Lullo a Leibniz* (Bologna, $_2$1983; first published in 1960), Cap. II, "Enciclopedismo e combinatoria nel Cinquecento," pp. 63–102, Cap. V, "La memoria artificiale e il metodo della nuova scienza: Ramo, Bacone, Cartesio," 1, "Pierre de la Ramée: la《memoria》come sezione della logica," pp. 155–162; Frances A. Yates, *The Art of Memory* (Chicago, 1966), Chap. VIII, "Lullism as an Art of Memory," pp. 173–198, Chap. X, "Ramism as an Art of Memory," pp. 231–242; Ong, *Op. cit.* (n. 6).

[36] Paolo Rossi, "The Legacy of Ramon Lull in Sixteenth-Century Thought," *Medieval and Renaissance Studies*, **5** (1961), pp. 184–185.

[37] In his *Clavis universalis*, Rossi wrote a suggestive comment on the Descartes of this period: "It is certain that the problem of the young Descartes—a man who has not 'caught his foundations of physics to be sided with'—may particularly appear to be close to the one

In the sixteenth century, Lullism was considered to be the symbolic ideal of the unification of the sciences, whether this goal was successfully attained or not. Around the spring of 1619 Descartes intended to create "an entirely new" mathematical science quite different from the Lullian art but possessing an aspect similar to it. It seems clear that the aspect common to the two was their character as the science of sciences. It was Descartes's intention that this character should contribute to the planned unification of arithmetic and geometry by means of his newly invented compasses and his reformed algebra.

At any rate, as a result of his meeting Beeckman, Descartes began to be awakened from his intellectual slumber. In his letter to Beeckman dated April 23, 1619, Descartes reconfirmed the certainty of his discoveries in the following words:

> As to other [discoveries] in which I took pride in my previous letters, I have really found them with new compasses and I am not deceived. But I am not going to write that piecemeal, for I will think over an entire work on this subject someday, in my judgment, which is new and not to be disregarded.[38]

Descartes's program was full of self-confidence. Nevertheless, we should not forget that at this time he had grasped only the first clue for his "entirely new science". This will be reconfirmed by a minute examination of the mathematical contents of the manuscript *Cogitationes privatae*.

In any event, on the occasion of meeting Beeckman, Descartes confirmed the importance of the mathematical approach to natural philosophy, and, with the idea of unifying the mathematical disciplines, left the Netherlands for Southern Germany, where the Thirty Years War began. In social status, he was then a soldier from the French elite, but his real intention was to reform the scheme of learning entirely as a mathematician-philosopher.

---

in the Lullian syntax and encyclo-pedia of the late sixteenth century: Behind the manifold of sciences there hides a profound unity, a law of connection, a common logic." *Op. cit.* (n. 35), p. 177 (my translation).

[38] *Op. cit.* (n. 1), in *AT*, X, p. 163; *Journal*, t. IV, p. 62: "Quod ad cætera quæ in superioribus me invenisse gloriabar, verè inveni cum novis circinis, nec decipior. Sed membratim non ad te scribam, quia integrum opus hac de re meditabor aliquando, meo judicio novum, nec contemnendum."

§ 2. The Mathematics in the *Cogitationes privatae*

A. The Plan of a Book Titled *Thesaurus mathematicus*

Descartes described his initial plan for "an entirely new science" in his letter of March 26, 1619 to Beeckman. That letter contained only a rough sketch of his ambitious program for reorganizing the mathematical disciplines, but did not mention its technical details. There is, however, a manuscript which Descartes is conjectured to have written during the years between 1619 and 1621 and makes known at least partly what and how he worked during this period which would be crucial for his mathematical career as a whole.

It is known that on June 1, 1676 the manuscript of Descartes was copied by Gottfried Wilhelm Leibniz, who "was interested in everything related to the French philosopher [i. e. Descartes]."[39] After his death, Leibniz's copy was deposited to the Royal Library of Hanover and first published in 1859 under the title *Cartesii Cogitationes privatae* by the French noble Foucher de Careil in the first volume of *Œuvres inédites de Descartes.*[40] Unfortunately, however, Foucher de Careil's transcription contained many errors. For example, he failed to comprehend what the cossic signs 𝔵, 𝔷, and 𝔠𝔢 in that copy signified. Still the worse, the original copy by Leibniz could not be found in the Library of Hanover when at the end of the nineteenth century Charles Adam and Paul Tannery were searching for it for inclusion in their *Œuvres de Descartes.* But due to the endeavours of Gustav Eneström, Henri Vogt, and Henri Adam (Charles's brother), Foucher de Careil's version was thoroughly emended into a mathematically meaningful form, which is now accessible to us in the tenth volume of the Adam-Tannery edition of *Œuvres de Descartes.* The *Cogitationes privatae* contains non-mathematical passages partly identical with Descartes's other writings of the same period under the various titles *Parnassus*, *Olympica*, *Democratica*, *Experimenta*, *Praeambla*, which were classified as C in the Stockholm inventory, made posthumously in 1650[41] and excerpted in Adrien Baillet's *La Vie de Monsieur Des-Cartes* of 1691. But here we focus our examination upon the mathematical part of the *Cogitationes privatae*, based upon the Adam-Tannery version of the Leibnizian copy.

Near the beginning part of the *Cogitationes privatae* we find the title of a mathematical book and the brief explanation of its contents. From this we know that at the time when he was writing down the collection of fragmentary notes *Cogitationes privatae* he entertained a plan to make his mathematical studies into a comprehensive book. The title of the planned book and its contents are explained as follows:

---

[39] Charles Adam, "Avertissement" to "Opuscules de 1619–1621, Ms. de Leibniz," *AT*, X, p. 207.

[40] Foucher de Careil, *Œuvres inédites de Descartes*, t. I (Paris, 1859), pp. 1–57.

[41] "Inventaire succinct des Écrits etc.," *AT*, X, pp. 7–8.

> The *Thesaurus mathematicus* of Polybius Cosmopolitanus.
>
> This work lays down the true means of solving all the difficulties in the science of mathematics, and demonstrates that the human intellect can achieve nothing further on these questions. The work is aimed at certain people who promise to show us miraculous discoveries in all the sciences, its purpose being to chide them for their sluggishness and to expose the emptiness of their boasts. A further aim is to lighten the agonizing toil of those who struggle night and day with the Gordian knots of this science, and who squander their intellectual resources to no avail. The work is offered afresh to learned men throughout the world and especially to the distinguished brothers of the Rose Croix in Germany.[42]

The author of the *Thesaurus mathematicus* Polybius Cosmopolitanus is no doubt a pseudonym of Descartes, as Henri Gouhier has suggested.[43] Descartes might have contemplated using the name Polybius Cosmopolitanus (Polybius, citizen of the world) after the well-known Greek historian Polybius Megalopolitanus, who was born in Megalopolis, Arcadia and who flourished in the second century B. C. The name of the book *Thesaurus mathematicus* was also the one which had been used at least once before as the title for a mathematical book. The *Thesaurus mathematicus* appeared in Frankfurt am Main in 1613. This work was mainly grounded upon George Joachim Rheticus's table of sines for every 10″ to fifteen decimal places with some additions by the editor Bartholomeo Pitiscus.[44] There is almost nothing common between the contents of the Rheticus-Pitiscus monograph on a trigonometric table and those of Descartes's planned work.

Judging from the correspondence in 1619 between Beeckman and Descartes, in the book titled *Thesaurus mathematicus* (The Mathematical Trea-

[42] *Cogitationes privatae*, in *AT*, X, p. 214: "*Polybij Cosmopolitani Thesavrvs Mathematicvs*, in quo traduntur vera media ad omnes hujus scientiae difficultates resolvendas, demonstraturque circa illas ab humano ingenio nihil vltra posse praestari: ad quorumdam, qui nova miracula in scientijs omnibus exhibere pollicentur vel cunctationem provocandam & temeritatem explodendam; tum ad multorum cruciabiles labores sublevandos, qui, in quibusdam hujus scientiae nodis Gordijs noctes diesque irretiti, oleum ingenij inutiliter absumunt: totius orbis eruditis & specialiter celeberrimis in G. (Germaniâ) F. R. C. denuo oblatus." The translation is taken from *The Philosophical Writings of Descartes*, translated by John Cottingham, Robert Stoothoff and Dugald Murdoch, Vol. 1 (Cambridge, 1985), p. 2.

[43] H. Gouhier, *Les Premières Pensées de Descartes: Contribution à histoire de l'Anti-Renaissance* (Paris, 1958), p. 109.

[44] Rheticus-Pitiscus, *Thesaurus mathematicus sive Canon sinuum ad radium 1.00000.00000.00000* (Francifurti, 1613). For this work, see R. C. A[rchibold], "Notes," *Mathematical Tables and Other Aids to Computations*, **3** (1948–1949), p. 558. See also Edward Rosen, "Rheticus, George Joachim," *DSB*, XI, pp. 395–398; H. L. L. Busard, "Pitiscus, Bartholomeo," *DSB*, XI, pp. 3–4. As to Pitiscus's position in the history of trigonometry, I have learn much from Nobuo Miura, "The Application of Trigonometry in Pitiscus: A Preliminary Essay," *Historia Scientiarum*, No. 30 (1986), pp. 63–78.

sure Trove) Descartes can be safely conjectured to have aimed at constructing a new mathematical science applicable to both discrete and continuous quantities. His claim that the book can provide the true means of solving all difficulties reminds us his sentences "almost nothing will remain to be found in geometry" and "all the thickest darkness can be dispelled" in his letter to Beeckman on 26 March, 1619 (see p. 102). The expression "true means" may suggest that his intended procedure for solving mathematical problems was not the usual one which was seen in Euclid's *Elements* or common textbooks of cossist algebra, but that of his compasses and/or his reformed algebra. By using these new devices it is supposed that he intended to cut the 'Gordian knots' of mathematics.

But his attempted monograph appears to have been quite different from treatises by some authors who boasted that they would "show us miraculous discoveries in all the sciences." Such writers may have had an actual reference and must have been the Rosicrucian brothers in Germany. About 1619, the Rosicrucian movement, it is known, swept across Germany, with the outbreak of the Thirty Years War in 1618. This rather mystical intellectual campaign sometimes promised to present people with a new science or a true wisdom. Descartes must have heard a rumor about the brothers of the Rose Croix,[45] and this hearsay may have reminded him of the old champion of Lullism he had met at an inn in Dordrecht. So, with Gouhier,[46] we sense an ironical nuance in the last sentence of the description of the planned book, "the book is offered [...] especially to the distinguished brothers of the Rose Croix in Germany." What Descartes was planning in 1619 must have been not a work like those by certain people who supported the Lullian art or the mysticism of the Rosicrucians, but a book on "an entirely new science" with solid foundations.

Just prior to the description on the *Thesaurus mathematicus*, Descartes delineates his intellectual experience up to his having grasped the idea of a new mathematical science:

> In my youth, when I was shown an ingenious invention, I used to

[45]On the relation between Descartes and the Rosicrusian movement, see Frances A. Yates, *The Rosicrusian Enlightenment* (London, 1972), Chap. VIII, "The Rosicrusian Scare in France," pp. 103–117. Yates's description is almost totally dependent upon A. Baillet, *La Vie de Monsieur Des-cartes* (Paris, 1691) and must be read critically. For criticism of Yates's general historiography emphasizing the continuity between the Renaissance and the Scientific Revolution, see P. Rossi, "Hermeticism, Rationality and the Scientific Revolution," in M. L. Righini Bonelli and William Shea, eds., *Reason, Experiment and Mysticism in the Scientific Revolution* (New York, 1975), pp. 247–273. Gouhier, on the other hand, places stress on the discontinuity between the two periods in his *Les Premières Pensées de Descartes: Contribution à l'histoire de l'Anti-Renaissance* (n. 43). He uses the term 'Anti-Renaissance' as the French synonym for 'Counter-Renaissance'. The latter nomination was first used by Hiram Haydn in his *Counter Renaissance* (New York, 1950) concerning the history of ideas of the Elizabethan period. Haydn defined the historical concept 'Counter-Renaissance' as "a protest against the basic principles of the classical renaissance, as well as against those of medieval Scholasticism." *Op. cit.*, Introduction, p. xi.

[46]H. Gouhier, *Op. cit.*(n. 43), p. 110.

> wonder whether I could work it out for myself before reading the inventor's account. This practice gradually led me to realize that I was making use of definite rules (*certae regulae*).[47]

From this statement we recognize that Descartes started to form his rules for directing the intellect at a considerably early period. He most likely cultivated these rules for solving mathematical and other problems at latest by 1619. Such rules will be fully developed, of course, in his later *Regulae ad directionem ingenii*.

At this stage Descartes seems to have begun to cease being a mere passive reader of books and to have made up his mind to write his own. The very first passage may be interpreted as an affirmation of his ambition:

> Actors, taught not to let any embarrassment show on their faces, put on a mask. I will do the same. So likewise, I am now about to mount the stage of the theater of the world, where I have so far been a spectator, and I come forward masked.[48]

The last sentence "I come forward masked" might have a certain relation to his use of the pseudonym Polybius Cosmopolitanus as the author of the *Thesaurus mathematicus*, if we are permitted to give free reign to our imagination. He is about to mount the stage of the theater of the world putting on a mask, that is to say, he begins to envision a plan for a book which will bring about a monumental reform in mathematics. As to the situation of the sciences, he writes the following observation, again in metaphorical language:

> The sciences are at present masked, but if the masks were taken off, they would be revealed in all their beauty. If we could see how the sciences are linked together, we would find them no harder to retain in our minds than the series of numbers.[49]

In Descartes's judgment, the true sciences have not been cultivated yet, and their mask should be taken off.

B. The Mesolabe Compass

To what extent and how did Descartes succeed in taking off the mask of mathematics when writing the *Cogitationes privatae*? What was the real content of "an entirely new science" proposed in the letter of March 26, 1619? To answer this question, we have to look at, before anything, the compasses which he considered to play a crucial role in his new discipline. He wrote about these compasses in the later part of the *Cogitationes privatae*. According to his letter

[47] *AT*, X, p. 214; Transtation, *Op. cit.* (n. 42), p. 2.

[48] *AT*, X, p. 214: "Vt comœdi, moniti ne in fronte appareat pudor, personam induunt: sic ego, hoc mundi theatrum conscensurus, in quo hactenus spectator exstiti, larvatus prodeo."

[49] *AT*, X, p. 215.; Translation, *Op. cit.* (n. 42), p. 3.

of March 26, 1619, he found four demonstrations with the help of the compasses. The first discovery was about the division of an angle into any number of equal parts, which will be discussed later. The others pertained to the solution of the three types of cubic equations: (1) $\pm a \pm bx = x^3$, (2) $\pm a \pm bx^2 = x^3$, (3) $\pm a \pm bx \pm cx^2 = x^3$ (see p. 101).

Let us first look at his attempt to solve the first type of cubic equation. He begins his explanation with the following words:

> I found equations between the following: 1𝔠𝔢 and 7𝔵+14, and the similar. I reduce to 1𝔵+2 equals $\frac{1}{7}$ 𝔠𝔢, and look for 1𝔠𝔢, which afterwards I will multiply by 7 [the first compas].
>
> Next we have to have another compass, two parts of which are the following. The first part has the line *bc* firmly annexed to the line *af* at a right angle; on the other hand, the line *de* is, of course, at a right angle to the line *fb* but movable along it. Moreover the line *fb* has a stylus fixed at the point *d*, in which it describes a line; likewise at the point *f* it has one [stylus] but movable, in which it describes another line in the following manner. The second part *dcegh*, remaining unchanged in the lines mutually firmly connected, flows over the line *ap*, where the first part is fixed at the immovable point *a*. The point *c* strikes against the line *dc* and so effects that the whole second part may come down. On the one hand, the line *cd* draws the line *de* along the space *fb* in accordance with the variation of the intersections, then the stylus d will describe the line of the first compass. On the one hand, the line *gh* will certainly intersect the line *de* and it will describe another curved line at the movable stylus *c*, of which the last line will cut *ap*, in which *ae* is the cube to be discovered, if *ab* of the first part is the unit, *ce* of the second is certainly the absolute number, which is two in the example.[50]

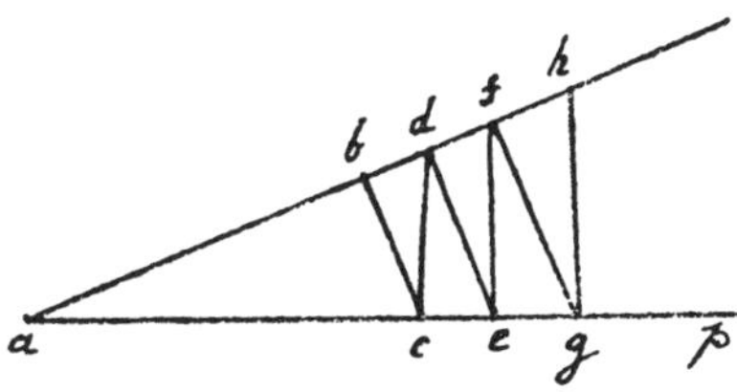

[Figure 3.3]

[50] *AT*, X, pp. 234–235: "Inveni æquationes inter talia: 1𝔠𝔢 & 7𝔵+ 14, & simile hoc. Reduco ad 1𝔵 + 2 æqu. $\frac{1}{7}$ 𝔠𝔢, & quæro 1𝔠𝔢, quem postea multiplicabo per 7 [primi circini]. Deinde alium circinum habere oportet, quorum duæ partes sunt tales. Prima habet lineam *bc* firmiter annexam *ad* angulos rectos lineæ *af*, lineam autem *de* ad angulos quidem rectos, sed mobilem per lineam *fb*. Linea *fb* habet præterea in puncto *d* stylum fixum, quo aliam lineam describit; in puncto *f* etiam vnum, sed mobilem, quo aliam lineam describit hoc pacto. Secunda pars *dcegh*, constans lineis firme invicem annexis, fluat supa lineam *ap*, vbi affixa est prima pars in puncto *a* immobili: punctum *c* impellit lineam *dc*, & ita efficiet vt tota secunda pars descendat, linea autem *cd* trahit lineam *de* per spatium *fb* juxta varietatem intersectionum, & tum stylus *d* lineam primi circini describet. Linea autem *gh* intersecabit etiam lineam *de*,

In this paragraph, Descartes tries to solve the cubic equation $x^3 = 7x+14$. Dividing both sides by 7, he gets $\frac{1}{7}x^3 = x + 2$. Then, he reduces this equation to $x^3 = x + 2$ by mistake, saying that "I look for **1𝔠𝔢**, which afterwards I will multiply by 7." But this error is not essential and is easily removed by a simple transformation of the root. The crucial problem is to solve the equation $x^3 + x = 2$ with the help of a compass. He calls this instrument "a mesolabe compass" or more literally "a compass of mesolabe" (*circinus mesolabi*) in a later passage of the *Cogitationes privatae*.[51] Hereafter we will use the former nomination.

Descartes's description on the compass quoted above is not necessarily elegant in the mathematical sense. Fortunately for us, however, in the *Géométrie* of 1637, he provides a clearer explanation of the same compass twice. He explains how to construct the mesolabe compass in Book Two of the *Géométrie* as follows:

> Consider the lines AB, AD, AF, and so on, which I assume to have been described with the aid of an instrument, YZ, composed of several rulers. These are so joined that when the one marked YZ is placed on the line AN, we can open and close the angle XYZ; and when it is completely closed, the points B, C, D, E, F, G, H are all assembled at point A. But to the extent that we open it, the ruler BC, which is joined at right angles to XY at point B, pushes the ruler CD toward Z; CD slides along YZ, always at right angles to it, and pushes GH, etc. And we can conceive of an

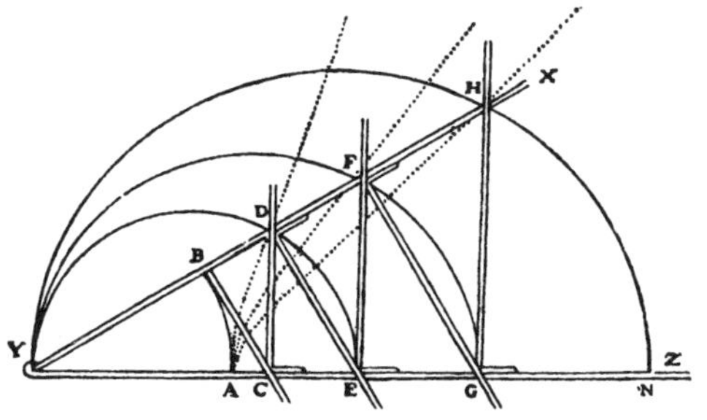

[Figure 3.4]

aliamque lineam curvam stylo *c* mobili describet, quæ vltima linea secabit *ap*, in quo *ae* est cubus inveniendus, si *ab* primæ partis sit vnitas, *ce* verò secundæ numerus absolutus, qui in exemplo est binarius." Here we draw attention to cossic symbols. Compared with Leibniz's autograph copy of the manuscript *De solidorum elementis* (see § 3), Leibniz is supposed to have used his own symbols ♃ (the astronomincal sign for Jupiter) and ꞓ (C with a little bar) for 𝔵 and 𝔠𝔢, respectively. See Descartes, *Exercices pour les éléments des solides*, éd. P. Costabel (Paris, 1987), Costabel's Introduction, p. x. But it is certain that Descartes's were the ordinary cossic ones 𝔵 and 𝔠𝔢 as can be imagined by his letters to Beeckman. Thus, we do not rewrite the cossic symbols into Leibniz's.

[51] *AT*, X, p. 239. On Descartes's compasses in general, see Michel Serfati, "Les compas cartésiens," *Archives de philosophie*, **56** (1993), pp. 197–230. See also *Idem*, "Le Développement de la pensée mathématique du jeune Descartes (l'éveil d'un mathématicien)," in *De la Méthode: Recherches en histoire et philosophie des mathématiques*, dirigé par M. Serfati (Besançon/Paris, 2002), pp. 39–104, which has provided a useful discussion on Descartes's compasses against the background knowledge on the mathematical career of young Descartes.

infinity of others, which are pushed consecutively in the same way, half of which always maintain the same angles with YX and the others with YZ. Now as we thus open the angle XYZ, the point B describes the line AB, which is a circle, and the other points D, F, H, where the intersections of the other rules occur, describe the other curved lines, AD, AF, AH, of which the latter are successively more complex than the first, and thus more complex than the circle.[52]

The essential function of the mesolabe compass is that one obtains the following continuous proportions because of the similar right angles:

$$\mathrm{YB/YC} = \mathrm{YC/YD} = \mathrm{YD/YE} = \mathrm{YE/YF} = \mathrm{YF/YG} = \mathrm{YG/YH}\ldots.$$

Putting YB = YA = 1 and YC = x, we obtain the continuous proportions of the geometrical progressions, $1, x, x^2$, etc.:

$$1 : x = x : x^2 = x^2 : x^3 = x^3 : x^4 = x^4 : x^5 = \ldots.$$

Returning to Figure 3.3 in the *Cogitationes privatae*, we have:

$$\frac{ab}{ac} = \frac{ac}{ad} = \frac{ad}{ae} = \frac{ae}{af} = \frac{ad^2}{ac} = \frac{ac^3}{ab^2}.$$

Therefore, $ce = ae - ac = \dfrac{ac^3}{ab^2} - ac$. Now putting $ab = 1$ and $ac = x$, it follows that

$$ce = x^3 - x, \text{ i. e. } x^3 = x + ce.^{53}$$

Consequently, if we open the compass so that the value of ce is exactly equal to 2, we get $ac = x$ as a positive real root of the cubic equation $x^3 = x + 2$.

In the next paragraph, Descartes discusses the cubic equation of the type $\pm a \pm bx \pm cx^2 = x^3$. This time, he does not use any compass:

> Moreover we obtain the equation between the following ce, ʒ, 2e, provided that there are as many ʒ as 2e in the following way:
>
> 1ce equals 6ʒ−62e+56.
>
> I reduce [it] to the ternary number of roots, and I will have
>
> ½ce equals 3ʒ−32e+28.
>
> Next I subtract the unit from N, make a cube from the remainder, add the unit to its root and produce a cube from that root. The result is ½ce. If it is multiplied by 2, it will make a cube required.[54]

---

[52] *AT*, VI, pp. 391–392; Olscamp, pp. 191–192.

[53] We refer to G. Eneström's note b. in *AT*, X, p. 235.

[54] *AT*, X, p. 236: "Fit præterea æquatio inter talia, ce, ʒ, 2e, dummodo quot sint ʒ tot 2e, & hoc modo:

Clearly here Descartes is trying to solve the cubic equation of which equal coefficients of the first and second degrees: $a_3x^3 = hx^2 - hx + a_0$, taking as an example, $x^3 = 6x^2 - 6x + 56$. Making the coefficient of the first degree 3, he gets $x^3 = 3x^2 - 3x + 28$. But at this point he commits an error similar to the one in the previous example. By mistake he reduces the above equation to $x^3 = 3x^2 - 3x + 28$. Certainly, following the procedures he indicates, we get

$$(x-1)^3 = 28 - 1, \quad x^3 = (\sqrt[3]{28-1} + 1)^3.$$

However, as the commentator Gustaf Eneström has noticed,[55] Descartes mistakenly writes,

$$\frac{1}{2}x^3 = (\sqrt[3]{28-1} + 1)^3.$$

Further Descartes considers the cubic equation in which the coefficients of ʒ and 𝔵 are not equal. Again he makes the same mistake as in the previous case in replacing $\frac{1}{4}x^3$ simply by x^3.

The last example Descartes discusses is the cubic equation $x^3 = -3x^2 - 3x + 26$. His directions are the following: "I add the unit to the absolute number; next I remove the unit from the root produced, and the cube required is produced from the root."[56] This time he does not make an error. In fact, the given equation is transformed into $(x+1)^3 = 27$. Consequently, $x = \sqrt[3]{26+1} - 1$ and $x^3 = (\sqrt[3]{26+1} - 1)^3$.

The remaining type of cubic equation is $\pm a \pm bx^2 = x^3$. He tackles this problem by using the mesolabe compass again. His attempt appears under the subtitle "Another compass for cubic equations 1𝔠𝔢 and Oʒ O𝔵." Then follows his 'solution' of the cubic equation $x^3 = x^2 + b$:

> If a cube equal to ON *dg* and one unknown square should be found, let the following compass be constructed: *dce* slides over *ap*, strikes *bc* at the point *c* by sliding, and moves so that *af*, to which *bc* is affixed at a right angle, may descend at the same time, and describes a line of the mesolabe compass at the intersection of *af* and *cd*.[57]

1𝔠𝔢 æqu. 6ʒ−6𝔵+56.
Reduco ad numerum radicum ternarium, habeboque
$\frac{1}{2}$𝔠𝔢 æqu. 3ʒ−3𝔵+28.
Deinde ex N tollo vnitatem, ex residuo cubum formo, cujus radici vnitatem addo, & quod cubice producitur ex illâ radice est $\frac{1}{2}$𝔠𝔢; quod si multiplicetur per 2, producet cubum quaesitum."

[55] *AT*, X, p. 236, note a.

[56] *Op. cit.*, p. 237: "Addo vnitatem numero absoluto; deinde ex radice producti vnitatem demo, & producitur ex radice cubus quæsitus."

[57] *AT*, X, pp. 238–239: "*Alius circinus ad æquationes* 1𝔠𝔢 *&* Oʒ *ON.* Si inveniendus sit cubus æqualis ON *dg* & quadrato vni incognito, talis circinus fabricetur: *dce* fluit supra *ap*, fluendo pellit *bc* in puncto *c* adigitque vt descendat simulque *af*, cui affixa est *bc* ad angulos rectos, describitque intersectione *af* & *cd* lineam circini mesolabi."

Thereafter Descartes continues to explain how to operate his compass for solving the equation. As the copyist Leibniz himself complained,[58] the sentences are very obscure and confused. Perhaps the original author Descartes did not attain a correct answer. At least he did not leave a precise description concerning how to construct the compass in his notes. But a reconstruction of a solution is possible, as Eneström has shown in a footnote of the Adam-Tannery edition.[59]

In Figure 3.3, used to solve the equation $x^3 = x + 2$, we put $ab = 1$, $ac = \sqrt{x}$, $fg = b$. Noting the similar triangles, we have $ad = \frac{ac^2}{ab} = x$, $ae = \frac{ad^2}{ac} = \frac{x^2}{\sqrt{x}} = \frac{x^3}{2}$, $af = \frac{ae^2}{ad} = \frac{x^3}{x} = x^2$, $ag = \frac{af^2}{ae} = \frac{x^4}{x^{\frac{3}{2}}} = x^{\frac{5}{2}}$, $ah = \frac{ag^2}{af} = \frac{x^5}{x^2} = x^3$. Since $ah = af + fh$, it follows that $x^3 = x^2 + fh$. Consequently, when we substitute the value of fh for b, by opening or closing the angle bac, we have ad as a positive root of the cubic equation $x^3 = x^2 + b$.

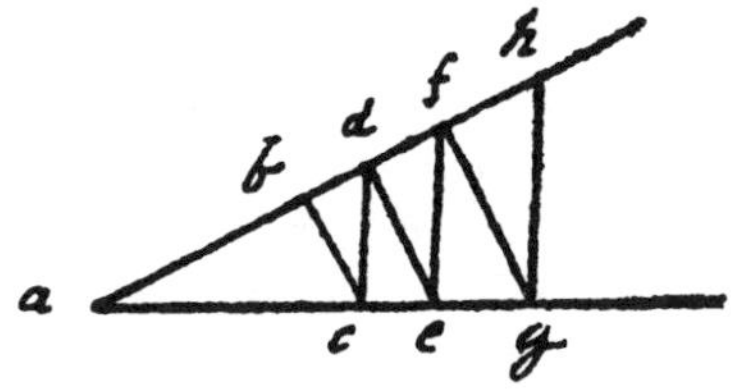

Descartes concludes his explanation on the cubic equations by expressing his conviction that the equations 1𝔠𝔢 = O𝔷−ON or ON−O𝔷, 1𝔠𝔢 = O𝔷−ON or ON−O𝔷 can be solved in a similar way.[60]

As far as the passages which we have examined above are concerned, Descartes's attempt at solving cubic equations is incomplete, or rather more systematically incorrect. He has not considered the general case of the following type of the cubic equation: $\pm a \pm bx \pm cx^2 = x^3$. But he himself seems to have noticed this omission. As a sign of this, he writes an addendum afterwards with the paragraph heading "A General Rule for Complete Equations with Four Terms."[61] In this paragraph, he indicates the procedure for eliminating the coefficient of the unknown of the second degree by the substitution of a root, thus reducing the original cubic equation finally to the form $\pm a \pm bx = x^3$. The latter type of the cubic equation is one of the first case which he has solved with the aid of his compass. Descartes thus might have believed that his attempt at solving all the types of cubic equations was finally accomplished.

Now here arises a historically important question: How did Descartes invent his mesolabe compass? To answer this question, we have to know, to begin with, what a mesolabe is. We approach this problem in two ways, i. e. by inquiring about the origin of the word 'mesolabe' and that of the instrument itself. The English word 'mesolabe' originated from the Greek noun '*μεσόλαβος*' (or '*μεσόλαβον*', sometimes '*μεσολάβιον*'), which in turn was formed by the combination of the two words '*μεσός*' (middle, mean) and '*λαβ-*' (i. e. '*λαμβάνειν*', to take, seize). The Latin terminologies corresponding to

[58]See Leibniz's remarks c, d, & f on p. 239.

[59]P. 239, note a.

[60]*AT*, X, p. 240.

[61]*AT*, X, p. 244: "Regula generalis ad æquationes quatuor terminorum completas."

'μεσόλαβος' and 'μεσολάβιον' are 'mesolabus' and 'mesolabium', respectively. As can be imagined from this etymological consideration, 'mesolabe' literally means 'mean-taking'. But in antiquity it had a more limited meaning, designating a mathematical instrument ascribed to Eratosthenes for finding two mean proportionals mechanically.

What was Eratosthenes's mesolabe? How was it transmitted to modern Europe? Pappus, in the first half of the fourth century A. D., provided a minute explanation of that instrument in Book III of his *Collection*, clearly mentioning the names of both the instrument and its inventor.[62] In Commandino's Latin translation of 1588, the names of the mathematician and his mathematical device appeared as "*in Eratosthenis mesolabo.*"[63] In his well-known commentary on Book II of Archimedes's *On the Sphere and Cylinder*, Eutocius of Ascalon, who flourished around 480 A. D., also asserted that Eratosthenes solved the duplication of the cube mechanically by inventing a mathematical apparatus for finding two mean proportionals betwen the two given lines.[64] But he did not refer to the name of that instrument. Due to Marshall Clagett it is known that as early as the thirteenth century William of Moerbeke made a very understandable Latin translation of the commentary by Eutocius, including the portion concerning Eratosthenes's device.[65] In his version of Pappus, Commandino did not forget to make mention of Eutocius's commentary on Archimedes.[66] In addition to Pappus's and Eutocius's explanation, Vitruvius in his *De Architectura* used the Latin word 'mesolabus' to denote Etatosthenes's instrument without any detailed mathematical explanation.[67] This is all that we generally know about the ancient sources of the description of Eratosthenes's mechanical device for finding mean proportionals.

Let us look at how Eratosthenes's instrument is constructed and used, following the explanation of Pappus *apud* Commandino. There the mesolabe is invoked "to find two mean proportionals in continued proportion when two straight lines are given."[68] This problem is important because the duplication

---

[62]Pappus, *Collectio*, ed. F. Hultsch, I (Berlin, 1876), pp. 54–56; *La Collection mathématique*, tr. Ver Eecke, t. I (Paris, 1933), p. 39.

[63]*Collectiones*, ed. F. Commandino (Pesaro, 1588), f. 4v.

[64]Archimedes, *Opera omnia cvm Commentariis Evtocii*, ed. I. L. Heiberg (corrigenda adiecit E. S. Stamatis), Vol. III (Stvtgardiae, 1972), pp. 88–97; Archimède, tr. Charles Mugler, t. IV: *Commentaires d'Eutocius et fragments* (Paris, 1972), pp. 64–69; *Les Œuvres complètes d'Archimède*, tr. Paul Ver Eecke, t. II (Paris, 1960), pp. 609–615.

[65]Marshall Clagett, *Archimedes in the Middle Ages*, Vol. II: *The Translations from the Greek by William of Moerbeke*, Part II: *Texts* (Philadelphia, 1976), pp. 246–248.

[66]Commandino, *Op. cit.*(n. 63), f. 5r.

[67]Vitruvius, *De Architectura*, Lib. 9, Prooem. 14: "Itaque Archytas cylindrorum descriptionibus, Eratosthenes organica mesolabi ratione idem explicaverunt." I. e. "Archytas solved the problem [the duplication of the cube] by a diagram with cylinders; Eratosthenes by means of an instrument, the mesolabium." See Vitruvius, *On Architecture*, Vol. II, tr. Frank Granger (London 1934), pp. 206–207.

[68]Commandino (n. 63), Lib. III. Problema I. Propositio V. f. 5r: "Duabus datis rectis lineis, duas medias proportiones in continua analogia inuenire."

of the cube can be accomplished once it has been solved. For, if x and y are two mean proportionals between the two given straight lines a and b, that is, if $a : x = x : y = y : b$, then $x^2 = ay$, $y^2 = bx$, $xy = ab$, and $x^3 = a^2b$. If $b = 2a$, we obtain an equation representing the problem of doubling a cube: $x^3 = 2a^3$.[69]

Let there be the rigid plinth ABCD and in it equal triangles AEH, MFK, NGL having right angles at E, F, and G (see Figure 3-5). Let the triangle AEH remain fixed and the triangle MFK and NGL be movable in parallel along the rulers AB and CD. To find two mean proportional lines between AC and XL, let MFK and NGL slide such that P and O on the lines EH and FK, respectively are in a straight line with A, X. Then, noticing the similar triangles, we have

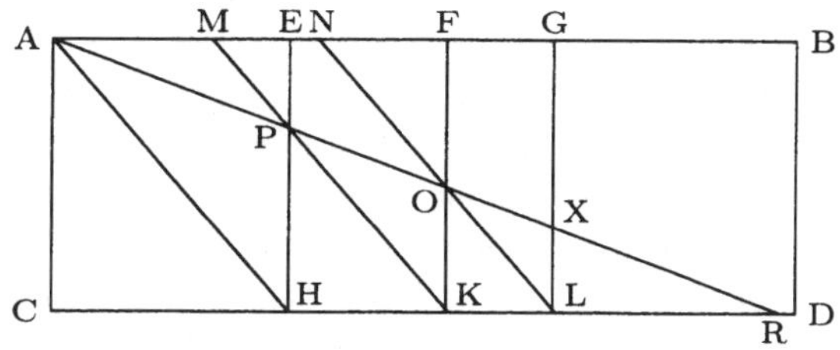

Figure 3.5

$$\frac{AC}{PH} = \frac{AR}{PR} = \frac{AH}{PK} = \frac{HR}{RK} = \frac{PH}{OK} = \frac{PR}{RO} = \frac{PK}{OL} = \frac{KR}{LR} = \frac{OK}{XL}.$$

Consequently

$$\frac{AC}{PH} = \frac{PH}{OK} = \frac{OK}{XL},$$

so that AC, PH, OK, and XL are in continued proportion. Therefore, if AC and XL are two given straight lines, then PH and OK are the required mean proportionals.[70]

There seems to exist a certain resemblance between Descartes's mesolabe compass used in the *Cogitationes privatae* and Eratosthenes's mechanical instrument for finding two mean proportionals. We have one more bit of evidence which might help to confirm Descartes's knowledge of Eratosthenes's mesolabe. In Book Three of the *Géométrie*, he employs his mesolabe compass, which he has already explained in Book Two (see pp. 114–115), for investigating many mean proportionals. In the paragraph with the margin subtitle "Exemple touchant l'inuention de plusieurs moyennes proportionelles," he, indicating to his mesolabe compass, states: "I do not believe there is any easier or more clearly demonstrated method for finding however many mean proportionals you might wish, than to use the curved lines described by [his mathematical instrument]."[71] To find many mean proportionals may have been Descartes's original aim for devising his own compass around the year 1619. As was seen in the case

[69] Heath, *A History of Greek Mathematics*, Vol. 1, pp. 252–253.

[70] Commandino, f. 5r-5v. Cf. Heath, Vol. 1, pp. 258–259.

[71] *AT*, VI, pp. 442–443; Olscamp, p. 228. Descartes's mathematical demonstration which follows is: "For if we wish to find two mean proportionals between YA and YE, we have only

of Eratosthenes, Descartes's use of his instrument devised originally for finding mean proportionals in order to solve the cubic equation would be a rather natural extension of its application.

Clavius also discussed the problem of finding two proportional lines in Book VI, Problem 10 (Proposition 15) of his *Geometria practica*. He understood this problem to be a natural extension of Euclid VI, 13: "To two given straight lines to find a mean proportional." As for the mathematicians who solved the problem of discovering two mean proportional lines, he referred to Eratosthenes, Pappus, Eutocius, Hero, Philoponus of Byzantium, Diocles, Nicomedes and many others.[72] But he made mention of neither the mesolabe nor Eratosthenes's mathematical demonstration concerning that instrument.

As we have stated above, Vitruvius and Pappus used the Latin and Greek words corresponding to 'mesolabe', respectively.[73] And Pappus and Eutocius provided a mathematical explanation of Eratosthenes's instrument. One might then venture to conclude that it was Book III of Pappus's *Collection* that inspired Descartes to invent his own mathematical compass for finding mean proportionals and for solving the cubic equation. This conjecture may be highly probable. But we cannot deny other possibilities as well, for mathematicians in the sixteenth and seventeenth centuries knew quite well not only the use of Eratosthenes's instrument through the various Renaissance editions of Archimedes with Eutocius's commentary but also the name of his apparatus, the 'mesolabe'. As to the term 'mesolabe', J. J. Scaliger used it as the title of his book published in 1594, and Viète used it in his *Variorum de rebus mathematicis responsorum liber VIII* of 1593 and his 1595 criticism of Scaliger's aforementioned work.[74]

to describe a circle whose diameter is YE; and because this circle cuts the curve AD at point D, YD is one of the required mean proportionals. The demonstration is obvious to the eye, merely by applying this instrument to the line YD: for, as YA, or YB which is equal to it, is to YC, so YC is to YD, and YD to YE." AT, VI, p. 443; Olscamp, pp. 228–229.

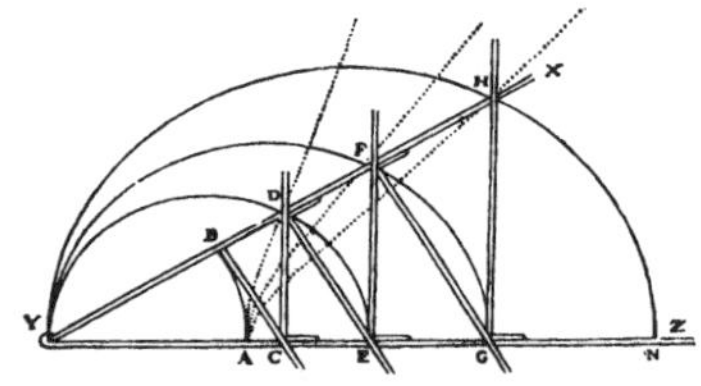

[72]Clavius, *Geometria practica* (n. 28), pp. 160–163. Almost all the mathematicians referred to here are examined in Heath, *A History of Greek Mathematics*, Vol. 1, pp. 244–270. As for what was known during the Renaissance on the problem of mean proportionals, see Marshall Clagett, *Archimedes in the Middle Ages*, Vol. III: *The Fate of the Medieval Archimedes, 1300 to 1565*, Part III: *The Medieval Archimedes in the Renaissance, 1450–1565* (Philadelphia, 1978), pp. 1163–1179.

[73]Principally only two ancient authors, Vitruvius and Pappus, are referred to in Henricus Stephanus, *Thesaurus Graecae Linguae*, Vol. VI (Graz, 1854), col. 616 and in Henry George Lidell and Robert Scott, *A Greek-English Lexicon* (Oxford, 1968), p. 1106. There could have been other cases, of course.

[74]J. J. Scaliger, *Mesolabium* (Leiden, 1594); Viète, *Variorum de rebus Mathematicis Responsorum Liber VIII* (Tours, 1593); *Idem, Pseudo-Mesolabum* (Paris, 1595), in *Opera mathematica*, ed. Frans van Schooten (Leiden, 1646), pp. 347–435 & 258–285. In his work, Viète mentioned Eutocius's commentary on Archimedes and Vitruvius.(*Opera mathematica*, pp. 348–350.)

Descartes clearly read Archimedes's two books *On the Sphere and Cylinder* (see p. 46), although we cannot determine whether this was before the years around 1619 or after that period. What can be ascertained at least is that he may have been well informed about the construction and use of Eratosthenes's mechanical device and certainly knew its name.

What Descartes accomplished with the mesolabe compass is the construction of the lengths of any geometrical progressions x^n (in today's notation) in order to solve the cubic equations. It is highly questionable whether or not he was finally successful in resolving them in a fully algebraic manner. The descriptions concerned were only fragmentary and mathematically incomplete. But the *Cogitationes privatae* shows that Descartes made some strides beyond the realm of Clavius's *Algebra* in his attempt to solve cubic equations.

C. The Other Mathematical Instruments

The *Cogitationes privatae* further shows that Descartes's invention of compasses was not restricted to the mesolabe. He also devised an instrument for dividing an angle into any number of equal parts. This is the very first compass that he referred to in his letter of March 26, 1619. His description of this mathematical instrument in the *Cogitationes privatae* begins with the subtitle "Compass to divide an angle into however many parts":

> Let there be the following compass: ab, ac, ad, ae are equal leaves divided equally at

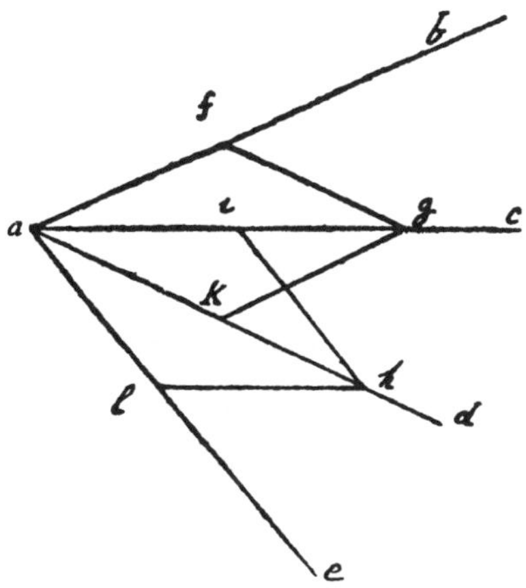

[Figure 3.6]

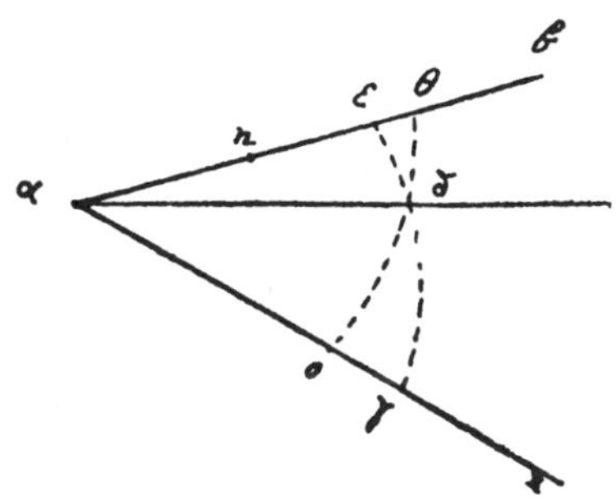

[Figure 3.7]

> points f, i, k, l; likewise fg is equal to af, etc. Then let the angles bac, cad and dae be always equal, but one be unable to be augmented or diminished without the others' being moved. Therefore let the

> angle $b\alpha x$ be divided: I apply the line ae over αx. By leaving them immobile, I raise the line ba toward the part b, which draws ac and ad with itself. And a line will be described by point g like $\gamma\delta\epsilon$. Then let $n\alpha$ be taken equal to af, and around point n the part of the circle $\theta\delta o$ be drawn so that $n\theta$ be certainly equal to fg: I say that the line $\alpha\delta$ divides an angle into three equal parts. In this manner an angle can be divided into many if the compass consists of many leaves.[75]

It should be noticed that in the compass of Figure 3.6 the points a, f, i, k and l are all fixed as pivots and that the points g and h are movable along the arms ac and ad, respectively. Descartes first places the compass on the given angle $b\alpha x$ so that ae may fall on αx (Figure 3.7). Then he raises the arm ab up to the side αb. The angle $b\alpha\delta$ identical with the angle bag of the compass will be the required one third angle. The curved line $\gamma\delta\epsilon$ indicates the locus of g while the arm ab being opened from ae lying exactly on αx[76]; θo is the arc of a circle of radius fg around n which indicates the locus of g when n is supposed to be fixed. Obviously the intersection point δ of the two curves will show the position of the side $\alpha\delta$ when the angle $b\alpha x$ is trisected by $\alpha\delta$.

Nevertheless, Descartes's interest in compasses extended beyond the mesolabe compass and the trisection compass. Remember that in the first paragraph in which he argued about the cubic equation with the aid of the mesolabe compass (see p. 113) he used the words "another compass" (*alius circinus*). As this expression suggests, before discussing the cubic equations and the trisection of an angle, he has written about a compass for describing a conic section:

> A conic section can be described by the following compass: Let AD be perpendicular and AB be an oblique surface. Let the leg of the compass be immobile and BC be turned about over the oblique plane yet such that CB can become shorter if it is imagined to ascend through C.[77]

[75] *AT*, X, pp. 240–241: "Circinus ad angulum in quotlibet partes dividendum. Sit talis circinus: ab, ac, ad, ae sunt æquales laminæ divisæ pariter in punctis f, i, k, l; item fg æquales af, &c. Vnde sit vt anguli, bac, cad & dae sint semper æquales, nec vnus possit augeri vel minui, quin alij etiam mutentur. Sit igitur angulus $b\alpha x$ dividendus: applico lineam ae supra αx; quâ ibi manente immobili, elevo lineam ba in partem b, quæ secum trahit ac & ad, lineaque describetur à puncto g talis $\gamma\delta\varepsilon$. Deinde sumatur $n\alpha$ æqualis af, ex puncto n ducatur pars circuli $\theta\delta o$, ita vt $n\theta$ sit etiam æqualis fg: dico lineam $\alpha\delta$ dividere angulum in tres partes æquales. Ita potest dividi angulus in plures, si circinus constet pluribus laminis."

[76] To use anachronistic conceptions and notation, the equation of the curve $\varepsilon\delta\gamma$ in polar coordinates is $\rho = 2a\cos(\varphi/2)$, where $\rho = \alpha\delta$, $\varphi = \angle\delta\alpha\gamma$, a $= af$, since $\rho = ag = ah = \alpha\gamma\cos\angle h\alpha\gamma$ ($\angle h\alpha\gamma$ is a right triangle.) Cf. *AT*, X, p. 241, note a by Eneström.

[77] *AT*, X, pp. 232–233: "Describi potest sectio conica tali circino: sit AD perpendicularis, superficies obliqua AB. Sit pes circini immobilis, volvatur BC supra planum obliquum, ita tamen vt CB possit brevior fieri, si imaginetur per C ascendere."

Descartes considers the compass (Figure 3.8) of which one leg CA is immobile; around the axis CA revolves the other leg CB which can ascend or descend, adjusting the inclined plane AB. The conic section described by point B will be identical to the one which is the intersection of the cone generated by the leg of the compass CB with the plane AB.

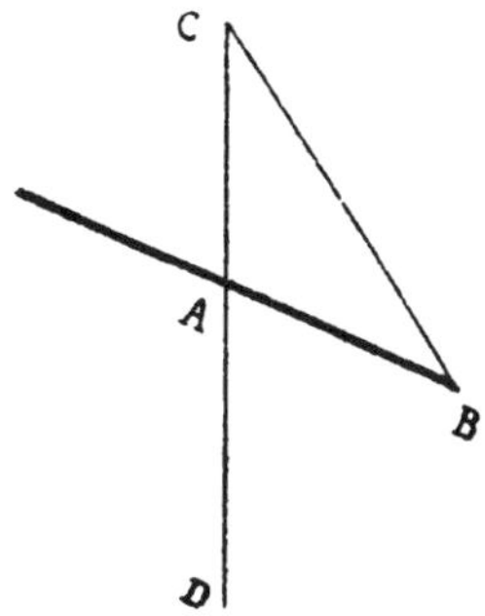

Figure 3.8

Further, he devises another compass to draw a cylindrical section. His explanation is the following:

> A section of a cylinder can be drawn by the following compass in the same manner: Let ACDE be a compass of which a leg is immobile; the line DE descends or ascends freely through point D as it will differ from the plane.[78]

The compass in Figure 3.9 has an immobile axis CA, a constant hand CD, and a leg DE which can descend or ascend through the joint D. The stylus E will describe a figure on AB, which is the intersection of the cylinder generated by DE with the inclined plane AB.

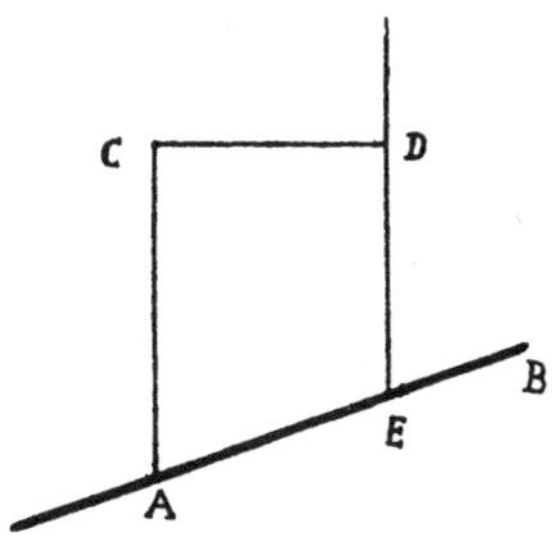

Figure 3.9

In addition to the compasses examined above, for a time at least Descartes seems to have been fascinated by any kind of compass whatever. His interest in compasses was never passing. He wrote down the following note: "I saw an instrument suitable for transferring all pictures: it consists of a foot with a two-headed compass."[79] There is some ambiguity in Descartes's description. But this compass is supposed to have been a certain kind of proportional compass for enlarging or reducing drawings, or more probably the pantograph, which was invented between 1603 and 1605 by the German Jesuit astronomer Christoph Scheiner,[80] for transferring pictures faithfully with the ratio required.

His interest eventually came to extend over more general mathematical or mechanical instruments than compasses. He continues the previous sentences on the two-headed compass, stating:

> Another [instrument] is also for depicting all clocks, which I can

[78] *AT*, X, p. 233: "Sectio cyclindri, eodem pacto, circino duci potest ita: sit ACDE circinus, cujus pes immobilis est; linea DE descendet vel ascendet libere per punctum D prout à plano distabit."

[79] *AT*, X, p. 241: "Vidi commodum instrumentum ad picturas omnes transferendas: constat in pede cum circino bicipiti."

[80] See G. L'E. Turner, *Op. cit.* (n. 27), p. 57.

> invent by myself. The third is for measuring solid angles. The fourth is of silver for measuring planes and pictures. The other most beautiful one is for transferring pictures. Another is affixed to a tibia of an orator for measuring moments [time]. Another is for deploying war artilleries.—Peter Roth, *Arithmetica philosophica.*—Benjamin Bramer.[81]

Here Descartes specifies various kinds of mathematical instruments which were being invented and used in the early seventeenth century.

The two mathematicians Descartes mentions draw our attention. The Nurembergian mathematician Peter Roth is referred to as the author of the *Arithmetica philosophica*, which belonged to the tradition of the German cossist algebra. Descartes may have been interested in this work because Roth discussed the cubic equation following the manner of Cardano (Part I), or perhaps because he calculated polygonal and pyramidal numbers after the example of the Ulm mathematician Johannes Faulhaber (Part II).[82] The calculation of polygonal or pyramidal numbers will be a topic of Descartes's manuscript *De solidorum elementis*, which we will examine in the next section.

The other mathematician, Benjamin Bramer, was a German mathematical practitioner who had wide-ranging interests in mathematics, especially in inventing mathematical instruments. As we touched upon very briefly in the previous section, like Faulhaber, he was known as a propagandist and reformer of the sector invented by Galileo around the turn of the century. His booklet *Beschreibung und Underricht, wie allerley Theylungen zu den Mathematischen Instrumenten zu verfertigen: Neben dem Gebrauch eines newen Proportional Instruments* (1615) was concerned with sectors having more sophisticated proportional scales than those of Galileo's model. (Figure 3.10, taken from the above-mentioned booklet, shows Bramer's sector with various proportional scales connected with a movable ruler at the fixed pivot.) This treatise on mathematical compasses referred to Eutocius's commentary on the second book of Archimedes's *On the Sphere and Cylinder*, without any detailed explanation of Eratosthenes's mesolabe, in relation to the problem of the duplication of the cube.[83] Descartes's reference to Bramer might have been simply an expression of his general interest in mathematical instruments, such as the sector or the proportional compass.

---

[81] *AT*, X, p. 242: "Aliud quoque ad omnia horologia depingenda, quod per me possum invenire. Tertium ad angulos solidos metiendos. Quartum argenteum ad plana & picturas metiendas. Pulcherrimum aliud ad picturas transferendas. Aliud affixum oratoris tibiæ ad momenta metienda. Aliud ad tormenta bellica noctu dirigenda.—Petri Rothen *Arithmetica philosophica.*—Benjamin Bramerus."

[82] Peter Roth, *Arithmetica philosophica, oder* [...] *Kunstliche Rechnung der Coß oder Algebrae* (Nürnberg, 1608). Cf. *AT*, X, p. 242, note a. Charles Adam has recorded the year of publication as 1607 instead of 1608 by mistake.

[83] B. Bramer, *Op. cit.*(n. 26), p. 39. As for the other works of Bramer, see Paul Kirchvogel, "Bramer, Benjamin," *DSB*, II, p. 419.

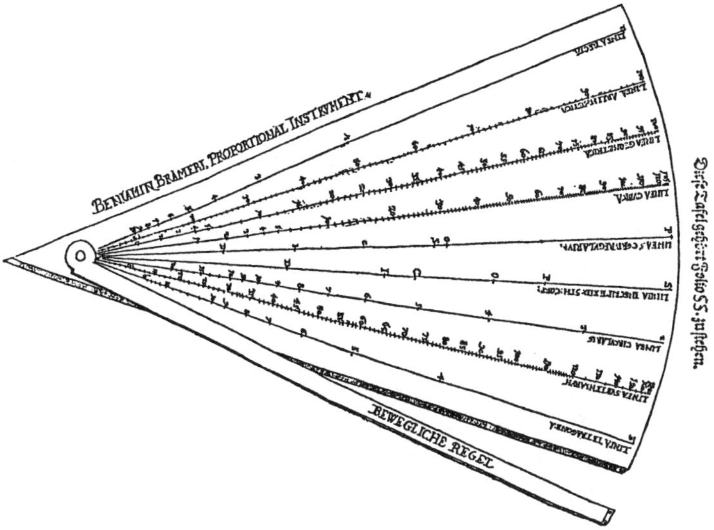

Figure 3.10

An observation should be made about Descartes's attempt to solve mathematical problems by means of his newly invented compasses. He not only tried to tackle cubic equations with the aid of his mesolabe compass and to divide an angle into any number of equal parts by devising a mechanical apparatus, but also showed his interest in mechanical instruments in general. He was certainly participating within the same intellectual atmosphere as the other mathematicians interested in mathematical compasses, such as Galileo, Faulhaber and Bramer.

Nevertheless, Descartes's interest in mathematical compasses was only one side of his mathematical activities. The mathematical practitioners mentioned above who invented and reformed the sector or the proportional compass tended almost without exception to neglect the theoretical development of the mathematical sciences. Their main concerns were fundamentally practical, that is to say, to fabricate useful mathematical devices for practical purposes. They, of course, called theoretical knowledge into service to make instruments as accurately as possible. This practical intention can be seen most clearly in Galileo's treatise on the geometric and military compasses mentioned in Section 1. As Stillman Drake has appropriately pointed out, Galileo achieved a mechanical approximation to practical problems "that lay beyond the scope of university mathematics."[84] It was not his main aim to construct a theoretical mathematics itself by the aid of mechanical tools.

[84] S. Drake, *Op. cit.* (n. 23), Introduction, p. 22.

Descartes's direction of study seems to have been quite the opposite. He surely resorted to mechanical devices, but did so in order to create his own respectable theoretical discipline. We have to admit that his attempts at justifying the use of compasses in his *Cogitationes privatae* were awkward and sometimes mistaken. This notwithstanding, he never forget to make efforts to translate the meanings of geometrical figures expressed by his compasses into the theoretical language that he had previously learned. But we should emphasize again that he tried to form "an entirely new" mathematical science not in a fully algebraic language of the written form but by the help of his compasses, as far as his notes *Cogitationes privatae* were concerned. Nevertheless, it is undeniable that that science was certainly an algebraic discipline although helped by compasses.

We should remember here that in his letter of March 26, 1619, Descartes told Beeckman that his "new science" should not be like the art of Lull, but should have solid mathematical foundations. A similar tendency in his thought appeared when he criticized Lambert Schenkel's book *De Arte memoriae* in the *Cogitationes privatae*. According to Frances A. Yates, Schenkel was well-known for his powers of memory in the late sixteenth and early seventeenth centuries.[85] Descartes's attitude toward this person seems to have been similar to his attitude toward the old man of Dordrecht who was proud of his art which originated from Ramon Lull. We encounter the following remark of Descartes in the *Cogitationes privatae*:

> On reading through Schenkel's profitable trifles (in the book *De Arte memoriae*) I thought of an easy way of making myself master of all I discovered through the imagination. This would be done through the reduction of things to their causes. Since all can be reduced to one it is obviously not necessary to remember all the sciences. When one understands the causes all vanished images can easily be found again in the brain through the impression of the cause. This is the true art of memory and it is plain contrary to his nebulous notions. Not that his is without effects, but it occupies the whole space with too many things and not in the right order. The right order is that the images should be formed dependent upon one another. He omits this which is the key to the whole mystery.[86]

Descartes's negative stance toward the vulgar art of memory is apparent here. The same stance was enunciated in his letter on April 29, 1619. There he was moreover denying the efficacy of the Lullian art and simply expressed his interest in knowing whether or not that art contained a useful key. But

---

[85]F. A. Yates, *The Art of Memory* (n. 35), p. 300. Cf. *Biographie universelle (Michaud) ancienne et moderne*, nouvelle éd. t. 38 (Paris, 1843), "Schenkel (Lambert/Thomas), 1547–*c.* 1603," pp. 283–284.

[86]*AT*, X, p. 230. The translation has been taken from Yates, *Ibid.*, p. 373.

this time, he tends to establish his rather constructive evaluation of the art of memory as Rossi has demonstrated in his *Clavis universalis*.[87] Descartes even proposes the key to what he regards as the true art of memory. According to his own criterion, "the right order" of the art "is that the images should be formed dependent upon one another."

A similar principle according to which Descartes thought the images should be formed might have been at work when he speculated on how he was to create a new mathematical discipline through the unification of arithmetic and geometry. But at least in the *Cogitationes privatae*, the principles of the unification thereof were still the mathematical compasses, not matured algebraic tools. Nevertheless, it is also certain that at the time of the composition of the *Cogitationes privatae* he started his journey toward a new science in which problems can be expressed very distinctly in our imagination by means of symbols which are easily memorized. This program will be fully stated in his *Regulae ad directionem ingenii*, especially in Rule XIV.

D. The Interest in Figurate Numbers, the Extension of the Pythagorean Theorem and Other Results

In the *Cogitationes privatae* there are a number of notes discussing mathematical sciences outside of pure mathematics. Descartes wrote, for example, memoranda on the mathematical law of free fall, hydrostatics, music and gnomonics. But here we look at only a few more problems from mathematics.

According to a note, one day "Isaac of Middelburg asked me if the cord ACB affixed at the nails a and b may describe a part of a conic section."[88] The problem was to determine the form of a chain suspended from two points and had been known since Galileo. (The name of the 'catenary' comes from the Latin word for chain, 'catena'.) Galileo mistakenly thought that the curve sought was a parabola. The catenary problem was successfully resolved by Christiaan Huygens, Johann Bernoulli and Leibniz in the late seventeenth century, and the 'catenary' curve is found to be able to be expressed by an exponential function. This problem contributed to the formation of a new mathematical field later named 'calculus of variations' (*calculus variationum*) by Leonhard Euler in 1766 and played an important role when the latter two mathematicians showed the effectiveness of their new differential calculus.[89] Descartes simply added a sentence to the preceding problem

Figure 3.11

[87]Rossi, *Clavis universalis* (n. 35), pp. 174–175.

[88]*AT*, X, p. 223: "Petijt à me Isaacus Middelburgensis an funis ACB affixus clavis *a*, *b*, sectionis conicæ partem describat."

[89]For the history of the catenary problem, see H. J. M. Bos, "Newton, Leibniz and the Leibnizian Tradition," Chap. 2 of I. Grattan-Guiness, ed., *From the Calculus to Set Theory,*

proposed by Beeckman: "Now I cannot permit myself the free time to examine that in detail."[90]

Between the notes arguing about mathematical compasses, we find the following observations on the nature of some characteristic numbers:

> If a square of a triangular number is subtracted from the square of the subsequent triangular [number], there remains a cube. For instance, 10, 15: Remove 100 from 225. 125 remains. From the progression 1 | 2 || 4 | 8 || 16 | 32 ||, the perfect numbers 6, 28, and 496 are obtained.[91]

The first topic is concerned with figurate numbers. The concept of figurate numbers, for example triangular numbers, is ascribed to the Pythagorians.[92] A triangular number is one of the series $\{n(n+1)/2; n = 1, 2, \cdots\}$ of successive sums of the natural numbers from 1 to n (Figure 3.12). It is so named because that number of points can be arranged into the form of a triangle. If the given triangular number is $n(n+1)/2$, then the next one may be written as $(n+1)(n+2)/2$. The difference between the square of $n(n+1)/2$ and that of $(n+1)(n+2)/2$ equals $(n+1)^3$, i. e. the cube of $(n+1)$, according to our method of calculation. For instance, we get the example which Descartes discusses: $15^2 - 10^2 = 5^3$.

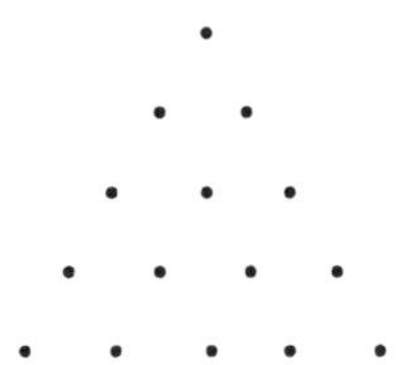

Figure 3.12

Perfect numbers which appear in the second observation of Descartes are also supposed to have belonged to the Pythagorean arithmetic.[93] The definition of those numbers is seen in Definition 22 of Euclid VII: "A perfect number is that which is equal to its own parts." We also find a general theorem on perfect numbers in Euclid IX, 36. In modern notation, it can be rewritten as follows: "If the sum of any number of terms of the series $1, 2, 2^2, \cdots, 2^{n-1}$ $(= S_n)$ is prime, then $S_n \cdot 2^{n-1}$ is a perfect number."[94] Descartes's cases above are the most primitive, and can be derived from when $n = 2, 3, 5$:

$$6 = 1+2+3;\ 28 = 1+2+4+7+14;$$
$$496 = 1+2+4+8+16+31+62+124+248.$$

1630–1910 (London, 1980), pp. 80–82; See also Hermann H. Goldstine, *A History of the Calculus of Variations from the 17th through the 19th Century* (New York/Heidelberg/Berlin, 1980), Index "Heavy chain problem."

[90] *AT*, X, p. 223: "Quod non licet per otium nunc disquirere."

[91] *AT*, X, p. 241: "Si subtrahatur numeri triangularis quadratus ex quadrato sequentis triangularis, restat cubus. Vt 10, 15: tolle 100 ex 225, restat 125. Ex progressione 1 | 2 || 4 | 8 || 16 | 32 || habentur numeri perfecti 6, 28, 496."

[92] Heath, *A History of Greek Mathematics*, Vol. 1, pp. 76–84.

[93] *Ibid.*, pp. 74–76.

[94] Cf. Heath, *The Thirteen Books of Euclid's Elements*, Vol. 2, (New York, $_2$1956), pp. 421–426; Clavius, *Euclidis Elementa*, pp. 378–379.

Toward the end of the *Cogitationes privatae*, Descartes begins to write his computations on the rectangular tetrahedron. The rectangular tetrahedron is defined as a tetrahedron of which each of the three angles at a vertex is a right angle. In other words, it is a solid obtained by cutting a rectangular parallelopiped near a corner. His notes begin with the following description:

> *In the rectangular tetrahedron, the power of the base is equal to the powers of three faces together.*
>
> For example, let three sides of the basis be $\sqrt{8}, \sqrt{20}, \sqrt{20}$; On the other hand, three sides above the basis 4, 2, 2: the area of the basis will be 6; [the area of] the three faces 2, 4, 4; of which the squares are 36, [and] 4, 16, 16, the latter three are equipotent to the former [see Figure 3.13 (a)]. Similarly, let sides of the basis be $\sqrt{13}$, $\sqrt{20}$, 5; and above the basis 2, 3, 4: The area of the basis will be $\sqrt{61}$; on the other hand, the surfaces are 3, 4, and 6, and the square of them are 61, and 9, 16, 36, [and the latter three] are equal to the former [see Figure 3.13 (b)].

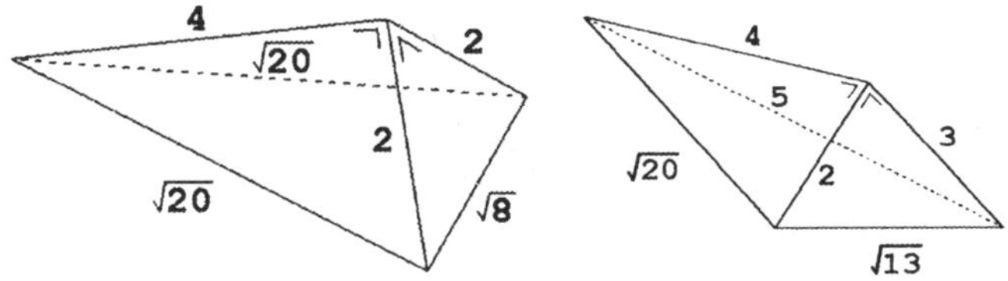

(a) (b)

[Figure 3.13]

> For this reason, very many unknown questions can be solved about the rectangular tetrahedron and about the non-rectangular tetrahedron through the relation to the rectangular one. This demonstration comes from the Pythagoreans and can be also extended to quantities of four dimensions: There the square of the solid opposed to the right angle is the squares from the other four solids altogether. To this let there be the example (*paradigma*) of progression in numbers 1, 2, 3, 4; in symbols 2e, ʒ, ce; in right angles of two, three, four lines.[95]

[95] *AT*, X, pp. 246–247: "*In tetraedro rectangulo, basis potentia æqualis est potentijs trium facierum simul.* V. g., sint basis tria latera, $\sqrt{8}, \sqrt{20}, \sqrt{20}$; tria verò latera supra basin, 4, 2, 2: area basis erit 6; trium facierum, 2, 4, 4; quorum quadrata sunt, 36, <&> 4, 16, 16, quæ tria æquipollent priori. Item, sint latera basis $\sqrt{13}, \sqrt{20}, 5$; & supra basin, 2, 3, 4: area basis erit $\sqrt{61}$; facierum verò, 3, 4, 6, quorum quadrata sunt 61, & 9, 16, 36, æqualia priori. Hinc plurimæ quæstiones ignotæ solvi possunt circa tetrædra rectangula & non rectangula per relationem ad rectangula. Hæc demonstratio ex Pythagoricâ procedit, & ad quantitatem quoque quatuor dimensionum potest ampliari; in quâ quadratum solidi angulo recto oppositi æquale est quadratis ex 4 alijs solidis simul. Sit ad hoc paradigma processionum in numeris 1, 2, 3, 4; in figuris, 2e, ʒ, ce; in angulis rectis duarum linearum, trium, quatuor."

Clearly here Descartes intends to generalize the ordinary Pythagorean theorem for right-angled triangles in a plane to those in the spaces of three and four dimensions! But in order to do so, he advances his speculations by taking numerical examples and by calculating algebraically (as will be shown in the subsequent notes). It is in the mode of algebraic thinking that he can extend his imagination even beyond solid geometry. In today's notation, the Pythagorean theorem in two dimensions can be represented by the formula "$a^2 + b^2 = c^2$," where c is the side subtending the right angle. In the three-dimensional space, he formulates it in the form "$A^2+B^2+C^2 = D^2$," where A, B, and C are the areas of the three sides and D is the area of the base which is opposite to the right-angled corner. Finally, he seems to proceed to the case of four dimensions and to imagine the algebraic formula "$\alpha^2 + \beta^2 + \gamma^2 + \delta^2 = \varepsilon^2$," where α, β, γ, and δ are the volumes of the four solids and ε is the volume of the solid "opposed to the right angle." If we are allowed to make an anachronistic observation, geometers of the later period who think of points and spaces in the abstract set-theoretical manner may be able to justify the extended Pythagorean 'theorem' formulated by Descartes in four-dimensional Euclidean space. But the extension is not so simple as Descartes thought. He does not state anything about how to define the right angle in four dimensions.[96] So his attempt is not sufficient. What is certain and significant, however, is that in some sense he is doing geometry in quite a 'formalistic' or algebraic way.

We do not know for certain what he means in the last sentence beginning with the words "to this." But we could conjecture the sentence means that the extended Pythagorean 'theorems' can be exemplified and ascertained sometimes by concrete numbers such as 1, 2, 3, or 4, sometimes by formulae containing unknown quantities 𝔵, 𝔷, 𝔠𝔢, and that "the right angle" can be reinterpreted as that between two lines in the case of two dimensions, and as that made by three or four lines in the cases of three or four dimensions, respectively. This interpretation may be a mere conjecture without solid foundations. But the following notes might corrobarate our conjecture to some measure and give a clue as to how Descartes calculated concrete values of the sides, the areas of the base, and of the sides in the passage quoted previously:

> *If the base of a rectangular pyramid is given, the sides over the base are easily found.*
>
> Let, for example, the sides over the base be $\sqrt{13}, \sqrt{20}, 5$. Let 1𝔵 be put for the first side over the base; for another $\sqrt{13 - 1𝔷}$.; and for the third $\sqrt{20 - 1𝔷}$.; the power of these two is equal to 33−2𝔷 (for it is equal to the power of the side), or 1𝔷 equals 4. Therefore,

[96]For a discussion of this problem, see Pierre Costabel, "L'Initiation mathématique de Descartes," *Archives de philosophie*, **46** (1983), pp. 643–644. It is known that in the middle of the 19th century Ludwig Schläfli successfully extended the Pythagorean theorem to the multidimensional case. See his "Theorie der vielfachen Kontinuität," *Gesammelte mathematische Abhandlungen*, Bd. 1 (Basel, 1950), p. 193.

if the base and the opposed angle are known, we can know the total pyramid, as Euclid demonstrates on the triangle.[97]

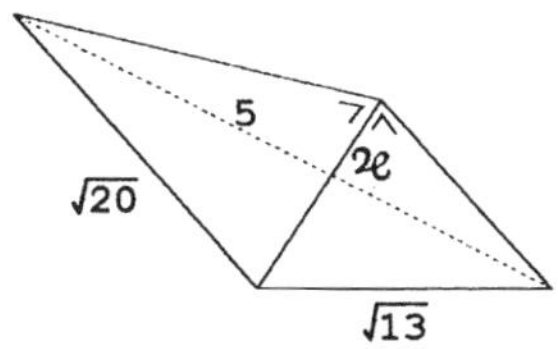

Figure 3.14

The term "pyramid" as used here is synonymous with tetrahedron. The rectangular tetrahedron the base of which has three sides $\sqrt{13}, \sqrt{20}, 5$ is identical to the second example in the preceding paragraph (see Figure 3.13 (b)). In modern symbols, Descartes sets one of the unknown sides to x (Figure 3.14). The other two sides will be calculated as $\sqrt{13-x^2}$ and $\sqrt{20-x^2}$ by the Pythagorean theorem. By applying the same theorem again to the remaining right-angled triangle, we obtain $(13-x^2)+(20-x^2) = 33-2x^2 = 25$, i. e. $2x^2 = 8$. To take a positive root, we have 2 for the first side.

In the next paragraph Descartes proceeds further with the general calculations about the rectangular tetrahedron, the base of which has the sides α, β, γ.[98] The three sides over the base abg are calculated as

$$\sqrt{\cdot}\ \tfrac{1}{2}\alpha q + \tfrac{1}{2}\gamma q - \tfrac{1}{2}\beta q.;$$
$$\sqrt{\cdot}\ \tfrac{1}{2}\alpha q + \tfrac{1}{2}\beta q - \tfrac{1}{2}\gamma q.;$$
$$\sqrt{\cdot}\ \tfrac{1}{2}\beta q + \tfrac{1}{2}\gamma q - \tfrac{1}{2}\alpha q..$$

Let the three sides required be x, y, and z. By the Pythagorean theorem, $y^2+z^2=\alpha^2, z^2+x^2=\beta^2, x^2+y^2=\gamma^2$ (see Figure 3.15). Then,

$$2x^2 = \beta^2+\gamma^2-\alpha^2,$$
$$2y^2 = \alpha^2+\gamma^2-\beta^2,$$
$$2z^2 = \alpha^2+\beta^2-\gamma^2.$$

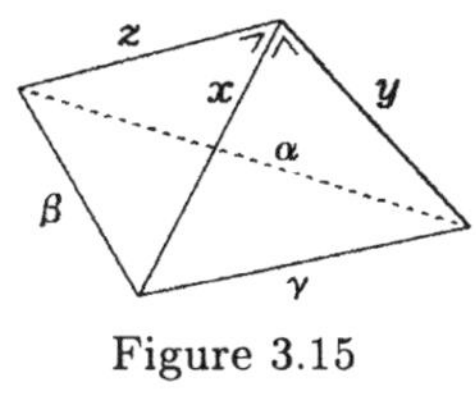

Figure 3.15

Thus we obtain Descartes's results. He probably writes αq for our α^2, probably following Clavius's alternative sign q (*quadratus*) for 𝔷.[99] He then calculates the areas of the three sides, the area of the base, and finally the volume of the tetrahedron.[100] He obtains the volume of the rectangular tetrahedron by

[97] *AT*, X, p. 247: "*Datâ basi pyramidis rectangulæ, facilè inveniuntur latera super basin.* Sint, v. g., latera basis, $\sqrt{13}, \sqrt{20}$, &5. Pro primo latere supra basin ponatur 1ℭ; pro altero, $\sqrt{.13 - 1𝔷.}$; & pro tertio, $\sqrt{.20 - 1𝔷}$; quorum duorum potentia, quia æqualis potentiæ lateris, est æqualis 33—2𝔷, vel 1𝔷 æq. 4. Ergo notâ basi & angulo opposito, totam pyramidem possumus agnoscere, vt de triangulo Euclides demonstrat."

[98] *AT*, X, pp. 247–248.

[99] Clavius, *Algebra*, p. 5.

[100] *AT*, X, p. 248. In our notation (or precisely speaking, in the later Descartes's!), the area of the three sides are $\sqrt{(\frac{1}{16}\alpha^4 + \frac{1}{8}\beta^2\gamma^2 - \frac{1}{16}\beta^4 - \frac{1}{16}\gamma^4)}$; $\sqrt{(\frac{1}{16}\beta^4 + \frac{1}{8}\alpha^2\gamma^2 - \frac{1}{16}\alpha^4 - \frac{1}{16}\gamma^4)}$;

calculating $xyz/6$ in the above notation. Further, he gives an example of which the three given sides of the base are $\sqrt{13}, \sqrt{20}, 5$. He first calculates $\alpha^2 + \beta^2 + \gamma^2 = 29$; then $(\alpha^2 + \beta^2 + \gamma^2) - \gamma^2 = 29 - 13 = 16 = x^2$, $(\alpha^2 + \beta^2 + \gamma^2) - \beta^2 = 29 - 20 = 9 = y^2$, $(\alpha^2 + \beta^2 + \gamma^2) - \alpha^2 = 29 - 25 = 4 = z^2$; so $x^2y^2z^2 = 576$ and $xyz = 24$. He finally gets the volume of the pyramid $xyz/6 = 4$. This is the very last result which Descartes records in the *Cogitationes privatae.*

The extension of the Pythagorean theorem to multi-dimensional 'spaces' is impressive, and certainly in advance of its time. But it is clearly denied that the result was historically transcendent. As Ivo Schneider has shown in his comprehensive biography of Johannes Faulhaber, similar attempts were found, for example, in Faulhaber's *Miracula arithmetica* of 1622. As to this problem which is closely related to the overall mathematical thought of Descartes in this period, we will touch upon it at the end of this chapter.

Descartes's calculations on the rectangular tetrahedron are impressive in that he very naturally reforms the cossic symbols used in Clavius's *Algebra* into the ones which we have just observed. If Leibniz and Foucher de Careil copied Descartes's signs in this part faithfully, they may be judged to be considerably close to the ones of his later years. This is significant for the development of his reformed algebra. Nevertheless, the overall meaning of his mathematical thought of this period will be discussed in detail in Section 4.

§ 3. The *De Solidorum Elementis*

A. A History of the Manuscript

Besides the fragmentary notes *Cogitationes privatae*, Descartes is supposed to have drafted a short piece, which Charles Adam dates to the period from 1619 to 1621. In 1676, Leibniz copied this manuscript then in the hands of Clerselier. The title which Leibniz gave to the manuscript was *Progymnasmata de solidorum elementis excerpta ex manuscripto Cartesii* (Preliminary Exercises on the Elements of Solids Extracted from a Manuscript of Descartes). This work is identical to the one found in the Stockholm inventory under the letter M.[101]

The original manuscript by Descartes had disappeared before Foucher de Careil published Leibniz's copy in the second volume of his *Œuvres inédites de Descartes* of 1860.[102] And unfortunately Foucher de Careil could not under-

---

$\sqrt{(\frac{1}{16}\gamma^4 + \frac{1}{8}\alpha^2\beta^2 - \frac{1}{16}\alpha^4 - \frac{1}{16}\beta^4)}$; the area of the basis is $\sqrt{\frac{1}{8}(\alpha^2\beta^2 + \alpha^2\gamma^2 + \beta^2\gamma^2) - \frac{1}{16}(\alpha^4 + \beta^4 + \gamma^4)}$; the volume is $\sqrt{\{\frac{1}{288}(\alpha^4\beta^2 + \alpha^4\gamma^2 + \beta^4\alpha^2 + \beta^4\gamma^2 + \gamma^4\alpha^2 + \gamma^4\beta^2) - \frac{1}{144}\alpha^2\beta^2\gamma^2 - \frac{1}{288}(\alpha^6 + \beta^6 + \gamma^6)\}}$. We reproduce Descartes's notation only for the last formula:
$\sqrt{}.\{.\frac{1}{288}\alpha qq\beta q + \frac{1}{288}\alpha qq\gamma q + \frac{1}{288}\beta qq\alpha q\ [+\frac{1}{288}\beta qq\gamma q] + \frac{1}{288}\gamma qq\alpha q + \frac{1}{288}\gamma qq\beta q - \frac{1}{144}\alpha q\beta q\gamma q - \frac{1}{288}\alpha qc - \frac{1}{288}\beta qc - \frac{1}{288}\gamma qc\}$.

[101] *AT*, X, p. 10.

[102] Foucher de Careil, *Œuvres inédites de Descartes*, t. 2 (Paris, 1860), pp. 214–234.

stand the mathematical contents in the copy, mainly because he was unable to figure out cossic symbols. Since the publication of the *editio princeps*, not a few mathematicians have contributed to showing how to revise the text and to interpret its mathematical content. Among those contributors, the French mathematicians C. Mallet, E. Prouhet,[103] and Ernest de Jonquières[104] are notable. The 1908 Adam-Tannery edition provided a Latin text much more reliable than previous ones. But even this version was not totally free from minor errors.

Fortunately, the manuscript by Leibniz is still extant in the Niedersächsische Landesbibiliothek in Hanover. Two significant studies on it appeared: Pasquale J. Federico, *Descartes on Polyhedra—A Study of the De Solidorum Elementis*[105] and Descartes, *Exercises pour les éléments des solides—Essai en complément d'Euclide*, edited and translated by Pierre Costabel.[106] Federico's posthumous work provides an accurate English translation and precise mathematical commentaries. But his interpretation of Descartes's text does not always seem to be beyond dispute, in particular, as we will show in Subsection D, concerning the date of Descartes's original manuscript. Costabel's monograph, on the other hand, can be said to be almost authoritative in its Latin text, French translation, notes, and commentaries. A few historical remarks on Costabel's work, however, should be made, despite its almost complete definitiveness, about Descartes's discussion on solid geometry (see Subsection B). It is for this reason that in this section we reexamine historically the *Progymnasmata de solidorum elementis.* (Hereafter as is customary we refer to it simply as the *De solidorum elementis.*)

B. Solid Geometry

Although the Leibniz copy is not partitioned, we can divide the *De solidorum elementis* into two parts. The first part is concerned with solid geometry, and the second part with polyhedral numbers.[107] First we will examine Part I on solid geometry. We confine our discussion to passages which may help us understand Descartes's characteristic way of thinking about mathematics. An almost exhaustive examination of this text can be found in Federico's and Costabel's

[103]For C. Mallet's and E. Prouhet's critical reviews, see Pasquale Joseph Federico, *Descartes on Polyhedra: A Study on the De Solidorum Elementis* (New York/Heidelberg/Berlin, 1982), pp. 6–7.

[104]E. de Jonquières, "Écrit posthume de Descartes. *De Solidorum Elementis.* Texte latin (original et revu) suivi d'une traduction française avec notes," *Mémoires de l'Académie des Sciences de l'Institut de France*, 2e série, **45** (1890), pp. 325–379.

[105]Federico, *Op. cit.* (n. 103).

[106]René Descartes, *Exercices pour les éléments des solides: Progymnasmata de solidorum elementis—Essai en complément d'Euclide*, Edition critique avec introduction, traduction, notes et commentaires par Pierre Costabel (Paris, 1987). See my review of this book in *Isis*, **79** (1988), pp. 731–732.

[107]This partition which we follow below is due to Federico. In his edition (n. 106), Costabel calls the table of formulae at the end of the treatise "la troisième partie," but does not consider it independently of Part II on polyhedral numbers.

monographs mentioned above.

Descartes begins Part I with a definition of the solid right angle. The solid angle is defined in Definition 11 of Euclid XI.[108] Descartes seems to presuppose this definition in the following discussion:

> A solid right angle is one which embraces the eighth part of the sphere, even though it is not formed by three plane right angles. But all the angles of the planes by which it is bounded, taken together, equal three right angles. As in a plane figure all the exterior angles, taken together, equal four right angles, so in a solid body all the exterior solid angles, taken together, equal eight solid right angles. By exterior [solid] angle I mean the mutual bending or inclination of the planes, which is to be measured with the help of the plane angles which comprise the solid angle. For the part by which the sum of all the plane angles forming a solid angle is less than the four right angles which form a plane, designates the exterior solid angle.[109]

Descartes first defines the solid right angle and then states the proposition that "all the exterior solid angles, taken together, equal eight solid right angles." Importantly, he provides no proof.

In his popular book ***Induction and Analogy in Mathematics***, George Pólya has presented Descartes's statement on the exterior solid angles as an exercise for analogical proof.[110] After having introduced Pólya's proof, Federico stated that "Descartes may of course have had a proof, but he may not have bothered. The analogy with plane figures is so close and direct that he may have thought further proof unnecessary."[111] Federico further continued his observation, and thought it curious that the word 'analogy' did not appear in any of the major philosophical writings of Descartes.

As we have seen in the previous section, however, Descartes used analogical reasoning when he attempted to extend the Pythagorean theorem to three and four dimensions. And it is memorable that his interest in solid geometry was not limited to the notes in the *De solidorum elementis*. In his *Cogitationes privatae*, he reported that he had seen a compass for measuring solid angles and carried out calculations on the rectangular tetrahedron, a corner of which is a solid right angle. There he even discussed polygonal (triangular) numbers. From these facts we do not necessarily want to conclude that the *De solidorum elementis* was composed exactly during the same period as the *Cogitationes privatae*, but simply to point out that there is a certain similarity and continuity between the two opuscula. This observation will surely influence the determi-

[108] Heath, *The Thirteen Books of Euclid's Elements*, Vol. 3 (New York, $_2$1956), pp. 261 & 267–268. Cf. Clavius, *Euclidis Elementa*, p. 479.

[109] *AT*, X, p. 265; Federico, *Op. cit.*, pp. 43–44.

[110] G. Pólya, *Induction and Analogy in Mathematics*, Volume 1 of *Mathematics and Plausible Reasoning* (Princeton, 1954), pp. 57 & 226.

[111] Federico, p. 45.

nation of the date of Descartes's original manuscript, about which we will argue later (Subsection D).

The next paragraph which attracts our attention is Descartes's algebraic attempt to prove that there cannot be more than five regular 'Platonic' polyhedra. His statement reads as follows:

> As follows it is proven that there cannot be more than 5 regular bodies: since if one takes α for the number of solid angles, and $\mathfrak{c}$ for the number of faces, one must be able to divide
> $$\frac{2\alpha-4}{1\mathfrak{z}} \text{ and } \frac{2\mathfrak{z}-4}{1\alpha}$$
> in a manner which no fraction occurs; otherwise it is certain and evident that a regular body is not possible. But this can only be so if α is 4, 6, 8, 12, 20 and respectively $\mathfrak{c}$ is 4, 8, 6, 20, 12, by which are generated the 5 regular bodies.[112]

Suppose that the given regular polyhedron has faces of the regular n-polygon. From the last sentence of the paragraph quoted above (p. 178), Descartes infers that $\Sigma = 2\pi\alpha - 4\pi$, where Σ signifies the sum of all the plane angles.[113] On the other hand, the sum of the plane angles on each face of the regular polyhedron may be written $(n-2)\pi$. Therefore, $\Sigma = 2\pi\alpha - 4\pi = (n-2)\pi \times \mathfrak{z}$. Consequently, we obtain Descartes's first formula $(2\alpha - 4)/1\mathfrak{z} = n - 2$. Next note that each solid angle of the polyhedron has the same number m of plane angles. Let the number of the plane angles be P. Then, $\mathrm{P} = m\alpha$. Therefore, $\Sigma = \mathrm{P}\pi - 2\mathfrak{z} = m\alpha\pi - 2\pi$. We have already had the formula $\Sigma = 2\pi\alpha - 4\pi$. Accordingly, $m\alpha - 2 = 2\alpha - 4$. So, we acquire Descartes's second formula $(2\mathfrak{z} - 4)/1\alpha = m - 2$. Now we put $n - 2 = a, m - 2 = b$. Then, $\alpha = (8+4a)/(4-ab)$, $\mathfrak{z} = (8+4b)/(4-ab)$. From the definitions of a and b, they must be integers such that $1 \leqq ab < 4$. Then $(a, b) = (1,1), (1,2), (2,1), (3,1)$, and $(1,3)$. Consequently $(\alpha, \mathfrak{z}) = (4,4), (6,8), (8,6), (12,20)$, and $(20,12)$. Thus we have Descartes's conclusion that there cannot be more than five regular bodies.[114]

Again Descartes never presents a detailed proof. But our reconstruction does not seem to deviate essentially from his way of reasoning. The result itself was known in ancient Greece. We can see its geometrical demonstration in Euclid XIII, 18, the final proposition of Euclid's *Elements*. What Descartes's words quoted above suggest is that he is supposed to have inferred the same result from a basically algebraic point of view. In fact, his reasoning is due to the fact that both $(2\alpha - 4)/1\mathfrak{z}$ and $(2\mathfrak{z} - 4)/1\alpha$ cannot be fractions.

---

[112] *AT*, X, pp. 266–267; Federico, p. 50. As we have already stated in our note 50 above, throughout his manuscript copy of *De solidorum elementis*, Leibniz systematically used his own symbols ♃ and ♀ for $\mathfrak{z}$ and $\mathfrak{c}$, respectively. We safely suppose that Descartes's original symbols were the usual cossic ones. So in our texts, we use $\mathfrak{z}$ and $\mathfrak{c}$, not those of Leibniz.

[113] *AT*, X, p. 265, *l.* 20–p. 266, *l.* 4; Federico, p. 46.

[114] See Federico, pp. 50–51.

The third topic to be examined is related to the theorem which has been usually ascribed to Leonhard Euler: S − A + F = 2, where S is the number of vertices (or that of solid angles), A is the number of edges, and F is the number of faces of a convex polyhedron.

Even quite recently, for example, Pierre Costabel argued on the priority of the discovery of that theorem and concluded categorically that Descartes essentially discovered the relation F − A + S = 2.[115] It may be said that the fundamental point in Costabel's argument has been repeated several times since the *editio princeps* was published in 1860. Costabel stated that "L'affirmation que F + S = A + 2 apparaît explicitement dans le texte cartésien vers la fin de la première partie que nous considérons ici."[116] In the footnote attached to this sentence, he referred to the Adam-Tannery edition, Volume X, p. 269, *ll.* 15–16. In fact there we find the sentence "the actual number of plane angles is $2\varphi + 2\alpha - 4$, where φ indicates the number of faces and α the number of solid angles.

Unfortunately, however, we nowhere encounter an explicit statement of the relation F + S = A + 2, which Costabel insisted existed in the text. Certainly, Descartes states in another place in the manuscript that "there are always twice as many plane angles as sides (*latera*) on the surfaces of a solid body; for one side is always common to two faces."[117] If we combine this with the above statement that the number of plane angles is $2\varphi + 2\alpha - 4$, we certainly obtain the Euler formula. This is indeed a small step. But such a conceptual error is what Imre Lakatos in his excellent essay *Proofs and Refutations: The Logic of Mathematical Discovery* has warned the historian of mathematics not to commit. The following is Lakatos's observation, which seems to us just:

> The priority should not be granted to Descartes without a minor qualification. It is true that Descartes states that the number of plane angles equals $2\varphi + 2\alpha - 4$ where by φ he means the number of faces and by α the number of solid angles. It is also true that he states that there are twice as many plane angles as edges (*latera*). The conjunction of these two statements of course yields the Euler formula. But Descartes did not see the point of doing so, since he still thought in terms of angles (plane and solid) and faces, and did not make a conscious revolutionary change to the concepts of 0-dimensional vertices, 1-dimensional edges and 2-dimensional faces as a necessary and sufficient basis for the full topological characterisation of polyhedra.[118]

[115] P. Costabel, "Le théorème de Descartes-Euler," *Studia Cartesiana*, **1** (1979), p. 34; Reprinted in *Idem, Démarches originales de Descartes savant* (Paris, 1982), p. 24.

[116] *Ibid.*, p. 31; Reprint, p. 21.

[117] *AT*, X, p. 267, l. 31–p. 268, l. 1; Federico, p. 54.

[118] I. Lakatos, *Proofs and Refutations: The Logic of Mathematical Discovery*, ed. by John Worrall and Elie Zahar (Cambridge, 1976), p. 6, footnote 1.

In fact, in Descartes's text the concept of side (*latus*) does not play an essential role, and moreover the number of sides is not denoted by a specific sign. Lakatos has stressed in the same text that Euler had to invent the term 'acies' (edge) for use instead of the old 'latus' (side) which is basically a polygonal concept. Nevertheless, Costabel still insisted the relation $F + S = A + 2$ was at least implicitly used in Descartes's discussion and stated that "Le lecteur moderne peut donc légitimement croire que la relation $F + S = A + 2$ est ici sous-jacente au texte cartésien."[119] In another place, he wrote: "La querelle faite à Descartes est vraiment byzantine. Et s'il [i. e. Descartes] n'a pas écrit: Théorème $F - S - 2 = R$ [in our present discussion, A], il a *bien vu* que, lorsqu'on se donne S, il *faut* que F soit tel que $F + S - 2$ soit plus petit que $3S - 6$; c'est-à-dire que F est soumis à une condition."[120] (Italics are in the original.)

It is true that Costabel differed from his predecessors who had insisted that Descartes obtained the Euler formula on several points. First, Costabel even suggested the possibility that Euler had studied Leibniz's manuscript before composing his treatises on polyhedra. But the evidence which Costabel presented to affirm his conjecture is, as he himself might admit, only circumstancial.[121] Second, he reexamined the Leibniz copy and conjectured, by comparing its size with that of the Descartes original recorded in the Stockholm inventory, that Leibniz had excerpted the original and suppressed some part of it. But from this alone the affirmation that Descartes possessed the formula $F + S = A + 2$ is a *non sequitur*. Descartes might have reached it. But such a conjecture is speculation, not one derived from a critical analysis of the text.

Third, Costabel further observed that Euler himself had not obtained his relation with a satisfactory demonstration. It was in Proposition 4 of the "Elementa doctrinae solidorum" of 1758 that Euler proposed his formula: "In every solid body bounded by plane faces the sum of the number of solid angles and the number of faces exceeds the number of edges by two."[122] As Costabel insisted, Euler's proof for this proposition was in fact applicable only to particular convex bodies, that is to say, Euler's was simply inductive. For these reasons, Costabel drew the following conclusion in his paper: "Objectivement, d'après le contenue des textes, [...] la palme revient à Descartes sans aucun partage. Bien entendu, du point de vue de l'histoire, revient à Euler le mérite d'avoir reconnu et publie le 'theoreme' tandis que le texte de Descartes demeurait inconnu et inedit dans les papiers de Leibniz."[123]

---

[119]Costabel, *Loc. cit.*(n. 116).

[120]Costabel, *Exercices pour les éléments des solides* (n. 106), p. 101.

[121]Costabel, *Ibid.*, pp. 93–96. Cf. Costabel, "Euler lecteur de Descartes," *Revue du XVIIIe siècle*, **18** (1986), pp. 281–288.

[122]Leonhard Euler, "Elementa doctrinae solidorum," *Novi commentarii academiae scientiarum Petropolitanae*, **4** (1752/3), 1758: *Opera omnia*, Ser. prima, XXVI (Lausanne, 1953), p. 78: "In omnia solido hedris planis incluso aggregatum ex numero angulorum solidorum et ex numero hedrarum binario excedit numerum acierum."

[123]Costabel, *Op. cit.* (n. 115), p. 34; *Démarches originales de Descartes savant*, p. 24.

Notwithstanding Costabel's statement, until Descartes's original manuscript (not the Leibniz copy) has not been discovered, we may not affirm with certainty that Descartes knew Euler's formula. What Euler did was, strictly speaking, to propose the explicit formula $F - A + S = 2$ with an inductive demonstration. To determine who was the first discoverer of that relation with a satisfactory proof is a subtle philosophical problem, as Lakatos's insightful essay *Proofs and Refutations* has taught us. Of course, our critical assessment does not completely negate Descartes's role as a forerunner of Euler's proposition and his significant place in the history of polyhedral studies. What Descartes accomplished in Part I of his *De solidorum elementis* pertaining to solid geometry is that he presented, for example, an algebraic demonstration of Euclid XIII, 18 on 'Platonic' solids and some important formulae, a combination of which may lead to the Euler formula. He is memorable for such achievements alone.

C. Figurate Numbers

Part Two of the *De solidorum elementis* is about polygonal and polyhedral numbers. As we have noted in § 2, D, since the time of Pythagoras the numbers which can be arranged into geometrical figures have been called 'figurate numbers'. Polygonal numbers belong to a representative category of figurate numbers of which the corresponding geometrical figures are polygons. Polyhedral numbers can be also defined in a manner similar to polygonal numbers, but are of a higher level of sophistication than the latter because their correlative figures are three-dimensional.

The ultimate goal of Part Two is to show how to calculate polyhedral numbers and their related quantities. But Descartes first gives directions for finding polygonal numbers:

> Solids (*solida*) are best of all formed by superimposing gnomons having always one angle vacant, and from which the whole figure can be resolved into triangles. From which it is easily recognized that the weights of all the polygonals are obtained by multiplying the triangulars by the numbers 2, 3, 4, 5, 6, etc., and subtracting from the product 1, 2, 3, 4 radices, etc.[124]

As to the text, Federico has suggested the word *solida* (solids) in the first sentence is a copying error for *plana* (planes).[125] Costabel, on the other hand, argued that *solida* should be read as it stands on the ground that *solida* can be interpreted as "des configurations pondérées [...] qui sont aussi bien à deux dimensions qu'à trois."[126] Certainly unless we interpret *solida* strictly as exclusive

[124] *AT*, X, p. 269; Federico, p. 92; Costabel, *Exercices* (n. 106), p. 33.

[125] Federico, p. 93.

[126] Costabel, *Exercices*, p. 36: "Le mot *solida*, qu'il faut bien traduire par *solides* pour éviter de faire trop de place dans la traduction à une interprétation, est sans aucun doute appelé

of plane figures, are the sentences quoted above mathematically meaningful.

What Descartes shows here are two ways of counting polygonal numbers. For simplicity, take as an example the pentagon. Pentagonal numbers are calculated, according to Descartes's directions, either by adding dots on gnomons (which make an arithmetical progression) or by dividing the given polygon into triangles. The expression "by superimposing gnomons" in the first sentence is related to the first method (Figure 3.16 (a)). In the second sentence, Descartes indicates the method of calculating polygonal numbers corresponding the latter case by triangulation. The pentagonal numbers in Figure 3.16 (b) consist of 3 (= 5 − 2) triangular numbers, but here 2 (= 5 − 3) common sides are counted twice. So the number of dots in Figure 3.16 (b) is (3 × the number of the corresponding triangular number) minus (2 × the number of dots in the common side). Following Federico, we can generalize these procedures to the n-th a-gonal number $p(a, n)$. Federico has obtained the formula

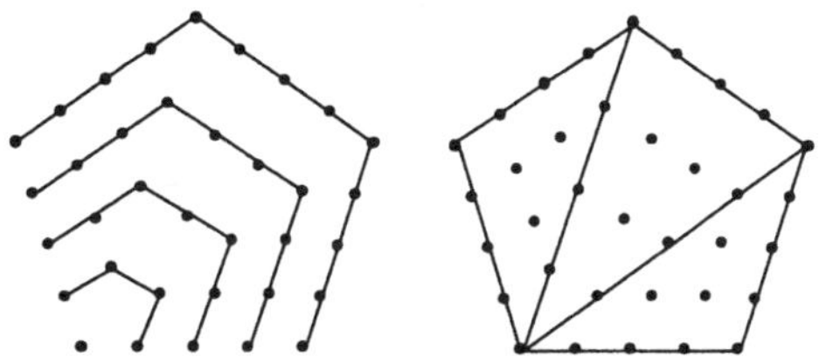

(a) (b)

Figure 3.16

$$p(a, n) = (a - 2).p(3, n) - (a - 3)n.$$ [127]

This is what Descartes indicates in the passage quoted above. The word "root" (*radix*) in the text signifies n in the above formula. Descartes refers to the number of dots as the "weight" (*pondus*) throughout the manuscript.

The calculation of polygonal numbers is for Descartes only a preliminary for counting polyhedral numbers, first for five regular polyhedra, and second for semi-regular polyhedra. Here we will not simply repeat Federico's (and Costabel's) comprehensive mathematical analyses about how the polyhedral numbers are calculated. Instead we will concentrate our examination on some passages which may be supposedly related to the determination of the date when Descartes composed the *De solidorum elementis*.

As polygonal numbers can be divided into traiangles, so a polyhedron can be divided into pyramids. Judging from the following algebraic directions for finding polyhedral numbers, Descartes already seems to have been well versed in how pyramidal numbers are calculated:

> The algebraic expressions for these figurate [polyhedral] numbers are found by multiplying the exponent (*exponens*) of the face plus $\frac{1}{2}\mathfrak{X}$

par cette idée sous-jacente." "Les *solida* sont essentiellement des configurations pondérées, qui s'opposent aux formes pures et immatérielles, mais qui sont aussi bien à deux dimensions qu'à trois."

[127] Federico, pp. 89 & 94.

> by $\frac{1}{3}\mathfrak{x} + \frac{1}{3}$, then by the number of faces, which is done as many times as there are different types of face in the given body; then add to or subtract from the result the number of radices multiplied by $\frac{1}{2}\mathfrak{z} + \frac{1}{2}\mathfrak{x}$, and the number of angles multiplied by $\mathfrak{x}$.[128]

Descartes's words here are analogous to the case of the polygonal numbers. In the case of the plane figures, he has assumed that the triangular numbers are known. At any rate, it is quite simple to calculate them, for $p(3,n) = n(n+1)/2$.

To count the pyramidal numbers which are supposed to have been given, on the other hand, is not so simple as the case of the triangular numbers. He gives easy directions for obtaining the pyramidal numbers. "The exponent of the face" in the passage above stands for the formula for the polygon of the face, i. e. $p(a,n)$. Descartes's procedure for calculating the pyramidal numbers can be formulated as follows:

$$P(a,n) = (p(a,n) + n/2).(n+1)/3.$$

Concerning this formula for the pyramidal numbers, Federico has stated: "How Descartes arrived at this expression for the general formula for a pyramidal number is not evident."[129]

How did Descartes acquire his knowledge about the pyramidal numbers? Of course, he could have derived the above formula $P(a,n)$ recursively from $P(a,n) = P(a,n-1) + p(a,n)$. That he may have done it this way cannot be irrevocably refuted, but this interpretation seems a little anachronistic. However, according to L. E. Dickson, a historian of the theory of numbers, the concept of the pyramidal numbers had been known since ancient Greece and many mathematicians in the late sixteenth and early seventeenth centuries tried to formulate them as generally as possible. Among others, he has referred to Johannes Faulhaber and Peter Roth.[130] We speculate that Descartes simply took for granted the formula for the pyramidal numbers because he drafted the *De solidorum elementis* on the basis of a treatise or some treatises about figurate numbers.

Since the year 1860 when mathematicians began to argue about the Descartes manuscript, a booklet under the title *Numerus figuratus, sive Arithmetica analytica arte mirabili jnavdita nova constans* (Figurate Number, or Analytical Arithmetic Consisting of an Admirable Unheard of and New Art) published in 1614,[131] which had been once ascribed to Faulhaber, has been suggested as a source. Above all, Gaston Milhaud pointed out that the terms

---

[128] *AT*, X, p. 275, *ll.* 4–10; Federico, p. 108.

[129] Cf. Federico, p. 99.

[130] Leonard Eugene Dickson, *History of the Theory of Numbers*, Vol. 2: *Diophantine Analysis* (Washington, D. C., 1919; New York, ${}_{2}1966$), p. 5.

[131] There is no indication of the place of publication, but it was probably in Frankfurt am Main. We refer to a copy at the Stadtbibliothek Ulm.

'pondus' and 'radix' in the *De solidorum elementis* are the same as the aforementioned *Numeris figuratus.*[132] Federico followed Milhaud's lead and stated that "Faulhaber's enthusiasm may have inspired him [Descartes] to think about the subject and to evolve some new ideas. Furthermore, Faulhaber's equations of pyramidal numbers may well have been the starting point for Descartes's treatment of pyramids."[133]

Johannes Faulhaber was a mathematical practitioner in Ulm who inclined toward mysticism and published many books, especially on practical mathematics.[134] It has been traditionally reported since Daniel Lipstorp and Adrien Baillet, who propagated Lipstorp's description widely, that Descartes visited him in Ulm and discussed some mathematical problems.[135] In the tract *Numerus figuratus*, which had been ascribed to Faulhaber, treated polygonal and pyramidal numbers from the viewpoint of number mysticism. What the author of this tract added to these concepts of figurate numbers is the notion of 'hyperpyramidal' numbers, to use Federico's terminology.[136] The hyperpyramidal numbers are obtained from successive summations of the pyramidal numbers. The process of summation can be repeated infinitely so that these numbers have a stratified structure. The tract gave only 'cubi-cossic' (i. e. third degree) equations for the pyramidal numbers and several equations of higher degrees for the hyperpyramidal numbers. But it did not venture to proceed to the more sophisticated polyhedral numbers.[137] It may be possible that Descartes didn't read the tract, but in the second part of the *De solidorum elementis* he went far beyond it, ignoring its mysticism and showing how to calculate the polyhedral numbers.

Nethertheless, it does not follow from earlier considerations that the *Numerus figuratus* was the only source for Descartes's knowledge on the pyramidal numbers. To give only one other possible source, Peter Roth's *Arithmetica philosophica* was explicitly cited in the *Cogitationes privatae* (p. 124). The subject of the second part of Roth's book is polygonal and pyramidal numbers. There Roth especially made efforts to calculate pyramidal numbers up to those with a hectagon as the base. The table of 98 pyramidal numbers occupies eight pages

---

[132]Gaston Milhaud, *Descartes savant* (Paris, 1921), pp. 86–87.

[133]Federico, p. 118.

[134]Faulhaber's books have been examined in Ivo Schneider, *Johannes Faulhaber* (n. 158, below), Kurt Hawlitscheck, *Johann Faulhaber 1580–1635: Eine Blütezeit der mathematischen Wissenschaften in Ulm* (Ulm, 1995), and A. G. Kästner, *Geschichte der Mathematik*, Bd. III (Göttingen, 1799; Hildesheim/New York, 1970), pp. 111–152. Cf. Paul A. Kirchvogel, "Faulhaber, Johann," *DSB*, IV, pp. 549–553.

[135]D. Lipstorp, *Specimen Philosophiae Cartesianae* (Lugduni Batavorum, 1653), pp. 78–80. Cf. *AT*, X, pp. 252–253; A. Baillet, *La Vie de Monsieur Des-Cartes* (Pais, 1691), Premiere Partie, pp. 67–70. Baillet's description was totally dependent on Lipstorp's.

[136]Federico, p. 117.

[137]Without examining the tract of *Numerus figuratus*, Prouhet and Milhaud mistakenly thought that this tract treated some polyhedral numbers. This error has been corrected by Federico, pp. 116–118.

of his book.[138] According to Roth's calculation, the first 'cubicossic' equation for the trigonal pyramidal numbers is (1𝔠𝔢+ 3𝔷+ 2ꝶ)/6 (in our notation, $P(3,n) = (n^3+3n^2+2n)/6$). The last one for the hectagonal pyramidal number is calculated as (98𝔠𝔢+ 3𝔷−95ꝶ)/6, i. e. $P(100,n) = (98n^3 + 3n^2 - 95n)/6$. These laborious calculations may have been sufficient for one to have obtained the general formula $P(a,n) = \{p(a,n) + n/2\}.(n+1)/3$, whether by induction or with a sufficient proof. This is exactly what Descartes indicates in the *De solidorum elementis* without presenting a detailed explanation (see p. 140).

By our previous observations, we don't want to claim that Descartes derived all his knowledge on the pyramidal numbers from Roth's *Arithmetica philosophica*. Part Two of Roth's treatise itself was not necessarily original. Its shortened title is "Der andere Theil von der Resolution deß Faulhabers gantzen Cubicossischen Lustgartens." It is without doubt that Faulhaber's book which Roth's book of 1608 referred to was *Arithmetischer Cubicossische Lustgarten* published in Tübingen in 1604. we will touch upon this problem later. In turn, the *Numerus figuratus* referred to Peter Roth as one of his predecessors.[139] What is certain is that Descartes was inspired by some mathematicians such as Faulhaber, Roth, or the author of the *Numerus figuratus* to treat figurate numbers, which were a widely argued topic in the early seventeenth century.

Descartes made calculations of the polyhedral numbers which the author of the *Numerus figuratus* and Roth never had tackled. It may have been very natural for one to have a plan to generalize the pyramidal numbers to the polyhedral numbers. But note that Descartes's treatment is confined to certain particular polyhedral numbers, i. e. those corresponding to regular and semi-regular polyhedra. Five regular polyhedra had been well known and were discussed by Descartes in Part One. But semi-regular polyhedra were unfamiliar and had been known generally through Book V of Pappus's *Collection*. Pappus ascribed these polyhedra to Archimedes.[140] It is known that it was Kepler's *Harmonice mundi*, published in 1619, that made known the Archimedean semi-regular polyhedra to mathematicians of the early seventeenth century.[141] The Commandino version did not have any illustrations of the Archimedean solids, while Kepler's book provided beautifully drawn figures of them. (Figures 3.17, 3.18 have been taken from Kepler's *Harmonice mundi*.) There is no evidence in Descartes's text that he consulted Kepler's book. Perhaps we may reasonably suppose that the semi-regular solids were known to mathematicians of the early seventeenth century not only through Kepler's *Harmonice mundi* but also through other writings. We will treat Descartes's probable sources below.

---

[138] Peter Roth, *Arithmetica philosophica*, ff. 126r–129v.

[139] *Op. cit.* (n. 131), p. 15.

[140] Pappus, ed. Hultsch (n. 62), pp. 350–358; Commandino, ff. 33v–34r; Ver Eecke, pp. 272–277.

[141] Johannes Kepler, *Harmonice mundi* (Linz, 1619), in *Gesammelte Werke*, Bd. VI (München, 1940), pp. 83–87; *L'Harmonie du monde*, tr. Jean Peyroux (Paris, 1979), pp. 74–80.

In all events, Descartes calculated the polyhedral numbers for five regular and eleven semi-regular polyhedra. Two of the thirteen Archimedean solids were not touched upon, but these are the only two which cannot be derived by appropriate truncations (cutting off corners) of the regular polyhedra. Federico shows in detail how Descartes calculated the polyhedral numbers. Here it will be sufficient for us to reconstruct his typical method of calculation only for the dodecahedron (out of 5 regular solids) and the icosidodecahedron (out of 11 semi-regular solids).

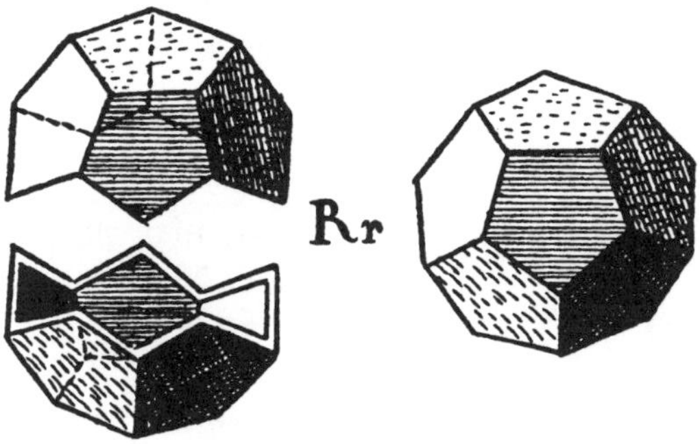

Figure 3.17

Let us first calculate the dodecahedral numbers. (see Figure 3.17.) Descartes obtained them by adding the gnomons constituting the dodecahedron. He provided no sufficient explanation for this method, but we can conjecture it in a manner analogous to the polygonal numbers roughly as follows: The dodecahedron has 12 pentagonal faces, 30 sides, and 20 vertices (solid angles). With one vertex and its neighboring sides and faces vacant, the dodecahedral surface loses 3 faces, 12 sides (among them 9 are reduced to single sides), and 10 vertices. The resultant surface is the gnomon which possesses 9 (= 12 − 3) pentagonal faces, 18 (= 30 − 12), and 10 (= 20 − 10) vertices. Descartes calculates F − R + A (F, R and A stand for the number of faces (*facies*), sides (*radices*) and solid angles (*anguli*), respectively), when the side (*radix*) of the polyhedron n (= 𝔵) equals 1, 2, 3 and 4 as the following figure indicates.[142]

		F − R + A	
		F − R + A	0
Gnomon	$n = 1$	9 − 18 + 10	1
	2	45 − 36 + 10	20
	3	108 − 54 + 10	84
	4	198 − 72 + 10	220

He seems to make calculations of F − R + A for the following reason: When the dodecahedron is divided into pyramids with pentagonal bases, we obtain 9 pentagonal pyramids. But 18 sides are counted twice and hence $28n$ must be subtracted; in this procedure exactly 10 dots too many are subtracted and hence 10 must be added. So in general the dots in the n-th gnomon are $9p(5,n) - 18n +$ 10. Summing these from $n = 1$ to n, we obtain the formula of the dodecahedral numbers: $9P(5,n) - 18p(3,n) + 10n$. In the end, this amounts to $(9n^3 - 9n^2 +$

[142] *AT*, X, p. 270; Federico, p. 96; Costabel, *Exercices*, p. 4.

2)/2. Descartes may have obtained this formula by induction. But he, at the same time, appears to have calculated the polyhedral numbers as generally as possible. This is confirmed by the following case of the icosidodecahedron.

The icosidodecahedron is characterized as the semi-regular Archimedean solid having 20 triangles and 12 pentagons as faces. (see Figure 3.18.) As to this polyhedron, Descartes writes as follows:

> The body formed of 20 triangles and 12 pentagons [has] 60 sides, 30 algles, 32 faces, and its gnomon has 18 triangular and 10 pentagonal faces, minus 48 radices, plus 21 angles.

Gnomon	F +	F −	R + A
	18 +	10 −	48 + 21
	54 +	50 −	96 + 21
	108 +	120 −	144 + 21.[143]

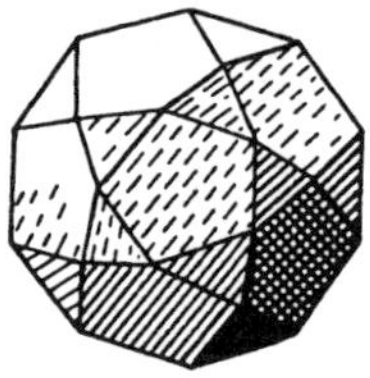

[Figure 3.18]

From the above description, we may suppose that Descartes possesses the formula for the gnomon of the icosidodecahedral numbers: $18p(3,n) + 10p(5,n) - 48n + 21$ in our notation.

Further from this we calculate the icosidodecahedral numbers $18P(3,n) + 10P(5,n) - 18p(3,n) + 21n$, or $8n^3 - 10n^2 + 3n$ in a way analogous to the case of the dodecahedron. Descartes certainly knows this formula, since this is exactly what he provides toward the end of Part Two on figurate numbers.[144]

In his conclusion, Descartes presents a table of formulae for five regular polyhedra and semi-regular polyhedra (out of the eleven semi-regular solids mentioned). The table has five columns: the first column gives the names or the main characteristics of the solids; the second provides the weights (*pondera*) of the polyhedra, i. e. the polyhedral numbers; the third the geometrical weights (*pondera geometrica*), or the volumes; the fourth the major axes (*axes majores*), or the diameters of the circumscribing spheres; the fifth and last the algebraic expresssions for the side of the original solid from which the polyhedron is derived.

These results of figurate numbers appear quite original prima facie. But as we saw above, he probably acquired his knowledge of the pyramidal numbers, for example, from Faulhaber and/or Roth. Further, as a probable source of Descartes's calculations on certain formulae of the regular and semi-regular polyhedra placed at the end of the treatise, Costabel suggests the French mathematician François de Foix-Candale's Latin edition of Euclid's *Elements*, published in Paris in 1566, and its enlarged edition, published in 1578 and 1602. The first 1566 edition comprised Book XIV, which is usually ascribed to Hyspcles,

[143] *AT*, X, p. 274; Federico, p. 106; Costabel, *Exercices*, pp. 5–6.
[144] *AT*, X, p. 276; Federico, p. 108; Costabel, pp. 7 & 35.

and Book XV, just as many Renaissance versions did. These additional books are about the regular solids. Further, the second augmented edition contained Books XVI, XVII and XVIII, which were the editor's own products. According to Costabel, this second edition has detailed discussions on the semi-regular polyhedra, and more importantly not a few resemblances can be found between the last numerical table of Descartes's opusculum and Foix-Candale's version.[145] For this reason, Costabel has added to his monograph the subtitle "Essai en complément d'Euclide." In doing so, he suggests that the *De solidorum elementis* was a work which belonged to the basic trend of the Euclidean tradition and which complemented this intellectual current. Foix-Candale's editions are now almost forgotten, but during the late sixteenth and early seventeenth centuries they were sometimes referred to: For example Clavius borrowed the entire Book XVI of Foix-Candale's for his 1589 edition of Euclid and Mersenne mentioned these versions in 1626.

Admitting this, however, it was Descartes's *De solidorum elementis* in which the polyhedral numbers were calculated, and certain formulae concerning the regular and semi-regular solids were presented systematically for the first time. Certainly, as Costabel informs us, Descartes may have started from a Renaissance version of Euclid's *Elements* and a work or works of Faulhaber and/or Roth. But he went beyond such starting points. And what should be emphasized is that the basic tone of the memoir is algebraic. In the memoir Descartes endeavored to prove the theorem on 'Platonic' solids through a certain algebraic means and to obtain algebraic formulae for solids. It is no accident, then, that cossic symbols appear frequently throughout the tract. It is on metrical aspects and their algebraic formulations that Descartes focused his mathematical mind. He may have attempted to complement the contemporary Euclidean tradition, but it was through fundamentally algebraic tools that he did so.

The *De solidorum elementis* can be also said to be a rather esoteric memoir. It may be entirely free from the mysticism of, for instance Faulhaber. But it treats even the sophisticated polyhedral numbers, especially those of some semi-regular Archimedean solids. We may allow ourselves to call the treatise an esoteric mathematical masterpiece of Descartes's youth.

D. The Date of the Original Manuscript

What is most questionable in Federico's almost definitive work appears to us to be his conjecture concerning the date of Descartes's original manuscript. When did Descartes compose his treatise *De solidorum elementis*? In the Adam-Tannery edition, Adam has placed the opusculum between 1619 and 1621. Milhaud has conjectured more specifically that Descartes wrote it in the winter of 1619–1620 when he is supposed to have visited Faulhaber.[146]

[145] Costabel, *Exercices*, pp. 87–92.
[146] G. Milhaud, *Op. cit.* (n. 132), pp. 84–88.

On the other hand, Federico has suggested the possible date of about 1630. Federico's conjecture is based mainly on the fact that various propositions used by Descartes in Part One about the area of spherical triangles and the measure of solid angles originated from some textbooks of his predecessors. Federico, moreover, has hypothesized that the sources were, for example, Willebrord Snel's *Doctrinae triangulorum canonicae libri quatuor*, published in Leiden in 1627, or Albert Girard's *Invention nouvelle en l'algebre* of 1629.[147] This conjecture on the date of Descartes's manuscript constitutes an important difference between Federico and me.

It is certain, for instance, that Girard treated solid geometry in a satisfactorily rigorous manner and gave such formulae as the spherical excess formula which asserts that the area of a spherical triangle is equal to the sum of its three angles less two right angles.[148] As Federico has pointed out, such a formulation would be necessary for a rigorous and thorough treatment of such concepts as solid and exterior solid angles.[149] But Descartes's reference to the technical terms in solid geometry was only for special cases like the right solid angle. Moreover, his demonstrations were in general not thorough, but rather analogical and inductive. It might be said that the background knowledge which Federico has ascribed to Descartes is necessary only in the eyes of today's methematicians. Some elementary textbooks of solid geometry may have been sufficient for Descartes to write the manuscript of the *De solidorum elementis*.

Federico's possible error about the date of the original draft seems to be caused by his minimal grasp of Descartes's overall mathematical career. Around the year 1630, to which Federico has ascribed the date of the original manuscript, Descartes's main interest was no longer mathematics but studies in natural philosophy (even though their main characteristic was mathematical) and metaphysics. At this juncture, Descartes's letter to Mersenne on March 31, 1638, which we have already quoted in the Introduction (p. 3), becomes crucial. In that letter, he stated that in principle he had ceased to be fully occupied 15 years before with mathematical studies undertaken at his own initiative. In other words, if his statement is trustworthy, his essential mathematical activities had become marginal by about 1623. However, a remark should be made here. There still remains the problem of determining whether or not Descartes's autobiographical statement is correct. This question can be answered, strictly speaking, only after we have examined Descartes's activities after 1623, more concretely the mathematical contents of the *Regulae ad directionem ingenii* and the *Géométrie*. In fact, as we will see in Chapters 4 and 5, Descartes exhibited considerable mathematical progresses in these two works. But, if we are

[147]Federico, p. 32.

[148]Federico, p. 39. Cf. Albert Girard, *Invention nouvelle en l'algebre* (Amsterdam, 1629); Reprinted edition by D. Bierens de Haan (Leiden, 1884), "De la mesure de la superface des triangles & polygones spheriques, nouvellement inventé," ff. Gv-[H4]v.

[149]Federico, pp. 32 & 39.

permitted to state in advance the results of our examination of these works the developments which he made in them may be interpreted as fundamentally conceptual, in other words, they are changes (even though radical) with respect to basic notions, and the Pappus problem which he solved in 1631–1632 was one which had been proposed by another mathematician and was not undertaken on his own initiative. On the other hand, the *De solidorum elementis* is a considerably lengthy treatise and contains sophisticated results. And this can be safely said to be a work undertaken on Descartes's own initiative. Accordingly, it would not be reasonable to assume that he gave his time to such a specific field of mathematics after 1623. Of course, our final answer should be postponed to the end of Part II of this book.

One could raise another possibility as to the date of the manuscript, that is to say, around the year 1638. Around the summer of 1638, Descartes wrote a short note under the title "Numeri polygoni" (Polygonal Numbers),[150] in which he developed his arguments on number theory by using cossic symbols. By composing such a memorandum, whose topic is surely related to the subject of Part Two of the *De solidorum elementis*, he seems to have intended to respond to challenges of about 1638 by contemporary mathematicians such as Fermat. This can be ascertained from Descartes's letters to Mersenne in June and July of 1638.[151] But, judging from his aforementioned statement in 1638 on his mathematical career, it would be unwarranted to suppose that he wrote the *De solidorum elementis* about the same time as the memorandum "Numeri polygoni." He might have used cossic signs in the note "Numeri polygoni" of about 1638 while thinking over his treatise of his youth on the polygonal numbers.

Thus, the most probable date of the original manuscript can be safely restricted to the period between 1619 and 1623. We have already pointed out that there are some resemblances between the *Cogitationes privatae* and the *De solidorum elementis*, and that Peter Roth, whose book *Arithmetica philosophica* was cited in the former, treated the same subject as Part Two of the latter treatise. Thus we can find a certain continuity between the two manuscripts.

Furthermore, in the text of the *De solidorum elementis* we encounter an immature passage on the dimensionality of 𝔵, 𝔷, and 𝔠𝔢. After giving some examples of the polygonal weights, Descartes makes the following observation:

> If one considers these figures as measurable, then all the units are

[150]This note was first published posthumously as a part of 'Excerpta ex Mss. R. Des-Cartes," in *Opuscula posthuma, physica et mathematica* (Amsterdam, 1701), p. 4. Cf. *AT*, X, pp. 297–299. Notice that the title "Numeri polygoni" was an addition by the editor of the *Opuscula*. In the 1701 *Opuscula* version, the cossic signs 𝔵 and 𝔷 were rewitten as x and xx, respectively.

[151]"Descartes, à (Santpoort), à Mersenne, à Paris. (3 juin 1638)," *AT*, II, pp. 158–168; *AM*, II, pp. 278–285; *CM*, VII, pp. 257–267. "Descartes, à (Santpoort), à Mersenne, à Paris. 27 juillet 1638," *AT*, II, pp. 25–257; *AM*, II, pp. 354–355; *CM*, VII, pp. 405–407. Cf. L. E. Dickson, *Op. cit.*(n. 130), pp. 6–7.

> understood as being of the same kind as the figures themselves: that is for triangles a triangular unit; pentagons are measured by a pentagonal unit, etc. Then the proportion between a plane and its radix is the same as the square to its radix; and a solid as a cube: so if the radix is 3, the plane will be 9, the solid 27, etc., for example. This holds also for the circle and the sphere, and all other figures. For if the circumference of a circle is three times larger than another, the area of the first will be nine times larger. From which it is observed that our mathematical progression, 2e, ʒ, ce, etc., is not attached to linear, square, cubic, figures, but is designated generally by the diverse species of measure.[152]

What Descartes means here is a little obscure, but his intention in writing this statement may be clarified by considering a concrete example. Let us take the case of the pentagonal weights. Following Descartes's directions for finding the polygonal numbers, we obtain $p(5,n) = (3n^2 - n)/2$ as the pentagonal weights. In this formula, he seems to say that n^2 and n should not be interpreted as attached to square and linear figures, respectively. Instead, according to him, the formula for the pentagonal numbers $p(5,n) = (3n^2 - n)/2$ must be measured as a whole by a pentagonal unit. The logical implication of what he intends must be as we have just interpreted it. However, is such a conception suitable for the mature Descartes, who had already published his *Géométrie*, at least, or who had composed even his *Regulae ad directionem ingenii*? His thought on dimensionality had become quite simplified in his later days. Thus the *De solidorum elementis* does appear to be a work of his youth. Consequently, we maintain that Adam's conjecture is very probable. Milhaud has ascribed the date of the composition to the winter of 1619–1620 during which Lipstorp reported that Descartes met Faulhaber. This hypothesis may be possible but seems to be restrictive. We at least tentatively hold the date of the *De solidorum elementis* to have been the years from 1619 to 1623, for safety a longer period than Adam, when Descartes was studying mathematics very earnestly and looking for the solid foundations of his "entirely new science". Costabel's considerations basically accord with ours.[153] Be that as it may, it can be said that he, by composing the *De solidorum elementis*, shows that his exceptional ability in mathematics should be ranked with other leading mathematicians of that day.

---

[152] *AT*, X, p. 271; Federico, p. 95.

[153] Costabel, *Exercices*, pp. 103–109. I have arrived at my conclusion independently of Costabel.

§ 4. Descartes's Mathematical and Philosophical Dream of 1619

So far we have pursued the development of Descartes's mathematical thought following his encounter with Isaac Beeckman in 1618. The aim of this section is to recapitulate his mathematical and philosophical thought of the period from 1619 to 1623, especially in relation to that of his later days.

The year 1619 is known to have been extremely important not only for the turn of Descartes's mathematical career but also for his conception of a significant philosophical idea based on his previous mathematical research. According to his autobiographical statement in Part II of the *Discours de la méthode*, at twenty-three years of age he conceived a definitive program of an intellectual reform which would be fully developed later. At times this overall plan has been interpreted in connection with the three dreams which he is supposed to have had on the night of November 10–11, 1619. Although his manuscript *Olympica* in which he is thought to have written his account of these dreams is not extant, we know their content from Adrien Baillet's *La Vie de Monsieur Des-cartes*, which excerpted part of the *Olympica*.[154] We may discuss the three dreams using Baillet's biography of Descartes. But here we are concerned not with the dreams themselves but simply with his mathematical project closely interrelated to the philosophical one, which he seems to have had in 1619. So the word 'dream' in the title of this section is only metaphorical.

Now, the place where Descartes dreamed at a *poèle* in the late autumn of 1619 is believed to have been the Duchy of Neuburg an der Donau rather than Ulm. This theory has become an almost established fact from the discovery of a copy of Pierre Charron's *De la Sagesse* on skepticism dedicated to "the most learned grateful friend and younger brother René Descartes" with the date of "the end of 1619" by a Jesuit living in Neuburg.[155]

But, what about a story on Descartes's visit to Ulm which was at times supposed to connect him with Faulhaber? Recently an attention has been paid to the letter of Johann Baptist Hebenstreit in Ulm, once rector of a gymnasium, to Johannes Kepler in Linz, which is conjectured to have been composed on February 1, 1620. The beginning part of that letter can be read as follows:

> I am not sure whether a certain Cartelius, a man of extreme erudition and of extraordinary humanity, will revisit you with mine. Surely, sometimes I suffer from troubles caused by unpleasant and

[154] Baillet, *La Vie de Monsieur Des-Cartes*, t. 1, pp. 80–86.

[155] Frédéric de Buzon, "Un exemplaire de la *Sagesse* de Pierre Charron offert à Descartes en 1619," Bulletin cartésian XX: *Archives de philosophie*, **55** (1992), pp. 1–3: "Doctissiomo Amico grato et minori fratri Renato Cartesio d. d. ded. P. Johannes B. Molitor S. J. exeunte Anno 1619 JBM." Cf. G. Rodis-Lewis, *Descartes* (Introduction, n. 1), pp. 71 & 330, n. 42. The theory to ascribe the location of the *poèle* to Neuburg began with A. Baillet, *Vie de Monsieur Descartes*, rédite et abregé (Paris, 1692): Collection "Grandeurs" (Vanves, 1946), p. 33.

> impludent beggars through friends. But, Cartelius seems to be different, and very refined, so you should support him by advice.[156]

We have no further evidence to determine that "Cartelius" in this letter is identical with Cartesius, the Latin equivalent of Descartes. Moreover, the letter had a date of "February 1", but the year "1620" is a conjecture of the editor. Thus, the identification of "Cartelius" with Descartes isn't definitive, but there are reasons for thinking so from the content of the letter.[157]

Even if "Cartelius" in the aforementioned letter isn't Descartes himself, it may be certain that Descartes was strongly influenced by Southwest German mathematicians around Ulm from the *Cogitationes privatae* and the *De solidorum elementis*. Ulm was known a place where produced good mathematicsans since the late Middle Ages, as evidenced the saying that "Ulmenses sunt mathematici." If we consider it to be a fact that Descartes was staying in Neuburg from the late autumn to the end of 1619, we may conjecture naturally that immediately afterwards he also visited Ulm, about 100 kilometers from Neuburg. According to Ivo Schneider's *Johannes Faulhaber, 1580–1635: Rechenmeister in einer Welt des Umbruchs*, published in 1993, Roth's *Arithmetica philosophica* was written to respond the *Arithmetischer Cubicossische Lustgarten*, published by Faulhaber in 1604, and the *Numerus figuratus* of 1614 and Faulhaber's *Miracula arithmetica* of 1622 were published under the influence of Roth's monograph.[158] Today, the author of the *Numerus figuratus* is ordinarily thought not as Faulhaber himself but his disciple Johannes Remmelin in Ulm.

In any event, it is clear that having been stimulated by writings of mathematicians near Ulm for studying rather sophisticated mathematical studies Descartes drafted some parts of the *Cogitationes privatae* and the *De solidorum elementis*. There we witness elements different from what he had acquired when he encountered with Beeckman. Remember that Descartes referred to the Rosicrucian brothers in his planned book entitled *Thesaurus mathematicus*. It was known that Faulhaber, representative *Rechenmeister* in Ulm, composed a writing related to the Rose Croix.[159] And Roth's *Arithmetica philosophica* was referred to explicitly in the *Cogitationes privatae*. In his *Miracula arith-*

[156]Kepler, *Gesammelte Werke*, Bd. XVII: *Briefe 1612–1620* (München, 1955), p. 416: "Nescio utrum Cartelius quidam, vir sané eruditus, et humanitate singularj, meas tibj reddiderit: Per equidem invitus amicos molestiâ illâ afficio, quam interdum ipsis afferunt ingratj et impudentes aeruscatores. Cartelius tamen alius videbatur, quemque consilio juvares dignissimus." A photograph copy of the original letter can be seen in K. Hawlitscheck, *Johann Faulhaber* (n. 134), p. 59.

[157]Hawlischeck, *Loc. cit.* (n. 134). Cf. Schneider (n. 158, below), the Edior's note to the letter in *Gesammelte Werke*, Bd. XVII, p. 516, and Kenneth Manders, "Descartes et Faulhaber," *Archives de philosophie*, **58** (1995), Bulletin cartésien XXIII, pp. 1–12.

[158]Ivo Schneider, *Johannes Faulhaber 1580–1635: Rechenmeister in einer Welt des Umbruchs* (Basel/Boston/Berlin, 1993).

[159]F. A. Yates, *The Rosicrusian Enlightenment* (n. 45), pp. 114–115. Cf. Hawlischeck, *Op. cit.* (n. 134), pp. 50–54.

metica Faulhaber made attempts to extend the theorems of Pythagoras and of Hero.[160] From these only we might not determine hastily that Descartes visited Ulm and met Faulhaber there. Nevertheless, to imagine so may be a quite natural conjecture, and it is certain that at least he met mathematicians surrounding Faulhaber, or even read books by them. It might be possible that he was taught the abovementioned mathematical results by Faulhaber. It might be also possible that some results in the *Cogitationes privatae* were simply copies from Faulhaber. In reverse, Descartes could have taught Faulhaber. What is certain is that from the autumn of 1619 Descartes had the same mathematical interest as the German cossists or *Rechenmeister* (a representative of whom was Faulhaber) who absorbed and developed the knowledge of higher algebra, including the solution of cubic equations since Girolamo Cardano, and that he obtained his own good results. This seems to have been an event in his career comparable his encounter with Beeckman.

Now, we return to our discussion on the significance of Descartes's mathematical studies during the years of 1619–20. He certainly produced mathematical results of a higher degree stimulated by his encounter with books by Faulhaber and Roth. Representative examples of these may have been a new challenge to solve cubic equations and an attempt to extend the Pythagorean theorem in the *Cogitationes privatae* and the *De solidorum elementis* as a whole. But his mathematical fruits were different from those by the German mathematical practitioners. Toward the end of his biography on Faulhaber, Schneider paid attention to that Descartes's style had a radical originaity sometimes seen only in amateur mathematicians and contrasted to the esoteric style of the German professional *Rechenmeister*. But we shouldn't forget that his innovative and plain style was simply in a germinative stage in this period, and that it would become evident only in later years. Still, it cannot be denied that its germ appeared in these years. The mathematics of Faulhaber and others around him was colored by a kind of esoterism typically seen in the Renaissance mathematics. In contrast, in Descartes's mathematics we sense a critical consciousness of this kind of esoterism. We have a book entitled *The Counter-Renaissance* by Hiram Haydn, publsihed in 1950.[161] In Descartes's thought as a whole we can surely sense a rather strong will for "the Counter-Renaissance" or "the Anti-Renaissance." This difference in the styles of mathematics seems to have been related to the fact that Descartes was deeply interested in philosophy as well as in mathematics. Accordingly, we consider the significance of the mathematical research for several years beginning with 1619 in the entire mathematical-philosophical career of Descartes.

To begin with, let us consider Descartes's own statement of 1637. According to it, at the onset of winter of 1619 in Germany he stayed all day shut up

[160] Schneider, *Op. cit.* (n. 158), pp. 123–129.

[161] Hiram Haydn, *The Counter Renaissance* (n. 45). See also H. Gouhier, *Les Premières Pensées de Descartes: Contribution à l'histoire de l'Anti-Renaissance* (n. 43).

alone in a *poèle* and reviewed his own ideas. He contemplated his need to reform his thought completely, just as an architect planned and remodelled buildings beginning from their foundations. It was to this end that he contrived the well-known four rules of his method, which was comprised of the advantages of logic, geometrical analysis, and algebra, but free from their defects. The rules consisted of, in sum, (1) never to accept anything as true if it is not judged as clear and distinct, (2) to divide a problem into parts which can be better understood, (3) to recompose the parts into knowledge of the most complex in an orderly manner, (4) to complete enumeration.[162]

It was his previous mathematical studies—Descartes continues in his statement on his career—from which he acquired a clue for these four rules. He had no difficulty in finding the most simplest and most easily known science: "Mathematicians alone have been able to find demonstrations—that is to say, certain and evident reasonings—I had no doubt that I should begin with the very things that they studied."[163]His answer was that "the simplest and most easily known science" is mathematics and that he started from it. But the mathematics suitable for his ideal was not to be found among the existing mathematical disciplines. Then he explains what he thought to be the essential matters in mathematics:

> From this, however, the only advantage I hoped to gain was to accustom my mind to nourish itself on truths and not to be satisfied with bad reasoning. Nor did I have any intention of trying to learn all the special sciences commonly called 'mathematics' (*mathématiques*). For I saw that, despite the diversity of their objects, they agree in considering nothing but the various ratios or proportions (*rapports ou proportions*) that hold between these objects. And so I thought it best to examine only such proportions in general, supposing them to hold only between such items as would help me to know them more easily. At the same time I would not restrict them to these items, so that I could apply them the better afterwards to whatever others they might fit. Next I observed that in order to know these proportions I would need sometimes to consider them separately, sometimes merely to keep them in mind or understand many of them together. And I thought that in order the better to consider them separately I should suppose them to hold between lines, because I did not find anything simpler, nor anything that I could represent more distinctly to my imagination and senses. But in order to keep them in mind or understand several of them together, I thought it necessary to designate them by the briefest possible symbols (*chiffres*). In this way I would take over all that is best in

[162] *Discours de la méthode*, in *AT*, VI, p. 18.
[163] *AT*, VI, p. 19.

> geometrical analysis and in algebra, using the one to correct all the defects of the other.[164]

This passage indicates that Descartes had to cultivate a normative mathematical discipline to attain his ideal goal of evident reasoning. This normative science, according to his description, treats proportions. What was Descartes's normative discipline like? Proportions are mathematical relations which cover both numbers and continuous magnitudes. He clearly thought of a common discipline applicable not only to the arithmetical theory of proportion exemplified in Euclid VII but also to the geometrical one in Euclid V. Proportions can easily be rewritten into algebraic equations by those who are familiar with algebra. So if the normative science of Descartes was algebra, we can well understand the quoted passage as a whole.

The identification of Descartes's normative discipline with algebra can be confirmed by a fact which we have cited in Chapter 2, § 5 (p. 92). What he endeavored to study in earnest among the mathematical sciences before 1618 was algebra. This is reconfirmed by Descartes's assertion in his conversation with Frans Burman in 1648. The mature Descartes commented on his *Discours de la méthode* and emphatically told the young Dutchman Burman, then only twenty years old: "[O]ne needs a mathematical mind which must then be polished by actual practice. Now this mathematical knowledge must be acquired from Algebra."[165] In Descartes's judgment, a mathematical mind must be formed above all by practicing algebra. This is truly an important statement.

But the benefit which can be obtained from his normative discipline was not to be derived from vulgar algebra. Descartes's ideal science must be a reformed algebra. How then can reformed algebra be characterized? As can be comprehended from the later part of the passage cited above, this discipline uses "the briefest possible symbols" and importantly is capable of playing the part of geometrical analysis. Thus Descartes's reformed algebra was to be furnished with an orderly symbolism and, as a tool, may have been called algebraic analysis. It must also have been identical to "the true mathematics" which in Rule IV of his *Regulae ad directionem ingenii* he stated he had cultivated from the vestiges left by Pappus and Diophantus (see p. 91).

Descartes tried to apply his newly acquired rules, at the outset, to difficult mathematical problems. He continued his narration on these rules immediately after the quoted paragraph:

> In fact, I venture to say that by strictly observing the few rules I had chosen, I became very adept at unravelling all the questions which fall under these two sciences [i. e. geometrical analysis and

[164] *Ibid.*, pp. 19–20.

[165] *AT*, V, pp. 176–177: "[...] requiritur ad id ingenium mathematicum, quodque usu poliri debet. Ea autem haurienda ex Algebrà." See *Descartes's Conversation with Burman*, translated with Introduction and Commentary by John Cottingham (Oxford, 1976), pp. 47–48.

> algebra]. So much so, in fact, that in the two or three months I spent in examining them—beginning with the simplest and most general and using each truth I found as a rule for finding further truths—not only did I solve many problems which I had previously thought very difficult, but also it seemed to me towards the end that even in those cases where I was still in the dark I could determine by what means and to what extent it was possible to find a solution.[166]

By the winter of 1619–1620, if his statement is trustworthy, Descartes had already conceived at least the germ of his reformed algebra, and from that time on he was determined to begin to consider the possibility of generalizing his rules to problems of the other sciences. He gives all the details of this intellectual situation:

> But what pleased me most about this method was that by following it I was sure in every case to use my reason, if not perfectly, at least as well as was in my power. Moreover, as I practised the method I felt my mind gradually become accustomed to conceiving its objects more clearly and distinctly; and since I did not restrict the method to any particular subject-matter, I hoped to apply it as usefully to the problems of the other sciences as I had to those of algebra. Not that I would have dared to try at the outset to examine every problem that might arise, for that would itself have been contrary to the order which the method prescribes. But observing that the principles of these sciences must all be derived from philosophy, in which I had not yet discovered any certain ones, I thought that first of all I had to try to establish some certain principles in philosophy. And since this is the most important task of all, and the one in which precipitate conclusions and preconceptions are most to be feared, I thought that I ought not try accomplish it until I had reached a more mature age than twenty-three, as I then was, and until I had first spent a long time in preparing myself for it.[167]

Here Descartes expresses his will to apply what he had gained from his research of mathematics, particularly of algebra, to the other fields of knowledge in a more general manner. But in the winter of 1619-1620 he had only just begun to understand the core of his general philosophical project, as he himself asserts.

Reflecting on how his philosophical thought developed, can we say something similar about his view of mathematics? We should notice that Descartes's account in the *Discours de la méthode*, written after about 18 years, cannot perhaps be taken to give the exact state of his thought of 1619. At least, it would be

---

[166] *AT*, VI, pp. 20–21.
[167] *Ibid.*, pp. 21–22.

safe to say that hindsight must be mixed with his statement of 1637. According to his account of 1637, by the winter of 1619–1620 he had already obtained the essential core of his algebra in which proportions, or equations, holding between lines play the principal part and which is to be carried out using "the briefest possible symbols." Here we have to discriminate between what he claimed in 1637 and what can be said about the actual state of his mathematical thought from the extant literature. It seems to us that the above characterization of his algebra of 1619 is an overstatement. What Descartes meant by his algebra in 1637 appears to be not the algebra of 1619 but rather the one in the *Géométrie* of 1637.

Descartes's mathematical thought of the period immediately following his meeting Beeckman, as recorded explicitly in the extant documents, was somewhat more moderate than his description of later years. It is certain that he started to cherish a rather ambitious program which would unify arithmetic and geometry as early as March 26, 1619. In order to accomplish this goal, he tried to solve the cubic equation in a tolerably successful manner (albeit along with systematic mistakes) with the aid of his newly devised mesolabe compass. The symbolism of his algebra in this period was fundamentally cossic and can be viewed to have been under the influence of the 1608 Clavius *Algebra*, although some novel elements can be detected toward the end of his *Cogitationes privatae*. Unfortunately, so far as the available documents inform us, he did not leave any indication as to how he would formulate his solution of the cubic equation algebraically, other than the one using the mathematical instrument, or how he would overcome the dimensionality of geometrical magnitudes.

But at the same time his mesolabe compass can be considered to have been in some way related to his plan for reforming algebra. As Jules Vuillemin has emphasized in his *Mathématiques et métaphysique chez Descartes*, Descartes's compass constitutes a very crucial step in his theory of proportion, that is to say, his algebra.[168] It was by this instrument that he could make any of the geometrical progressions signified by the cossic symbols $\mathfrak{x}$, $\mathfrak{z}$, $\mathfrak{ce}$, etc. give a clear visual representation. Furthermore in this period, we conjecture that he established himself as a creative mathematician to the extent that he ingeniously took on problems of solid geometry and of figurate numbers to obtain unprecedented results of high quality.

It is certainly undeniable that by the winter of 1619–1620 he could have already possessed his plan of a reformed algebra capable of serving as an algebraic analysis, at least in a germinal form in his mind. It may simply not have been recorded in an explicit form. This is why we may justly call his 1619 program for the unification of the mathematical disciplines a 'mathematical dream'.

According to a general observation on Jesuit mathematical thought of the period of our concern, algebra was underdeveloped among the mathematical

---

[168] Jules Vuillemin, *Mathématiques et métaphysique chez Descartes* (Paris, 1960), Chapitre IV: "La théorie des proportions," pp. 115–117.

sciences in the early seventeenth century, even though the Jesuits had a textbook of algebra by Clavius.[169] Descartes, who began with Clavius's *Algebra* as the starting point, travelling in the Netherlands and Germany, came out of the mathematical domain of the Society of Jesus and was to make great progress. Just as he was parting with the scholastic system of learning by establishing the universal method in philosophical thought in general, in the field of mathematical thought he was bidding farewell to the school mathematics of the Jesuit Order. This is the context for his idea of "an entirely new science" and the treatise entitled *De solidorum elementis*. What we have is a preliminary program, rather than the fruit of mature thought.

As a commentator on the dreams which Descartes is supposed to have had on November 10–11, 1619 has pointed out, they can be persuasively interpreted as prophetic.[170] Prophetic dreams are claimed by audacious young men who possess an ambitious plan but whose goal is remote. Descartes might have been such an intellectual youth. In the *Cogitationes privatae*, he writes the following note:

> In November 1619, I had a dream involving the Seventh Ode of Ausonius, which begins *Quod vitae sectabor iter?* ['What road in life shall I follow?'].[171]

In November 1619, Descartes might have questioned "Quod vitae sectabor iter?" above all concerning his philosophical plan. He seems to have asked this question about his mathematical career to some extent as well. It is certain that at the time of the winter of 1619–1620, the foundations of his mathematical program had become more solid than those of his philosophical program. Because Descartes's program of reforming mathematics and his philosophical 'dream' were closely interrelated, his mission or ambition to accomplish that 'dream' may have led him to transform dreams into substance near Neuburg.

The germ of the program of learning which Descartes later brought to flower emerged in 1619 while travelling from the Netherlands to Germany in the midst of the turmoils of the Thirty Years War, which had just started. His ideas grew, and the central part of the first comprehensive treatise of his reformed algebra was conceived, perhaps before 1623. His *Regulae ad directionem ingenii*, the first monograph on his general rules, was drafted much later, probably by

[169] Jean Dhombres, "Une Mathématique baroque en Europe: réseaux, ambitions at acteurs," in *L'Europe mathématique: Histoires, mythes, identités*, sous la direction de Catherine Goldstein, Jeremy Gray et Jim Ritter (Paris, 1996), pp. 144–181, at p. 159.

[170] Alice Browne, "Descartes's Dream," *Journal of the Warburg and Courtauld Institutes*, **40** (1977), pp. 256–273. In this regard, Sebba fundamentally agrees with Browne. See Gregor Sebba, *The Dream of Descartes*, Assembled from Manuscripts and Edited by Richard A. Watson (Carbondale and Edwardsville, Ill., 1987), p. 7: "What Descartes discovered was his mission, not his philosophy." Cf. John R. Cole, *The Olympian Dreams and Youthful Rebellion of René Descartes* (Urbana/Chicago, 1992).

[171] *AT*, X, p. 216; Translation (n. 42), p. 4.

about 1628. These will be the subject of the next chapter.

Chapter 4

The Mathematical Background of the *Regulae ad Directionem Ingenii*

§ 1 The Old *Algebra*: The First Fruit of "An Entirely New Science"

Descartes's program of 1619 for reforming mathematics (see Chapter 3, § 1) was embodied in his manuscript titled *Algebra* . Descartes referred to this writing for the first time, as far as we know, in October of 1628 when he visited his old friend Beeckman, who was then in Dordrecht. Beeckman took notes in his *Journal* describing the content of his discussions with Descartes, just as he had done in 1618–1619. This document is extremely important for understanding Descartes's mathematical thought during the period between 1619 and 1628. In it Descartes told of his nine year trek which began in 1619, the year when the two scholars bade farewell to one another. Under the margin subtitle "History of Descartes and of His Relationship with Me," Beeckman writes as follows:

> M. René Descartes du Perron, who in 1618 in Breda of Brabant wrote for me the *Compendium of Music*, in which he revealed his view on music and which has been inserted into this *Journal*. He, I say, came to Dordrecht to call on me on October 8, 1628, after he had gone first to Middleburg from Holland in order to look for me there. He told me that insofar as arithmetic and geometry were concerned, he had nothing more to discover; that is, in these branches during the past nine years he had made as much progress as was possible for the human mind. He gave me perspicacious specimens of this and promised to send me his *Algebra* a little later from Paris, which, he said, was finished and by which not only had he arrived at a perfect knowledge of geometry but also he claimed to embrace all human knowledge. Perhaps he will come here to perfect and publish it so that together we may complete the study of what remains to be

> discovered in the sciences. Indeed, after travelling through Germany, France and Italy, he affirms that he has not found anybody with whom he is able to discuss from the bottom of the heart and from whom he can hope for aid in his researches besides me. So he says that everywhere there is a lack of true philosophy, which he calls the work of devotees. In truth I prefer him to all the arithmeticians and geometers whom I have ever seen and read.[1]

Beeckman's note indicates that during the nine years from 1619 to 1628, Descartes claimed that he had "arrived at a perfect knowledge of geometry" and that he would eventually encompass the whole of human knowledge. Significantly it was in a completed treatise called *Algebra* that Descartes claimed to have "had nothing more to discover" as to arithmetical and geometrical knowledge. According to Beeckman's note, Descartes promised to send this treatise from Paris to the Dutch scholar.

Descartes seems to have kept his promise at least in part, for in a letter to Beeckman of October 17, 1630, he referred to his treatise as "the *Algebra*, which I have given to you."[2] But, it isn't necessarily clear that the he gave to his elder friend the entire manuscript of his *Algebra*. This monograph was mentioned again as "ma vieille Algebre"[3] in his letter to Mersenne which was written in January of 1638 (see below p. 174). Descartes's old *Algebra* was no doubt identical to the manuscript listed in the Stockholm Inventory as the article "D. Un petit registre in octavo, contenans cent cinquante cinq pages, ou il semble avoir escrit pour son usage une introduction contenans les fondemens de son algebra."[4] Descartes's autograph is not to be found on this treatise, nor have

---

[1] *AT*, X, p. 331; *Journal*, t. III (La Haye, 1945), pp. 94–95: "{Historia Des Cartes ejusque mecum necessitudo.} D. Renatus Des Cartes Du Peron, qui anno 1618 in meam gratiam Bredæ Brabantinorum *Musicæ Compendium* conscripsit, quo suam sententiam de musicâ mihi aperuit quodque huic operi insertum est, is, inquam, die 8^0 mensis Octobris 1628 ad me visendum venit Dortrechtum, cùm priùs frustra ex Hollandiâ Middelburgum venisset, ut me ibi quæreret. Is dicebat mihi se in arithmeticis et geometricis nihil ampliùs optare, id est se tantum in ijs his novem annis profecisse quantum humanum ingenium capere possit. Cujus rei non obscura mihi specimina reddidit, paulò post Parisijs suam *Algebram*, quam perfectam dicit quàque ad perfectam geometriæ scientiam pervenit, imò quâ ad omnem cognitionem humanam pervenire potest, propediem ad me missurus aut ipsemet huc ad eam edendam et limandam venturus, ut communi operâ id quod restat in scientijs, perficiamus. Galliâ enim, Germaniâ et Italiâ peragratâ, dicit se non invenisse alium, cum quo secundùm animi sui sententiam disserere et â quo adjumentum in studijs suis sperare possit, quàm per me. Tantam dicit esse ubique inopiam veræ philosophiæ, quam vocat operam navantium; ego verò illum omnibus, quos unquam vidi aut legi, arithmeticis et geometris præfero." I refer to Jack Rochard Vrooman's partial translation in his *René Descartes: A Biography* (New York, 1970), p. 78.

[2] "Descartes à Beeckman. Amsterdam, 17 octobre 1630," *AT*, I, p. 159; *AM*, I, p. 151: "[...] *Algebram*, quam tibi dedi, [...]."

[3] 'Descartes, à (Santpoort), à Mersenne, à Paris. (25 janvier 1638)," *AT*, I, p. 501; *AM*, II, p. 84; *CM*, VII, p. 39.

[4] *AT*, X, p. 8.

any copies of it been discovered.[5] But fortunately a part of its content was recorded under the title "A Certain Specimen of Descartes's Algebra" a little after the above quoted paragraph in Beeckman's *Journal*. We will examine it below.

According to Beeckman's statement cited above, the two men meeting after the nine year interregnum recognized one another as distinguished scholars. The elder especially praised the younger as the best mathematician he had ever met. Beeckman's praise of Descartes continues with the margin subtitle "Why the Learned Are Few":

> I certainly believe that the reason why there are so few really learned men is that those who are gifted for the sciences, as soon as they have made a discovery, are eager to write and are not content to publish what they have discovered, but seizing the occasion write their recent works together with scientific discoveries from the beginning to the extent that they oppress their intelligence, which is perfectly capable of many discoveries, by undertaking labor which is neither useful nor new. This man, on the contrary, has still written nothing, but meditating until the age of thirty-three, he seems to have found what he was looking for better than the others. This should be said so that one will imitate him rather than the host of scribblers.[6]

Beeckman was impressed by the mature Descartes of thirty-three years, who had been making steady progress forging his idea related to "an entirely new science" about which the March 26, 1619 letter described. Descartes seemed to him one of those ideal scholars who kept original ideas until they were fully developed. Nine years had passed since the last communication between the two "physico-mathematicians." In the fall of 1628, the younger man must have talked with self-confidence about his ideas on mathematics and of natural philosophy. He might have appeared to the elder man as if he were adhering to Horace's maxim: "Keep back till the ninth year."[7]

Now let us examine Descartes's "old *Algebra*" as it is recorded in Beeck-

---

[5]The editors of the *AT* edition states, "Cette Algèbre, qui datait de 1618–19, n'a pas été retrouvée." *AT*, I, p. 168. But they do not give the reason why this manuscript has been ascribed to the years 1618–1619.

[6]*AT*, X, p. 332; *Journal*, t. III, p. 95: "{Docti cur pauci.} Causam verò cur tam pauci hîs versatissimi sint, esse existimo quia omnes qui ingenio tali pollent, ubi se aliquid invenisse autumant, statim scripturiunt, nec tantùm ad quod invenere edunt, verùm eam occasionem arripientes, nova opera scientiasque ab ovo conscribunt, atque ita suum ingenium, ad plurima perfectè invenienda aptissimum, multitudine laboris non utilis aut novi obruunt. Ille verò necdum quicquam scripsit, sed usque ad 33 ætatis suæ annum meditando, eam rem quam quæsivit, perfectiùs quàm reliqui invenisse videtur. Hæc dicta sunto ne quis potiùs numerum scripturientium quàm illum imitetur."

[7]Horatius, *Ars poetica*, l. 388: "Nonumque prematur in annum." Descartes later cited this saying in his letter toward the end of November of 1633 to Mersenne. *AT*, I, p. 272; *AM*, I, p. 242; *CM*, III, p. 559.

man's *Journal* under the title "A Certain Specimen of Descartes's Algebra." What attracts our attention concerning the beginning part of that memorundum is that Descartes seems to have taken pains to unify arithmetical numbers and geometrical magnitudes. The latter quantities have dimensions, so a crucial problem for him in his treatise must have been treating their dimensionality. How does Descartes handle the most fundamental object of his algebra? Beeckman writes that he argues about how quantities in general are to be represented as follows:

> The same man [Descartes] states that he discovered a general algebra, and for it he uses not solid figures but simply planes, for the mind penetrates them more easily. And thus the other [mathematical] things, in addition to geometry, are best represented by these.
>
> He denotes unity by a small square; in this manner he denotes a point as well. He denotes a line or root by reducing the breadth of a square into a parallelogram. His square is formed by the same number of roots; and a cube, made by as many squares as numbers indicate, reduced to an oblong form; a biquadratic in the same manner etc. Or rather, he further explains all these by lines so that *a* may represent a point, *b* a line, *c* a square, and *d* a cube. In this manner *f* also represented a cube constructed from the multiplication of a square *e* by the number of roots [Figure 4.1].

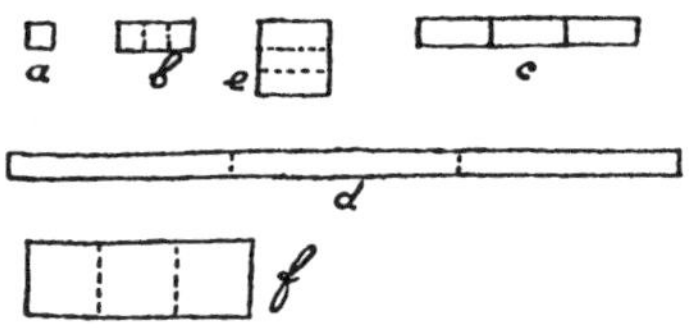

[Figure 4.1]

> And with greater trouble, he solved the same [problem] by simple lines just as is seen here in the margin, where cossic signs are added to individual lines, lines signifying those quantities which are prefixed [Figure 4.2].[8]

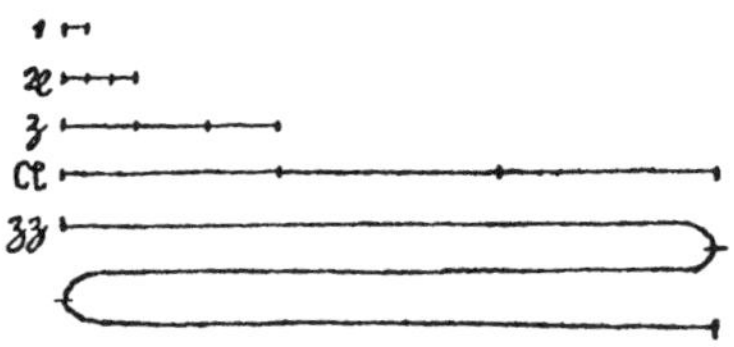

[Figure 4.2]

[8] *AT*, X, pp. 333–334; *Journal*, t. III, pp. 95–96: "{Algebræ Des Cartes specimen quoddam.} Dicit Idem se invenisse algebram generalem, ad eamque se non uti corporum figuris, sed planis duntaxat, quia eæ facilius mentibus insinuantur. Atque ita es aliæ præter geometriam, ijs optimè exprimuntur. Concipit unitatem per quadratum exiguum; ita etiam punctum concipit. Lineam verò aut radicem concipit per parallelogrammum, ex uno istius quadrati latere et longitudine debitâ conflatum. Quadratum concipit ex tot talibus radicibus factum; cubum, ex tot quot numeri indicant quadratis ad formam oblongam redactis factum;

Descartes proposes three different ways of representing algebraic quantities. The three manners depend on how the unit is represented. (1) If we think of the unit as having length and breadth, we depict it by means of a square, □; (2) by a point, •, if we regard it as the element which makes up a discrete number; and finally (3) by a line, —, if we view it as having only length. In the first case, Descartes presents the two ways e and c for expressing a square and similarly two ways d and f for a cube. He does not venture to explain the second case in detail, simply suggesting that the unit can also be represented by a point. The third way of representation draws our attention. According to Beeckman, Descartes has arrived at this method of expression with some toil. Using this representation, all algebraic quantities can be reduced to a one-dimensional line segment. These modes of representing quantities are essentially identical with those in Rule XV of the *Regulae ad directionem ingenii.*[9]

However, even after contriving the third way by means of simple lines, Descartes does not seem to have thought that he had finally resolved the problem of geometrical dimensionality. According to Beeckman, he further proceeds to speculate on dimensions higher than the third degree:

> He particularly denotes a cube by three dimensions as do other [people]. But he regards a biquadratic as if a stone cube were made from a simple cube which is considered to be wooden: for exactly one dimension is added as a whole; but if another dimension is to be added, he thinks of an iron cube; then golden etc. It happens not only in gravity but also in colors and all the other qualities. Therefore, distinguishing three squares from the wooden cube, he at last thinks also of classifying the cube brought together only by wood, stone, etc. so that the iron cube may be transformed into the wooden cube in such a manner that the simple cube [may be tranformed] into the squares, which are to be observed in each genus.[10]

Descartes thus tries to justify the meaningfulness of dimensions greater than three. To accomplish this, he first identifies an ordinary cube of three dimensions

---

biquadratum eodem modo etc. Imò hæc omnia etiam lineis explicat, ita ut *a* punctum, *b* lineam, *c* quadratum, *d* cubum representet. Eo modo quoque *f* cubum representabat ex multiplicatione quadrati *e* per numerum radicis confectum. Nec minori negotio eadem absolvit per nudas lineas, quemadmodum hîc ad marginem videre est, ubi notæ cossicæ singulis lineis adjectæ sunt, lineis eas quæ præfixæ sunt quantitates significantibus."

[9] See *Journal*, t. III, p. 95, note 4.

[10] *AT*, X, p. 334; *Journal*, t. III, p. 96: "Particulariter verò concipit cubum per tres dimensiones, ut etiam alij faciunt; at biquadratum concipit ac si ex cubo simiplici, qui consideratur ut ligneus, fieret cubus lapideus; ita enim per totum additur una dimensio; at si altera dimensio sit addenda, considerat cubum ferreum, tum aureum etc., quod non solùm fit in gravitate, sed etiam in coloribus et omnibus alijs qualitatibus. Secans igitur ex cubo ligneo quadrata tria, concipit tandem se secare cubum ex ligneitate, ferreitate etc. solâ conflatum, ita ut ferreus cubus ad ligneum perducatur eo modo quo cubus simplex ad quadrata in unoquoque genere observanda." We have used the text of the *Journal* rather than that of the *AT* edition.

with a wooden cube. Then he signifies four dimensions by a stone cube, five dimensions by an iron cube and so on. The greater the number of dimensions, the more weight the material of the cube gains. This idea may appear immature, awkward, and doubtful, but Descartes's purpose in giving multidimensional quantities a concrete meaning is at least understandable.

In the next paragraph, Beeckmen presents Descartes's way of solving the quadratic equation, taking as an example: **1ʒ∞6𝔛+7**, or in our notation, $x^2 = 6x + 7$. Beeckman's note reads as follows:

> In this method, the same man reduces a binomial to one term, as you see. In fact, wishing to take away 6 roots of the unknown square AB, he divides 6 by 2. But, because both FC and GB contain 3 roots, the square DC is taken away twice when FC and GB are removed; accordingly, **6𝔛** and a square of the half, i. e. 9, will be taken away. Therefore, one who wants to remove **6𝔛** must add 9 so that the lesser square DE may remain. Once this has been known, its side is also known, since the half of the root being added, the root of the first square is obtained. Thus the lesser square is taken out from the larger, through which the larger root is found.[11]

1 ʒ ∞ 6 𝔛 + 7
— 6 𝔛 + 9
1 ʒ ∞ DE . 16
DF . 4
DG . 3
AC . 7

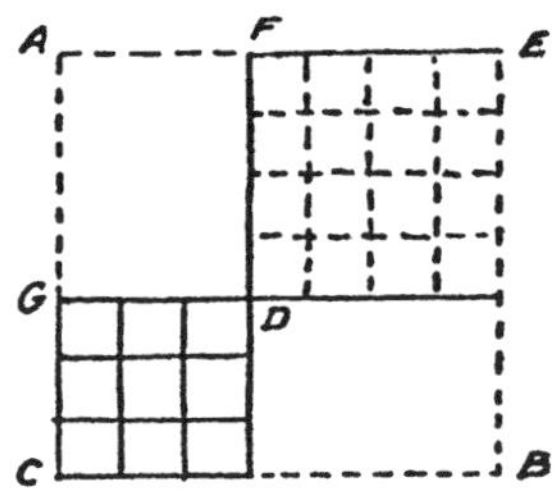

[Figure 4.3]

In this account, Descartes solves the quadratic equation $x^2 = 6x + 7$. The expression "to reduce a binomial to one term" implies that one transforms the given equation into a squared form from which a simple equation can be derived. To remove $6x$, he changes the original equation into $x^2 - 6x + 32 = 16$. Then, $(x - 3)^2 = 42$. $(x - 3)^2 = y^2$ signifies the lesser square. Descartes (*apud* Beeckman) takes only the positive root of the equation $y^2 = 16$. Hence, $y = 4$. Consequently, $x = y + 3 = 7$, which is the larger root. Descartes seems to treat the problem always referring to the figure (see Figure 4.3). As presented, his method is elementary and containing almost nothing surpassing the cossist tradition.

[11] *AT*, X, pp. 334–335; *Journal*, t. III, p. 96: "Idem hoc pacto, ut vides, minuit binomium uno nomine. Cupiens enim auferre 6 radices quadrati *AB* incogniti, dividit 6 per 2. At, quia *FC* et *GB* continent utrumque 3 radices, cùm *FC* et *GB* auferuntur, aufertur quadratum *DC* bis; auferentur igitur 6𝔛 et quadratum ex dimidio, videlicet 9. Idcirco qui auferre vult 6𝔛, debet addere 9, ut restet minus quadratum *DE*. Quo cognito cognoscitur etiam ejus latus, quod, addito dimidio radicum, habetur radix quadrati primi. Ita ex majore quadrato excipitur minus, quo mediante invenitur majoris radix."

The last paragraph on Descartes's algebra discusses irrational numbers and various kinds of the roots of equations:

> By a parabola he explains irrational numbers which cannot be explained otherwise. On the other hand, he names some roots 'true', some 'implicit' (that is, less than nothing), and some 'imaginary' (that is, not explicable at all); and, moreover, from a common table he looks at how many roots a certain equation can have, one of which is required.[12]

By "irrational numbers", Descartes seems to mean in particular the numbers explicable in the form $\sqrt{a}$, where a is a rational number which cannot be a square. Certainly the number $b = \sqrt{a}$ can be rewritten into the form $b^2 = a$, which represents a parabola, one of the conic sections. For irrational numbers, he uses the words "numeri surdi" (surd numbers) in the letter of March 26, 1619,[13] and "nombres sours" in Book III of the *Géométrie.*[14] In the latter work where surd numbers are discussed, he only deals with a cubic equation whose coefficients contain $\sqrt{3}$. Descartes's terms "true" (*verus*), "implicit" (*implicitus*) and "imaginary" (*imaginarius*), respectively, correspond to 'positive', 'negative' and 'imaginary' in today's terminology. In Book III of the *Géométrie*, he still uses the terms "true" (*vray*) and "imaginary" (*imaginaire*),[15] but "implicit" is replaced by "false" (*faux*).[16] "True" and "false" there, however, are used as applicable not only to real but also to imaginary roots.[17] So, precisely speaking, "true" or "implicit" (or "false") may simply mean "having a positive or negaive sign," respectively.

What the last sentence in the paragraph quoted above means is somewhat ambiguous and debatable. But if the idea incorporated into this sentence had been developed and integrated into Book III of the *Géométrie* as the other concepts in this paragraph were, it could indicate the well-known "Rule of Signs" ascribed to Descartes.[18] If this is the case, the statement will be paraphrased as "Descartes sees, observing the table of the signs of the coefficients in the equation, how many positive roots there can be." Or it could be a simple criterion for ascertaining the number of roots by means of the degree of the equation. In Book III of the *Géométrie*, this criterion is stated as follows: "Understand that in each equation there can be as many different roots (that is, values of the un-

---

[12] *AT*, X, p. 335; *Journal*, t. III, pp. 96–97: "Irrationales numeros, qui aliter explicari non possunt, explicat per parabolam. Nominat autem quasdam radices *veras*, quasdam *implicitas* (id est minores quàm nihil), quasdam *imaginarias* (id est omninò inexplicabiles); ac videt ex tabulâ vulgari quot aliqua æquatio radices habere possit quarum una sit quæsita."

[13] *AT*, X, p. 157. See Chapter 3, p. 102.

[14] *AT*, VI, p. 452; Olscamp, p. 235.

[15] *AT*, VI, pp. 446 & 453

[16] P. 445.

[17] P. 453.

[18] *AT*, VI, p. 446: The paragraph under the margin subtitle "Combien il peut y auoir de vrayes racines en chaque Equation." Cf. Olscamp, p. 230.

known quantities), as the unknown quantity has dimensions."[19] One difficulty of the latter case is that the words "from a common table" in Beeckman's note make almost no sense.

What can be asserted from our examination of Beeckman's record "A Certain Specimen of Descartes's Algebra" is that before October of 1628, Descartes had already possessed his own system of algebra, which can be thought to be a product of "an entirely new science." A problem which had troubled him by that time was the dimensionality of geometrical magnitudes. He surely had the idea (even if unsuccessful) that all quantities should be regarded as a one dimensional line. But parallel to this idea, he seems to have retained alternative ways of representing quantities, that is, by means of small squares or dots. The assertion that he considered himself to have arrived at a final solution to the problem of dimensionality by 1628 would thus be debatable. What is noteworthy is that in his old *Algebra*, in so far as it was documented in Beeckman's Journal, Descartes still used cossic notation. His attempt at constructing a new algebra in the *Cogitationes privatae* was, as our analysis in Chapter 3, § 2 has shown, by means of certain mathematical instruments rather than in a purely written form. But in the old *Algebra*, Descartes did not rely on any mathematical instruments, simply manipulating cossic algebraic symbols in order to solve problems. It might appear curious that a treatise which is fundamentally confined within the cossist tradition would be thought to be connected to "an entirely new science." But Descartes's phrase "an entirely new science" seems to have been literally "entirely new" when compared to classical mathematics rather than cossist mathematics. But even compared with typical textbooks of cossist algebra, his purely conceptual attempt to unite arithmetical numbers and geometrical magnitudes is a salient point of contrast, although it was surely awkward. Thus, "A Certain Specimen of Descartes's Algebra" can be characterized not only as an embodyment of "an entirely new science," but also a preliminary formulation for the mathematical part of the *Regulae ad directionem ingenii*.

The topics which Descartes talked about with Beeckman in October of 1628 were not confined to the principles of his algebra. The French mathematician seems to have been eager to discuss, among other things, music and optics with the older Dutch scholar. Several pages of Beeckman's *Journal* from this period are fully occupied with these subjects. Their contents are indicated to some extent by the following margin subtitles: "Angle of Refraction Explained by Descartes," "The Ratio of Thickness of Musical Strings," and "To Burn Very

---

[19] *AT*, VI, p. 444; Olscamp, p. 229: The paragraph under the margin subtitle "Combien il peut y auoir de racines en chaque Equation." This proposition is, of course, a form of the fundamental theorem of algebra. In 1629 Albert Girard claimed the same in his *Invention nouvelle en l'algebre* (Amsterdam, 1629); Reprinted edition (Leiden, 1884), ff. [E 4r]–F3v. Girard in his turn referred to François Viète's *De Recognitione aequationum* of 1615. Cf. Viète, *Opera mathematica* (1646), p. 158; Witmer's translation, pp. 309–310.

Remote Places by Solar Rays."[20]

In Beeckman's *Journal*, we also find notes under the following titles which can be ascribed to the period between October 8, 1628 and February 1, 1629: "An Ellipse in Which All Parallel Rays Converge in a Point of a Denser Medium," "A Hyperbola through Which Rays Converge into One Point," and "The Part of the Ellipse through Which Rays in the Air Exactly Converge."[21] Beeckman's note at the beginning of these passages reads "Described word for word from the writings of the earlier oft-mentioned M. Descartes."[22] This comment suggests that the mathematical theory of refraction presented in these notes was excerpted from writings of Descartes which Beeckman obtained from him sometime between October 8, 1628 and February 1, 1629. The question may arise as to whether or not these excerpts were part of the old *Algebra*. The editor of Beeckman's *Journal* Cornelis de Waard has conjectured in the affirmative.[23] This question, however, can be answered only after we have examined them.

Here we analyze in detail only one example which has the same mathematical characteristics as the above-mentioned dioptrical excerpts, as to what Descartes's theory of refraction using conic sections was like. In Beeckman's *Journal* we find a paragraph under the margin subtitle: "A Hyperbola through Which All Parallel Rays Converge into One Point."[24] Beeckman begins this paragraph with the following statement: "On February 1, 1629 in Dordrecht. M. Descartes had left this proposition on hyperbola without demonstration and required me to look for a demonstration; as I have discovered it, he was pleased and judged that it was authentic."[25] As De Waard has conjectured, the author seems to have found a proof for the presented proposition some days after

[20] *AT*, X, pp. 335–338; *Journal*, t. III, pp. 97–99.

[21] *AT*, X, pp. 338–340; *Journal*, t. IV, pp. 135–136: "Ellipsis in quà omnes radij parallele concurrent in puncto medij densioris." "Hyperbola per quam radij in unicum punctum concurrunt." "Ellipsis pars per quam radij in aere exactè concurrunt." De Waard has ascribed these excerpts to the period between December 15, 1628 and February 1, 1629 on the ground that Beeckman is supposed to have received Descartes's writings which he later excerpted by mail. We suppose for safety that Beeckman made the excerpts during the wider period between October 8, 1628 and February 1, 1629.

[22] *AT*, X, p. 338; *Journal*, t. IV, p. 135: "Ex scriptis D. Des Chartes [*sic!*] antè sæpè dicti ad verbum descripta."

[23] See the following statements by C. de Waard, the editor of Beeckman's *Journal*: *Journal*, t. III, p. 109, note 1): "Le reste de la feuille est occupé par des extraits ("*ex scriptis* D. Des Chartes anté sæpe dicti ad verbum descripta"), emprunté probablement à l'*Algèbre* que Descartes avait promis de vouloir enyoyer de Paris;" and the editor's explanation on Beeckman's excerpts, *Journal*, t. IV, p. 135: "En visitant Beeckman, le 8e octobre 1628, à Dordrecht, Descartes lui avait promis de lui envoyer son *Algèbre* après son retour à Paris. C'est était sans doute de cet écrit que Beeckman tira les extraits suivants."

[24] *AT*, X, p. 341; *Journal*, t. III, p. 109: "Hyperbola, per quam omnes radij paralleli in unum punctum exactè incidant, demonstrata."

[25] *Ibid*: "10 Feb. 1629 Dortrechti. Hanc de hyperbolâ propositionem D. Des Chartes indemonstratam reliquerat, ac me rogavit ut ejus demonstrationem quærerem; quam cùm invenissem, gavisus est ac genuinam esse judacavit."

Descartes proposed it on October 8, 1628 and to have written the note about his demonstration on February 1, 1629, soon after the French mathematician had approved of it in a letter.[26]

The following detailed demonstration was certainly not Descartes's own, but Beeckman's. It seems, however, to have been done analogously to the former's method for an ellipse. Descartes's own attempts recorded by Beeckman were all more or less elliptic. Thus, Beeckman's description modeled on Descartes's proof deserves a full examination. He presents the following demonstration:

> Now it is as follows: *let* A *and* E *be the two foci*, GB *and* UC [=VC] *be parts of a hyperbola*, WG *be a ray parallel to* AE, *meeting* GF *perpendicularly, and be refracted into* E; *or from* E *meeting at* G, *be refracted parallel into* W. *And let* AG *be another line by which together with* GE *the hyperbola is described, and let* QR *and* ST *be the sines of the ray going out and into to the perpendicular* HQ, *which cuts the tangent at a right angle. But*, GM *has been produced by the bisection of the angle* AGE. *It must be shown that* ST *is to* QR *as* BC *to* AE.

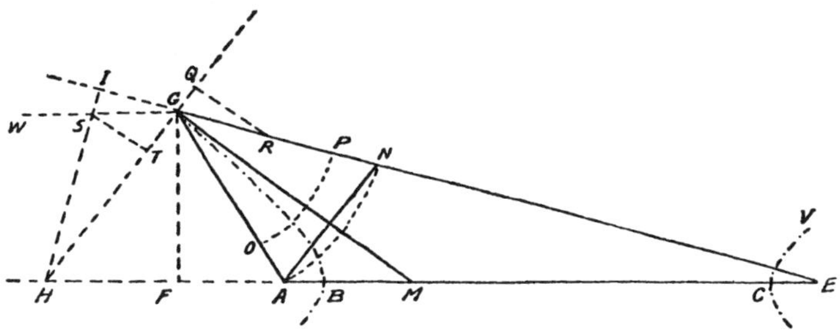

[Figure 4.4]

> But since QRG and HIG are similar triangles, as STG and GHF, it is certain that ST is to GR as GF is to HI; and since IHE and GFE are also similar, it will be
>
> as GF to HI so GE to HE.
>
> Now let GN be equal to GA, and OA and PN be equal to AB, which is also equal to CE. But by removing the equals GP and GO from GE and GA, PE and BE will be equal from the construction of the hyperbola; in fact the vertex B is marked when AO about the center A and EP about the center E, being moved, form a straight line and tangents to B mutually. But, since NP is equal to AB and EC, NE will be less than AE by the double of AB, that is AB

[26]See *Journal*, t. III, p. 109, notes 2) and 3).

> and EC, therefore equal to BC. And since the straight line AN is at a right angle to the line GM by Proposition 9 of Euclid I, GH and AN will be parallel and the trialgles ANE and HGE will be similar; accordingly
>
> > [*as*] *NE to AE so GE to HE*, therefore *as BC to AE* and this is *as ST to QR*.
>
> That which was to be demonstrated.[27]

The point of the demonstration is to show that the ratio of ST to QR, that is $\sin SGT/\sin RGQ$, is equal to the ratio of BC to AE, which is given. Once this relationship (the law of refraction) has been demonstrated, it is easy to see that all the parallel rays through the hyperbola converge into one point E. It should be noticed that the difference between GE and AG is BC by the main characteristic of the hyperbola. Accordingly, $PE = GE - GP = GE - GO = (GE - AG) + AO = BC + CE = BE$. This formula is obtained "from the construction of the hyperbola," to use Beeckman's phrase. We speculate from our analysis of Beeckman's discussion above that Descartes had obtained the law of refraction by the fall of 1628 at latest and used it for his dioptrics. At that time Descartes left Beeckman the above problem, which could only be solved relying on the dioptrical law.

This idea will be incorporated into Discours VIII of Descartes's *Dioptrique* of 1637. The main property of the hyperbola which Beeckman refers to will be illustrated there by means of gardener's instrument[28] (see Figure 4.5). Also the form of the refraction through the hyperbolic lens will be represented by an illustration (Figure 4.6).[29] We will further examine Descartes's dioptrical research during his Paris years of the 1620s below.

---

[27] *AT*, X, pp. 341–342; *Journal*, t. III, p. 110: "Ea autem talis est: *Sint* A, E *duo foci, partes hyperbolarum* GB *et* UC, WG *radius parallelus* AE, *perpendiculariter* GF *incidens, et refringatur in* E; *vel ex* E *in* G *incidens, refringatur parallelus in* W. *Sitque* AG *altera linea, ex quâ cum* GE *hyperbola describitur, sintque* QR *et* ST *sinus radij egredientis et ingredientis ad perpendicularem* HQ, *quæ tangentem* GM *secat ad angulos rectos.* GM *verò ex bisectione anguli* AGE *nata est. Ostendendum est* ST *se habere ad* QR *ut* BC *ad* AE.

At cùm *QRG* et *HIG* triangula similia sint, ut et *STG* et *GHF*, certum est *ST* esse ad *QR* sicut *GF* est ad *HI*; cùmque *IHE* et *GFE* etiam similia sint, erunt

ut GF *ad* HI, *sic* GE *ad* HE.

Fiat jam *GN* æquale *GA*, et *OA* et *PN* æqualia *AB*, quod etiam æquale est *CE*. At ablatis æqualibus *GP* et *GO* ex *GE* et *GA*, erunt *PE* et *BE* æqualia ex constructione hyperbolæ; vertex enim *B* notatur, cùm *AO* super *A* et *EP* super centro *E* motæ unam rectam efficiunt se invicem tangentes ad *B*. Cùm autem *NP* æquale sit *AB* et *EC*, erit *NE* minor quàm *AE* duplici *AB*, ad est *AB* et *EC*, ergo æqualis *BE*. Cùmque *AN* recta per 9 *primi* Euclidis sit ad angulos rectos ad lineam *GM*, erunt *GH* et *AN* parallelæ et triangula *ANE* et *HGE* similia, ideòque

⟨*ut*⟩ NE *ad* AE, *sic* GE *ad* HE, *ergo etiam ut* BC *ad* AE *et hæc ut* ST *ad* QR.

Quod erat demonstrandum."

[28] *AT*, VI, p. 176.

[29] *AT*, VI, p. 181.

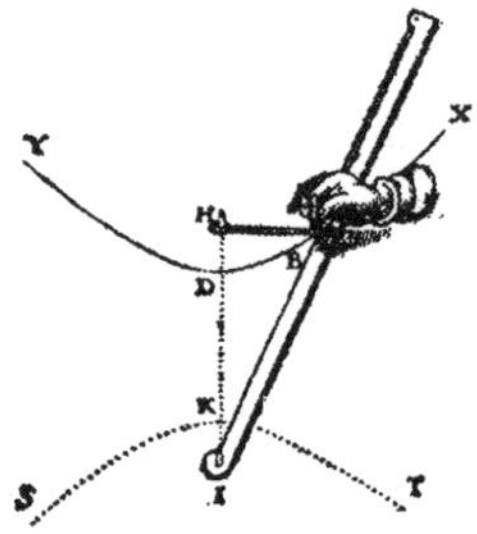

Figure 4.5

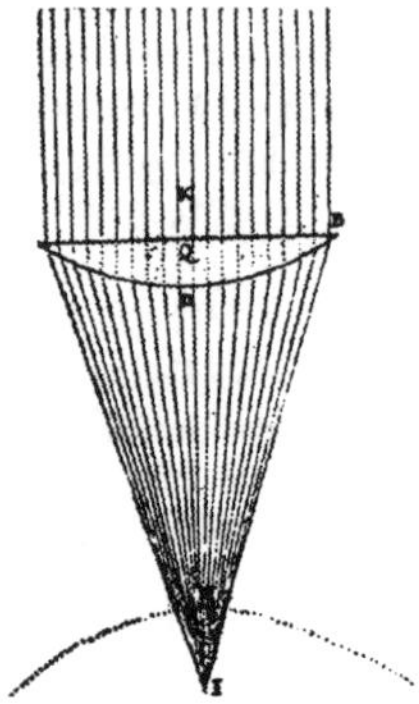

Figure 4.6

Beeckman's notes ascribed to the period around the winter of 1628–1629 were not confined to the basic concepts of algebra and the mathematical theory of refraction, which were mentioned above, for Beeckman, after having written the previous demonstration, inserted into his *Journal* copies of two mathematical problems. De Waard has supposed again that they had some relation to Descartes's algebraic treatise sent from Paris.[30] The first problem has the margin subtitle: "It Is Demonstrated That One Can Find Two Mean Proportionals by Means of a Parabola."[31] The second one is to solve a quadratic equation, again with the aid of a parabola.

Beeckman writes about the first problem: "As soon as Descartes had discovered that two mean proportionals are found by means of a parabola, a certain French mathematician of Paris demonstrated this geometrically in the following manner. That I have described word for word."[32] The problem of discovering proportional means is, as we have seen in Chapter 3, § 2, closely connected with the duplication of the cube and was a challenge to mathematicians of the sixteenth and seventeenth centuries. It is known that Mersenne was informed about Descartes's solution in the summer of 1625 and incorporated it in his *Harmonicorum libri*, published in Paris in 1636.

According to Mersenne's report,[33] Descaartes solved the problem by intersecting a parabola and a circle. Let two lines be given such that $m = 2n$. Draw a part of a parabola whose vertex is A (see Figure 4.7). Take the focus O such

[30] *Journal*, t. III, p. 110, note 1).

[31] *AT*, X, p. 341; *Journal*, t. IV, p. 136: "Parabolâ duo media proportionalia inveniri posse demonstratur."'

[32] *AT*, X, p. 342; *Journal*, t. IV, p. 136: "Cùm D. Des Chartes invenisset per parabolam duo media proportionalia inveniri, hoc mathematicus quidam Gallus Parisijs geometricè demonstravit hoc modo. Quod ad verbum descripsi."

[33] Marin Mersenne, *Harmonicorum Libri* (Paris, 1636), pp. 146–147; "(René Descartes, à Paris), à Mersenne, à Paris. (été de 1625)," *CM*, I, pp. 156–159.

that $AO = Nm$ and B on the axis such that $AB = Nm$. Erect the straight line $BC = n$ at a right angle to the axis. Then describe a circle about the center C and through A. Let the intersection of the two curved lines be D and the perpendicular DI be drawn from D to the axis of the parabola. DI and IA will be the larger and lesser mean proportionals between n and m, respectively.

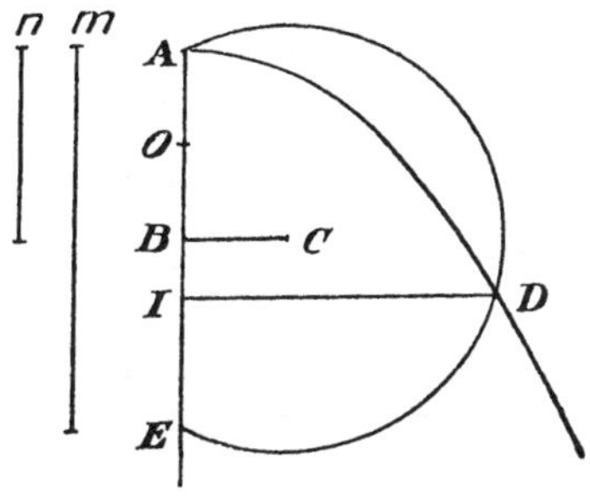

Figure 4.7

Descartes's solution was communicated through Mersenne to several other mathematicians. Claude Mydorge was one such mathematician who responded to Descartes by sending him a more general solution with a detailed demonstration in the same year, 1625. Accordingly, "a certain French mathematician of Paris" in Beeckman's comment can be identified as Mydorge.[34]

It can be safely said that it was Mydorge's letter to Descartes that Beeckman transcribed word for word. Mydorge's solution was generalized in the sense that the two given lines were arbitrary. Mydorge (*apud* Beeckman) begins his solution with the following words: "A Solid Problem Constructed Surely. *When two straight lines are proposed, to assign two means in continued proportion.* Let there be the two proposed, the lesser GB, the larger BH. And now it is necessary to find two means in continued proportion between them."[35]

Interestingly, Mydorge described his soution quite formally, following three steps in his demonstration: (1) analytically (Ἀναλυτικῶς), (2) synthetically (Συνθετικῶς), and (3) by proof (Ἀπόδειξις). The analysis is roughly as follows: Let it be done already and the two means be ED the lesser and EA the larger (Figure 4.8). From the presupposition, $GB : ED = ED : EA = EA : BH$. Then,

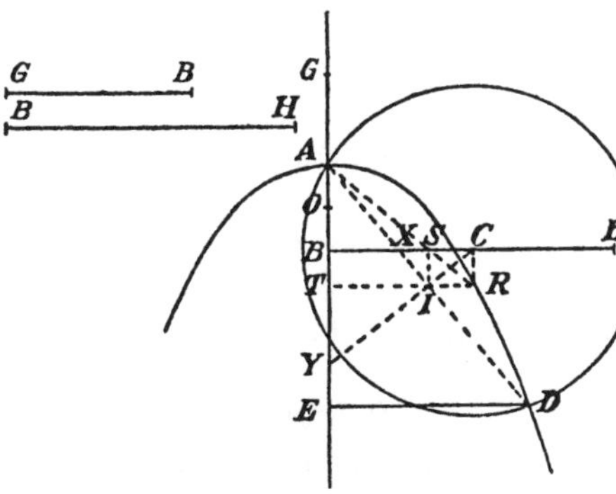

Figure 4.8

[34]See *Journal*, t. IV, p. 136, note 4). Claude Mydorge was interested in the theory of conic sections and made use of it to construct optical lenses. Cf. Pierre Speziali, "Mydorge, Claude," *DSB*, IX, pp. 598–599.

[35]*AT*, X, p. 342; *Journal*, t. IV, p. 136; "Claude Mydorge, à Paris, à René Descartes (à Paris). (1625)," *CM*, I, pp. 269–273: "PROBLEMA SOLIDUM SOLIDE CONSTRUCTUM. ***Propositis duabus lineis rectis, binas medias in continuâ proportione assignare.*** Sunto binae propositae, minor GB, major BH. Oporteat autem inter eas binas medias in continuà proportione invenire."

$ED^2 = GB.EA$. Draw a parabola whose vertex is A and the *latus rectum* is GB. Let the straight line AD be connected and be cut into two equal parts at I. And let TI be extended to R (R is not always on the parabola) so that TR is equal and parallel to the half of BH, i.e. BC. Then, $AB : BS \; (= TI = NED) = TI : TA \; (NEA) = TA : TR \; (= BC)$. Therefore, the triangles ATI and ATR are similar and the angle TAI is equal to the angle ART. But, $AT : TR = SI : IR \; (= IS : SC) = YT : TI$. Accordingly, the triangles ATR, ISR, YTI, and ITA are all similar and the angles ART, ICS, YIT, and TAI are all equal one to another. Therefore, $AT : TI = TI : TY$. Consequently, AIY is an angle in a semicircle, thus, a right angle. Likewise, AIC is also a right angle. Therefore, the triangles AIC and DIC are equal. Consequently, AC is equal to CD and both are the radius of a circle whose center is C.

In the synthesis, Mydorge constructs a parabola and a circle and arrives at the continued proportion, $BG : ED = ED : AE = AE : BH$. In the proof, he finally demonstrates the previous relationship. Descartes will rewrite the Mydorge solution into a simple algebraic formula and incorporate it into Book III of the *Géométrie* of 1637 under the margin subtitle "L'inuention de deux moyenes proportionelles."[36]

The second problem which Beeckman transcribed is titled "To resolve cossic equations [formed] by lens by means of a parabola."[37] Some elucidation may be necessary to comprehend this description properly. Let us examine Beeckman's comment preceding Descartes's own explanation so as to find a clue as to what Descartes dealt with. Beeckman introduces Descartes's solution in the following words:

> With the help of a parabola to construct all solid problems by a general method. That M. Descartes in another place calls a universal secret to resolve all complicated equations of third and fourth dimension [formed] by geometrical lines. That I describe word for word from his writings.[38]

According to Beeckman's transcription which follows the above explanation, Descartes tackles biquadratic equations of the type $z^4 = \pm az^2 \pm bz \pm c$ (in his later notation) by intersecting a parabola with a circle in various ways (see Figure 4.9). This idea will be developed and constitute a rather long section in Book III of the *Géométrie* under the title "Façon generale pour constuire tous

[36] *AT*, VI, p. 469.

[37] *AT*, X, p. 344; *Journal*, t. IV, p. 138: "Parabolâ æquationes cossicas lineis exponere."

[38] *AT*, X, p. 344; *Journal*, t. IV, p. 138: "AUXILIO PARABOLAE OMNIA SOLIDA PROBLEMATA GENERALI METHODO CONSTRUERE. *Quod alio loco vocat D. Des Chartes secretum universale ad æquationes omnes tertiâ vel quartâ dimensione involutas lineis geometricis exponendas. Quod ex illius scriptis ad verbum describo.*"

les problesmes solides, reduits a vne Equation de trois ou quatre dimensions."[39] In the 1637 version, Descartes presents his solution in the classical manner of analysis-synthesis, that is, describing first the way to construct a parabola and a circle, and then its demonstration. What Beeckman transcribed in 1629 were the directions for constructing the two curved lines which correspond to the analytical part of the 1637 version.

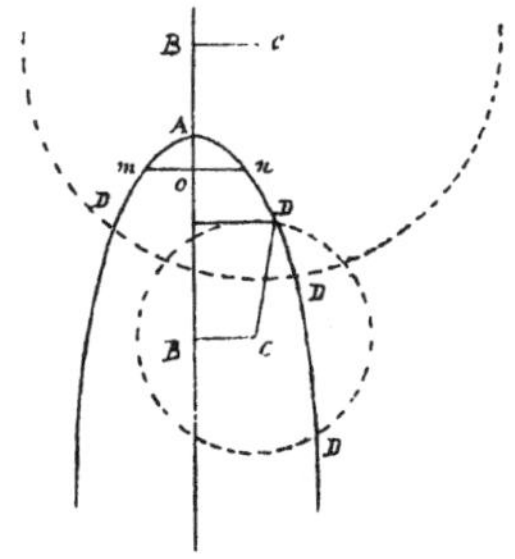

Figure 4.9

Descartes seems to have thought much of his solution of the cubic and biquadratic equations by means of a parabola. At the end of his transcription, Beeckman writes his observation on Descartes's discovery:

> M. Descartes considers this discovery so important that he grants that he has never discovered anything more outstanding and that no one has been ever discovered anything more outstanding.[40]

In the *Géométrie* of 1637, Descartes will comment on this same solution as follows: "This rule is as general and as complete as one could wish."[41]

With this method for solving the biquadratic equation, Beeckman concluded his excerpt from Descartes's writings. Judging from the entries in Beeckman's *Journal*, we can ascertain that by the fall of 1628, Descartes had possessed in an already written form at least (1) the basic algebraic rules applicable to both arithmetical and geometrical quantities, (2) the mathematical theory of refraction through lenses with the shapes of conic sections, (3) the method for finding two mean proportionals by means of a parabola, and (4) the general method for solving the cubic and quartic equations by means of a parabola.

After having examined Beeckman's exerpt from Descartes's writings, we have to ask the important question: what was the content of Descartes's old *Algebra* and when did he draft it? In order to answer this question we should first consider the relationship between the old *Algebra* and Beeckman's excerpts. It is beyond doubt that the first extract with the title "A Certain Specimen of Descartes's Algebra" belonged to the old *Algebra*. On the other hand, it is somewhat problematical whether or not we can ascribe the other excerpts to Descartes's monograph of the *Algebra*. According to the description of the Stockholm Inventory, Descartes's manuscript of the *Algebra* consisted of 155 pages in octavo and its content concerned the foundations of his algebra (see

[39] *AT*, VI, pp. 464–469; Olscamp, pp. 244–247. Cf. J. F. Scott, *The Scientific Work of René Descartes* (London, 1952), pp. 147–150.

[40] *AT*, X, p. 346; *Journal*, t. IV, p. 139: "*Hanc inventionem tanti facit D. Des Chartes, ut fateatur se nihil unquam præstantiùs invenisse, imò a nemine unquam præstantiùs quid inventum.*"

[41] *AT*, VI, p. 467; Olscamp, p. 246.

p. 226). This appears to suggest prima facie that only the first excerpt originated in Descartes's old *Algebra*.

But we should not draw too hasty a conclusion. There exists Descartes's important testimony about his algebraic monograph's relation to his mature *Géométrie* of 1637. Descartes wrote to Mersenne probably at the beginning of 1638:

> I would not see any difficulty in sending him [Mydorge] my old *Algebra* (*ma vieille Algebre*), except that it is a work which, in my opinion, is not worth reading. And because nobody has, as far as I know, any copy of it, I would be glad if it did not part from my hands anymore. But if he can take the trouble to examine the third book of my *Géométrie*, I hope he will find it rather easy and that he will thereafter proceed to the end of the second book.[42]

Descartes wrote these words in order to answer questions which Mydorge (through Mersenne) might have raised about "many things which I (Descartes) wrote on vision,"[43] i. e. optics. He requested the Parisian optics expert Mydorge to read Book Three (and the end of Book Two) of the *Géométrie* probably as a mathematical preparation to optical studies. Importantly, he was thinking at the moment of the possibility of sending his old *Algebra* to Mydorge instead of recommending appropriate parts of the *Géométrie*. We do not know exactly what parts of Books Two and Three Descartes referred to. But his pronouncement here seems to indicate that his old *Algebra* had a part which he thought to have been almost totally superseded by the last two books of his final mathematical treatise of 1637. On the other hand, we frankly admit that even this testimony is not sufficient to determine precisely what the content Descartes's *Algebra* included.

And we should pay attention that Descartes wrote in the letter "nobody has any copy of it." Didn't he really send anybody the manuscript of *Algebra*? Didn't he keep his promise that he would send it to Beeckman in October 1628? Or, did he say so, simply because Beeckman wasn't residing in Paris? Further, did he write so, since he had at least once become estranged from Beeckman. At last, we are not certain as to what was the fact about the old *Algebra* being mere conjectures.

Our consideration above suggests, in sum, that the old *Algebra* contained some mathematical arguments other than Beeckman excerpted in his *Journal* under "A certain specimen of Descartes's algebra" explaining only basic notions of algebra.

When, then, did Descartes write his old *Algebra*? The answer to this question naturally depends on which parts of Beeckman's extracts were as-

[42] "Descartes, (à Santpoort), à Mersenne, à Paris. (25 janvier 1638)." *AT*, I, pp. 501–502; *AM*, II, p. 84; *CM*, VII, p. 39.
[43] *AT*, I, p. 501; *AM*, II, p. 84; *CM*, VII, pp. 38–39.

cribed to Descartes's algebraic treatise. First of all, the mathematical discussions recorded in "A Certain Specimen of Descartes's Algebra" seem to have constituted the core of that treatise. We know that Descartes began to engage in such discussions in 1619. But the discovery of the principal ideas of his algebra is one thing and the date of the composition of his treatise is another. The latest supposed date must be 1628, the year around which he is conjectured to have drafted the *Regulae ad directionem ingenii*.

As to the second group of Beeckmen's extracts related to dioptrics, we may safely ascribe them to the period around 1626. As John A. Schuster's extensive analysis has shown,[44] during this period in Paris Descartes conducted dioptrical research in collaboration with Claude Mydorge. According to Schuster, in 1626 Mydorge and Descartes almost certainly succeeded in formulating the mathematical law of refraction and, moreover, used a lens having a surface which was hyperbolic in shape.[45] So it was in 1626, to be exact, that Descartes established his basic idea which resulted in Beeckman's mathematical argument shown above. And we have already seen that it was in 1625 that Descartes found two mean proportionals by intersecting a parabola with a circle (p. 171). This was the topic of the third group of Beeckman's excerpts.

Further what we should pay attention to in relation to these two dioptrical or mathematical topics that problems were solved by means of geometrical tools, not algebraic ones. Thus, if Descartes's *Algebra* was limited strictly to algebraic themes, the second and third groups of Beeckman's excerpts would not have been part of it. To put it the other way around, if the former was not the case, they were possibly part of it.

Now when did Descartes discover his solution of the cubic and biquadratic equations using a parabola, which belonged to the fourth group of Beeckman's excerpts? Baillet, following Lipstorp, has ascribed this discovery to the year 1620, soon after Descartes's alleged meeting with Faulhaber.[46] This date is possible, but there is no evidence for it outside of Lipstorp's biography of Descartes, published in 1653. We do not deny the possibility of the year 1620 but rather suggest the alternative, that the discovery was made about the same time as that of two proportional means, namely, around 1625. It should be noticed that Descartes solved this problem algebraically.

---

[44]John A. Schuster, "Descartes and the Scientific Revolution 1618–1634: An Interpretation," (Ph. D. Dissertation, Princeton University, 1977), Chapter 4, "The Early Scientific Career: The Optical Triumph of the Parisian Years," pp. 268–368, esp. pp. 299–329.

[45]See, in particular, "Claude Mydorge, à Paris, à Mersenne (à Paris). (février-mars 1626?)," *CM*, I, pp. 404–415. Descartes wrote to Golius on February 2, 1632: "[...] il y a enuiron cinq ans, dont M. Mydorge traça luy-mesme le modelle [...]." *AT*, I, p. 239; *AM*, I, p. 218, note. And about the hyperbolic lens, he wrote to Constantin Huygens on December 11, 1635: "Il y a desia huit ou neuf ans que ie fis aussi tailler vn verre par lemoyen du tour, et il reüssit parfaitement bien [...]." *AT*, I, pp. 335–336; *AM*, I, p. 298.

[46]Baillet, *La Vie de M. Des-Cartes*, t. I, p. 70. Cf. Lipstorp, *Specimena philosophiae cartesianae* (Leiden, 1653), pp. 79–80; *AT*, X, p. 253.

We may conclude the following: If the old *Algebra* was extensive enough to include purely geometrical subjects containing the second and third groups of Beeckman's extracts, we can ascribe the date of its composition to the period between 1626 and 1628, although some parts of it may have been written before these years. Otherwise, it would be possible that Descartes wrote his old *Algebra* as early years as from 1619 to 1623, for example. The year 1623 was the *terminus ad quem* by which Descartes thought his main mathematical discoveries secured on his own initiative had been essentially accomplished. One thing should be stated here about 1623. It was around this year that Descartes is commonly supposed to have returned to Paris from his foreign travels. His main interest during his Paris years after 1623 seems to have shifted from mathematics to natural philosophy, especially optics. It is highly probable that the central content of Descartes's algebra as it appears in the old *Algebra* had been already acquired by about 1623. At least Descartes himself thought that his career had progressed in this way.

At all events, if the *Regulae ad directionem ingenii* represents Descartes's philosophical speculation during the period between 1619 and the late 1620s in comparison with the more mature philosophical part of his *Discours de la méthode* of 1637, his old *Algebra* was a product of his mathematical studies during the years between 1619 and 1628. It must be admitted that the algebraic monograph contained parts which were immature in comparison with the *Géométrie*. But it certainly served as a significant preliminary for the later definitive form of his algebra. Moreover, it seems to have played an important role not only for his mathematical treatise but also for his philosophical speculations of the late 1620s. In other words, it appears to have functioned as the mathematical foundations of the *Regulae ad directionem ingenii.*

§ 2. The Mathematics in the *Regulae ad Directionem Ingenii*

A. Pure Mathematics as the Paradigm for Certainty

Descartes seems to have begun to cherish his idea of a general method in the fall of 1619 (see Chapter 3, § 4). If it is credible that the night of November 10-11, 1619 was a private turning point of Descartes's career, the public recognition of his philosophical thought may be said to have taken place on the occasion of a meeting at the Paris palace of the Papal Nuncio, probably in the fall of 1627.[47] Recalling this meeting, Descartes later wrote to Villebressieu, one of the participants:

> You saw these two results of my fine Rule or Natural Method (*ma belle règle ou Méthode naturelle*) in the discussion which was forced

[47]See *AM*, I, p. 30, note 1. Adrien Baillet ascribed the date of this meeting to 1628, in his *La Vie de M. Des-Cartes*, t. I, p. 160. But this raises a difficulty to conciliate with Descartes's and Bérulle (a participant)'s presence in Paris.

> on me in the presence of the Papal Nuncio, Cardinal de Bérulle, Father Mersenne and all that great and learned company assembled at the Nuncio's palace to hear M. de Chandoux lecture about his new philosophy. I made the whole company avow what great power the art of reasoning well (*l'art de bien resonner*) has over the minds of those who have no learning beyond the ordinary. I showed them that my principles are more certain, more true, and more natural than any of those which are currently received in the learned world. You were as convinced as everybody else, and you were all good enough to beg me to put them in writing and to publish them.[48]

Indeed, the 1627 assembly seems to have been the first occasion upon which Descartes revealed to the public his "natural method" which had been fostered essentially by mathematical studies. Chandoux's lecture is conjectured to have been directed against the entire Aristotelian philosophy from a standpoint influenced by Renaissance naturalism. The Cardinal de Berulle, the founder of the French Oratory and an attendant of the meeting, is reported to have been very much impressed by Descartes's rejoinder to Chandoux and to have encouraged him to devote himself to establishing his own philosophy founded on the aforementioned "natural method." This time Descartes must have felt a firm sense of mission as a creative thinker.[49]

We can obtain important information concerning the process of the formation of his method from an autobiographical statement in Part Two of his *Discours de la méthode*. But the *Regulae ad directionem ingenii* provides at times more information about how it was formed and even about what it consisted in, albeit the rules for his method as stated in the *Regulae* must be admitted to have been generally immature. It was in the *Regulae*, which is conjectured to have been composed around 1628, that his "natural method," made public in the fall of 1627, was explicitly formulated for the first time.

As the author himself declares toward the end of Rule XII, at first the *Regulae* was intended to consist of three parts: Part I on the theory of knowledge in general (Rules I–XII), Part II on mathematics (Rules XIII–XXI), and Part III on natural philosophy. Part III was never written and even Part II is incomplete. Because the text of the *Regulae* as a whole was left unfinished, textual problems concerning it have been raised in recent years.[50] We will discuss this problem in the next section. We consciously confine our discussion to the mathematics, although we include some discussion of the relation between mathematics and

[48] "Descartes à Villebressieu. Amsterdam, été de 1631?," *AT*, I, p. 212; *AM*, I, p. 198. Cf. Baillet, *Op. cit.*, p. 163.

[49] See Baillet, *Op. cit.*, p. 165; Vrooman, *René Descartes* (n. 1), pp. 72–75.

[50] We may not be able to discuss the contents of the *Regulae ad directionem ingenii* without having solved the textual problem proposed by Jean-Paul Weber, *La Constitution du texte des Regulae* (Paris, 1964). But we postpone this problem until the next section. The basic argument of this section will not be influenced by this problem.

philosophy as necessary.

The ultimate goal of the *Regulae* is to reorganize the entire scheme of learning and to establish a solid method for pursuing "the truth of things." The first striking point of the *Regulae* is that Descartes's most important criterion which a cognition must satisfy to be admitted as true knowledge is certainty (*certitudo*). Rule II clearly tells us this: "We should attend only to those objects of which our minds seem capable of having certain and indubitable cognition."[51]

Descartes compares the two disciplines, dialectic and mathematics, as to whether they meet the rule of certainty. It is crucial in this comparison that he disdains the supposed certainty of dialectic, on the one hand, and praises that of mathematics, on the other. He certainly admits that dialectic (*dialectica*), which is also called logic (*logica*) following the medieval and Renaissance tradition, as the method of philosophizing and probable syllogisms "exercise the minds of the young, stimulating them with a certain rivalry."[52] Nevertheless, by using dialectic, he claims, one cannot acquire any knowledge, but merely increase the number of doubts. So he accuses dialecticians:

> Those chains with which dialecticians (*Dialectici*) suppose they regulate human reason seem to me to be of little use here, though I do not deny that they are very useful for other purposes. In fact none of the errors to which men—men, I say, not the brutes—are liable is ever due to faulty inference; they are due only to the fact that men take for granted certain poorly understood observations, or lay down rash and groundless judgments.[53]

The criticism of dialectic will be a frequently repeated theme in the *Regulae*.

It is arithmetic and geometry, the two subjects of pure mathematics, that Descartes extols instead of dialectic or logic. He declares that of all the sciences arithmetic and geometry alone are free from any taint of falsity or uncertainty and satisfy Rule II. He states the reason why these two subjects are much more certain than the other disciplines:

> They alone are concerned with an object so pure and simple that they make no assumptions that experience might render uncertain; they consist entirely in deducing conclusions by means of rational arguments. They are therefore the easiest and clearest of all sciences and have just the sort of object we are looking for.[54]

But he does not necessarily insist that only these two branches of pure mathematics should be studied. He entertains one reservation at the conclusion of the rule of certainty:

---

[51] *AT*, X, p. 362.
[52] *Ibid.*, p. 363.
[53] P. 365.
[54] *Ibid.*

> Now the conclusion we should draw from these considerations is not that arithmetic and geometry are the only sciences worth studying, but rather that in seeking the right path of truth we ought to concern ourselves only with objects which admit of as much certainty as the demonstrations of arithmetic and geometry.[55]

This reservation is essential in order to understnad rightly the natural method in the *Regulae.* In all events, the reform of existing sciences must be carried out with the criterion of certainty at its center. All sciences have to be reorganized to attain the same level of certainty as the demonstrations of pure mathematics.

In order to accomplish this goal, Descartes proposes training our mind or more accurately our intelligence (*ingenium*) with the view of forming true judgments. In Rule III, he analyzes the acts of our intellect by means of which we are able to arrive at true knowledge, not merely probable conjecture, and admits only two intellectual acts which are suitable for such an end: intuition (*intuitus*) and deduction (*deductio*).[56] He first defines the former concept as follows:

> By 'intuition' I do not mean the fluctuating testimony of the senses of the deceptive judgment of the imagination as it groups things together, but the conception of a clear and attentive mind, which is so easy and distinct that there can be no room for doubt about what we are understanding. Alternatively, and this comes to the same thing, intuition is the indubitable conception of a clear and attentive mind which proceeds solely from the light of reason (*rationis lux*). Because it is simpler, it is more certain than deduction, though deduction [...] is not something a man can perform wrongly.[57]

By means of this function of the mind, everyone can mentally intuit that "he exists, that he is thinking, that a triangle is bounded by just three lines, and a sphere by a single surface, and the like."[58]

Descartes then defines 'deduction': By deduction "we mean the inference of something as following necessarily from some other propositions which are known with certainty."[59] He states summarily as follows as to how we can judge truly and soundly through the most certain routes to knowledge:

> We are distinguishing mental intuition from certain deduction on

[55] *AT*, X, p. 366.

[56] *AT*, X, p. 368, *l.* 12: "intuitus et inductio." But this pair concepts should be read "intuitus et deductio," following the Dutch version of 1684, which reads "d'inzieninge en d'afleiding," that is "intuition and deduction." Cf. R. Descartes, *Regulae ad directionem ingenii: Texte critique établi par Giovanni Crapulli avec la version hollandaise du XVIIème siècle* (La Haye, 1966), p. 8*. See also *AT*, X, "Appendice," p. 694.

[57] *AT*, X, p. 368.

[58] *Ibid.*

[59] *AT*, X, p. 369.

> the grounds that we are aware of a movement or a sort of sequence in the latter but not in the former, and also because immediate self-evidence (*praesens evidentia*) is not required for deduction, as it is for intuition; deduction in a sense gets its certainty from memory. It follows that those propositions which are immediately inferred from first principles (*prima principia*) can be said to be known in one respect through intuition, and in another respect through deduction. But first principles themselves are known only through intuition, and the remote conclusions only through deduction.[60]

Certainly Descartes's model from which the two concepts are derived is synthetic proof in mathematics. We should note that the process of attaining first principles through intuition is not further scrutinized in the text of *Regulae*. Instead we are told that we need a method (*methodus*) for obtaining true knowledge.

Descartes defines the method in Rule IV as consisting in having reliable rules for distinguishing what is false from what is true and being such that one can constantly increase one's knowledge. According to him, the method is above all useful for intuition and for finding deduction which alone lead to true knowledge. He puts it: "If our method properly explains how we should use our mental intuition to avoid falling into the opposite error [rejecting what is true]and how we should go about finding the deductive inferences that will help us attain this all-embracing knowledge, then I do not see that anything more is needed to make it complete; for as I have already said, we can have no knowledge (*scientia*) without mental intuition or deduction."[61] Dialectic contains no such function. Dialectic, claims Descartes, is of no use, "or rather should be reckoned a positive hindrance."[62]

But how do we secure a model for the method? We find an extremely important statement from which we can learn the origin of the Cartesian method in the *Discours de la méthode*. This statement has been already quoted *in toto* in Chapter 2, § 5 (p. 110). There Descartes has told us that in forming his method, the geometrical analysis of the ancients and the algebra of the moderns played a crucial role. In referring to these two disciplines, he has stated that they "are simply the spontaneous fruits which have sprung from the innate principles of this method." In the same place, we are told that "a sort of arithmetic called 'algebra'" is achieving for numbers what the geometrical analysis of the ancients did for figures. So to be more exact, we contend that it was algebraic analysis *par excellence* that played the central role when he devised his natural method. In other words, the model discipline for his method was, among other things, algebra playing the role of analysis, that is, a tool for theorem-finding or problem-solving.

If the paradigmatic science for the Cartesian method was algebraic anal-

---

[60] *Ibid.*, p. 370.
[61] P. 372.
[62] *Ibid.*

ysis, it is quite natural that in the rules which follow Rule IV we find passages which discuss the analytic method. Commentators on the *Regulae* at times have justly stated that in Rules V and VIII we witness a clear avator of the analytic method.[63] In fact Descartes states in Rule V: "We shall be following this method exactly if we first reduce complicated and obscure propositions step by step to simpler ones, and then, starting with the intuition of the simplest ones of all, try to ascend through the same steps to a knowledge of all the rest."[64] In this passage, we can easily detect the regressive characteristic of analysis and its reverse which marks synthesis. In the same rule he compares his method to the thread of Ariadne which Theseus held in the Labyrinth.[65]

The way of determining the 'linea anaclastica' in Rule VIII is another example of the analytic method, which is applied this time to dioprtics.[66] The 'linea anaclastica' is characterized as "a line from which parallel rays are so refracted that they intersect at a single point." This line is exactly the hyperbola about which Descartes and Beeckman communicated in the winter of 1628–1629 (see pp. 235–237).

Furthermore, throughout the text of the *Regulae* Descartes takes examples from the theory of proportion in order to explain several rules of his method. For example, Rule VI, which he considers to be one of the principal keys to his method, is elucidated by using this theory of proportion. Rule VI asserts that "we should attend to what is most simple in each series of things in which we had directly deduced some truths from others."[67] Everybody knows, Descartes states, the simple truth that the number 6 is twice 3. But repeating such a question as to what twice 6 is, we finally notice that the numbers 3, 6, 12, 24, 48, etc. are continued proportionals. All of this, he says, is so clear as to seem almost childish.

"Nevertheless," he continues, "when I think carefully about it, I can see what sort of complications are involved in all the questions one can ask about the proportions or relations (*habitudines*) between things, and in what order the questions should be investigated. This one point encompasses the essential core of the entire science of pure mathematics."[68] Descartes notes in this example that after having become familiar with the continued proportion beginning with the ratio of 3 to 6, he can relatively easily tackle problems which look prima facie rather complicated, such as finding the mean proportional when the extreme terms 3 and 12 are given. He further proceeds to raise the question of finding three mean proportionals between the extreme terms 3 and 24. Rule VI commands that this problem should be split up and made easier. Thus he

[63]For example, see Boyce Gibson, "The Regulae of Descartes (II.)," *Mind*, **7** (1898), p. 339; L. J. Beck, *The Method of Descartes: A Study of the Regulae* (Oxford, 1952), pp. 155–167.

[64]*AT*, X, p. 379.

[65]*Ibid.*, p. 380.

[66]Pp. 393–394.

[67]P. 381.

[68]Pp. 384–385.

suggests that first we look for the single mean proportional between 3 and 48, i. e. 12; then a further mean proportional between 3 and 12, i. e. 6; then another between 12 and 48, i. e. 24. He gives the procedure for discovering the single mean proportional between the two extreme terms, stating that "we must attend at the same time to the two extreme terms and the ratio between them, in order to obtain a new ratio by dividing this one."[69] We may translate this procedure into algebraic language, for example, $a : x = x : b$. But here Descartes does not do so. It was common sense at that time that the language of the theory of proportion could be written in the language of algebra. Then, it seems clear for us that he had the method of analysis, particularly algebraic analysis, at hand as a model for his general method.

On the whole, in Book One of his *Regulae* Descartes asserts first of all that all reliable knowledge should possess the certainty of mathematics. He then examines the nature of mathematical reasoning and splits it up into two components: intuition and deduction. What he looks upon as important is the method for discovering the truth of things. In the domain of mathematics, this amounts to the method of analysis being used as a tool for finding theorems and for solving problems. It is the mathematical method of analysis that he endeavors to reformulate into a much more generalized version.

B. Algebra as the Method of Analysis

That algebraic analysis is an incomparable paradigm for the method becomes clearer in Book Two of the *Regulae*, that is Rule XIII and the rules following it, which treat mathematics more directly than Book One.

Rules XIII and XIV instruct us that once we understand a problem we should analyze it by dividing it into the simplest parts and re-expressing it in unambiguous terms. In these rules, Descartes discusses how we inquire about what we do not know on the basis of what we already know. He formulated the rules as generally as possible. But at the same time it seems undeniable that he considers them on the analogy of algebraic analysis. We can observe this in some of the rules in Book Two. In the introductory part of Rule XIV he notes the following:

> [W]henever we deduce something unknown from something already known, it does not follow that we are discovering some new kind of entity, but merely that we are extending our entire knowledge of the topic in question to the point where we perceive that the thing we are looking for participates in this way or that way in the nature of the things given in the statement of the problem.[70]

He asks here how we can extend what we know to what we do not know. He

[69] P. 385.
[70] P. 438.

answers that this heuristic procedure can be accomplished by means of various kinds of comparison:

> We should note that comparisons are said to be simplest and straightforward only when the thing sought and the initial data participate equally in a certain nature. The reason why preparation is required for other sorts of comparison is simply that the common nature in question is not present equally in both, but only by way of other relations or proportions which imply it. The chief part of human endeavour is simply to reduce these proportions to the point where an equality (*æqualitas*) between what we are seeking and what we already know is clearly visible.[71]

In order to look for something unknown, Descartes says, we should establish certain proportions which contain both the things sought and the things given. And under the condition that the terms of the proportions which represent a given problem have been abstracted from every subject so as to be expressible in the language of magnitudes, we can reduce these proportions to the form of equalities. The last point is emphasized and restated as follows: "Our intention is simply to reduce the proportions, however complicated, to the point where we can discover some equality (*æqualitas*) between that which is unknown and something known."[72] We should pay attention to the fact that Descartes's terminology here consists of that of the theory of proportion or theory of equations.

A similar procedure is the topic of the entire Rule XVII. Rule XVII tells that "we should make a direct survey of the problem to be solved, disregarding the fact that some of its terms are known and others unknown, and intuiting, through a train of sound reasoning, the dependence of one term or another."[73] Here Descartes is particularly concerned with the complicated question of discerning certain intermediate propositions on the ground that we know the first and last propositions in the series of our deductive inference. In this case, we have to follow, we are told, the order which is completely indirect and reverse of the usual deductive process. He presents a key to resolving such problems:

> So the trick here is to treat the unknown ones as if they were known. This may enable us to adopt the easy and direct method of inquiry even in the most complicated of problems. There is no reason why we should not always do this, since from the outset of this part of the treatise [i. e. from Rule XIII, the opening rule of Book Two] our assumption has been that we know that the unknown terms in the problem are so dependent on the known ones that they are

[71] P. 440.
[72] P. 447.
[73] P. 459.

> wholly determined by them. Accordingly, we shall be carrying out everything this Rule prescribes if, recognizing that the unknown is determinded by the known, we reflect on the terms which occur to us first and count the unknown ones among the known, so that by reasoning soundly step by step we may deduce from these all the rest, even the known terms as if they are unknown.[74]

The opening sentence reminds us of the well-known definition of analysis in Book VII (Domain of Analysis) of Pappus's *Collection*: "Analysis is the path from what one is seeking, as if it were established, by way of its consequences, to something that is established by synthesis."[75] Of course, we may be able to list the other possible sources for Descartes's discussion of the analytic method.[76] In any case, the quoted passage teaches us crucial procedures which mathematical analysts, particularly algebraic analysts, follow when they form proportions or equations from both known and unknown quantities under certain conditions. To treat the unknown quantities as if they were known is just to represent them by certain letters which can be distinguished from other letters denoting known quantities.

From the material that we have just examined above, it has become evident that the method of analysis as used in mathematics underlies Descartes's discussion in Book Two of his *Regulae*. Let us see very briefly how the idea of algebraic analysis emerged and how Descartes's description of analysis in the *Regulae* was developed in the later *Géométrie*.

As is well known, the idea of algebraic analysis was explicitly formulated for the first time in early modern Europe by Francois Viète in his *In artem analyticem isagoge*, published in Tours in 1591. His forerunners in Arabic and medieval and Rensaissance European mathematics will be discussed thematically in Chapter 5, § 3, C. Viète defined analysis in general as "assuming that which is sought as if it were admitted [and working] through the consequences [of that assumption] to what is admittedly true."[77] He cited this sentence as having been proposed by Theon of Alexandria.[78]

In Viète's analytic art, the theory of proportion tended to be rewritten in an algebraic manner, that is to say, in the language of the theory of equations.

---

[74]Pp. 460–461.

[75]Pappus of Alexandria, *Book 7 of the Collection*, Part 1: *Introduction, Text, and Translation*, edited with Translation and Commentary by Alexander Jones (New York/Berlin/Heidelberg/Tokyo, 1986), p. 82. Cf. Commandino, f. 157r; Hultsch, Vol. II (Berlin, 1877), pp. 634–635; Ver Eecke, t. II, p. 477.

[76]See, among others, André Robert, "Descartes et l'analyse des anciens," *Archives de philosophie*, **13** (1937): *Autour du "Discours de la méthode" (1637–1937)*, pp. 221–245. [pp. 301–325.]

[77]Viète, *In artem analyticem isagoge* (Tours, 1591), f. 4r; *Opera mathematica* (Leiden, 1646), p. 1: "Adsumptio quæsiti tanquam concessi per consequentia ad rerum concessum." See also T. Richard Witmer's translation, p. 11.

[78]We will discuss Viète's sources for the definition of analysis in Chapter 5, §3.

For instance, Book VI, Propositions 16 and 17 of Euclid's *Elements*, which are formulated geometrically, could be transformed into the following proposition:

> If there are three or four magnitudes, and as the first is to the second, so that second, or else some third, is to another, that which is under the extreme terms will be equal to that which is under the means. And so, a proportion can be called the composition (*constitutio*) of an equation, an equation the resolution (*resolutio*) of a proportion.[79]

Thus Viète reduced proportional relationships under certain geometrical conditions to algebraic equations. He also reinterpreted the definition of analysis itself into algebraic terms. For example, 'zetetics' (*Zetetice*), a sort of analysis, was defined as the art "by which is found the equation or proportion between the magnitude that is being sought and those that are given."[80] In ancient Greece, the heuristic characteristic of analysis was emphasized as was seen in Geminus's concise definition: "Analysis is the heuristic of demonstration."[81] In Viète's analytic art it was the newly invented symbolic algebra, i. e. algebraic analysis, that took the role of the heuristic art.

For the moment we will not discuss whether or not Descartes knew Viète's works when he composed the text of *Regulae*. But it is certainly algebraic analysis as the art of discovery that he considers to be the principal basis of his method in Book Two. What is known and what is unknown are usually mixed up in a complicated manner. This manner is first to be expressed in the language of the thory of proportion. Then the proportions representing some relations with magnitudes are translated into algebraic equations which have both known and unknown magnitudes. It remains only to solve the algebraic equations to aquire the unknown magnitudes.

In the *Géométrie* of 1637, the method of analysis will be beautifully reformulated in the paragraph under the margin subtitle: "Comment it faut venir aux Equations qui seruent a resoudre les problesmes." There the object of the analytic method appears as "lines":

---

[79] Viète, *Isagoge*, f. 4v; *Opera mathematica*, p. 2. The translation is by the Reverend J. Sinfree Smith, in Jacob Klein, *Greek Mathematical Thought and the Origin of Algebra*, tr. by Eva Brann (Cambridge, Mass., 1968; New York, 1992), p. 320. Cf. Witmer's translation, p. 15. Compare with the geometrical counterpart in the *Elements*: (Book VI, Proposition 16) "If four straight lines be proportional, the rectangle contained by the extremes is equal to the rectangle contained by the means; and the reverse." Heath, *The Thirteen Books of Euclid's Elements*, Vol. II (New York, $_2$1956), p. 221. For a more general discussion on the theory of proportion rewritten in the modern analytic tradition, see my paper, "The Acceptance of the Theory of Proportion in the Sixteenth and Seventeenth Centuries—Barrow's Reaction to the Analytic Mathematics," *Historia Scientiarum*, No. 29 (1985), pp. 93–126.

[80] Viète, *Isagoge*, f. 4r; *Opera mathematica*, p. 1. J. W. Smith's translation, p. 321. Cf. Witmer's translation, p. 12.

[81] "᾿Ανάλυσις ἐστιν ἀποδείξεως εὕρεσις." Ammonius's commentary on Aristotle's *Prior Analytics*, in Aristoteles, *Organon Graece*, ed. Theodor Waitz, Pars Prior (Leipzig, 1844; Aalen, 1965), p. 44. Analysis as the heuristic art in mathematics is fully discussed in Michael S. Mahoney, *Op. cit.*(Ch. 2, n. 50).

> Thus, if we wish to solve some problem, we should first of all consider it solved, and give names to all the lines—the unknown ones as well as the others—which seem necessary in order to construct it. Then, without considering any difference between the known and the unknown lines, we should go through the problem in the order which most naturally shows the mutual dependency between these lines, until we have found a means of expressing a single quantity in two ways. This will be called an equation, for the terms of one of the two ways [of expressing the quantity] are equal to those of the other. And we must find as many such equations as we assume there to be unknown lines.[82]

In this later description of algebraic analysis, the language of the theory of proportion recedes into the background and that of the theory of equations comes to the fore. The *Regulae* thus takes a definitive step in the history of the analytic method.

It is the method of analysis stated in the algebraic form that Descartes adopts in order to pursue the relations expressed in magnitudes. The main concern of Rule XIV is how to abstract various mathematical and physical problems and reduce them to certain proportions or equations in which magnitudes alone appear. It is exactly to obtain these equations successfully that we are told to be concerned exclusively with an extended object. Descartes deliberately chooses the notion of 'extension' (*extensio*) as the abstract field in which algebraic analysis works, refraining from using the term 'quantity' (*quantitas*), "for there are some philosophers so subtle that they have even distinguished quantity from extension."[83] From this statement we understand that 'extension' as the field of algebraic analysis is so general a conception as to encompass any mathematically imaginable objects.

Let us see, then, how Descartes develops his 'algebra of extension'. In considering proportional relationships, we certainly have to take account of such characteristics as dimension, unity, and shape. But, in Descartes's judgment, "it is enough for our purposes if we consider all the characteristics of *extension itself* (*ipsa extensio*) which may assist us in elucidating differences in proportion"[84] (The emphasis is mine). For example, there can be countless different dimensions within the same subject in question, such as length, weight, and speed. Nevertheless, according to the author of the *Regulae*, it is clear that these dimensions "add absolutely nothing to the things which possess them, and that they are understood in the same way whether they have a real basis in the objects themselves or are arbitrary inventions of our mind. [...] Yet these all function in the same way from the point of view simply of dimension, which is how they ought to be viewed here in the mathematical disciplines. Whether

[82] *AT*, VI, p. 372; Olscamp, p. 179.
[83] *AT*, X, p. 447.
[84] *Ibid.*

dimensions have a real basis is something for the physicists to consider."[85] The same thing can be said about the unity and shape of a figure.

Descartes thus attaches importance to "extension itself." In doing so, he intends to unify arithmetic and geometry. Sometimes continuous magnitudes, the object of geometry, can be counted as discrete numbers by using the unit. He puts it:

> [W]e should realize that, with the aid of the unit we have adopted, it is sometimes possible completely to reduce continuous magnitudes to a set (*multitudo*) and that this can always be done partially at least. The set of units can then be arranged in such an order that the difficulty involved in discerning a measure (*mensura*) becomes simply one of scrutinizing the order (*ordo*). The greatest advantage of our method lies in this progressive ordering.[86]

In order to make his algebraic analysis work for both arithmetic and geometry, Descartes further scrutinizes the concept of extension and introduces specific figures by means of which we can treat not only quantitative aspects of geometrical figures, but also numbers. These figures are called "rectilinear and rectangular surfaces, or straight lines." At the end of Rule XIV, he explains why these notions are necessary:

> We have as much reason to abstract proportions from geometrical figures, if the problem has to do with these, as we have from any other subject matter. The only figures that we need to reserve for this purpose are rectilinear and rectangular surfaces, or straight lines, which we also call figures, because, as we said above, these are just as good as surfaces in assisting us to imagine an object that is really extended. Lastly, these same figures must serve to represent sometimes continuous magnitudes, sometimes a set or a number. To find a simpler way of expressing differences in relation would be beyond the bounds of human endeavour.[87]

These figures, rectangular surfaces or straight lines, are, respectively, oblongs, the unit being understood as a small square, or straight lines, the unit being a short line. These have already observed in Beeckman's note titled "A Certain Specimen of Descartes's Algebra" (see pp. 162–165). They will be a subject of Rule XV. Indeed as Lüder Gäbe has pointed out in his study of Rule XIV of the *Regulae*, this is not the vulgar geometry as a theory of lines and angles as exemplified by the elementary part of Euclid's *Elements*, but the new geometry "comme une theorie ou mesures de corps ou de surfaces sont placees a la base

[85] *AT*, X, p. 448.
[86] *Ibid.*, p. 452.
[87] *Ibid.*

et ont pour consequences des rapports de grandeurs linéaires."[88] We should add to this observation that the manuscript of the old *Algebra* formed the backdrop for Descartes's description of Rules XIV and XV.

After introducing in Rule XV fundamental figures representing both continuous magnitudes and discrete numbers, Descartes develops his symbolic algebra in Rule XVI, which functions as the organon for algebraic analysis. In this algebra, the figures discussed above are represented by very concise symbols (*brevissimae notae*). For example, the letters a, b, c, etc. denote figures which are unknown. Moreover, he writes "$2a^3$" for a magnitude which means "twice the magnitude symbolized by the letter a, which contains three relations,"[89] where the "number of relations" is to be understood as the dimensions of the magnitudes.

By assigning very concise symbols to express the dimensions of the magnitudes, Descartes seems to begin to surpass even his own old *Algebra*. According to him, moreover, "we must note above all that the root (*radix*), the square (*quadratus*), the cube (*cubus*), etc. are nothing but magnitudes in continued proportion which, it is always supposed, are preceded by the arbitrary unit mentioned above."[90]

As far as the extant text of the *Regulae* is concerned, Descartes continues to think that any magnitudes should be represented by either surfaces or lines. Even the four basic algebraic operations, addition, subtraction, multiplication, and division, are elucidated in Rule XVIII by means of these devices. Addition and subtraction are explained simply as the juxtaposition and deduction of two lines respectively. "In multiplication," he writes:

> [W]e also conceive the given magnitudes in the form of lines; though in this case we imagine them as forming a rectangle, for if we multiply a by b,
>
> 
>
>
> we fit one line at right angles to the other, thus:
>
> 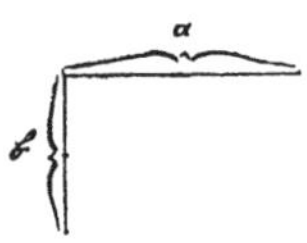
>
>
> to make the rectangle:
>
> 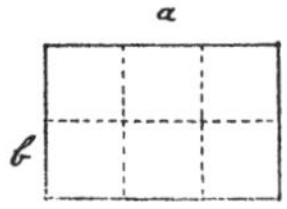
>

[88] Lüder Gäbe, "La Règle XIV: Lien entre géométrie et algèbre," *Archives de philosophie*, **46** (1983), p. 658.

[89] *AT*, X, p. 455.

[90] *Ibid.*, p. 457.

Again, if we wish to multiply ab by c,

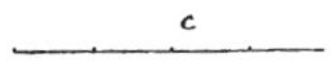

we ought to conceive ab as a line, viz.,

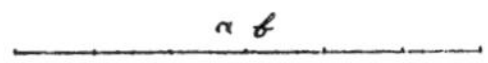

in order to obtain for abc following figure.[91]

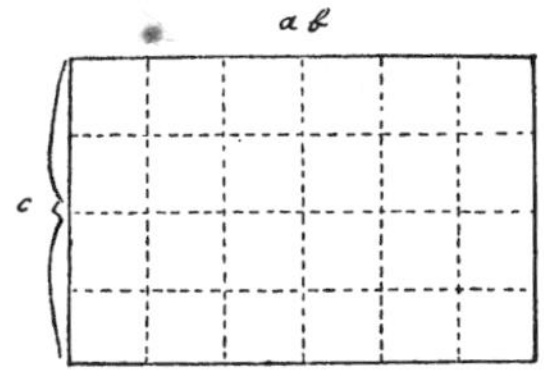

This is a rather complicated procedure for one who knows the later 'algebra of line segments' in the *Géométrie.* He clearly knows how a magnitude of the second dimension can be reduced to one of the first dimension with the help of the unit, the existence of which Descartes has assumed here.[92] And he follows this procedure when he conceives ab as a line in the passage cited above. Without stating any particular reason, he cleaves to algebra using surfaces and straight lines.

Nevertheless, such is the algebra of the *Regulae*, which, together with the old *Algebra*, will be surpassed by the later simpler one of the *Géométrie*, just as the method of the *Regulae* will be overcome and reformulated in an orderly manner by the *Discours de la méthode*. Descartes's epistemological position in the *Regulae* that pure mathematics serves as the ideal for certainty in knowledge will be questioned at least once as well in the later philosophical writings (see our discussion in Chapter 8, § 1). The contents of the *Regulae* remained incomplete just as the text itself was left unfinished.

§ 3. *Mathesis Universalis*

Descartes's concept of 'mathesis universalis', which appears exclusively in Rule IV of the *Regulae ad directionem ingenii*, has been a focus of vigorous debate among recent Cartesian scholars, particularly since the publication in 1964 of Jean-Paul Weber's *La Constitution du texte des Regulae.* Even before the appearance of Weber's monograph, that concept was variously interpreted, for example, as algebra, the system of mathematico-mechanical physics, universal science as the ideal of the unification of the sciences, or even the method itself.[93]

---

[91] Pp. 465–466.

[92] P. 463.

[93] To make up an exhaustive bibliography would be almost impossible. But for various representative interpretations of 'mathesis universalis', see, for example, John A. Schuster, "Descartes' *Mathesis Universalis*: 1619–28," in Stephan Gaukroger, ed., *Descartes: Philoso-*

But after Weber's textual critique and historical examination appeared, more careful analyses have been published. Among others, extremely important contributions, I maintain, are Giovanni Crapulli's Latin text of the *Regulae* with the 1684 Dutch version (1966)[94] and his extensive historical examination of the idea of 'mathesis universalis' in the sixteenth century (1969);[95] Jean-Luc Marion's philosophical study titled *Sur l'ontologie grise de Descartes* (1975);[96] and Marion's and Pierre Costabel's French translation and annotation of the *Regulae* (1977).[97] In addition to these studies, John A. Schuster's interpretation that the 'mathesis universalis' for Descartes meant his program of mathematical physics has gained a certain influence. The views of these scholars will be discussed thematically in the appropriate place. These indispensable works notwithstanding, it seems that a synthetic study is absolutely necessary, especially from the point of view of the history of mathematics.

Before examining what the concept of 'mathesis universalis' in fact was in Descartes's philosophico-mathematical thought, let us touch upon a textual problem of the *Regulae*, especially of Rule IV, which discusses the 'mathesis universalis'. This problem was first presented by Jean-Paul Weber in 1964. Weber contended that Rule IV of the *Regulae* consists of two mutually heterogeneous texts: Rule IV-A (*AT*, X, p. 368–p. 374, *l.* 15) and Rule IV-B (*AT*, X, p. 374, *l.* 16–p. 380). In the Amsterdam manuscript, which was used as the principal part of the *editio princeps* of 1701, the two parts were combined together as the unitary Rule IV, whereas in the other Hanover manuscript Rule IV-B was set at the end of the entire text of the *Regulae.* Moreover, Rule IV-B, Weber maintained, concerns the 'mathesis universalis' as a mathematical concept, while Rule IV-A concerns a more general method possibly applicable to the entirety of human knowledge. He conjectured, then, that the text of Rule IV-B must represent an earlier stage in the intellectual development of the young Descartes than that of Rule IV-A. He further suggested that Descartes's notion of 'mathesis universalis' may have originated from Adriaan van Roomen's usage of the same term.[98]

Weber's noting that the term 'mathesis universalis' appears only in Rule IV-B attracts our attention. Indeed, although the item "mathesis universalis" in

---

phy, Mathematics and Physics (Brighton, Sussex/Totowa, New Jersey, 1980), pp. 41–96, esp. notes 1 and 2 on pp. 80–81; Frederick Van de Pitte, "Descartes' *Mathesis Universalis*," *Archiv für Geschichte der Philosophie*, **61** (1979), pp. 154–174, esp. pp. 154–155.

[94] R. Descartes, *Op. cit.*(n. 56), éd. par Giovanni Crapulli.

[95] Giovanni Crapulli, *Mathesis universalis: Genesi di una idea nel XVI secolo* (Roma, 1969).

[96] Jean-Luc Marion, *Sur l'ontologie grise de Descartes: Science cartesien et savoir aristotélicien dans les Regulae* (Paris, 1975).

[97] René Descartes, *Règles utiles et claires pour la Direction de l'esprit et la recherche de la vérité*: Traduction selon le lexique cartésien, et annotation conceptuelle par Jean-Luc Marion avec notes mathématiques de Pierre Costabel (La Haye, 1977).

[98] Jean-Paul Weber, *Op. cit.* (n. 50), Chapitre I, "La Règle IV," pp. 3–17 and "Appendice A," pp. 247–249; *Idem*, "Sur la composition de la Regula IV de Descartes," *Revue philosophique de la France et de l'étranger*, **154** (1964), pp. 1–20.

the "Index des Matières" of the Adam-Tannery edition is cited in several places outside of Rule IV-B, actually only Rule IV-B contains the term 'mathesis universalis' explicitly.[99] But Weber's claim that the Hanover text is identical or closer to the original does not necessarily seem to be definitive, because as in the Amsterdam version, Rule IV-A and Rule IV-B are unified in the Dutch translation by J. H. Glazemaker published in 1684 as well. This was the text which Crapulli reedited in a critical manner in 1966.[100] Of course, neither do we assert that this vernacular edition is closer to the original manuscript than the Hanover version. Other than the original, there could have been a Latin manuscript to which Glazemaker referred. So it is in a sense understandable that the Philosophische Bibliothek edition of the *Regulae* published in 1973 has used the Hanover manuscript as the basic text and has therefore recognized Weber's claim.[101] Accordingly, neither Weber's position nor the traditional view has been supported by any further objective textual evidence.

Interpretative coherence and circumstantial evidence alone may warrant the correct reading of Rule IV. It is from this angle that Marion challenged Weber. Let us briefly scrutinize Marion's interpretation of both the notion of 'mathesis universalis' and the text of Rule IV. After our critical examination of Marion's thesis, we will present our own explanation.

Marion admitted that Rules IV-A and IV-B are in a sense heterogeneous. But he saw at the same time a parallelism between the two sections and found similarities in the structures of their arguments. Furthermore, in his judgment, the term 'mathesis universalis' has a different extension from the concept which Marion refers to by 'mathématique universelle'. The latter furnishes principles for the particular branches of mathematics and is limited to quantity alone (either discrete or continuous). It is a field within mathematics. The former, on the other hand, is concerned with 'order' (*ordo*) and 'measure' (*mensura*), and is one degree more abstract than mathematics proper. It is not mathematics any longer, but a "méta-mathématique non-mathématique" which is about the "mathématicité non-mathématique des mathématiques,"[102] as it were.

In Marion's opinion, the 'mathesis universalis' is a universal method which involves a second degree of abstraction, i. e. order and measure. Thus, "l'unique science, productrice d'universelle certitude, équipollente en d'infinis objets indifférents," Marion asserted, "Descartes la nomme soit Mathesis universalis (à

---

[99]See *AT*, V, "Index général," p. 812. Cf. J.-R. Armogathe et J.-L. Marion, *Index des Regulae ad directionem ingenii de René Descartes*, avec des listes de leçons et conjectures établies par G. Crapulli (Roma, 1976), pp. 73 & 146.

[100]Crapulli's edition (n. 94), "Introduction," esp. p. XIV.

[101]Descartes, *Regulae ad directionem ingenii*, Kritisch revidiert, übersetzt und herausgegeben von Heinrich Springmeyer, Lüder Gäbe, und Hans Günter Zekl (Hamburg, 1973). Cf. Heinrich Springmeyer, "Eine Neue Kritische Text-Ausgabe der *Regulae ad directionem ingenii* von René Descartes," *Zeitschrift für philosophische Forschung*, **24** (1970), pp. 101–125.

[102]Marion, *Op. cit.* (n. 96), p. 64. Throughout his French translation (n. 97) Marion distinguishes the term 'mathesis' from 'mathematica(e)'.

partir de la mathématicité non-mathématique des mathématiques, IV-B, 378, 8), soit méthode générale (à partir de la production de la certitude)."[103] As the 'mathesis universalis' is not a mathematics but a method according to Marion's opinion, it is natural that there be a continuity between IV-A and IV-B. Weber's thesis which emphasized discontinuity is, then, unwarranted. Marion's interpretation was later supported by Frederick P. Van de Pitte.[104]

Marion's and Van de Pitte's reading is certainly ingenious. But one thing which we cannot fully support is their distinction between 'mathesis' and 'mathematica(e)'. It is from this point that we have to start our own argument. In fact, it might be admitted that Descartes employs different terms, such as 'mathesis' and 'mathematica(e)', deliberately. The original Greek term 'μάθησις' meant 'the process of learning' or simply 'learning', and 'μάθημα' meant 'that which is learnt' or 'learning'. 'Μάθησις' surely emphasized the process of learning. But throughout Rule IV Descartes appears to use the word 'mathesis' as synonymous with mathematics in the modern sense. For example, we find a sentence in Rule IV-B: "I wondered why the founders of philosophy would admit no one to the pursuit of wisdom who was unversed in *Mathesis*."[105] As the footnote added to this sentence in the Adam-Tannery edition suggests, it is a reference to the motto "*οὐδεὶς ἀγεωμέτρητος εἰσίτω*" (A person ignorant of geometry must not enter.), which is said to have been inscribed at the entrance to Plato's Academy. Accordingly, 'mathesis' in this context can be interpreted as a notion which is not different from mathematics in the modern sense. Further, Descartes refers to 'vera Mathesis' frequently and states, "one can even see some traces of this *vera Mathesis*, I think, in Pappus and Diophantus who, though not of that earliest antiquity, lived many centuries before our time."[106] This time again, 'vera Mathesis' is translatable into 'true mathematics'. Even if the meaning of 'mathesis' differs from that of 'mathematica(e)', any difference obtaining between the two seems to be a difference of nuance and within mathematics.

Marion and Van de Pitte at times mention Crapulli's invaluable study on the concept of 'mathesis universalis' in the sixteenth century and particularly refer to Van Roomen's use of that term. According to the conclusion of Crapulli's book, which we can support, 'mathesis universalis' in the sixteenth century was understood in general as "the discipline common to all mathematical sciences, principles and properties."[107] And Van Roomen himself in his *Apologia pro Archimede* of 1597 called "mathesis universalis" "prima mathematica, seu prima mathesis."[108] Thus he also employed 'mathematica' and 'mathesis'

[103] Marion, *Op. cit.* (n. 96), p. 62.
[104] Frederick P. Van de Pitte, *Op. cit.* (n. 93).
[105] *AT*, X, p. 375, *ll.* 22–25.
[106] *AT*, X, p. 376, *ll.* 21–24.
[107] Crapulli, *Op. cit.*(n. 95), p. 145.
[108] Adriaan van Roomen, *Apologia pro Archimede* (Würzburg, actually published in Genève, 1597), p. 23.

indifferently.

Marion insisted, as we have already pointed out, that 'mathesis universalis' is more abstract than mathematics, the science of quantity, because it is concerned with order and measure. But order and measure can be construed as principal characteristics of mathematics, not necessarily of method. These terms appear not only in Rule IV-B but also in some of the rules of Book Two. To mention only one example, in Rule XVI Descartes refers to the dual function of number and states that it is "sometimes to express order (*ordo*), sometimes measure (*mensura*)."[109]

At the same time, we can understand the reason why Marion and Van de Pitte had to distinguish between the terms 'mathesis' and 'mathematica(e)'. As Crapulli has suggested, 'mathesis universalis' was understood as the discipline which treats common principles of mathematics. Also Van Roomen sometimes identified 'mathesis universalis' with 'prima mathesis'. According to Van de Pitte, an advocate of Marion's interpretation, 'prima mathesis' or 'mathesis universalis' "must be logically prior to (*i. e.* outside) mathematics,"[110] because it provides mathematics with its fundamental principles. Thus Van de Pitte stated the following:

> It is perfectly natural that he [Descartes] should employ the term "mathesis universalis" to mean the universal principles of learning, or universal method. And the term "mathesis" (when used alone) may mean the underlying principles which render mathematics a science, or simply the principles of learning (*i. e.* method)—or indeed both, since Descartes understands these to be the same principles.[111]

There is no other suggestion as to the reason both Marion and Van de Pitte placed the 'mathesis universalis' outside mathematics other than the statements quoted above. But here we have to ask Marion and Van de Pitte *to whom 'mathesis universalis' is "logically prior* to (*i. e.* outside) mathematics" (The emphasis is mine). In general, Marion interpreted the text of *Regulae* against the background of the works of Aristotle. To Aristotle or other ancient and medieval thinkers, the first principles which imply mathematical propositions may have been logically prior to and outside mathematics. However, is the same statement true for Descartes as well? For the Descartes of the *Regulae* at least, "arithmetic and geometry prove to be much more certain than other disciplines,"[112] and importantly "the first principles themselves are known only through intuition."[113] To him arithmetic and geometry, at least in their idealized form, are the normative disciplines after the example of which the other

---

[109] *AT*, X, p. 457, *l.* 19.
[110] Van de Pitte (n. 93), p. 159. Cf. *Ibid.*, p. 160, note 18.
[111] *Ibid.*, p. 161.
[112] *AT*, X, p. 365.
[113] *Ibid.*, p. 370.

disciplines should be reorganized or newly formed.

'Mathesis universalis' must be considered to refer to the common mathematical discipline placed prior to the main branches of pure mathematics. Of course, it may simultaneously function as a model for the universal method. In this sense and in this sense alone, we can acknowledge Marion's and Van de Pitte's insight that 'mathesis universalis' can be regarded as a universal method. However, it is a part of mathematics, not outside mathematics. In fact, such a dual function of 'mathesis universalis' was also recognized by Van Roomen in his *Universae mathesis [sic!] idea*, published in 1602. He stated in Chapter II "The Division of Mathematics" of the *Universae mathesis idea*:

> The *Universalis [mathematica]* is that which is concerned with quantity of every kind, namely, *Logistice* and *prima Mathesis*, the former as the organon of science and the latter as a science.[114]

Needless to say, the analysis and algebra which Descartes claims to have used to form his method have the same dual function and are viewed sometimes as the organon of discipline and sometimes as a discipline itself. Note that Van Roomen classified pure mathematics into two categories, 'universalis' and 'specialis'.[115] So in his use 'mathesis universalis' was a term applied to a discipline within pure mathematics. Whether or not there is a logical or philosophical problem in Descartes's regarding 'mathesis universalis' as a domain of mathematics, as Marion and Van de Pitte thought, is an issue to be treated separately. In fact we discuss this problem later in Chapter 8, § 1. In any case, we must not project our own interpretation on the historical Descartes.

Now what precisely was referred to by 'mathesis universalis'? In Rule IV-B we are told that at first Descartes was completely satisfied neither with the ancient nor contemporary writers of the mathematical disciplines. But he tells us subsequently he felt that in ancient times there must have been a kind of mathematics (*mathesis*) quite different from the one which prevailed in his day. He calls this discipline 'true mathematics' (*vera Mathesis*).[116] According to him, some traces of this true mathematics could be seen in the works of Pappus and Diophantus. Then he switches the period of his topic of argument from ancient Greece to his own day and states that some mathematicians tried to revive the true mathematics, "for it seems to me to be none other than the art which goes by the outlandish name (*barbaro nomine*) of 'algebra'."[117]

It seemed for Descartes that algebra or, at least, a certain kind of algebra, was an example of the true mathematics. After noticing this, he turned, we

[114] Adriaan van Roomen, *Vniversæ mathesis idea, qua mathematicæ universim sumptæ natura, præstantia, usus et distributio brevissimè proponuntur* (Herbipoli, 1602), p. 14: "Vniversalis est qvæ circa omnem versatur qvantitatem nempe Logistice & prima Mathesis, illa ut organum scientiæ, hæc ut scientia."

[115] *Ibid.*: "Pura iterum duplex universalis & specialis."

[116] *AT*, X, p. 376. See the full passage quoted on p. 91.

[117] *Ibid.*, p. 377. See our quotation on p. 92.

are told, "from the particular studies of arithmetic and geometry to a general investigation of mathematics (*Mathesis*)."[118] He began this investigation by asking what is generally meant by the term 'Mathesis'. He knew that in addition to arithmetic and geometry, astronomy, music, optics, mechanics and so on were considered to be parts of mathematics (*mathematica*), and he paid attention, he tells us, to the fact that almost everybody can easily distinguish what relates to 'Mathesis' from what is related to other disciplines. Then we encounter the crucial statement in which the term 'mathesis universalis' appears for the first time:

> When I considered the matter more closely, I came to see that the exclusive concern of *Mathesis* is with questions of order (*ordo*) and measure (*mensura*) and that it is irrelevant whether the measure in question involves numbers, shapes, stars, sounds, or any other object whatever. This made me realize that there must be a kind of general science (*generalis quaedam scientia*) which explains all the points that can be raised concerning order and measure irrespective of the subject matter, and that this science should be termed *Mathesis universalis*—not a far fetched designation but a venerable term with a well established meaning (*non ascititio vocabulo, sed jam inveterato atque vsu recepto*)—for in this science is contained everything on account of which the other sciences are called parts of *Mathematica*.[119]

Two things should be noticed concerning this passage. First, the term 'mathesis universalis' is not Descartes's invention but a "venerable" one "with a well established meaning." Nor is this term "a far fetched designation." This reminds us of the fact that he was previously attracted to the art which was called "by the outlandish name" algebra and which seemed to him the revival of the true mathematics of ancient times. Clearly, the expression "a far fetched designation" is deliberately used to suggest "an outlandish name." Accordingly, what Descartes indicates by 'mathesis universalis' has a close relationship to algebra. Further, he tells us that "everyone knows the name (i. e. *mathesis universalis*) of this subject and without even studying it understands what its subject matter is."[120] This comment implies that Descartes's concept of 'mathesis universalis' was basically identical with the one which was in wide use among mathematicians in the late sixteenth and seventeenth centuries. Crapulli's study of 'mathesis universalis' in the sixteenth century offers as examples of mathematicians who employed the term 'mathesis universalis' such figures as Cunradus Dasypodius, Johann Heinrich Alsted, and most notably Adriaan van Roomen. According to Crapulli, a model for 'mathesis universalis' during the

[118] *Ibid.*
[119] *AT*, X, pp. 377–378.
[120] *Ibid.*, p. 378.

initial period of its use was the theory of proportion understood as applicable to both continuous magnitudes and discrete numbers.[121] Throughout the *Regulae* Descartes himself invokes not a few examples from the theory of proportion to illustrate his method. And importantly, describing his mathematical career in Part Two of his *Discours de la méthode*, he states: "I did not have any intention of trying to learn all the special sciences commonly called 'mathematics' (*Mathematiques*), for I saw that, despite the diversity of their objects, they agree in considering nothing but the various ratios and proportions (*rapports ou proportions*) that hold between these objects."[122] Descartes viewed the theory of proportion as the core of mathematics. As we have observed in Section 2, in Book Two of the *Regulae*, he attempts to translate the language of the theory of proportion into that of the theory of equations. The latter is nothing but algebra.

As will be argued in detail in Chapter 7, § 2, in his *Universae mathesis idea* Van Roomen expressed 'mathesis universalis' with different names, such as 'supputatrix', 'logistice', and 'prima mathesis'.[123] 'Supputatrix' is a Latin word which designates the art of calculation. Clearly Van Roomen thought this art to be almost synonymous with practical arithmetic.[124] In the section of "the Division of the Mathematical Disciplines" of the "Prolegomena" to his version of Euclid's *Elements*, Clavius used the word 'supputatrix' as it meant the art of calculation.[125] And his disciple Biancani wrote a mathematical textbook in which he viewed practical arithmetic as concerned with 'supputatrix' or 'algebra'.[126] Descartes himself used not necessarily the word 'supputatrix' itself, but 'supputatio' (calculation) very close to the former in Rule XVI of the *Regulae*.[127] Because the word 'algebra' never appears in his *Universae mathesis idea*, Van Roomen might have included algebra there among the constituents of 'mathesis universalis', and his readers a fortiori could have reinterpreted it thus. So there is an increasingly great probability that 'mathesis universalis' can be interpreted as the reformed algebra.

Second, the two principal characteristics, order and measure, of 'mathesis universalis' are key concepts on the basis of which Descartes develops his 'algebra of extensions' in Rule XIV. There he claims that "there are but two kinds of things which are compared with each other: sets (*multitudines*) and magnitudes (*magnitudines*)"[128] in order to express all the various relations and proportions, and then reinterprets these two concepts which represent given

[121] Crapulli, *Mathesis universalis* (n. 95), p. 145.

[122] *AT*, VI, pp. 19–20.

[123] Van Roomen, *Vniversæ mathesis idea* (n. 114), pp. 17–20.

[124] *Ibid.*, p. 17: "Posset non incommodè si vox recepta esset Arithmopraxia vocari."

[125] Clavius, *Euclidis Elementa, Opera mathematica*, t. 1 (1611), p. 4.

[126] See William A. Wallace, *Galileo and His Sources: The Heritage of the Gollegio Romano in Galileo's Science* (Princeton, 1984), p. 145. The author and his book were G. Biancani's *Sphaera mundi seu cosmographia demonstrativa* (Bologna, 1620).

[127] *AT*, X, p. 456.

[128] *AT*, X, p. 450, *ll.* 15–17.

figures by another pair of concepts, i. e. order and measure:

> Moreover, if we are to explain which of all the available figures we are going to make use of here, we should know that all the relations which may possibly obtain between entities of the same kind should be placed under one or other of two categories, viz. order and measure.[129]

The terms, order and measure, are not metamathematical concepts as Marion and Van de Pitte have concluded, but can be explained as mathematical notions. In other words, they are contained within the basic mathematical vocabulary in the same way as, for example, the two important concepts, number and magnitude. They are not necessarily more abstract than the latter ideas, albeit they have wider extension.

These considerations make it reasonable to conjecture that Descartes's 'mathesis universalis' is exactly the reformed algebra which can function as the normative discipline for all certain knowledge. It is noteworthy that he states retrospectively at the end of Rule IV-B, "Up to now [...] I have devoted all my energies to *mathesis universalis*, so that I think I shall be able in due course to tackle the somewhat more advanced sciences, without my efforts being premature."[130] "The somewhat more advanced sciences" may be interpreted as the physics and/or metaphysics for whose sake Descartes tries to reform the universal method. We know that he had drafted his treatise on algebra by October of 1628. While we might not be able to identify his old *Algebra* as an embodiment of his 'mathesis universalis', it seems at least certain that that treatise assumed a crucial role as a concrete model for 'mathesis universalis'.[131]

At the same time, however, does this identification necessarily exclude the other interpretation that 'mathesis universalis' was identifiable with the method? Throughout the *Regulae*, as we have seen in Section 2, algebra plays the function of an analytic method. It is not without reason that some commentators of the *Regulae* have made 'mathesis universalis' equal to the universal method itself. However, we claim that 'mathesis universalis' is not necessarily identical with the method. In this sense, it is justfiable that Daniel Garber cast a question some interpreters who identified 'mathesis universalis' with the

---

[129] *AT*, X, p. 451, *ll.* 5–8.

[130] *Ibid.*, p. 379.

[131] There have been several historians of philosophy and of science who regard the *mathesis universalis* to be symbolic algebra. Among others, see Louis Liard, "La méthode et la mathématique universelle de Descartes," *Revue philosophique de la France et l'Étranger*, **10** (1880), pp. 569–600; *Idem, Descartes* (Paris, ₃1911); J. Klein, *Greek Mathematical Thought and the Origin of Algebra* (n. 79), pp. 197–211, esp. p. 294, note 310; The late Japanese historian of mathematics Nakamura Koshiro identified 'mathesis universalis' with a kind of 'meta-algebra', in his "*Sūgakuteki Hōhō no Keisei nitsuite*" (On the Formation of Mathematical Methods) (in Japanese), *Kagaku Kisoron Kenkyu* (Journal of the Japan Association for Philosophy of Science), Vol. 1, No. 4 (1955), pp. 20–26, at p. 26.

method.[132] Within the domain of the mathematical disciplines, they may be identifiable. 'Mathesis universalis' works fully only within the mathematical sciences. On the other hand, the method is applicable to any knowledge whatever. Rule IV tells us that "we need a method if we are to investigate the truth of things."[133] The method is needed not only for warranting the truth of mathematics but also for that of much more general things. The universal method is applicable to "the more advanced sciences" as well, to use Descartes's own words. It is useful for acquiring certain knowledge in general. Thus we can formulate the following proportional relationship: As 'mathesis universalis' is to mathematics so the universal method is to certain knowledge. It is in this sense that 'mathesis universalis' functions as the normative discipline for discovering the truth of things. If certain knowledge were restricted only to mathematics, the method would be equivalent to 'mathesis universalis'. But Descartes definitely negates this possibility at the end of Rule II, that is to say, he maintain there that certain knowledge is not restricted simply to mathematics (see p. 250).

In order to attain the universal method, Descartes suggests that studies outside of mathematics are required as well. To him mathematics is important only in the sense that it provides him with a way to familialize himself with the universal method. In Rule XIV he elaborates his conception as to the relation between the study of pure mathematics and the method:

> At this point we should be delighted to come upon a reader favourably disposed towards arithmetic and geometry, though I would rather that he had not yet embarked upon these studies than that he had been taught them in the usual manner. For the Rules which I am about to expound are much more readily employed in the study of these sciences (where they are all that is needed) than in any other sort of problem. Moreover, these Rules are so useful in the pursuit of deeper wisdom that I have no hesitation in saying that this part of our method was designed not just for the sake of mathematical problems; our intention was, rather, that the mathematical problems should be studied almost exclusively for the sake of the excellent practice which they give us in the method. I shall not assume anything drawn from the aforementioned disciplines, save perhaps certain facts which are self-evident and accesible to everyone.[134]

Indeed 'mathesis universalis' works as the normative science for one who intends to provide the method for certain knowledge.

There remains the textual problem in Rule IV which Jean-Paul Weber advanced in 1964. Is Rule IV-B an integral part of Rule IV-A or independent of the latter? Marion and Van de Pitte argued that there is a continuity between

---

[132] Daniel Garber, *Descartes' Metaphysical Physics* (Chicago/London, 1992), p. 319.
[133] *AT*, X, p. 371.
[134] *Ibid.*, p. 442.

the two sections of Rule IV. On the other hand, Pamela A. Kraus recently answered them, asserting that Rules IV-A and IV-B are heterogeneous in both intention and theme whereas they have stylistic and thematic parallels, and that Descartes composed Rule IV-B first and proceeded to write down Rule IV-A next, keeping IV-B at hand as a model.[135] Her attempted solution appears unusually persuasive. But unfortunately we do not have any definitive evidence which may either support or refute her thesis. And we have to point out one element in her conjecture with which we cannot be satisfied. She generally emphasizes a discontinuity between IV-A and IV-B and, in our opinion, fails to explain sufficiently the role of 'mathesis universalis' in the formation of the Cartesian universal method.

Now let us review what Descartes states toward the end of Rule IV-A. After noting that the geometrical analysis of the ancients and the algebra of the moderns are "the spontaneous fruits which have sprung from the innate principles" of his method,[136] Descartes comments on the rule which constitutes his method:

> I would not value these Rules so highly if they were good only for solving those pointless problems with which calculators (*Logistae*) and geometers are inclined to while away their time, for in that case all I could credit myself with achieving would be to dabble in trifles with greater subtlety than they. I shall have much to say below about figures and numbers, for no other disciplines can yield illustrations as evident and certain as these. But if one attends closely to my meaning, one will readily see that ordinary mathematics is far from my mind here, that it is quite another discipline I am expounding, and that these illustrations are more its outer garment (*integumentum*) than its inner parts.[137]

At the start Descartes rejected, he recollects, "ordinary mathematics" (*vulgaris Mathematica*) and began to praise another discipline. Mathematical illustrations about figures and numbers are, according to him, "its outer garment," not "its inner parts."

What then is this outstanding discipline which constitutes the central core of the mathematical disciplines? Descartes continues his explanation after the passage which we have just cited:

> This [discipline] should contain the primary rudiments of human reason and extend to the discovery of truths in any field whatever.

[135] Pamela A. Kraus, "From Universal Mathematics to Universal Method: Descartes's 'Turn' in Rule IV of the *Regulae*," *Journal of the History of Philosophy*, **21** (1983), pp. 159–174: "Evidently Descartes thought it useful to keep IV-B at hand as a model, making alternations as he went along, while composing IV-A." (p. 171.)

[136] *AT*, X, p. 373. See our citation on p. 92.

[137] *AT*, X, pp. 373–374.

> Frankly speaking, I am convinced that it is a more powerful instrument of knowledge than any other with which human beings are endowed, as it is the source of all the rest. I have spoken of its 'outer garment', not because I wish to conceal this science and shroud it from the gaze of the public; I wish rather to clothe and adorn it so as to make it easier to present to the human mind.[138]

This statement concludes Rule IV-A. Then begins the autobiographical narration about his mathematical career which we have already quoted in Chapter 2, § 5 (p. 110) and which forms the beginning part of Rule IV-B.

Now we ask again: What is the discipline "which should contain the primary rudiments of human reason and extend to the discovery of truths in any field whatever"? Weber interpreted it as the method.[139] Marion similarly viewed this sentence as a statement which explains the universality of the method.[140] We do not completely deny their explanations. Nevertheless, we should remind ourselves that "another discipline" is compared with ordinary mathematics. Moreover, the cited passage which contrasts that discipline with ordinary mathematics comes after his description of the spontaneous fruits of the method, i.e. geometrical analysis and algebra. If this discipline means the method, is it not the form confined to mathematics? If the answer is in the affirmative, it must have an extremely close relationship with 'mathesis universalis', for the latter conception is the corresponding part of the universal method restricted to the mathematical sciences. And so is the expression "another discipline" more conformable with 'mathesis universalis' than the universal method? At least we surmise that the last paragraph of Rule IV-A can be interpreted as an introductory remark to Rule IV-B. Then Rule IV-B may be a retrospective autobiographical sketch of how Descartes first discovered the method in the field of mathematics and how the universal method is to be understood accurately.[141]

In this way we can recover a natural connection between Rules IV-A and IV-B which might have been perceived by the copyist of the Amsterdam manuscript and the translator of the Dutch 1684 edition. We do not necessarily assert that ours is the only legitimate intepretation, but simply claim it has the same degree of validity as, or possibly greater validity than other previously proposed elucidations.

In sum, Descartes's 'mathesis universalis' was a very significant concept which functioned as a central model for the universal method. It meant above all a reformed algebra applicable first of all to arithmetic and geometry and then the other mathematical sciences. Descartes may seem to have seen its concrete realization in his old *Algebra*, although the latter monograph might

---

[138] *Ibid.*, p. 374.

[139] Weber, *La Constitution du texte des Regulae* (n. 50), p. 5.

[140] Descartes, *Règles utiles et claires etc.*(n. 97), p. 141, note (16).

[141] In this regard, Van de Pitte's observation on the text of Rule IV is suggestive. See *Op. cit.* (n. 93), pp. 172–173.

have been an insufficient specimen of his ideal of 'mathesis universalis' and ultimately of the universal method. The reformed algebra was to function as the method of analysis or the organon of discovery in mathematics. The universal method, on the other hand, was devised with 'mathesis universalis' as a model, so that Descartes could acquire certain knowledge in a broader field than the mathematical disciplines. That reformed algebra or 'mathesis universalis' was not ordinary algebra but a thoroughly 'methodized' ideal science which could be counted as one of the mathematical disciplines.

Here we should compare our interpretation just mentioned with that by Schuster and his followers which undertands that 'mathesis universalis' is identified with the program of the application of mathematics to natural philosphy, or of mathematical physics.[142] Certainly it was quite important for Descartes's program of the reform of learning to establish mathematical physics. Thus we should consider this problem independently of the construction of reformed algebra and of the reformation of metaphysics. Nevertheless, it may not be permitted to confuse the concept of 'mathesis universalis' with the program of mathematical physics. Schuster's errors are essentially two. One is that he has failed in interpreting the *Regulae*, especially Rule IV. Descartes's 'mathesis universalis' was a core part of the mathematical sciences and was to serve them as their normative discipline. It should not be confused with applied sciences of mathematics, of an "outer garment" in Descartes's own words. However, Schuster confused the two. There existed a large volumes entitled *Elementa matheseos universae* (first published in 1713–15, in 2 volumes) by Christian Wolff, a 'vulgar' disciple of Leibniz. Its most extensive version published in 1730–41 in 5 volumes was an encyclopedic writing on the mathematical sciences of the early eighteenth century beginning with the methodology of mathematics, and proceeding to various fields of pure mathematics such as arithmetic, geometry, trigonometry, finite and infinite analysis, and to 'mixed mathematics' such as mechanics, statics, hydrostatics, optics, catoptrics, dioptrics, astronomy, geography, hydrogeography, chronology, and even military and civil architecture.[143] The name of the sciences which Wolff called "mathesis universa" or mathematical sciences as a whole was nothing other than the "outer garment" of the mathematical disciplines, and the concept clearly different from 'mathe-

[142] J. A. Schuster, "Descartes and the Scientific Revolution 1618–1634: An Interpretation" (n. 44) & "Descartes' *Mathesis Universalis*: 1619–28" (n. 93). Giorgio Israel, "The Analytical Method in Descartes' *Géométrie*," in M. Otte and M. Panza, eds., *Analysis and Synthesis in Mathematics* (Ch. 2, n. 50), pp. 3–34, discussing Descartes's mathematics comprehensively, has been influenced by Schuster's interpretation partially and made its entire argument unclear. See especially the following: "In fact, Descartes proposes a quantitative interpretation of the Universe, centering on mathematics, i. e. *Mathesis universalis*, completely different from "vulgar" mathematics of his time, and a universal knowledge that permits reducing the analysis of any phenomenon to problems of "order" and "relations"." (pp. 17–18.)

[143] Christian Wolff, *Elementa matheseos universæ*, t. I–V (Halle, 1730–41): *Gesammelte Werke*, II. Abteilung, Lateinische Schriften, Bde. 29–33, edidit et curavit J. E. Hofmannus (Hildesheim, 1968–71).

sis universalis'. It may be certainly distinguished from Descartes's program of mathematizing natural philosophy. Descartes's was a project, whereas Wolff's was a fruit of it.

The other error which Schuster has committed is that he didn't pay sufficient attention to the contemporary use of the notion of 'mathesis universalis'. In the seventeenth century it generally meant a discipline common to mathematical sciences. This is exactly what Crapulli's comprehensive monograph has taught. In the meantime, Wolff was surely familiar with the concept of 'mathesis universalis' which was supposed to be different from the 'mathesis universa'. He referred to the 'mathesis universalis' as an independent heading of his *Mathematisches Lexicon* of 1716. There "mathesis universalis" is translated into the German "die Universal-Mathematick oder allgemeine Mathematick," and further interpreted to be "die Buchstaben-Rechen-Kunst" by Erasmus Bartholin, "die Rechen-Kunst mit Ziffern und Buchstaben" by John Wallis, or an attempt by Leibniz to extend these traditional uses to a discipline of quantity in general as examples. As for the last attempt, Wolff observes as follows: "Diese Mathesis universalis ist zur Zeit noch nicht erfunden," that is to say, this kind of 'mathesis universalis' is not yet found at present.[144] To confuse 'mathesis universalis' with 'mathesis universa' from the middle of the sixteenth to the early eighteenth centuries is amount to an error to mistake an intersection of sets for a union of them in set theory today. We will argue these uses again in Chapter 8, § 1, F.

In any case, the *Regulae ad directionem ingenii* was a treatise in which Descartes attempted to explicate the universal method, and was superseded by the more 'methodized' *Discours de la méthode* and other essays of 1637 and was left in the extant incomplete form. The *Regulae* seems to have been superseded, as we will show in what follows, by Descartes's first published book of 1637 principally through two paths, mathematical and philosophical.

What separated more strictly the mathematics of the *Regulae* and that of the *Géométrie* of 1637 was, for example, that only the latter could totally succeed in solving the problem of geometrical dimensionality by establishing the ingenious idea which can appropriately be called 'algebra of line segments'. We may state that it was only by means of this conceptually crucial notion that Descartes could accomplish successfully the 1619 program of "an entirely new science" in the literal sense of the words. As we saw above in the sections on the old *Algebra* and the *Regulae*, there were certain mathematical insufficiencies in Descartes's attempt to reduce multi-dimensional geometrical magnitudes to one-dimensional lines. This problem was resolved for the first time in a mathematically persuasive manner in the beginning part of the *Géométrie*. The con-

[144] Ch. Wolff, *Mathematisches Lexicon* (Leipzig, 1716): *Gesammelte Werke*, I. Abteilung, Deutsche Schriften, Bd. II, herausgegeben und bearbeitet von J. E. Hofmann (Hildesheim, 1965), col. 869. He referred to 'mathesis univessalis' in the above-quoted *Elementa matheseos universæ*, t. 1, p. 21: "Præfatio" ad *Elementa Arithmeticæ*.

ceptually matured mathematics of algebraic analysis first appeared with this essay. To show this is the task of the next chapter.

Furthermore, our considerations in this section lead to the second path as well concerning the philosophy of mathematics. In the *Regulae*, Descartes was modeling his general method on 'mathesis universalis'. But, with the deepening of his earnest metaphysical speculation which started in the winter of 1628–1629 in Holland, the *Regulae*'s epistemological assumption that pure mathematics provides the ultimate paradigm for certainty seems to have begun to be abandoned in some sense or other. Importantly, Descartes never used the concept of 'mathesis universalis' in his later philosophical writings after the *Regulae*. We do not know exactly whether this shift of his metaphysical position was related to the disuse of that concept. This is certainly a significant problem which should be discussed thoroughly. Indeed, the *Regulae ad directionem ingenii* and the idea of 'mathesis universalis' occupied a crucial place not only in Descartes's intellectual career but also in the entire history of modern thought, even in Western thought as a whole. The historical and philosophical significance of this pivotal concept will be argued in detail in Part II, and Descartes's 'mathesis universalis', in particular, again in Chapter 8, § 1 against the background of his philosophy of mathematics after the *Regulae*.

Chapter 5

The *Géométrie* of 1637

§ 1. The Pappus Problem

By the time that he composed his old *Algebra* and *Regulae ad directionem ingenii*, Descartes had established his basic program of algebraic analysis, which, even though immature, may be compared with that of François Viète. By that period his main interest, too, had already shifted from pure mathematics to natural philosophy (especially optics) and metaphysics. His studies carried out in the Netherlands during the years around 1630 first crystallized as *Le Monde, ou Traité de la lumière*, published posthumously in 1664, and the *Dioptrique*, one of the three essays added to the *Discours de la méthode* of 1637. The latter treatise exploited the theory of conic sections, which was excerpted partly in Beeckman's *Journal* in 1628–1629 (see Chapter 4, § 1), whereas the former was a writing which can be evaluated to be "the first attempt to construct an entire physical universe on mechanical foundations."[1]

We have already pointed out that Descartes claimed his mathematical activities undertaken at his own initiative essentially terminated before the year 1623. And on April 15, 1630 he wrote to Mersenne, who had requested him to propose some appropriate mathematical problems to the Parisian mathematicians, the following: "As for problems, I shall send you a million of them to propose to others if you like; but I am so tired of mathematics and now hold them in so little esteem that I can no longer take the trouble to solve them myself."[2]

Descartes had probably carried out his mathematical studies with enthusiasm before 1623. And surely about 1630 his main interest was no longer mathe-

[1] Michael S. Mahoney's Introduction to, Descartes, *The World* (New York, 1979), p. xv.

[2] "Descartes, à Amsterdam, à Mersenne, à Paris. 15 avril 1630," *AT*, I, p. 139; *AM*, I, p. 131; *CM*, II, p. 425: "Pour des Problesmes, je vous en envoyeray un milion pour proposer aus autres, si vous le desirés; mais je suis si las des mathematiques, et en fais maintenant si peu d'estat, que je ne sçaurois plus prendre la peine de les soudre moymesme."

matics. But he never completely lost interest in solving mathematical problems, especially at the earnest request of his friends. He even left an indication that he considered himself to be a student of mathematics. On April 16-26, 1629, he enrolled at the University of Franecker, where his signature has been preserved as "RENATUS DES CARTES GALLUS, PHILOSOPHUS" (René Descartes Frenchman, Philosopher).[3] But on June 27, 1630, the year when he admitted being tired of solving mathematical problems, he registered at the University of Leiden as "RENATUS DESCARTES PICTO, studiosus matheseos, annos natus XXXIII" (René du Perron Descartes, student of mathematics, 33 years of age).[4]

Descartes seems to have enrolled at the University of Leiden to study mathematics with Jacobus Golius and astronomy with Martinus Hortensius. Jacobus Golius, or Jacob Gool, was well known primarily as an Orientalist. At the University of Leiden he first took office as professor of Oriental languages in 1624. In that year, he left for the Near East, travelling there for four years. He was nominated in absentia for the additional position of professor of mathematics upon the death of Willebrord Snel in 1626, and assumed the position on his return in 1629. As a mathematician, he had a reputation for being well versed in the ancient theory of conic sections, especially that of Apollonius of Perga.[5]

How Descartes was viewed by others in the early 1630s can be seen to some extent from his friend Balzac's letter addressed to him in the spring of 1631: "Your efforts to conquer truth with plenty of force and courage seem to me something more noble than everything done with much noise and turmoil in Germany and in Italy."[6] What was happening in Germany and in Italy at that time must be political and religious conflicts, notably the Thirty Years War based on religious confrontation. The intellectual mission "to conquer truth" attempted by Descartes since 1619 was attaining a new stage in the 1630s.

It was on the occasion at which Golius proposed a problem stated in Pappus's *Collection* to mathematicians such as Descartes and Mydorge in 1631 that Descartes applied his method of algebraic analysis to ancient geometrical problems and gained the self-confidence that his mathematical method could surpass the ancient one. In his letter of April 5, 1632 to Mersenne, he wrote that it took five or six weeks for him to discover a solution to the problem Golius had posed.[7] And according to Leibniz's note "Remarques sur l'abrégé de la Vie

---

[3] Gustave Cohen, *Écrivains français en Hollande dans la première moitié du XVIIe siècle* (Paris, 1920), p. 436.

[4] *Ibid.*, p. 452.

[5] For the biography of Golius, see *Nieuw Nederlandsch Biografisch Woordenboek*, X (Leiden, 1937), cols. 287–289; *Nouvelle Biographie générale*, XXI (Paris, 1857), cols. 120–123.

[6] "Balzac à Descartes, [Paris, 25 avril 1631]," *AT*, I, pp. 200–201; *AM*, I, p. 188.

[7] "Descartes, à Amsterdam, à Mersenne, à Paris. 5 avril 1632," *AT*, I, p. 244; *AM*, I, p. 221; *CM*, III, p. 292.

de Monsieur des Cartes," he consumed six weeks on the problem.[8]

Descartes's solution to Pappus's problem and his way of sorting curves appear in Books One and Two of the *Géométrie*, and what state these were in 1631 can be reconstructed by means of a letter which he sent to Golius indubitably during January of 1632. From the very first sentence of that letter, it is clear that the letter is meant to supplement a previous writing which contained an entire solution: "I was extremely pleased that you gladly took pains to examine the writing which I sent to you."[9]

What was, then, the Pappus problem which Golius proposed to Descartes? Golius took the problem from Book VII ('Domain of Analysis') of Pappus's *Collection*. There Pappus counted Apollonius's Conics to be among the many treatises which constituted the 'Domain of Analysis'. It was in a commentary to Apollonius's *Conics* that Pappus mentioned a problem of the locus which he claimed that neither Euclid nor Apollonius nor any other mathematician had been able to solve entirely.

Pappus begins his explanation of the problem with the case of three or four lines (As Descartes did, we refer to Commandino's Latin version of 1588 as the basic text):

> If three straight lines are given in position, and if straight lines be drawn from one and the same point, making given angles with the three given lines, and if there be given the ratio of the rectangle contained by two of the lines so drawn to the square of the other, the point lies on a solid locus given in position, namely, one of the three conic sections. And, if lines be drawn making given angles with four straight lines given in position, and if the rectangle of two of them bears a given ratio to the rectangle of the other two; then, in the same way, the point lies on a conic section given in position.[10]

Apollonius may be said to have given, as Pappus suggests, a partial solution to the problem of the locus with reference to three or four lines in Book Three of his *Conics*.[11] But Apollonius's attempt stopped here. Pappus further

[8]See G. W. Leibniz, "Remarques sur l'abrégé de la Vie de Monsieur des Cartes," *Die philosophischen Schriften*, Bd. IV, herausgegeben von C. I. Gerhardt (Berlin, 1880), p. 312. Leibniz heard this story from Claude Hardy during his sojourn in Paris during the years between 1672 and 1676. Leibniz's observation is the following: "J'ay cela de M. Hardy qui me l'a couté autrefois à Paris. M. Boyle s'etonne dans quelque endroit que M. des Cartes a pu se resoudre à employer six semaines à un seul probleme. Mais il faut considerer que ce probleme contient une grande suite d'autres."

[9]"Descartes, à Amsterdam, à [Golius]. [Janvier 1632]," *AT*, I, p. 232; *AM*, I, p. 212.

[10]Commandino, f. 165r; *AT*, VI, pp. 377–378; Olscamp, pp. 182–182. I basically rely on Olscamp's translation. Cf. Alexander Jones, *Op. cit.* (Ch. 2, n. 56), pp. 120–121; Hultsch, II, pp. 678–679; Ver Eecke, II, p. 508.

[11]*Apollonii Pergaei quae exstant cum commentariis antiquiis*, edidit et latine interpretatus est J. L. Heiberg (Leipzig, 1893), Vol. I, Propositiones LIII-LVI, pp. 438–451; *Les Coniques d'Apollonius de Perge*, traduites par Paul Ver Eecke, Nouveau Tirage (Paris, 1963), pp. 273–

proceeds to the case of the loci when five or six lines are given. In these cases, a ratio is made out of a rectangle contained by two lines or a solid by three lines, these figures being somehow explainable in geometrical language. The point which makes a locus lies on a line which transcends the solid locus.

Pappus, then, attempts to generalize the problem to the cases when there are more than six lines. But in these events, a difficulty occurs, for, in his own words, "we cannot say whether a ratio of something contained by four lines is given to that which is contained by the rest, since there is no figure of more than three dimensions."[12] But he tries to overcome the difficulty by using an ingenious and sophisticated concept in the theory of proportion, namely compound ratio. Employing this concept he continues his explanation (*apud* Commandino): For in this are agreed those who formerly interpreted these things, in that they hold that a figure contained by these lines is not understandable in any way. This is permissible, however, both to say and to demonstrate generally in the aforesaid proportions by compound ratios and in the latter cases in this way."[13] Thus, the locus problem with reference to more than six lines can be formulated as follows:

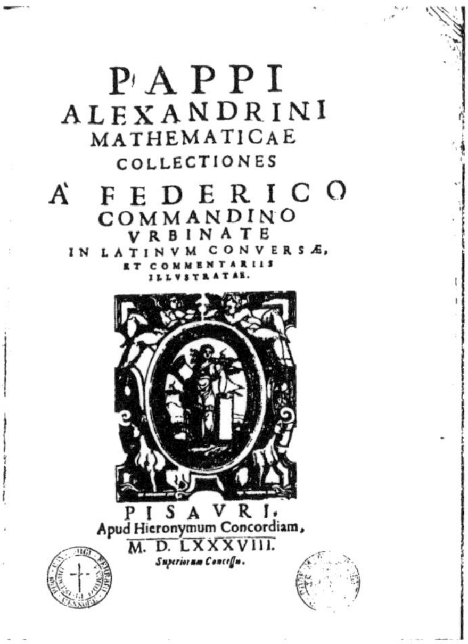
PAPPI
ALEXANDRINI
MATHEMATICAE
COLLECTIONES
A FEDERICO
COMMANDINO
VRBINATE
IN LATINVM CONVERSÆ,
ET COMMENTARIIS
ILLVSTRATAE.
PISAVRI,
Apud Hieronymum Concordiam,
M. D. LXXXVIII.
Superiorum Concessu.

Figure 5.1 The Title Page of Latin Translation by Federico Commandino of Pappus's *Collection*, Published in 1588 (The author's possesion)

280; T. L. Heath, *Apollonius of Perga: Treatise of Conic Sections* (Cambridge, 1896), "Locus with Respect to Three Lines &c.," pp. 119–125; Heath, *A History of Greek Mathematics*, Vol. 2 (Oxford, 1921), p. 157; H. G. Zeuthen, *Die Lehre von den Kegelschnitten im Altertum* (Kopenhagen, 1886; Hildesheim, 1965), pp. 126–150. Zeuthen and Heath interpreted the work of Apollonius by using the concept of 'geometrical algebra'. We are critical of their reconstruction, however.

[12] Commandino, f. 165v; *AT*, VI, p. 378; Olscamp, p. 183. Cf. Jones, pp. 122–123; Hultsch, II, pp. 680–681; Ver Eecke, II, p. 509.

[13] Commandino; Olscamp; Jones; Hultsch, *Ibid.* *AT*, VI, pp. 378–379; Ver Eecke, pp. 509–510: "Acquiescunt autem his qui paulo ante talia interpretati sunt, neque vnum aliquo pacto comprehensibile significantes quod his continetur. Licebit autem per coniunctas proportiones haec dicere demonstrare vni-uerse in dictis proportionibus, atque his in hunc modum." Note that Commandino's text is corrupted, so "in the aforesaid proportions" (in dictis proportionibus) should be read as "in the aforesaid propositions" (in dictis propositionibus; ἐπὶ τῶν τροειρημένων προτάσεων).

> If from any point straight lines be drawn making given angles with straight lines given in position; and if there be given a ratio compounded of them, that is the ratio that one of the lines drawn has to one, the second has to a second, the third to a third, and so on to the given line if there be seven lines, or if there be eight lines, of the last to a last, the point lies on the lines that are given in position. And similarly, whatever may be the odd or even number, since these, as I have said, correspond in position to the four lines; therefore they have not set forth any method so that a line may be known, etc.[14]

Let us elucidate the problem of Pappus in our own terminology. To explain the locus problem in a mathematically precise way, we shall use H. J. M. Bos's symbols for the given lines $(L_1, L_2, \cdots)$ and lengths $(d_1, d_2, \cdots)$ of straight lines drawn from a point (P) to meet the lines at given angles $(\varphi_1, \varphi_2, \cdots)$[15] (see Figure 5.2). (Figure (a) is taken from Descartes's *Géométrie* and (b) is from Bos's article.) And let $\alpha : \beta$ be a given ratio and a given line segment. The problem for three lines is to find points such that $(d_1 d_2) : (d_3^2) = \alpha : \beta$, and for four lines $(d_1 d_2) : (d_3 d_4) = \alpha : \beta$. The characteristic properties for five and six lines can be formulated as $(d_1 d_2 d_3) : (d_4 d_5 a) = \alpha : \beta$ and $(d_1 d_2 d_3) : (d_4 d_5 d_6) = \alpha : \beta$, respectively. These formulations are permissible without danger of anachronism since all symbols in the formulae above have a certain geometrical meaning. But in Pappus expressions like $d_1 d_2 d_3 d_4$ were not admitted because they could not be explained geometrically.

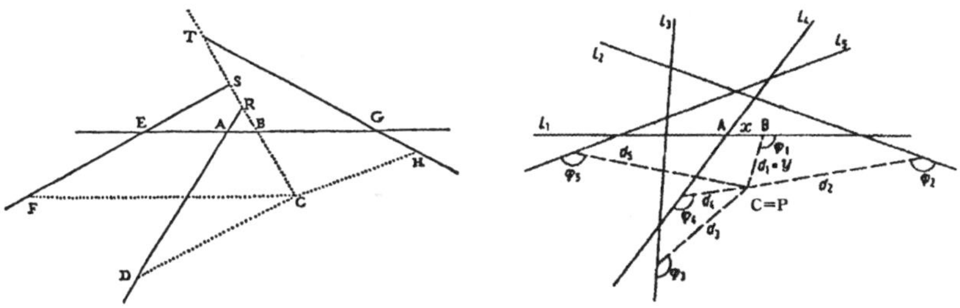

(a) (b)

Figure 5.2

It was the concept of compound or compounded ratio (ὁ συνημμένος λόγος or συγκίμενος λόγος; in Commandino's translation, *conjuncta proportio* or *composita proportio*, respectively) that helped Pappus overcome this difficulty. Some manuscripts of Euclid's *Elements* contain the following definition at the beginning of Book VI:

---

[14]Commandino; Jones; Hultsch, *Ibid.*; *AT*, VI, p. 379; Olscamp, pp. 183–184; Ver Eecke, p. 510.

[15]H. J. M. Bos, "On the Representation of Curves in Descartes' *Géométrie*," *Archive for History of Exact Sciences*, **24** (1981), pp. 295–338, esp. p. 299.

> A ratio is said to be compounded of ratios when the sizes of the ratios multiplied together make something.[16]

The definition is not presumed to have been authored by Euclid himself, as will be discussed later, and suggests that compounding ratios should be viewed as an analogue of multiplying fractions. Clearly Pappus adopted this definition of compounded ratio when he formulated the problem of the locus with respect to more than six lines. Let the compounding of the ratios $(\alpha : \beta)$ and $(\gamma : \delta)$ be denoted by $(\alpha : \beta) \cdot (\gamma : \delta)$. Then, the locus problem for any number of lines $(n \geqq 2)$ amounts to finding the points such that:

$$(d_1 : d_{n+1}) \cdot (d_2 : d_{n+2}) \cdots (d_n : a) \quad [\text{or } (d_n : d_{2n})] = \alpha : \beta.$$

Pappus could have identified $(\alpha : \beta) \cdot (\gamma : \delta)$ with $\alpha\gamma : \beta\delta$ in a geometrically legitimate manner through Euclid VI, 23: "Equiangular parallelograms have to one another the ratio compounded of the ratios of their sides," and then $(\alpha : \beta) \cdot (\gamma : \delta) \cdot (\varepsilon : \eta)$ with $\alpha\gamma\varepsilon : \beta\delta\eta$ by considering appropriate parallelepipeds. But we should note that he would never have permitted the above formulation by means of compound ratios to be explicated as

$$\begin{aligned}&(d_1 d_2 \cdots d_n) : (d_{n+1} d_{n+2} \cdots a)\\&[\text{or } (d_1 d_2 \cdots d_n) : (d_{n+1} d_{n+2} \cdots d_{2n})] = \alpha : \beta,\end{aligned}$$

for $d_1 d_2 \cdots d_n (n \geqq 4)$ does not make sense geometrically.

Let us, then, examine how Descartes attempts to solve Pappus's locus problem in Book One of the *Géométrie* before reconstructing his solution of the winter of 1631–1632. He takes up the ancient locus problem as an example which shows that the ancients had no true method for discovering solutions exhaustively, but that they had simply gathered together those propositions they had stumbled upon. "And we can also see this very clearly," he adds to the above observation, "from what Pappus put at the beginning of his seventh book."[17] Introducing Pappus's explanation of the problem and presenting his attempt to solve it up to six lines, he interrupts the quotation and comments on it as follows:

> Here I beg you to note in passing that scruple which caused the ancients to use the terms of arithmetic in geometry, which could proceed only from the fact that they did not see clearly enough the relation between the two, caused much obscurity and hindrance in the manner in which they explained them.[18]

[16]Thomas L. Heath, *The Thirteen Books of Euclid's Elements*, Vol. 2 (New York, ${}_2$1956), p. 189. Cf. Ian Mueller, *Philosophy of Mathematics and Deductive Structure in Euclid's Elements* (Cambridge, Mass., 1981), pp. 87–88; Ken Saito, "Compounded Ratio in Euclid and Apollonius," *Historia Scientiarum*, **31** (1986), pp. 25–59

[17]*AT*, VI, p. 377; Olscamp, p. 182.

[18]*AT*, VI, p. 378; Olscamp, p. 183.

By saying so, Descartes censures Pappus for the latter's obscure manners of explanation, especially in those which utilized the concept of compound ratio. After quoting a passage which contains Pappus's solution to the problem for more than six lines by means of compound ratios, he commences to expound his own solution of the ancient problem. His response to the problem of Pappus, up to six lines, is fundamentally the same as the ancients'. He continues with a quite similar explanation for the problem with reference to more than six lines:

> Or, if there are seven [straight lines given in position], what is produced (*ce qui se produist*) when we multiply four of them [straight lines drawn from a point to the given straight lines] by each other must be the given ratio to what is produced by the multiplication of the three others, and again of another given line. Or, if there be eight, the product (*le produit*) of the multiplication of four must have the given ratio to the product of the four others. And thus this problem can be extended to any number of lines. Then, since there are always an infinite number of points which can fulfill the requirement here it is also necessary to know and to trace the line in which they are to be found.[19]

Pappus certainly recognized, Descartes further asserts, that when only three or four lines are given in position, this line is one of the three conic sections. But he again accuses the ancient geometrical analyst:

> He [Pappus] does not undertake to determine it [the conic section], nor to describe it, no more than [he tries] to explain those where all these points must be found, when the problem involves a greater number of lines. He only adds that the ancients had recognized one of them which they had shown to be useful, but which seemed the most clear, and was not, however, the most important. This gave me occasion to try and discover whether, through my method, I could go as far as they had gone.[20]

Descartes's solution of the locus problem is methodically very coherent and applicable to all cases whatever number of straight lines are given. But it is by cutting a Gordian knot, as it were, that he secures the unitary character of his solution. In other words, he has acquired his unified solution by giving a legitimate meaning to the expressions such as $d_1 d_2 \cdots d_n (n \geqq 4)$. The expression $d_1 d_2 \cdots d_n (n \geqq 4)$ was thought in ancient times to be refractory to any geometrical interpretation since it was more than three dimensions. But for Descartes, who had been exerting himself to unify arithmetic and geometry with the help of his reformed algebra since 1619, it simply indicates a one-dimensional line segment. As we have seen in Chapter 4, § 1, in his old *Algebra* he tried to solve

[19] *AT*, VI, p. 380; Olscamp, p. 184.
[20] *AT*, VI, p. 380; Olscamp, pp. 184–185.

the problem of dimensionality for geometrical magnitudes. His attempts were rather awkward and in the end unsuccessful.

At the beginning of the *Géométrie*, however, Descartes has attained a deceptively simple solution to the dimensionality problem. In the following manner he explains that the product of the two line segments is not of two dimensions but simply of one dimension:

> [B]y taking one line which I shall call unity, in order to relate it as closely as possible to numbers (*nombres*), and which usually can be chosen arbitrarily, and then by taking two others, [we may] find a fourth line which is to one of these two lines as the other is to the unity—which is the same as multiplication.[21]

We should note that here Descartes takes the unit arbitrarily so that geometrical lines may be considered to be numbers. He further elucidates this operation geometrically.

> For example, let AB be unity [Figure 5.3 (a)], and let it be necessary to multiply BD by BC; I have only to join the points A and C, then to draw DE parallel to AC, and BE is the product of this multiplication.[22]

Let BC and BD be two given line segments. Construct a triangle, of which one side is BC and the other side is the unit. Since AC is parallel to DE by construction, AB is to BC as BD is to BE.

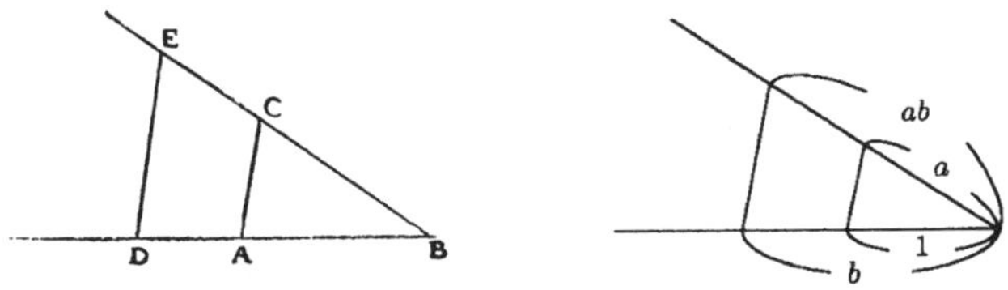

(a) Figure 5.3 (b)

BE, a line segment, is the product of multiplying BD by BC, for AB is the unit. Namely, in today's notation (Figure 5.3 (b)), from the proportional relation $1 : a = b : ab$, a line segment ab can be obtained. Division can be similarly interpreted as an operation on line segments. Together with the idea from Rule XVIII of the *Regulae*, which explains addition and subtraction by means of corresponding geometrical operations of two line segments, he has established

[21] *AT*, VI, p. 370; Olscamp, p. 177. The historical significance of Descartes's algebra of line segments is emphasized in M. S. Mahoney, "The Beginnings of Algebraic Thought in the Seventeenth Century," Chapter 5 of Stephan Gaukroger, ed., *Descartes: Philosophy, Mathematics and Physics* (Brighton, Sussex/Totowa, New Jersey, 1980), pp. 141–155, esp. p. 145.

[22] *AT*, VI, 370; Olscamp, pp. 177–178.

the 'algebra of line segments'. On introducing the notations a^2, b^3 and so on, he further declares:

> Here it is to be noted that by a^2 or b^3 or the like, I ordinarily mean simple lines, although, in order to make use of the names in algebra, I call them squares, cubes, etc.[23]

Thus by building the 'algebra of line segments', Descartes can relate any of the arithmetical calculations to geometrical operations. In this regard it is no accident that he states at the very beginning of the *Géométrie* that "all the problems of geometry can easily be reduced to such terms that thereafter we need to know only the length of certain straight lines in order to construct them."[24] The program for unifying arithmetic and geometry which he began to entertain as early as the spring of 1619 has now been successfully accomplished. He can then proclaim: "I shall not hesitate to introduce these arithmetical terms into geometry, in order to make myself more intelligible."[25] In this sense, nothing expresses the principal characteristic of his mathematics more appropriately than his own term "le calcul Geometrique" in Book Two.[26] With Descartes geometry has become mere calculation. Thereby, Pappus's formulation by means of compound ratios which represents the locus problem

$$(d_1 : d_{n+1}) \cdot (d_2 : d_{n+2}) \cdots (d_n : a) \text{ [or } (d_n : d_{2n})] = \alpha : \beta$$

can be translated into an algebraic one

$$(d_1 d_2 \cdots d_n) : (d_{n+1} d_{n+2} \cdots a) \text{ [or } (d_1 d_2 \cdots d_n) : (d_{n+1} d_{n+2} \cdots d_{2n})] = \alpha : \beta$$

in a theoretically justifiable manner.

Of course, Pappus himself might have thought at least implicitly that the formulation for $n \geqq 4$ through compound ratios was a natural extension of the ones for $n = 2$ and 3 interpreted geometrically as, for example, $(d_1 d_2 d_3) : (d_4 d_5 d_6) = \alpha : \beta$. It is reported that Pappus made frequent use of compound ratios which neither Euclid nor Apollonius treated.[27] In this sense, Pappus, a leading Alexandrian mathematician in the silver age of Greek mathematics who flourished during the first half of the fourth century A. D., was

[23] *AT*, VI, p. 371; Olscamp, p. 178.
[24] *AT*, VI, p. 369; Olscamp, p. 177.
[25] *AT*, VI, p. 370; Olscamp, p. 177.
[26] *AT*, VI, p. 390; Olscamp, p. 191.
[27] In this regard, Alexander Jones's remark is significant. He believes that Pappus used the notion of compound ratios which might be interpreted as that of 'products of ratios' under the influence of his predecessors who admitted powers of more than three dimensions in arithmetical cases. He counts Diophantus as one of these predecessors. See Jones's commentary on Pappus of Alexandria, *Book 7 of the Collection*, Edited with Translation and Commentary by Alexander Jones (New York/Berlin/Heidelberg/Tokyo, 1986), Part 1, p. 74 and Part 2, p. 404.

standing historically in a closer position to Descartes than the two predecessors who attempted a solution to the problem of the locus. Significantly, Pappus's formulation through compound ratios surely contributes to allowing Descartes to solve it easily in the sense that that notion may lead the latter rather straightforwardly to solve the problem of dimensionality.

Another noteworthy historical precondition which helped Descartes tackle the Pappus problem successfully was that by the time of Descartes the notion of compound ratio had tended to be generally interpreted in an arithmetical manner. Euclid VI, 23, which we have cited above (p. 210), had its arithmetical counterpart VIII, 5: "Plane numbers have to one another the ratio compounded [of the ratios] of their sides." In his excellent monograph titled ***Philosophy of Mathematics and Deductive Structure in Euclid's Elements***, Ian Mueller asserts about these two propositions that although Euclid did not define the compounding of ratios explicitly and did not understand it to be a representation of multiplication, the later scholiasts had a strong inclination to interpret the compounding as a multiplication, and in the case of numbers as a multiplication of numerical fractions.[28] This tendency was strengthened in the Middle Ages. By means of the notion of 'denomination' (*denominatio*), the compounding of two ratios was understood quite simply as the arithmetical operation of multiplying two fractions and simple ratio as a fraction.[29] But the concept of compound ratio continued to puzzle mathematicians until the time of Isaac Newton.[30]

Obviously in arithmetic, numbers can be multiplied as many times as one likes and the problem of dimensionality does not occur. It would be quite natural if Descartes, who had made efforts to remove the barrier between arithmetic and geometry, was inclined to understand the concept of compound ratio arithmetically. As we have already observed, he had to strive to clarify the difficulty of the dimensionality of geometrical magnitudes in a completely theoretical way, basically within the domain of geometry. But even if his attempt was admittedly geometrical, we should not overlook the fact that he successfully cleared up the difficulty by introducing the unit, primarily an arithmetical notion,[31] into a geometrical discussion. In Rule XVIII of the *Regulae*, he had already acknowledged the importance of the unit at least in arithmetical proportion, asserting that "if it is stated that as the unit is to a, say to 5, which is given, so is b, i. e. 7, which is given, to that which is required, which is ab, i. e. 35, then a and b are at the second place (*gradus*), and ab, which is produced by the

[28] Ian Mueller, *Op. cit.* (n. 16), pp. 87–88 & 154–155.

[29] See M. S. Mahoney, "Mathematics," Chapter 5 of David C. Lindberg, ed., *Science in the Middle Ages* (Chicago, 1978), p. 166.

[30] See Edith Sylla, "Compound Ratios: Bradwardine, Oresme, and the First Edition of Newton's *Principia*," in Everett Mendelsohn, ed., *Transformation and Tradition in the Sciences: Essays in Honor of I. Bernard Cohen* (Cambridge, England, 1984), pp. 11–43.

[31] Thomas L. Heath, *A History of Greek Mathematics*, Vol. 1 (Oxford, 1921), pp. 69–70; *Idem, The Thirteen Books of Euclid's Elements*, Vol. 2 (New York, $_2$1956), Book VII, Definition 1, pp. 277 & 279

former at the third."[32] What he achieved in the *Géométrie* was an ingenious translation of this relationship into a geometrical language. We may say that the unit plays a crucial role for his unification of the two principal fields of pure mathematics. The key with which he opened the door to an entirely new mathematics was, thus, not only technical but also deeply conceptual: That is to say, Descartes's mature thought in the *Géométrie* brought about drastic changes of such fundamental concepts as numbers and magnitudes by reorganizing the traditional mathematical disciplines.

Then, how does Descartes further proceed in his solution to the locus problem along with such a conceptual background and supersede the ancient mathematicians? In Book One of the *Géométrie* he begins to state his demonstration with the following words:

> Let AB, AD, EF, GH, etc., be several lines given by position, and let it be required to find a point, such as C, from which we can draw other straight lines (such as CB, CD, CF, and CH) on the given ones, in such a way that the angles CBA, CDA, CFE, CHG, etc., are given, and such a way that the product of the multiplication of one part of these lines is equal to the product of the multiplication of the others—or at least so that these products have some other given ratio, for this does not make the problem more difficult.[33]

Descartes thus sets the problem in a unified form. Then he tackles it with a method of analysis which is not geometrical but algebraic. The margin subtitle of the next paragraph shows this most clearly: "Comment on doit poser les termes pour venir a l'Equation en cet exemple."[34] His first goal is to obtain the algebraic equation with which points on the required locus are satisfied. So he states:

> First, I assume the thing already done (*ie suppose la chose comme desia faite*); and in order that I extricate myself (*me demesler*) from the confusion of all these lines, I consider one of the given lines and one of those which it is necessary to find, for example AB and CB, as the principal ones, to which I shall try to relate all the others. Let the segment of the line AB, which is between points A and B, be called x, and let BC be called y.[35]

"I assume the thing already done" is a typical expression frequently seen in the procedure of analysis. For example, in Pappus's definition of analysis at the beginning of Book VII of his *Collection*, we find the expression "in analysis we

---

[32] *AT*, X, p. 463.
[33] *AT*, VI, p. 382; Olscamp. p. 186.
[34] *AT*, VI, pp. 382–383.
[35] *AT*, VI, pp. 382–383; Olscamp, p. 186.

assume what is sought as if it has been achieved."[36] And we may be able to relate the clause "in order that I extricate myself from the confusion of all these lines" to the second rule of Descartes's universal method stated in Part Two of the *Discours de la méthode*: "To divide each of the difficulties I examined into as many parts as possible and as may be required in order to resolve them better."[37] The Greek verb 'ἀναλύειν' means "to undo, dissolve something into its elements, or resolve into its elements."[38]

But Descartes's method of analysis is not identical with Pappus's. The latter's analysis is geometrical whereas the former's is algebraic. Descartes supposes the two segments AB and BC, which vary according to the position of point C on the locus, to be known already and that they are set equal to x and y. These two line segments are the principal magnitudes which other magnitudes are to be related to. In Descartes's and Bos's notation, x is the distance along L_1 from a fixed point A to the intersection B of d_1 with L_1 (in the exceptional case when all lines L_1, L_2, ... are parallel, x cannot be taken) and $d_1 = y$. Then, all d_i can be represented in a linear equation of x and y, $d_i = a_i x + b_i y + c_i$, where a_i, b_i and c_i are all some given magnitudes. Consequently, the formula which characterizes points on the locus

$$(d_1 d_2 \cdots d_n) : (d_{n+1} d_{n+2} \cdots a)$$
$$[(d_1 d_2 \cdots d_n) : (d_{n+1} d_{n+2} \cdots d_{2n})] = \alpha : \beta$$

can be rewitten into the following algebraic equation:

$$\begin{aligned} &y(a_2 x + b_2 y + c_2) \cdots (a_n x + b_n y + c_n) \\ &= \frac{\alpha}{\beta}\Big(a_{n+1}x + b_{n+1}y + c_{n+1}\Big) \cdots \Big(a_{2n-1}x + b_{2n-1}y + c_{2n-1}\Big)a \\ &[\text{or } (a_{n+1}x + b_{n+1}y + c_{n+1}) \cdots (a_{2n}x + b_{2n}y + c_{2n})].^{39} \end{aligned}$$

This is a result of the algebraic analysis stated very generally.

In the usual manner once the algebraic equation has been obtained as the result of the analytic inquiry, the method of synthesis has to commence. According to the well-known definition of synthesis at the beginning of Book

[36] Jones, pp. 82–83: "ἐν [· · ·] τῇ ἀναλύσει τὸ ζητούμενον ὡς γεγονὸς ὑποθεμένοι[. . .]"; Commandino, f. 157r: "in resolutione [. . .] id quod quaeritur tanquam factù ponentes [· · ·]." Cf. Hultsch, II, p. 634; Ver Eecke, II, p. 477. More examples can be observed in Pappus's *Collection* such as Hultsch, I, p. 186, l. 14: "Let it be done" (γεγονότω) in Book IV, Proposition 4; Commandino, f. 38r: "Factum iam sit." Cf. Jaakko Hintikka and Unto Remes, *The Method of Analysis: Its Geometrical Origin and Its General Significance* (Dordrecht/Boston, 1974), p. 24.

[37] *AT*, VI, p. 18.

[38] Henry George Liddell and Robert Scott, *A Greek-English Lexicon* (Oxford, 1968), p. 112. See also our discussion on pp. 63–72.

[39] Bos, *Op. cit.* (n. 15), p. 300.

VII of Pappus's *Collection*, "in synthesis, by reversal, we assume what was obtained last in the analysis to have been achieved already, [...] we attain the end of the construction (*κατασκευή*, *compositio*) of what was sought."[40]

Following this method, Descartes, too, proceeds to describe lines characterized by the algebraic equations obtained above. Before having accomplished the construction, he makes general statements about the nature of curved lines. This is a task of the introductory part of Book Two.

Criticizing Pappus again, Descartes presents a new criterion by which curved lines can be classified exhaustively. In the paragraph with the margin subtitle "Quelles sont les lignes courbes qu'on peut receuoir en Geometrie," he states his observation:

> The ancients knew very well that there are three kinds of problems in geometry: the plane (*plan*), the solid (*solide*), the linear (*lineaire*). That is, some can be constructed merely by tracing straight lines and circles, while others cannot be constructed unless we use some conic section, and still others require some other, more complex, line. But I am astonished that they did not distinguish different degrees among these more complex curves besides this, and I am unable to understand why they called these curves mechanical (*Mechanique*), rather than geometrical. For if we say that they are so called because it is necessary to use a certain instrument (*machine*) to describe them, it is necessary to reject, for the same reason, circles and straight lines, seeing that they can only be described on paper with a compass and ruler, which we can also call instruments. Nor is it because the instruments that are used to trace them, being more complex than the ruler and compass, are therefore not as exact, for were this the case, it would be necessary to exclude them from mechanics, where exactness of works made by hand is desired, rather than from Geometry, where we seek only exactitude of reasoning, which can certainly be as exact with reference to these lines as to simpler ones.[41]

Apparently, in writing this statement Descartes was referring to Pappus's way of classifying problems and then curves. To understand how Descartes surpasses one of the greatest Greek mathematicians, let us compare the paragraph quoted above with Pappus's description in Book III of his *Collection*. Pappus put it:

> The ancients held that there were three genera of geometrical problems: some were called plane (*ἐπίπεδα*, *plana* in Commandino's

[40] Jones, *Ibid.* Cf. Hultsch; Commandino; Ver Eecke, *Ibid.* Mahoney conjectures that this part is an interpolation by a later scholiast. Mahoney, "Another Look at Greek Geometrical Analysis," (Ch. 2, n. 50), pp. 324–325

[41] *AT*, VI, pp. 388–389; Olscamp, p. 190.

> Latin version), some solid (στερεά, *solida*), and some linear (γραμμικά, *linearia*). Those that can be solved with straight lines and circumferences of circles are reasonably called plane, for the lines by which these problems are solved have their genesis in a plane. Problems that are solved by the use in their discovery of one or more sections of a cone are called solid, for in their construction it is necessary to use surfaces of solid figures, namely conic surfaces. There yet remains the third genus which is called linear, for lines other than those mentioned are used in the construction, which have a varied and more intricate genesis, such as the spirals, the quadratrices, the conchoids and the cissoids, which have many marvelous properties. As there is this difference between problems, the ancient geometers did not construct the aforementioned problem of the two straight lines, which is solid by nature, following geometrical reasoning, because it was not easy to draw the sections of the cone in a plane, but by using instruments (ὄργανα, *instrumenta*) then brought it to a manual construction and fit preparation, as is seen in the *Mesolabe* of Eratosthenes and the *Mechanics* of Philo and Hero.[42]

Following a criterion similar to the one mentioned above in Book VII Pappus classified loci into three categories: plane, solid, and linear.[43] It should be pointed out that Descartes's understanding of Pappus's statement was not sufficiently accurate. For example, according to Descartes's observation the ancients denied that curves higher than lines and circles were to be accepted in geometry and tended to call them 'mechanical'. In fact in a paragraph below the one quoted above, he clearly states: "It is true that they [the ancients] never completely accepted conic sections in their geometry."[44] Such an interpretation is historically untenable as A. G. Molland has persuasively contended.[45] A short reflection on Apollonius's *Conics* would be enough to see Descartes's historical incorrectness.

On the other hand, Descartes's intention in criticizing Pappus is fully understandable. He proposes his own criterion as to what should be accepted in geometry:

> It seems to me that it is very clear that if (as we do) we understand by 'geometrical' (*Geometrique*) that which is precise and exact, and by 'mechanical' (*Mechanique*) that which is not; and if we consider

[42] Hultsch, I, pp. 54–47; Commandino, f. 4v; Ver Eecke, I, pp. 38–39. As to the translation we referred to A. G. Molland, "Shifting the Foundations: Descartes's Transformation of Ancient Geometry," *Historia Mathematica*, **3** (1976), pp. 27–28.

[43] Jones, pp. 104–107; Hultsch, II, pp. 662–663; Commandino, f. 162v; Ver Eecke, II, pp. 495–496.

[44] *AT*, VI, p. 389; Olscamp, p. 191.

[45] Molland, *Op. cit.* (n. 42), pp. 35–36.

> geometry as a science that teaches a general knowledge of the measures of all bodies, we must no more exclude complex lines from it than simple ones, provided that we can conceive them as being described by a continuous movement, or by several successive movements of which the latter are completely determined by those which precede: for by this means, we can always have an exact knowledge of their measure.[46]

Thus he intends to transform the definition of geometry by modifying or widening the domain of exact knowledge. But exactly how does he widen the territory of geometry? First of all, attention should be called to the fact that the use of mathematical instruments is condemned less by Descartes than by the ancients. For example, in Descartes's judgment lines which are obtained by the mesolabe compass are to be welcomed into geometry. As we have seen in Chapter 3, § 2, this instrument is supposed to have been devised as an improved form of Eratosthenes's mesolabe, the tool for finding mean proportionals. Such an attitude might have reflected his general inclination to respect the practical and utilitarian aspects of knowledge.[47] He conceives the description of these curves "as clearly and distinctly as that of the circle, or at least as that of the conic sections."[48] We should note that the lines drawn by means of the mesolabe compass describe curves represented by (algebraic) polynomials with respect to the magnitude understood as varying some one-dimensional basic quantity (x

---

[46] *AT*, VI, pp. 389–390; Olscamp, p. 191.

[47] Descartes, for example, states on his physics in Part Six of his *Discours de la méthode*: "[A]s soon as I had acquired some general notions in physics and had noticed, as I began to test them in various particular problems, where they could lead and how much they differ from the principles used up to now, I believed that I could not keep them secret without sinning gravely against the law which obliges us to do all in our power to secure the general welfare of mankind. For they opened my eyes to the possibility of gaining knowledge which would be very useful in life, and of discovering a practical philosophy which might replace the speculative philosophy taught in the schools. Through this philosophy we could know the power and action of fire, water, air, the stars, the heavens and all the other bodies in our environment, as distinctly as we know the various crafts of our artisans; and we could use this knowledge—as the artisans use theirs—for all the purposes for which it is appropriate, and thus make ourselves, as it were, the lords and masters of nature. This is desirable not only for the invention of innumerable devices which would facilitate our enjoyment of the fruits of the earth and all the goods we find there, but also, and most importantly, for the maintenance of health, which is undoubtedly the chief good and the foundation of all the other goods in this life." *AT*, VI, pp. 61–62. Descartes's emphasis on the method of analysis itself may be interpreted as his preference for practice rather than mere speculation. In this regard, see the following literature. Gilson sees here an influence of Francis Bacon. Gilson, *Texte et commentaire* (Ch. 1, n. 1), pp. 443–446. Cf. Richard B. Carter, "Descartes' Methodological Transformation of Homo Sapiens into Homo Faber," *Sudhoffs Archiv*, **68** (1984), pp. 225–229. See also Harcourt Brown, "The Utilitarian Motive in the Age of Descartes," *Annals of Science*, **1** (1936), pp. 182–192. Incidentally, it was in this context that Karl Marx provided his single reference to Descartes in *Das Kapital*, Bd. 1 (1867): *Marx-Engels Werke*, Bd. 23 (Berlin, 1969), pp. 411–412, n. 111.

[48] *AT*, VI, p. 392; Olscamp, p. 192.

in our notation in Chapter 3, § 2).

Second, more importantly, Descartes criticizes Pappus because the latter did not distinguish different degrees among curves higher than the conic sections. This alleged fault of the ancients does not seem to have been internally theoretical to the framework of Greek geometry itself. In other words, it seems rather to be working together with the external reason that Descartes wants to introduce a new way of classifying curves, especially ones which appear as solutions of the locus problem of Pappus. He proposes his extended classification in the paragraph with the margin subtitile "La facon de distinguer toutes les lignes courbes en certaines genres. Et de connaistre le rapport qu'ont tous leurs poins a ceux des lignes droites."[49] It is exactly here that we can witness his real motive for questioning the ancients' way of classifying geometrical problems and curves. He insists that the most natural method of classifying curves is according to the degree of their equations:

> I could give here many other ways of drawing and conceiving curved lines whose complexity would increase, by degrees, to infinity. But in order to understand together all curves that are present in nature, and to classify them by order into certain classes (*genres*), I know of nothing better than to say that all points of those curves which can be called 'geometrical' (*Geometrique*)—that is, which fall under some precise and exact measure—necessarily have a certain relation to all the points of a straight line; and this relation can be expressed by a single equation of all the points. And when [no term of] this equation is higher than the rectangle of two undeterminate quantities, or else of the square of a single unknown quantity, the curved line is of the first and simplest class, which comprises only the circle, parabola, hyperbola, and ellipse. But when one or both of the equation's two quantities (for there must be two, in order to explain the relation between two points) reaches the third or fourth degree, the curve is of the second class; and when the equation reaches the fifth of sixth degree, the curve is of the third class; and so on the others, to infinity.[50]

The curves which can be accepted in geometry are fundamentally those represented by algebraic equations and classified in accordance with their degree.

Accordingly, it is with the help of the power of a new tool of mathematics, i. e. algebra, that Descartes tries to surpass Pappus. Thus some people today who tend to think rigorously might become skeptical about whether or not the modern mathematician surpasses the ancient at all. Indeed, as Pappus's way of classifying curves indicates, the ancient geometer sorted curved lines in terms of their genesis, not in terms of their properties. The ancients were not

[49] *AT*, VI, p. 392.

[50] *AT*, VI, pp. 392–393; Olscamp, pp. 192–193.

necessarily ignorant of a certain manner of characterizing geometrical curves by their properties. For example, Apollonius in his *Conics* argued about the conic sections (parabola, hyperbola, and ellipse) by demonstrating their properties ($\sigma\upsilon\mu\pi\tau\acute{\omega}\mu\alpha\tau\alpha$). But in agreement with Molland, it can be asserted that in antiquity genetic classifications of curves were usually viewed as more basic than those by properties.[51] Pappus was a typical example of this tendency. Descartes's method of classification, on the other hand, depends principally on the properties of curves, or to be precise, the algebraic equations which curves satisfy. In this sense, if he surpasses the ancients, he surpasses them especially with respect to the fundamental notions of his mathematical thought,which can be characterized as algebraic rather than purely geometrical.

With such a new conceptual, i. e. algebraic framework, Descartes can now state a quite general solution to the problem of Pappus. When only three or four straight lines are given, the curved line which has been obtained as the locus is necessarily one of the first class; and when no more than eight lines are given, the required line can be only of the second class, or lower in exceptional cases; and so on for the remainder.[52] His way of tackling the Pappus problem is quite exhaustive. In doing so, he follows the last rule of enumeration of his universal method stated in the *Discours de la méthode*: "Throughout to make enumerations so complete, and reviews so comprehensive, that I could be sure of leaving nothing out."[53]

After having demonstrated the solution of the locus problem with reference to three or four lines by exhaustively translating Apollonius's theory of conic sections into algebraic terms, Descartes declares with a triumphant voice the result of his method of synthesis:

> Since all equations which are no higher than the square are included in what I have just explained, not only is the problem of the ancients concerning three and four lines completely solved here, but also everything pertaining to what they called the composition (*composition*) of solid loci, and as a result also to that of plane loci, because these are included in the solid. For these loci are nothing but this: when the problem is to find some point lacking one condition for its complete determination (as in this example), any part of a straight line can be taken for the one which is required. And if this line is straight, or circular, we call it a plane locus. But if it is a parabola, or a hyperbola, or an ellipse, we call it a solid locus. And always, in any case, we can arrive at an equation which contains two unknown quantities, and which is analogous to some one of those I have just solved. But if the line which so determines the needed point is of a

[51] Molland, *Op. cit.* (n. 42), pp. 23–33.

[52] *AT*, VI, pp. 396–397; Olscamp, p. 196. For the mathematical details of Descartes's solution of the problem of Pappus, see Bos, *Op. cit.* (n. 15), pp. 298–303.

[53] *AT*, VI, p. 19.

> more complex degree than the conic sections, we can call it, in the same manner, a supersolid locus (*lieu sursolide*), and so for the others. And if two conditions are lacking for the determination of this point, the locus of the point is a surface, which may be completely flat, or spherical, or more complex. But the highest goal of the ancients in this matter was to achieve the composition of solid loci: and it seems that all that Apollonius wrote about conic sections was designed to search for this.[54]

It is by means of the method of problematical analysis in its algebraic form and of the method of synthesis (*composition*) in the form of representing algebraic equations that Descartes has successfully discharged the ancient problem of the locus. What we have emphasized is that his success was mainly conceptual and achieved by means of his methodological power, and that the triumph of his method was rooted in his longtime endeavor to reform algebra.

This mathematical achievement is not of the winter of 1631–1632, but of 1637. To what extent had Descartes solved the Pappus problem in the winter of 1631–1632? This problem is related to the questions about when and how Descartes became confident that his method of algebraic analysis superseded the ancient mathematical thought on the whole, and about the extent to which he was indebted to his contemporary mathematicians who possessed the idea of algebraic analysis. The latter question will be discussed in Section 3. The answer to the original problem will be given through a critical examination of Descartes's letter to Golius of January 1632, the only historical source related to the former's challenge to the ancient problem. Descartes gives his description of Pappus's problem and then his answer in that lettere as follows:

> If arbitrarily many straight lines are given, all points related to them according to the purport of the question will be on one of the lines which can be described by a single motion, continuous and determined entirely by some simple relations: indeed, by two or three at most, if straight lines given in position be not more than four; by three or four relations at most, if straight lines given in position be not more than eight; by five or six, if given straight lines be not more than twelve, and so on to infinity.[55]

From this statement, we know that by the beginning of 1632, Descartes had already acquired the principal part of the solution which is stated in the

---

[54] *AT*, VI, pp. 406–407; Olscamp, p. 203.

[55] *AT*, I, p. 233; *AM*, I, pp. 212–213: "Datis quotcunque rectis lineis, puncta omnia ad illas iuxta tenorem quæstionis relata, contingent vnam ex lineis quæ describi possunt vnico motu continuo, & omni ex parte determinato ab aliquot simplicibus relationibus; nempe, à duobus vel tribus ad summum, si rectæ positione datæ non sint plures quam quatuor; à tribus vel quatuor relationibus ad summum, si rectæ positione datæ non sint plures quam octo; à quinque vel sex, si datæ rectæ non sint plures quam duodecim, atque ita in infinitum."

Géométrie. The expression "simple relation" (*simplex relatio*) which appears in the letter will be explained later.

Descartes continues the description of his answer to the Pappus problem:

> And inversly, no line can be described such that one cannot find the position of some straight lines to which infinite points on it are related according to the purport of the question. In fact, these lines will not be more than four if the described curve does not depend on more than two simple relationships; nor more than eight if the curve does not depend on more than four relationships; and so on.[56]

This passage asserts that the inverse proposition of the previous statement holds, in other words, there is no curved line which cannot be represented by means of an algebraic equation which is obtained as a solution of some locus problem of Pappus. Descartes will expatiate further on this inverse statement in Book Two of the *Géométrie* as follows:

> Now because the position of the given straight lines can vary in every way, and as a result can change many of the known quantities, such as the signs + and − in the equation, in every imaginable way, it is evident that there is no curve of the first class which will not be useful to this problem, when it is proposed in four straight lines; nor any of the second which will not be useful when it is proposed in eight; nor any of the third when it is proposed in twelve; and so on for the rest. So that there is no curved line which is subject to calculation and can be included in Geometry which may not be used for some number of lines.[57]

To restate this passage more simply, he claims that all algebraic curves can occur as equations of the loci which become solutions of Pappus's problem. Unfortunately, this statement is too strong. Later, in the note titled "Errores Cartesij Geometriae," Isaac Newton will note that Descartes's assertion is incorrect and provide a proof for his criticism of Descartes.[58]

Further on in his letter to Golius, Descartes defines a simple relation which appears in the solution of Pappus's problem: "Now here I call simple

---

[56] *AT*, I, p. 233; *AM*, I, p. 213: "Et vice versâ nulla talis linea potest describi, quin possit inueniri positio aliquot rectarum, ad quas referantur infinita puncta, iuxta tenorem quæstionis, quæ illam contingent. Quæ quidem rectæ non erunt plures quam quatuor, si curua descripta non pendeat à pluribus quam duobus simplicibus relationibus; nec plures quam octo, si curua non pendeat à pluribus quam quatuor relationibus; & sic consequenter."

[57] *AT*, VI, p. 397; Olscamp, p. 196.

[58] See D. T. Whiteside, ed., ***The Mathematical Papers of Isaac Newton***, Vol. IV (Cambridge, 1971), Part II, 3, § 1: "Three Mistakes in Descartes's *Géométrie*," pp. 336–345, esp. pp. 340–345. This memorandum is supposed to have been composed in the late 1670s. Incidentally, Bos has recently given a modern and more elegant demonstration. See Bos, *Op. cit.* (n. 15), "Appendix: On the Curves that Occur as Loci of Problems of Pappus," pp. 332–338.

relations those which individually involve only single geometrical ratios."[59] This definition might appear to be rather obscure. But it would not be a serious mistake for us to interpret "simple relation" as a relation which can be expressed in the linear form $ax + by + c$, or in his own words of the *Géométrie* , a relation which "can always be expressed by three terms, of which one is composed of the unknown quantity y, multiplied or divided by some other known quantity; and another consisting of the unknown quantity x, also multiplied or divided by some other known quantity; and the third consisting of some known quantity."[60] In fact, this interpretation is not anachronistic, for Descartes in his letter to Golius restates his solution in terms of (algebraic) equations.[61] Our explanation is further corroborated by Descartes's terminology in the *Regulae*. In Rule XVI he notes that "those proportions which form a continuing sequence are to be understood in terms of a number of relations."[62] In this way accordingly, a^3 contains three relations, since it can be characterized by the three continued proportional relationships: $1 : a = a : a^2 = a^2 : a^3$.

In the sentences which follow his definition of a simple relation, he states that non-algebraic lines are to be excluded from the solutions of Pappus's problem:

> And this definition of the required lines is, unless I am mistaken, adequate and sufficient. For by saying that they are described by a single motion, I exclude the quadratrix, the spirals and others of the same sort, which are not described without using two or more movements independent of one another. And by saying that the motion should be determined by some simple relations, I exclude innumerable others to which no names which I would know are given. Finally, I define each class by the number of relations; and so the first class contains only the conic sections; on the other hand, the second [class], besides those which I have explained above, contains as many others as possible, and it would take long to enumerate them.[63]

---

[59] *AT*, I, p. 233; *AM*, I, p. 213: "Hîc autem simplices relationes illas appello, quarum singulæ non nisi singulas proportiones Geometricas inuoluunt."

[60] *AT*, VI, pp. 384–385; Olscamp, p. 188.

[61] *AT*, I, p. 234, *l.* 13, *AM*, I, p. 214.

[62] *AT*, X, p. 456: "Advertendum est etiam, per numerum relationum intelligendas esse proportiones se continuo ordine subsequentes [...]."

[63] *AT*, I, pp. 233–234; *AM*, I, pp. 213–214: "Atque hæc linearum quæsitarum definitio est, ni fallor, adæquata & sufficiens. Per hoc enim quod dicam illas vnico motu continuo describi, excludo Quadraticem & Spirales, aliasque eiusmodi, quæ non nisi per duos aut plures motus, ab inuicem non dependentes, describuntur. Et per hoc quod dicam illum motum ab aliquot simplicibus relationibus debere determinari, alias innumeras excludo, quibus nulla nomina, quod sciam, sint imposita. Denique per numerum relationum singula genera definio; atque ita primum genus solas Conicas Sectiones comprehendit, secumdum verò præter illas quas supra explicui, continet alias quam plurimas, quas longum esset recensere."

This concludes the theoretical part of Descartes's letter to Golius. From the last passage, we may be able to judge that by January 1632, Descartes had not only attained the equations which represent the loci of Pappus's problem, but had also acquired the basic idea as to how curves are to be sorted, which is fully developed at the beginning of Book Two of the *Géométrie*. But, note that nowhere in that letter to Golius do we find any explicit theoretical explanation of the dimensionality of geometrical magnitudes as appears at the beginning part of the *Géométrie* (see pp. 212–213).

Indeed it can be concluded from our consideration above that it was Pappus's problem proposed by Golius in 1631 that allowed Descartes's leap from the level of Greek mathematical thought and let him later pride of himself on having created a mathematical method which the ancients had never possessed and his contemporaries had never thought of. But, of course, the basis of this success was Descartes's own efforts conceptualizing his new algebra and historical conditions which made the formation of such a conceptual framework possible.

§ 2. The Composition of the *Géométrie*

It was essentially by means of his method of algebraic analysis that Descartes succeeded in discovering the solution of Pappus's locus problem in the winter of 1631–1632. We may say that this method, along with the writings he showed Beeckman in the winter of 1628–1629 (including the old *Algebra*), contained the principal materials for his *Géométrie* of 1637.

How was the *Géométrie* composed and what position did Descartes consider it to occupy with in the structure of his science as a whole? The *Géométrie* appeared as the last of the three essays of his *Discours de la méthode*. Descartes's desire to produce his system of knowledge, which would develop into the volume of 1637, was first announced, as far as we know, in 1635. In a letter of 1634, he told Mersenne that, having heard about Galileo's condemnation by the Roman Catholic Church for his support of the heliocentric system of the world, he had abandoned the publication of *Le Monde, ou Traité de la lumière*, which contained the same theory that had caused the Italian scientist's misfortune.[64] In that letter he wrote in a discouraging tone: "I desire to live in peace and to continue the life I have begun under the motto, to live well you must live unseen (*Bene vixit, bene qui latuit*)."[65]

In the next year, 1635, however, Descartes recovered from these adverse circumstances and completed the composition of the *Dioptrique*.[66] Further, in

[64] As to the history and the text of this tract, see Descartes, *The World*, Translation and Introduction by M. S. Mahoney (n. 1).

[65] "Descartes, à (Deventer), à Mersenne, à Paris. (fin de fevrier 1634)," *AT*, I, pp. 285–286; *AM*, I, p. 253; *CM*, IV, pp. 50–51. Ovid, *Tristia*, III, iv, 25.

[66] "Descartes, à Utrecht, à Mersenne, à Paris. (mars 1635)," *AT*, I, p. 322; *AM*, I, p. 284;

his letter to Constantin Huygens of November 1 of the same year, he revealed his intention to add the *Météores* to the intended volume.[67]

It was in his letter addressed to Mersenne in March 1636 that Descartes announced for the first time that his planned book would include the *Géométrie* and that he had provisionally entitiled the book: *Le projet d'une Science universelle qui puisse élever nostre nature à son plus haut degré de perfection. Plus la Dioptrique, les Meteores, et la Geometrie; où les plus curieuses Matieres que l'Autheur ait pû choisir, pour rendre preuve de la Science universelle qu'il propose, sont expliquées en telle sorte, que ceux mesmes qui n'ont point estudié les peuvent entendre.*[68] In the same letter, he wrote to Mersenne about the purpose for writing the *Géométrie*: "In the *Géométrie*, I try to give a general way of solving all the problems that have never yet been solved."[69]

According to a later statement of Descartes, the *Géométrie* was composed for the most part just as the *Météores* was being printed, that is to say, in November 1636, and he discovered a part of the content even during the period in which he was writing the treatise.[70] In drafting the *Géométrie*, Descartes must have recalled how he solved the Pappus problem, and then referred to his old *Algebra* and perhaps his other previous writings. The young Dutch mathematician Frans van Schooten drew beautiful illustrations for the three essays.[71]

It was with the date June 8, 1637 that the monumental book was published anonymously by the bookseller Ian Maire in Leiden under the title: *DISCOURS DE LA METHODE Pour bien conduire sa raison, & chercher la verité dans les sciences. Plus LA DIOPTRIQVE. LES METEORES. ET LA GEOMETRIE. Qui sont des essais de cete METHODE.* It is said that despite the concealing of the author's name, everyone in the intellectual world knew that Descartes had written the book. Therewith he was obliged to quit his life in accordance with the motto "Bene vixit, bene qui latuit."

CM, V, p. 125.

[67]"Descartes, à Utrecht, à Constantin Huygens. 1er novembre 1635," *AT*, I, pp. 329–330; *AM*, I, pp. 290–291.

[68]"Descartes, à (Leyde), à Mersenne, à Paris. (mars 1636), " *AT*, I, pp. 339–340; *AM*, I, p. 301; *CM*, VI, p. 44.

[69]*Ibid.*: "[E]n la *Geometrie*, ie tasche à donner vne façon generale pour soudre tous les Problemes qui ne l'ont encore iamais esté."

[70]"Descartes au P. Diriennes, 22 Février 1638," *AT*, I, p. 458 (where the date of the letter is attributed to October, 1637); AM, II, p. 140: "C'est vn traitte que ie n'ay quasi compose que pendant qu'on imprimoit mes *Météores* [...]." The Father Deriennes taught mathematics at the college of La Flèche from 1635 to the year of his death 1661. A. Romano, *La Contre-Reforme mathématique* (Ch. 1, m. 13), pp. 599–600; Cf. Élie Denisoff, "Les Étapes de la rédaction du «Discours de la méthode»," *Revue philosophique de Louvain,* **54** (1956), pp. 261–264; "Premier Essai" de *Idem, Descartes, Premier théoricien de la physique mathématique* (Louvain/Paris, 1970), pp. 15–18.

[71]See the explanation of Van Schooten in the "Table des noms propres," *AM*, I, 464. As wil be seen, Van Schooten edited the two Latin versions of the *Géométrie*. For his career, see Josef Ehrenfried Hofmann, *Frans van Schooten der Jüngere* (Wiesbaden, 1962).

The *Géométrie* differed from the other two essays and from the introduction to the three essays, *Discours de la méthode*. The difference is clearly set out in a note (*Advertissement*) prefixed to the text of the *Géométrie*:

> Thus for I have tried to make myself understandable to everybody; but as for this treatise, I am afraid that it will be able to be read only those who have already known that which is in the books of Geometry: For especially as they contain many truths demonstrated very well, I have believed that it would be superfluous to repeat them and have not hesitated about using them.[72]

By "the books of Geometry" did Descartes mean ordinary textbooks of geometry, in particular several essential books of Euclid's *Elements*? In any case, the *Géométrie* was written for those who possessed a sufficient knowledge of common mathematics. And as he claimed in a letter to Mersenne which probably dates from around the end of 1637, it was by this treatise, among others, that he thought he had successfully proved the preeminence of his universal method. He stated in that letter:

> I have simply tried by the *Dioptrique* and by the *Météores* to persuade [the reader] that my method is better than the ordinary, but I claim that I have proved that by my *Géométrie.*[73]

This statement indicates that Descartes's method was formed while he was studying mathematics and conversely that the *Géométrie* was the unequaled illustration of his method.

The treatise consists of three books. Like the monograph *Géométrie*, all the three books have titles hinting that they are closely related to classical geometry: Book I, "Of Problems that Can Be Constructed Using Only Circles and Straight Lines"; Book II, "Of the Nature of Curved Lines"; and Book III, "Of the Construction of Solid and Supersolid Problems." The title *Géométrie* for a treatise in which algebra played an enormously important role might appear *prima facie* curious, but it was entirely legitimate because in seventeenth century France the term 'géométrie' broadly pointed to pure mathematics,[74] and because the subject which he tried to deal with was nothing other than a classical discipline which had been traditionally called geometry. The approach

[72] *AT*, VI, p. 368: "Aduertissement. Iusques icy i'ay tasché de me rendre intelligible a tout le monde; mais, pour ce traité, ie crains qu'il ne pourra estre leu que par ceux qui sçauent desia ce qui est dans les liures de *Géométrie*: car, d'autant qu'ils contienent plusieurs verités fort bien demonstrées, i'ay creu qu'il seroit superflus de les repeter, & n'ay pas laissé, pour cela, de m'en seruir."

[73] "Descartes, (près d'Alemaer), à Mersenne, à Paris. (fin décembre 1637)," *AT*, I, p. 478; *AM*, II, p. 65; *CM*, VI, p. 345.

[74] É. Denisoff, *Op. cit.* (n. 70), p. 259, note 16. Denisoff refers to the item 'géométrie' in A. Lalande, *Vocabulaire de la philosophie* (Paris, 1932). See also Boyce Gibson, "La 《Géométrie》 de Descartes au point de vue de sa méthode," *Revue de métaphysique et de morale*, **4** (1896), pp. 386–398, esp. p. 390.

to that discipline, i. e. algebra, however, was quite new. He tried to reformulate the traditional field of mathematics by means of the novel method of algebraic analysis.

The very beginning sentence of Book I confirms our characterization of the *Géométrie*: "All the problems of geometry can easily be reduced to such terms that thereafter we need to know only the length of certain straight lines in order to construct them."[75] This is a kind of declaration of the establishment of a new paradigmatic discipline which is today called 'analytic geometry',[76] that is to say, geometry using the method of algebraic analysis. To solve geometrical problems through this approach, one has to provide oneself with an 'algebra of line segments' by means of which one can represent "the length of certain straight lines." The words "reduce" (*reduire*) and "construct" (*construire*) suggest the dual method of analysis and synthesis, which are, in fact, fully mobilized in Books I and II. What Descartes did in the *Géométrie* was to pose a revolutionary system of mathematics concerning how to reorganize and solve classical problems algebraically. In a word, he presents a new paradigm.

In the *Géométrie*, the 'algebra of line segments' was first set forth in a theoretically justifiable manner. The beginning of the *Géométrie* is devoted to a preparation for this organon of algebraic analysis. Descartes defines four or five algebraic (originally arithmetical) operations (addition, subtraction, multiplication, division, and the extraction of roots) among line segments, and shows that any result of their operations can be interpreted as a line segment. He never claims, strictly speaking, that a line segment is a number, the object of arithmetic. But by plotting the unit in a geometrical line segment, we may say that he takes a step toward the arithmetization of geometry, and after all, of mathematics as a whole, then toward the theory of real numbers.

As we saw in Chapter 3, §1, as early as the spring of 1619 Descartes began to entertain his program for reforming mathematics to be "an entirely new

[75] *AT*, VI, p. 369; Olscamp, p. 177.

[76] According to Boyer, the terminology 'analytic geometry' was used probably for the first time by Michel Rolle in his "De l'evanovissement des quantitez inconnues dans la géométrie analytique," *Mémoires de l'Académie royale des sciences* (Paris), Année 1709 (1709. 9. Aoust), pp. 419–450. See Carl B. Boyer, *History of Analytic Geometry* (New York, 1956), p. 155: "Incidentally, in Rolle's discussion of this problem the term 'analytic geometry' appeared in print, perhaps for the first time, in a sense analogous to that of today." On p. 141 of the same book Boyer also refers to Newton's treatise titled "Artis Analyticae Specimina vel Geometria Analytica," which was published in *Opera quae exstant omnia*, Vol. I, pp. 389-518, under the editorship of Samuel Horsley in London in 1779. The title is ascribed not to the author Newton but to a copyist William Jones. The tract which Horsley edited as *Artis analyticae specimina vel Geometria analytica* is basically identical to "De Methodus serierum et fluxionum," written in Winter 1670-1671, in D. T. Whiteide, ed, *The Mathematical Papers of Isaac Newton*, Vol. III (1670–1673), (Cambridge, 1969). Jones copied Newton's manuscript about 1710. See Whiteside's Introduction, *Op. cit.*, pp. 11–13. See also Whiteside's "General Introduction," in *The Mathematical Papers of Isaac Newton*, Vol. I (1664–1666), (Cambridge, 1967), pp. xxiii and xxviii, notes (26) and (34). I am skeptical about Newton's use of 'analytic geometry' in other works as well.

science". He was then aiming at the unification of arithmetic and geometry by emphasizing the importance of algebra for reorganizing the mathematical disciplines. In the manuscript titled *Algebra* he attemted to realize this program, but its form remained insufficient. A crucial difference between the previous versions of "an entirely new science" and the *Géométrie* is that only the latter was fully organized according to his 'natural method'. It was with this well 'methodized' tract that he not only attempted to demonstrate the preeminence and effectiveness of his method, but initiated an entirely new style, or more appropriately, paradigma of mathematics. The algebra of line segments may be said to be situated in the most fundamental position in his 'methodized' new mathematics. We can hardly overemphasize the historical significance of this conceptual tool. It was by means of the algebra of line segments that Descartes advanced the novel spirit of a thoroughly coherent 'methodized' algebraic analysis. In setting forth this revolutionary organon in the *Géométrie*, he gave the last vital touch to the program of "an entirely new science" which he had been continuously nurturing since 1619.

In order to carry out algebraic analysis, Descartes introduces letters of alphabet a, b, c, etc. and x, y, z, etc. to denote known and unknown quantities, respectively. This notation is exactly that which we are familiar with. Throughout the treatise, along with the method of analysis he tries to mobilize the method of synthesis in the form of the 'construction of equations'. According to H. J. M. Bos, in seventeenth century mathematics to construct an algebraic equation meant to perform "a geometrical construction yielding line segments with lengths equal to the roots of the equation."[77] So we may regard the construction of equations as the method of synthesis in its algebraic form. As we have already observed in Section 1, Descartes does not fail to apply consciously his method of enumeration to the classification of algebraic curves or equations. He states this clearly in Book Three on the theory of algebraic equations.[78] It is easily seen that the titles of the three books originated from the viewpoint of the construction of equations according to their degrees. The construction of equations was, then, a natural consequence of the author's methods of synthesis and enumeration. We emphasize again that all these procedures were carried out on the conceptual ground of the algebra of line segments.

Descartes concludes his treatise with the following remark: "And I hope that posterity will be grateful to me, not only for the things that I have explained, but also for those that I have voluntarily omitted, in order to leave the pleasure of inventing them to further generations."[79] In fact the *Géométrie* is

[77]H. J. M. Bos, "Arguments on Motivation in the Rise and Decline of a Mathematical Theory; the 'Construction of Equations', 1637–*ca.* 1750," *Archive for History of Exact Sciences*, **30** (1984), p. 331.

[78]In the section under the margin subtitle "Pourquoy les problesmes solides ne peuuent estre construits sans les sections coniques, ny ceux qui sont plus composés sans quelques autres lignes plus composées." *AT*, VI, p. 475; Olscamp, p. 251.

[79]*AT*, VI, p. 485; Olscamp, p. 259.

extremely outstanding concerning its method rather than its individual mathematical results. Thus what Descartes accomplished in his *Géométrie* of 1637 is, in fine, the definitive realization of his 1619 program of "an entirely new science."

So far we have pointed out that the *Géométrie* is based on the ingeniously discovered algebra of line segments. More precisely, however, it can be characterized as the algebra of finite line segments, and hence, his analytic mathematics can be characterized as the algebraic analysis of finite quantities. This characterization is conspicuous especially in his 'method of tangents', which Descartes sets forth in the latter part of Book Two. We consider this topic here, for we have yet to touch upon it. In the latter part of Book Two, Descartes first explains how to draw the normal and the tangent by the method of indeterminate coefficients. His manner is elucidated in a rather long section under the margin subtitle "Façon generale pour trouuer les lignes droites qui couppent les courbes données, ou leurs contingentes, a angles droits."[80] We will explicate here Descartes's method of drawing the normal and the tangent by using modern notation without distorting his essential line of thought.

Let the equation of the curve $F(x, y) = 0$. He first tries to determine the normal instead of the tangent by assuming that the curve is cut in a pont C and another point C′ by a circle the center of which is P (see Figure 5.4). Let OP be equal to v and the normal CP be s. So the subnormal MP $= v - y$. Then, we have the relation $s^2 = x^2 + (v-y)^2$, or $s^2 = x^2 + v^2 - 2vy + y^2$. By eliminating x in the above two equations, we get an equation $f(y, s, v) = 0$. When C and C′ coincide, i. e. CP is the normal, $f(y, s, v) = 0$ must have a double root equal to y_0. Accordingly, $f(y, s, v)$ can be written in the form $(y - y_0)^2 g(y, s, v)$. Comparing corresponding coefficients, we can determine the values of s and v.[81] This method of drawing the normal and the tangent also counts for Descartes as one of his most significant discoveries.

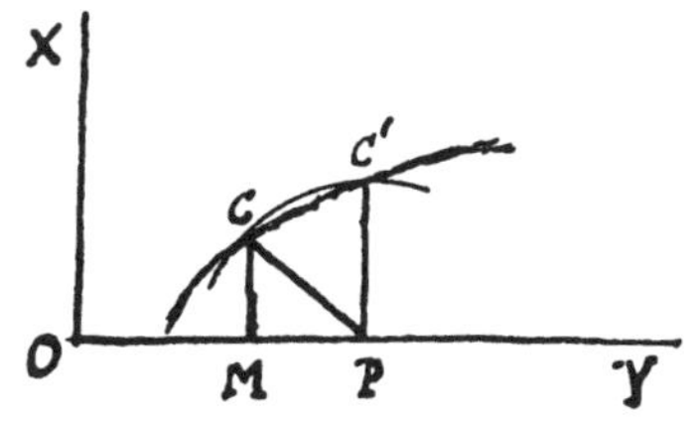

Figure 5.4

In the part which follows 'the method of tangents', Descartes attempts to introduce new curves which are to be useful in optics. These are ovals, which are viewed as a natural extension of ellipses and hyperbolas. Descartes's discussion of these curves also confirms that his mathematics in the *Géométrie* was basically finitist. At the same time it indicates that Descartes can be counted as a forerunner of the so-called 'inverse method of tangents.' Here, we show his historically important explanation in outline.

[80] *AT*, VI, pp. 413–423; Olscamp, pp. 207–215.

[81] Our simplified explanation owes much to the description by the editors of *CM*, VI, pp. 348–349.

There are four classes of ovals. The first oval, according to Descartes's description, can be drawn as follows.[82] Let two lines FA, AR be given, intersecting at A at a given angle (see Figure 5.5 (a)). Take an arbitrary point 5 on AG. Draw a circle with its center F and radius F5. Describe the straight line 56, which cuts the line AR at point 6, so that the ratio of A6 to A5 is given (We set this ratio $k < 1$). Mark AR equal to GA, and then describe a circle with its center G and radius R6. This circle cuts the other circle at the two points labeled 1, through which the oval passes.

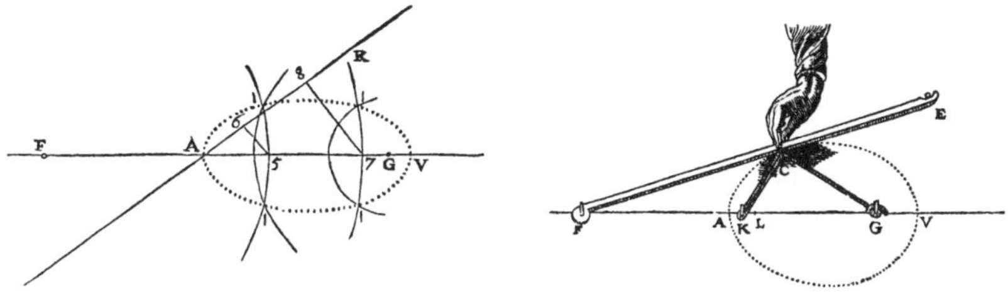

(a) (b)

Figure 5.5

Repeat this procedure by taking two points 7 and 8 instead of 5 and 6 respectively in such a manner that the straight line 78 is parallel to 56. Through this pointwise construction, we will have arbitrarily many points 1, which constitute an oval. This oval can be written in an analytical way by putting G1$= u$ and F1$= v$, for $u+kv = \mathrm{R6}+k.\mathrm{F5} = (\mathrm{AG} - \mathrm{A6})+k(\mathrm{FA} + \mathrm{A5}) = \mathrm{AG} - \mathrm{A6}+k.\mathrm{FA}+k.\mathrm{A5} = \mathrm{AG} + k.\mathrm{FA}$. Likewise, we obtain $u + kv = \mathrm{R8} + k.\mathrm{F7} = \mathrm{AG} + k.\mathrm{FA}$, for 78 is parallel to 56. So we may represent the first oval by the equation $u + kv = \mathrm{AG} + k.\mathrm{FA}$ $(k < 1)$.

Descartes also provides an alternative way of drawing the first oval, that is, one used by gardeners (see Figure 5.5 (b)). Assume the lines FA and AG to be equal and divide the entire line FG at point L, so that the ratio of LG to FL is equal to A6/A5 ($= k$). Let AL be bisected at K. A ruler FE pivoted at point F is capable of turning about at this point. "This cord, being attached to the end of the ruler at E, passes from C toward K, then from K again toward C, and from C to G, where its other end is attached, so that the length of this cord is composed of GA plus AL plus FE minus AF. And the movement of point C will describe this [first] oval in a manner similar to that in which the ellipse and the hyperbola have been descibed in the *Dioptrique*."[83] Thus, the equation of this curve is given as $\mathrm{EC} + 2\mathrm{CK} + \mathrm{CG} = \mathrm{GA} + \mathrm{AL} + \mathrm{FE} - \mathrm{AF}$. That it belongs to the first class of ovals is proved by noting the fact that EF is tangent to the curve at point C.[84]

[82] *AT*, VI, pp. 424–425; Olscamp, pp. 215–216.

[83] *AT*, VI, p. 428; Olscamp, p. 218.

[84] Descartes gives no demonstration. For a detailed proof, see J. F. Scott, *The Scientific Work of René Descartes* (London, 1952), pp. 127–128. Scott has provided his proof

Ovals are historically very remarkable since they seem to have been thought out for the first time in order to obtain aplanatic lenses or mirrors. After explaining how to draw ovals, Descartes begins to discuss the optical properties of ovals in a section with margin subtitle "Les proprietes de ces ouals, touchant les reflexions & les refractions." As to the first oval, he states the following:

> Besides this, for each of these ovals we must consider two parts, which have different properties: in the first oval, the part toward A causes rays passing through the air from F to converge toward point G, when they encounter the convex surface 1A1 of a lens which, in accordance with what was said in the *Dioptrique*, has refractions such that they can all be measured by the ratio between the lines A5 and A6, or similar ones, through the aid of which we described this oval. But the part that is toward V causes all the rays coming from point G to be reflected toward F, when they meet at G the concave surface of a mirror of the shape of 1V1, which is of a material such that it diminishes the lines A5 and A6: for from what has been proven in the *Dioptrique*, it is evident that as here pointed, the angles of reflection will be unequal, as will be the angles of refraction, and can be measured in the same way.[85]

Let us consider the first case, i. e. the refraction through the arc 1A1 of the first oval. The discussion about the reflection from the arc 1V1 is basically similar. The rays coming from point F are to be reflected at point C (= 1) toward G in such a way that the angles of refraction satisfy the condition $\sin\angle P'CF/\sin\angle PCG = A5/A6$ $(= 1/k)$, where the straight line PP′ is the normal of the curve at point C (see Figure 5.6). In the next section "Demonstration des proprietes de ces ouals touchant les reflections & refractions," Descartes actually shows by using his method of normals that with respect to the first oval

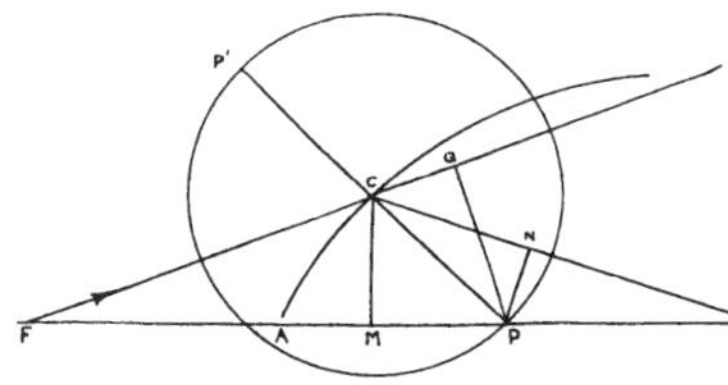

Figure 5.6

by representing the oval in an algebraic equation of the fourth degree. If the point M is taken to be the foot of the perpendicular line through C and we put $AM = y, CM = x, AF = c, AG = b$, and $A5 : A6 = d : e$, the equation will be $[(d^2 - e^2)x^2 + (d^2 - e^2)y^2 - 2(e^2c + bd^2)y - 2ec(ec + bd)]^2 = 4e^2(bd + ec)^2(x^2 + c^2 + 2cy + y^2)$. Scott has referred to *The Geometry of René Descartes*, translated by David Eugene Smith and Marcia L. Latham (New York, ${}_2$1954), p. 135, note [173]. Smith in turn referred to Claude Rabuel, *Commentaires sur la Geometrie de M. Descartes* (Lyons, 1730), p. 348. Incidentally, in his manuscript *Geometria* (*ca.* 1693), Newton discussed a Cartesian oval. See D. T. Whiteside, ed., *The Mathematical Papers of Isaac Newton*, Vol. VII (Cambridge, 1976), pp. 384–387, esp. Whiteside's footnote (85) on pp. 385–387.

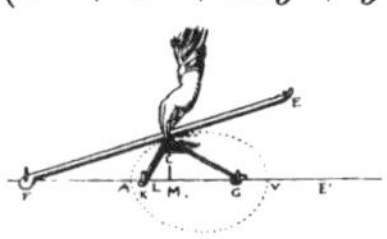

[85] *AT*, VI, pp. 429–430; Olscamp, p. 219.

the rays always reflect in such a manner that $\sin \angle P'CF / \sin \angle PCG = A5/A6$.[86]

Such results are obtained as a generalization of reflection or refraction by means or mirrors or lenses, respectively, the shapes of which are ellipses or hyperbolas. In Chapter 4, §1, we have seen that around 1626 Descartes with his Parisian colleagues studied refraction by optical instruments having a surface which was a conic section in shape. When they discussed optical properties of the conic section-shaped lenses, they were, of course, at the same time familiar with important characteristics of these classical curves. In contrast, in the case of ovals we should note that a property of the normal or tangent of the curve is known first, that is to say, the refractive law according to which the rays pass through is given first, then a curve is obtained afterward. In short, when a property of the tangent is given, the original curve is obtained. Thus Descartes's study of ovals is one of the very early examples of the 'inverse method of tangents,'[87] although his treatment is primitive and fundamentally finitist, like his method of tangents, not a method which involves some infinitesimal procedure.

When and how did Descartes make the study of these ovals? According a detailed examination by Hara, a Japanese historian of mathematics,[88] Descartes's description on ovals in the *Géométrie* is closely tied to his manuscript which was first published posthumously in the *Opuscula posthuma, physica et mathematica* (Amsterdam, 1701) and much later republished under the editorship of Paul Tannery in Volume X of the Adam-Tannery edition as the "Excerpta mathematica X–XII." The "Excerpta mathematica" may give a hint as to when and how the author of the *Géométrie* discovered the ovals. As to the date of the composition of these fragments, Tannery has conjectured: "Ce fragment doit remonter à une date où Descartes venait seulement d'imaginer sa méthode des tangents, et n'en possédait pas encore bien la pratique."[89] This

[86] *AT*, VI, pp. 431–433; Olscamp, pp. 220–222. Cf. Scott, *Op. cit.* (n. 84), pp. 129–130. Figure 5.6 is taken from Scott, p. 129.

[87] For the later development of the inverse method of tangents, see D. T. Whiteside, "Patterns of Mathematical Thought in the Later Seventeenth Century," *Archive for History of Exact Sciences*,**1** (1960), X. 3. "The concept of tangent," pp. 348–365; Christoph J. Scriba, "The Inverse Method of Tangents: A Dialogue between Leibniz and Newton (1675–1677)," *Archive for History of Exact Sciences*, **2** (1963), pp. 113–137.

[88] Kokiti Hara, "Comment Descartes a-t-il découvert ses ovales?", *Historia Scientiarum*, No. 29 (1985), pp. 51–82. Importantly, Hara points out that the text and figures in the Adam-Tannery version are not a faithful reproduction of those of the *Opuscula posthuma*. For example, the titles of the "Exerpta mathematica," X: "Ovales Opticae Qvatvor," XI: "Earvm Descriptio et Tactio," and XII: "Earvmdem Octo Vertices, Horvmqve Vsvs" are not Descartes's original but are taken from "Index Exerptorum" of the 1701 *Opuscula* version, namely by the editor of the *editio princeps*. (Cf. *AT*, X, "Avertissement" par Charles Adam, p. 280) The "Exerpta mathematica, X-XII" which we are discussing appear on pp. 9–17 in the 1701 edition and pp. 310–324 in the Adam-Tannery edition. Incidentally, Christiaan Huygens's reconstruction of how Descartes found ovals can be seen in his *Traité de la lumière* (Leiden, 1690): *Œuvres complètes de Christiaan Huygens*, publiées par la Société Hollandaise des Sciences, t. XIX (La Haye, 1937; Amsterdam, 1967), pp. 529–530.

[89] Paul Tannery, "Éclaircissements sur les ovales," *AT*, X, p. 325.

conjecture can be justly accepted. But the problem still remains as to when and how Descartes formulated his method of tangents. Reflecting upon the period of his dioptrical research, it is certain that Descartes studied his ovals after 1626.[90] Moreover, on reviewing the notation of the "Excerpta mathematica X-XII," which is very close to that of the *Géométrie*, we may safely assert that he found his ovals after having composed the *Regulae* and before deciding to write the *Dioptrique* and the *Géométrie*, thus probably between the years 1628 and 1635. The *terminus ad quem* is suggested from the fact that he drafted the fragments not in French but in Latin, so they may not be conjectured to have been an a preparatory manuscript for the *Dioptrique* or the *Géométrie.* Our examination above shows, in any case, that, insofar as the *Géométrie* is concerned, his 'method of tangents' and his study of ovals related to the 'inverse method of tangents' did not go beyond the algebraic analysis of finite quantities. We will discuss this issue in detail in Section 4.

After having identified main characteristics of his mature mathematical thought, we should reconsider here Descartes's crucial statement of 1638, in which he asserts that he terminated his essential mathematical research undertaken at his own initiative about 1623. Our considerations in the previous and present chapters indicate that he surely made some significant mathematical discoveries after 1623. But this does not seem to contradict what we stated in Chapter 3, § 3, D. They can be characterized as innovations related to basic notions of his algebra or ones which were made at the urging of others. It may be said, moreover, that after 1623 his main interest had shifted from solving mathematical problems, so that he thought his fundamental mathematical studies had been completed around 1623. Such reflections seem to lead to the following conclusion: In 1638 Descartes did not state anything that is essentially contradictory to historical evidence and he did not think that he had tackled those substantially new mathematical subjects which are exemplified by the *De solidorum elementis* on his own initiative after about 1623. Accordingly, we may assert, at least, that Descartes drafted the *De solidorum elementis* probably before 1623. We have to add promptly once again, however, that even after 1623 Descartes's mathematical thought developed dramatically and that it culminated conceptually with the algebra of line segments which was made public only in the *Géométrie* of 1637.

[90] We do not know for certain how Descartes acquired the knowledge of ovals in general. He does not seem to have mentioned these curves in his writings and letters during the 1620s, as far as I examined them. For mathematicians of the early seventeenth century, ovals of any kind might have been a natural extension of the conic sections, as can be seen in Kepler's *Astronomia nova* (1609).

§ 3. Descartes's Place in the Formative Period of the Modern Analytic Tradition

A. Plagiarism Debates with Jean Beaugrand in 1637–1638

With the *Géométrie*, Descartes established his mature system of algebraic analysis, which exerted great influence upon later generations of mathematicians. As we have frequently pointed out, he was not the first mathematician who engaged in building the modern analytic tradition of mathematics. The formation of this tradition was initiated by "Viète's analytic program," to use Michael S. Mahoney's term.[91] There arises, then, an important problem as to whether or not Descartes studied other analytic mathematicians, especially Viète, when constructing his own analytic mathematics. This has been a conundrum which historians of early modern mathematics have been challenged to resolve. In this section we will try to answer this question, having been furnished the background knowledge of Descartes's mathematical career up to the *Géométrie*. In the following discussion, the main line of which is in general accord with Mahoney's studies on the formation of algebraic analysis, we will focus on Descartes's relation to the analytic program and attempt to supplement previous research on this intellectual current.

At the end of the *Géométrie*, Descartes stated his expectation that later generations would promptly accept his style of mathematics and tackle problems which he voluntarily left unsolved. Despite his hope, with the clear exception of Florimond de Beaune, who wrote *Notae breves* for Frans van Schooten's Latin edition edition of the *Géométrie* published in 1649[92] it is doubtful that contemporary French mathematicians unconditionally welcomed this mathematical treatise.

One of the reasons that the prompt acceptance of Cartesian mathematics was retarded was Jean Beaugrand's accusation that the author of the *Géométrie* had plagiarized some of the writings of the forerunners of analytic mathematics, especially François Viète and/or Thomas Harriot. Beaugrand was renowned primarily as a member of the legal profession, but he had also established some fame as a mathematician by 1630. (For the details, see Subsection B.) He had some relation to the publication of the *Géométrie*. At the beginning of 1637 Descartes asked the French king for the Privilège to publish his *Discours de la méthode* and the three *Essais* through Mersenne. Beaugrand, who was then a *Sécrétaire du Chancelier*, was in the position responsible whether it should be offered or not.

[91] A part of the title of Chapter II "Nullum Non Problema Solvere: Viète's Analytic Program" of Michael S. Mahoney, *The Mathematical Career of Pierre de Fermat (1601–1665)* (Princeton, 1973; ${}_2$1994). We have also learned much from his doctoral dissertation at Princeton University: "The Royal Road: The Development of Algebraic Analysis from 1550 to 1650, With Special Reference to the Work of Pierre de Fermat" (1967) and his paper "The Beginnings of Algebraic Thought etc." (cited in n. 21 above).

[92] We will discuss the Latin version of 1649 in Section 4.

At that time, the *Chancelier*, Pierre Seguier, retained a copy of Descartes's first essay the *Dioptrique*, which was submitted for examination. It was the *Sécrétaire du Chancelier*, Beaugrand, who sent that copy without the author's permission to his able protégé Fermat to ask the latter's opinion of it. This brought about a well-known debate between Descartes and Fermat on the theory of light.[93]

Probably toward the end of 1637, moreover, after having examined Descartes's third essay the *Géométrie*, the same Beaugrand is supposed to have had a personal conversation with Mersenne to express his estimation of it. On that occasion, the *Sécrétaire du Chancelier* seems to have charged the author with plagiarising Viète and other mathematicians.[94] It is in order to reexamine whether Beaugrand's charge was just or not as well as to place Descartes in the appropriate position in the tradition of analytic mathematics that we have to argue about the historically important problem concerning Descartes's relation to the mathematicians of earlier generations.

The following letter of Descartes to Mersenne, which was probably written soon after his having been informed by Mersenne of Beaugrand's accusation is an exemplary document wherein Descartes stated in his own words the relationship between himself and other contemporary mathematicians, especially Viète. He begins this letter with the following statement, a part of which has been already quoted above (p. 324). "The author of the *Géostatique*" in the first sentence is Beaugrand,[95] of course:

> The judgment the author of the *Géostatique* formed about my writings does not affect me much. And I am not so glad of being obliged to talk advantageously of myself. But since only a few people could understand my *Géométrie* and you desire me to write to you the opinion I have of it, I think it is the right time to tell you that my opinion is such that I do not wish to add anything to it; and that I have simply tried by the *Dioptrique* and the *Météores* to persuade [the reader] that my method is better than the ordinary, but I claim that I have proved that by my *Géométrie*. For, from the outset, I solve in it a problem which, according to Pappus's statement, could

[93]The content and background of the debate is analyzed in Mahoney's book on Fermat (n. 91), Chapter IV, Section IV, "Looking under the Bed: Descartes vs. Fermat, 1637–38," pp. 170–193; A. I. Sabra, *Theories of Light from Descartes to Newton* (London, 1967; Cambridge, 1981), Chapter III, "Descartes' Explanation on Reflection. Fermat's Objections" & Chapter IV, "Descartes' Explanation of Refraction. Fermat's 'Refutations'," pp. 69–135.

[94]For the situation in which Beaugrand made his criticism of the *Géométrie* public, see *AM*, I, p. 65, note 1; *CM*, VI, p. 344, the forward by the editors. As to the intricate relation between Beaugrand and Descartes which led to their mutual discord, see "Beaugrand, (Jean de)," (Table des noms propres), *AM*, I, pp. 429–430.

[95]For Beaugrand's career and publications, consult Henry Nathan, "Beaugrand, Jean," *DSB*, I, pp. 541–542. His *Géostatique* was published in Paris in 1636 and its Latin version as *Geostatica* in Paris in 1637.

> never be resolved by any of the ancients; and we can say that it has never been by any of the moderns either since none of them have written anything on it and, nevertheless, the ablest ones have tried to find out those other things which Pappus states in the same place to have been searched for by the ancients, such as the *Apollonius redivivus*, the *Apollonius Batavus* and others, among whom we have to place M. your *Conseiller De maximis et minimis*; But none of those men have been able to achieve anything that the ancients had been ignorant of. After that, what I set forth in the second book concerning the nature and properties of curved lines and the way to examine them is, as it seems to me, beyond the ordinary geometry just as Cicero's rhetoric is beyond children's a, b, c. And I have such little confidence in that which your Geostatician promises that it does not seem to me less ridiculous than the braggards of Italian comedies, to say that he will give in a Preface better ways than mine for discovering the tangents of all curved lines.[96]

Descartes here praises himself for having solved the problem of Pappus, which in his judgment neither the ancients nor the moderns have successfully grappled with. He compares his method of solving that problem with some works of other contemporary mathematicians. The works he refers to are by Marino Ghetaldi, Willebrord Snel, and Fermat. Note that in the passage quoted above Descartes identifies the titles of the books with their authors, for example, the *Apollonius redivivus* with Ghetaldi. Let us briefly examine these works in order to understand well what Descartes wants to claim.

Ghetaldi's *Apollonius redivivus seu restituta Apollonii Pergaei inclinationum geometria* published in Venice in 1607 was, as the title itself suggests,

---

[96] *Op. cit.* (n. 73), *AT*, I, pp. 478–479; *AM*, II, pp. 65–66; *CM*, VI, pp. 344–346: "Le iugement que l'autheur de la *Geostatique* fait de mes écrits me touche fort peu. Et ie ne suis pas bienaise s'estre obligé de parler auantageusement de moymesme; mais pour ce qu'il y a peu de gens qui puissent entendre ma *Géométrie*, et que vous desirez que ie vous mande quelle est l'opinion que i'en ay, je croy qu'il est à propos que ie vous die qu'elle est telle, que ie n'y souhaitte rien dauantage; et que i'ay seulement tasché par la *Dioptrique* et par les *Météores* de persuader que ma methode est meilleure que l'ordinaire, mais ie pretens l'auoir demonstré par ma *Geometrie*. Car dés le commencement i'y resous vne question, qui par le témoignage de Pappus n'a pû estre trouuée par aucun des Anciens; et l'on peut dire qu'elle ne l'a pû estre non plus par aucun des modernes, puis qu'aucun n'en a écrit, & que neantmoins les plus habiles ont tasché de trouuer les autres choses que Pappus dit au mesme endroit auoir esté cherchées par les anciens, comme l'*Apollonius Redivivus*, l'*Apollonius Batavus*, et autres, du nombre desquels it faut mettre aussi M. vostre Conseiller *De maximis & minimis*; mais aucun de ceux-là n'a rien sceu faire que les anciens ayent ignoré. Apres cela, ce que ie donne au second liure, touchant la nature & les proprietez des lignes courbes & la façon de les examiner, est, ce me semble, autant au delà de la geometrie ordinaire, que la rhetorique de Ciceron est au delà de l'a, b, c des enfans. Et ie croy si peu ce que promet vostre Geostaticien, qu'il ne me semble pas moins ridicule de dire qu'il donnera dans vne Preface des moyens pour trouuer les tangentes de toutes les lig[nes] courbes qui seront meilleurs que le mien, que le sont les Catitans de Comedies Italiennes."

an attempt to reconstruct Apollonius's lost work *Neuses* (*νεύσεις*). As is clear from "Ad Lectorem" attached to his work, Ghetaldi tried to restore Apollonius's *Neuses*, following Pappus's account in Book VII of his *Collection*.[97]

The *Apollonius redivivus* of 1607 was supplemented by the same author's *Liber secundus*, which appeared in Venice in 1613, Ghetaldi composed the second book of *Apollonius redivivus*, having been stimulated by Alexander Anderson's *Supplementum Apllonii redivivi* (1612). Although Ghetaldi gave high praise to Anderson, an editor of Viète, and to the analytic art in general,[98] his way of restoring the lost work of the ancient geometer was not algebraic, but rather followed the classical manner. In Venice in 1607 he also published the *Supplementum Apollonij Galli* as a sister volume of his *Apollonius redivivus*.[99] The author thereby attempted to supplement, as the title clearly indicates, his master Viète's *Apollonius Gallus* (1600).

Willebrord Snel also joined in Viète's and Ghetaldi's program to reconstruct the lost works of Apollonius which were listed in Book VII of Pappus's *Collection*. Snel's writings on Apollonius consist of three parts. The first part has remained unpublished and its manuscript is now at the library of the University of Leiden.[100] The second was published in Leiden in 1607 under the title *Περὶ λόγου ἀποτομῆς καὶ τερὶ χωρίου ἀποτομῆς resuscitata Geometria*. In his work, Snel expressed his indebtness to Viète and defined the '*ἀνάλυσις*' to be "the art of finding well in mathematics" (*ars bené inveniendi in mathematicis*) following Ramus's and Viète's definition.[101] In composing this book, he apparently referred to Pappus's *Collection* and tried to restore Apollonius's two lost works, the *Cutting off of a Ratio* (*λόγου ἀποτομή*) and the *Cutting off of an Area* (*χωρίου ἀποτομή*).[102] Interestingly, Snel solved a problem (*Problema III*) following Pappus's classical dual method of analysis-synthesis.[103]

The third part appeared in Leiden in 1608 as the *Apollonius Batavus seu exsuscitata Apollonii Pergaei περὶ διωρισμένης τομῆς Geometria*, which is the

[97]Marino Ghetaldi, *Apollonius redivivus* (Venice, 1607), "Ad Lectorem": *Opera omnia*, Redactor Žarko Dadić (Zagreb, 1968), pp. 195–196. Pappus's description on Apollonius's *Neuses* can be seen in Alexander Jones, *Book 7 of the Collection*, Part 1, pp. 112–115 and 196–229 (Lemmas to the Neuses); Hultsch, II, pp. 670–673 and 770–821; Commandino, ff. 163v–164r & 202v–219v; Ver Eecke, II, pp. 501–503 and 596–633. Cf. Heath, *A History of Greek Mathematics*, Vol. 2 (Oxford, 1921), pp. 189–192 and 412–416.

[98]Ghetaldi, *Apollonius redivivus seu restitutae Apollonii Pergaei de inclinationibus geometriae*, liber secundus (Venice, 1613), "Ad Lectorem": *Opera omnia*, p. 228.

[99]With regard to Ghetaldi's general attempt to restore Apollonius's works, see Luigi Campedelli, "Ghetaldi (Ghettaldi), Marino," *DSB*, V, pp. 381–383, esp. p. 382.

[100]Dirk J. Struik, "Snel (Snellius or Snel van Royen), Willebrord," *DSB*, XII, p. 500.

[101]Wilebrordus Snellius, *Περὶ λόγου ἀποτομῆς etc.* (Leiden, 1607), pp. 3 & 7.

[102]For Pappus's account of the *Cutting off of a Ratio* and the *Cutting off of an Area*, see Jones, Part 1, pp. 86–89 and 126–141 (Lemmas to the *Cutting off of a Ratio and an Area*); Hultsch, II, pp. 64–643 and 684–705; Commandino, ff. 158r–159r and 166r–173r; Ver Eecke, II, pp. 480–482 and 512–530. Cf. Heath, *A History of Greek Mathematics*, Vol. 2, pp. 175–180; Jones, Part 2, pp. 510–514.

[103]Snellius, *Op. cit.* (n. 101), pp. 12–13 (p. 12 was numbered as p. 10 by mistake).

work Descartes mentions in his aforementioned letter to Mersenne. The aim of this book was to restore Apollonius's *Determinate Section* (*διωρισμένη τομή*), while referring to Pappus's description of it.[104] According to Pappus, problems in that work of Apollonius can be stated generally as follows: "Let four points A, B, C, D be given on a straight line, and let the ratio $\alpha : \beta$ also be given. It is required to cut the line in a point X such that

$$\text{AX.BX} : \text{CX.DX} = \alpha : \beta.\text{"}^{105}$$

This formulation is similar to that of Pappus's locus problem which Descartes solved. But the problem of determining X is quadratic and can be settled easily. Our examination of Snel's two published works demonstrates that the analytical method used in them was not algebraic and did not essentially go beyond the conceptual framework of the ancient geometers.

Noteworthy is that both Ghetaldi's *Apollonius redivivus* and Snel's *Apollonius Batavus* were works which came after Viète's *Apollonius Gallus seu exsuscitata Apollonii Pergaei περὶ ἐπαφῶν Geometrica, ad V. C. Adrianum Romanum Belgam*, published in Paris in 1600. Viète attempted in his *Apollonius Gallus* to reconstructing Apollonius's *Tangencies* (*ἐπαφοί*) by using Pappus's explanation in the seventh book of geometrical analysis.[106] A problem which particularly interested Viète was drawing a circle which touches three given circles. Viète's solution to that problem was by means of only the ruler and the compass and is said to have impressed the Belgian mathematician Adriaan van Roomen. The book was published under the editorship of Ghetaldi.[107] The aforementioned works by Ghetaldi and Snel were meant to realize Viète's overall project of reviving lost ancient mathematical works by means of the method of geometrical analysis. Unfortunately, however, these works together with the book of Viète were all purely geometrical.

In his letter cited above, Descates also refers to "M. your *Conseiller De maximis et minimis*" and to Beaugrand's method for drawing tangents to curved lines. "M[onsieur] your *Conseiller*" is undoubtedly the *Conseiller au Parlement de Toulouse* Fermat, and *De maximis et minimis* must be his short tratise *Methodus ad disquirendam maximam et minimam et de tangentibus linearum curvarum*, a copy of which Mersenne probably sent to Descartes a bit prior to the composition of the latter's reply which we are examining here. But judging from the context of the letter, Descartes here very likely refers to Fermat's manuscript *Apollonii Pergaei libri duo de locis planis restituti* rather than the

[104] Jones, Part 1, pp. 88–91 & 142–195; Hultsch, II, pp. 642–645 & 704–771; Commandino, ff. 159r & 173r–202v; Ver Eecke, II, pp. 482–483 & 530–596.

[105] See Jones, Part 2, p. 514. We modify only notation from the Greek to the Roman letters. Cf. Heath, Vol. 2, pp. 180–181.

[106] Jones, Part 1, pp. 90–95 & 230–249; Hultsch, II, pp. 644–649 & 820–853; Commandino, ff. 159r–160r & 219v–232r; Ver Eecke, II, pp. 483–485 & 633–658. Cf. Jones, Part 2, pp. 534–539; Heath, Vol. 2, pp. 181–185

[107] H. L. L. Busard, "Viète, François," *DSB*, XIV, p. 22.

Methodus ad disquirendam maximam et minimam. Descartes seems to have confused the titles of these manuscripts with one another.[108] Fermat's tract on plane loci was a work which took over what Viète, Ghetaldi, and Snel had left unfinished in their program to restore Apollonius's lost works. The task of the mathematician of Toulouse in his *De Locis planis* was to reconstruct the *Plane Loci* (τόποι ἐπίπεδοι).[109] According to Mahoney's detailed examination, this treatise shows a transition from the geometrical analysis of the ancients to the algebraic geometry exemplified by the same mathematician's well-known tract *Ad locos planos et solidos isagoge.*[110] But as can be easily judged, Fermat's attempt to revive Apollonius's *Plane Loci* (not the *Isagoge*), together with the works of Ghetaldi and Snel, cannot be compared with Books I and II of Descartes's *Géométrie.* Nevertheless, of course, we should not overlook the fact that at the end of 1637 Descartes was well versed in the mathematical works influenced by Viète's analytical method, albeit they were without exception fundamentally geometrical.

In the same letter to Mersenne, Descartes further insists that his method of drawing the normal or the tangent to a given curved line is superior to Beaugrand's (and perhaps Fermat's as well). We have already seen in Section 2 (p. 230) his own method of tangents by means of indeterminate coefficients.

What then were Fermat's an Beaugrand's methods for describing tangents? In his *Methodus ad disquirendam maximam et minimam* and *De tangentibus linearum curvarum*, Fermat tries to apply what he learned from Diophantus and Viète to his method for drawing a tangent to curved lines. He presents his method by taking the example of a parabola. Let BDZ be the given parabola, which has the vertex D and the diameter DC (Figure 5.7). Let B be the point at which the line BE is to be drawn tangent to the parabola. We choose on the segment BE a point O through which we draw the ordinate OI. Then, since O is exterior to the parabola, $CD/DI > BC^2/OI^2 = CE^2/IE^2$. Let CD be a given magnitude D. Put CE and CI equal to A and E, respectively. Thus we have

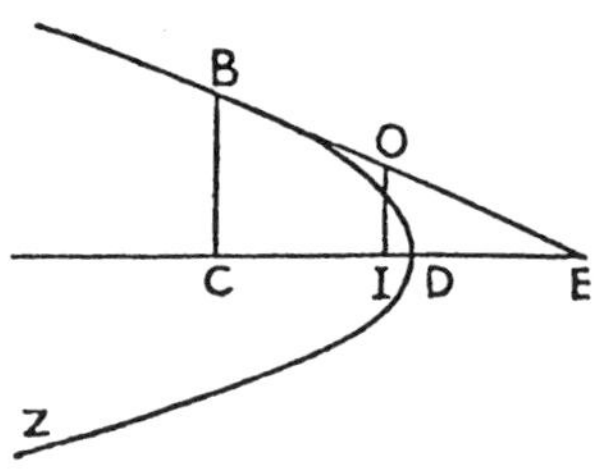

Figure 5.7

$$\frac{D}{D-E} > \frac{A}{A^2+E^2-2AE}$$

[108]The editors of the various editions of the letter of Descartes have all agreed that Descartes mentions Fermat's manuscript on Apollonius's *Plane loci. AT*, I, p. 478, note d; *AM*, II, p. 65, note 4; *CM*, VI, p. 345, note 5.

[109]As for Apollonius's *Plane loci*, see Jones, Part 2, pp. 539–546.

[110]See Mahoney, *The Mathematical Career of Pierre de Fermat* (n. 91), Chapter III, Section III, "The Origins of the Introduction: Apollonius' *Plane Loci* and *Conics*," pp. 92–117.

or $DA^2 + DE^2 - 2DAE > DA^2 - A^2E$. 'Adequating'[111] the previous formula, $DE^2 - 2DEA$ is 'adequal' to $-A^2E$. Dividing all terms by E, we have $DE + A^2$ to be 'adequal' to $2DA$. On taking out DE, there remains A^2 equal to $2DA$, i. e. $A = 2D$. "Thus we have proved that CE is the double of CD, which is in fact the case."[112]

The process corresponding to the word 'adequal', in Fermat's mind, was purely finitistic, but it tended to be interpreted by later mathematicians to involve something similar to the concept of a limit. Beaugrand doubtlessly learned his method of tangents from his protege Fermat. The only essential difference which distinguished his method from Fermat's was that Beaugrand used e instead of Fermat's E.[113]

What becomes clear from our analysis of the methods of tangents is this. Descartes's method of tangents is exceedingly complicated to handle, whereas with Fermat's (and Beaugrand's) it is relatively easy to get fast results, although the latter might involve a difficulty with regard to the notion of 'adæqualitas'. More importantly, Descartes is so confident of the success of the *Géométrie* that he would never condescend to learn something from others. Judging from the later disputes between Descartes and Fermat through their correspondence during 1638, the former at the end of 1637 perhaps had yet to examine closely a copy of Fermat's tract *Methodus ad disquirendam maximam et minimam etc.* His confusion of the *De locis planis* with *De maximis et minimis* might be said to have been caused partly because of his self-confidence and partly because of his self-conceit.[114]

The same letter continues and begins to concern itself with Book III of the *Géométrie* . Descartes directly compares his mathematics with a work by Viète on the theory of equations:

---

[111] For the special word 'adæquare', see Mahoney, *Fermat*, pp. 163-164. The term 'adæqualitas' was first coined by Xylander as the Latin word for Diophantus's 'παρισότης'. Heath translated it into 'near-as-possible equality'. See Heath, *Diophantus of Alexandria* (Cambridge, 1910; New York, ₂1964), pp. 95–98. Having examined various manuscript sources, the following paper has pointed out that Fermat thought 'adæquare' to be almost synonymous with 'æquare', meaning "to put equal." I don't believe this assertion to be definitive, but it must be certain that 'adæquare' had a very primitive sense close to "to put equal." Herbert Breger, "The Mysteries of Adaequare: A Vindication of Fermat," *Archive for History of Exact Sciences*, **46** (1993–1994), pp. 193–219.

[112] *Œuvres de Fermat*, éd. Charles Henry et Paul Tannery, t. I (Paris, 1891), pp. 134–136; *Varia opera mathematica*, ed. Samuel de Fermat (Toulouse, 1679), pp. 63–64. An English translation is in D. J. Struik, ed., *A Source Book in Mathematics, 1200-1800* (Cambridge, Mass., 1969), pp. 222–224. Cf. Mahoney, *Fermat* (n. 91), pp. 166–167.

[113] Beaugrand's technique can be seen in one of his *facta*, which is reproduced in *CM*, VIII, pp. 90–92.

[114] A more interesting story about the debate between Descartes and Fermat which developed in 1638 can be seen in Mahoney, *Fermat*, pp. 170–192. Mahoney states his observation on the personalities of the two French mathematicians: "Descartes learned nothing from it [the debate], even though he was in error; Fermat, even though he was correct, gained new mathematical insights that led him to revise his methods and sharpen his tools" (p. 171).

> And it is far from likely that what I wrote can be easily taken from Viète; on the contrary, the reason why my treatise is so difficult to understand is that I tried to put into it only what I thought to have been known neither by him nor by anyone else. As can be seen, if one compares what I wrote about the number of roots in each equation on page 372, which is the place where I began to present the rules of my *Algebra*, with what Viète wrote about it at the end of his book *De Emendatione aequationum*, one will see that I determine it in general for all equations, whereas, having provided some particular examples, of which he is so proud that he wanted to conclude his book with it, he showed that he could not solve it in general. Thus I started where he had stopped. And I have done this without knowing that, for I have much more perused Viète, since I received your last letter, than I had ever done before, having found it here in the hands of one of my friends. And between us, I do not think that he knew as much as I thought, although he was very skilful.[115]

The exact title of Viète's book which Descartes refers to is *De Æquationum recognitione et emendatione*, which was published in Paris in 1615 under the editorship of the author's Scottish friend Alexander Anderson. The friend from whom Descartes borrowed that tract is supposed to have been the Dutch mathematician Godefoid de Haestrecht.[116]

Let us glance at Viète's tract and compare it with Descartes's theory in the *Géométrie*. In the fourteenth and last chapter of his treatise, Viète presented four theorems on the theory of equations. Theorem I is the following:

> (Theorem I) If $\overline{B+D}$ in $A - A$ quad. be equal to B in D, A is explicable by either B or D.[117]

---

[115] *AT*, I, pp. 479–480; *AM*, II, p. 66; *CM*, VI, p. 346: "Et tant s'en faut que les choses que i'ay écrites puissent estre aisément tirées de Viète, qu'au contraire, ce qui est cause que mon traitté est difficile à entendre, c'est que i'ay tasché à n'y rien mettre que ce que i'ay crû n'auoir point esté sceu ny par luy, ny pas aucun autre. Comme on peut voir, si on confere ce que i'ay écrit du nombre des racines qui sont en chaque équation dans la page 372, qui est l'endroit où ie commence à donner les regles de mon Algebre, auec ce que Viète en a écrit tout à la fin de son liure *de Emendatione æquationum*; car on verra que ie le determine generalement en toutes équations, au lieu que luy n'en ayant donné que quelques exemples particuliers, dont il fait toutesfois si grand estat qu'il a voulu conclure son liure par là, il a monstré qu'il ne le pouuoit determiner en general. Et ainsi i'ay commencé où il auoit acheué; ce que i'ay fait toutesfois sans y penser, car i'ay plus fe'uilleté Viète depuis que i'ay receu vostre derniere, que ie n'auois jamais fait auparavant, l'ayant trouué icy par hazard entre les mains d'vn de mes amis; et entre nous ie ne trouue pas qu'il en ait tant sceu que ie pensois, nonobstant qu'il fust fort habile."

[116] *CM*, VI, p. 346, note 2.

[117] Viète, *Opera mathematica* (1646), p. 158. Cf. Witmer's translation, p. 309. Both *AT*, I, p. 479, note a & *AM*, II, p. 66, note 2 mistakenly refer to "Chap. XVI" instead of "Chap. XIV".

Translating this into our own terminology, if $(B+D)A - A^2 = BD$, then $(A-B)(A-D)=0$, i. e. $A=B$ or $A=D$. That is, Viète resolved the quadratic equation into its linear factors, $A-B$ and $A-D$. He used this procedure up to the quintic equation $(A-B)(A-D)(A-G)(A-H)(A-K)=0$, which was treated in the last theorem IV of the same chapter. One may argue that with these four theorems he partially understood the relations between the coefficients and the roots of an equation.[118] At any rate, he presented his results for only four cases and concluded his entire treatise with the following sentence: "I have treated [this problem][elsewhere] at length and in other respects with the elegant reasoning behind this beautiful observation, [so] this must be the end and crown [of this work]."[119]

Let us turn to Descartes. On page 372 (in the first edition) of the *Géométrie*, Descartes commences to argue the following problem: "Combien il peut y auoir de racine en chaque Equation." His very first statement of this paragraph asserts that each equation has as many roots as the unknown quantity has dimensions, as we have already seen in Chapter 4, § 1 (p. 166). Viète adopted only positive roots representing geometrical magnitudes. On the other hand, Descartes there accepts negative roots (*fausses racines*, in his own terminology)[120] as well; proposes a criterion to examine how many positive roots (*vrayes racines*) are in each equation, which is later called 'Descartes's Rule of Signs'[121]; and even asserts: "The true [positive] roots, as well as the false [negative] ones, are not always real (*reél*), but sometimes only imaginary (*imaginaire*)."[122] Although he does not notice the correlations between the coefficients and the roots because he takes examples with numerical coefficients (not general, symbolic ones), we must admit that his insistence that his formulation is more general than Viète's is basically correct. From this, one may state that Descartes is a pathfinder of purely abstract, structural mathematics. Mahoney, for one, contends that in Book III of the *Géométrie* "for the first time there appears new, purely abstract, non-intuitive objects in mathematics, which arise out of structural considerations."[123] Accordingly, Descartes was among the first to postulate n roots for an algebraic equation of n-th degree.[124]

---

[118] For example, see Frédéric Ritter, "François Viète, Inventeur de l'algèbre moderne 1540–1603: Essai sur sa vie et son œuvre," *Revue occidentale philosophique, sociale et politique*, seconde série, **10** (1895), p. 401; H. L. L. Busard, "Viète, François," *DSB*, XIV, pp. 23–24.

[119] *Opera mathematica*, p. 158; Witmer, p. 310.

[120] The first edition, p. 372; *AT*, VI, p. 445; Olscamp, pp. 229–230.

[121] The first edition, p. 373; *AT*, VI, p. 446; Olscamp, p. 230. See Smith and Latham, *Op. cit.* (n. 84), p. 160, note [196].

[122] The first edition, p. 380; *AT*, VI, p. 453; Olscamp, p. 236. See our discussion on p. 196.

[123] M. S. Mahoney, "The Beginnings of Algebraic Thought in the Seventeenth Century" (n. 21), p. 146.

[124] According to Joseph E. Hofmann, this fact, which is equivalent to the fundamental theorem of algebra, was known, for example, by P. Roth (1608), Albert Girard (1629), and Thomas Harriot (1631). Hofmann, *Geschichte der Mathematik*, Erster Teil (Berlin, $_2$1963), p. 166. Cf. Mahoney, *Fermat*, p. 150, note 15. On the development since Girard, see Christian Gilan,

Thus, we recognize that what Descartes wants to communicate to Mersenne is essentially justifiable. Insofar as we restrict our discussion to the mathematical matters Descartes mentions in his letter to Mersenne, he started, as he claims, where Viète had stopped.

But we must not disregard Descartes's statement in his letter quoted above that he "much more perused Viète" than he had even done before. He never denies that he read Viète before the composition of the *Géométrie*, or at the latest before the time he wrote the letter to Mersenne. When, then, did he read Viète? We will try to answer this important question in Subsection B.

Descartes seems to have considered the priority problem concerning mathematical discoveries stated in the *Géométrie* to be rather important. But he did not like to become entangled with debates about his publication of 1637. Unfortunately, however, counter to his hopes Beaugrand moved openly to attack him for the priority of Viète (sometimes of Harriot). Beaugrand sometimes accused the author of the *Géométrie* by circulating anonymous pamphlets[125] and sometimes sent letters to Mersenne claiming that his opponent plagiarized Viète's *De Aequatione recognitione et emendatione*.[126]

In replying to such charges against him, Descartes addressed letters to Mersenne, on one occasion by reiterating the following sentence, similar to one which he had previously stated (see p. 242): "I begin where Viète had finished,"[127] and on another occasion by asserting strongly that "I have no acquaintance with the geometer about whom you wrote to me and I am surprised with what he says, that we have studied Viète together in Paris; for this is a book of which I do not remember to have seen even the cover while I was in France."[128] The geometer to whom Descartes referred is conjectured to be Jean-Baptiste Chauveau.[129] But note that the last statement does not necessarily imply that Descartes had never studied any work by Viète because it was concerned only with a particular book of Viète and a particular period.

"Sur l'histoire du théorème fondamental de l'algèbre: Théorie des équaions et calcul intégral," *Archive for History of Exact Sciences*, **42** (1991), pp. 91–136.

[125] See Paul Tannery, "La Correspondance de Descartes dans les Inédits du Fonds Libri," esp., Chapter III, "Les pamphlets mathématiques contre Descartes," *Mémoires scientifiques*, publiés par J.-L. Heiberg et H.-G. Zeuthen, t. IV: *Sciences modernes*, éd. par Gino Loria (Toulouse/Paris, 1926), pp. 203–228.

[126] "(Jean Beaugrand), à (Paris), à Mersenne, à Paris. (début de mars 1638)," *CM*, VII, pp. 87–103, esp. p. 88. For Beaugrand's charge that Descartes plagiarized Harriot, see the editors' note to *CM*, VII, pp. 201-202.

[127] "Descartes, à (Santpoort), à Mersenne, à Paris. 31 mars 1638," *AT*, II, p. 82; *AM*, II, p. 213; *CM*, VII, p. 120: "[I]e commence en cela par ou Viète auoit fini."

[128] "Descartes, à (Santpoort), à Mersenne, à Paris. (20 février 1639)," *AT*, II, p. 524; *AM*, III, p. 195; *CM*, VIII, p. 325: "Ie n'ay aucune connoissance de ce Geometrie dont vous m'écrituez, et ie m'étonne de ce qu'il dit, que nous auons estudié ensemble Viète à Paris; car c'est vn livre dont ie ne me souuiens pas auoir seulement iamais vû la couuerture, pendant que i'ay esté en France."

[129] *CM*, VIII, p. 328: the editor's note for l. 17.

Further examination of the ample, related documents of this kind would not be of much importance for our study. But it would be undeniable that Descartes's inclination not to express his indebtness to his predecessors made matters worse. In fact, in the *Géométrie*, only the two mathematicians, Girolamo Cardano and Scipione Ferro, among the moderns, are referred to as discoverers of the solution of the cubic equation.[130]

Another important factor which complicated the disputes between Beaugrand and Descartes was that it took place in the seventeenth century. This may need a little elucidation. A. Rupert Hall in his *Philosophers at War: The Quarrel between Newton and Leibniz* enumerates the causes which made any priority debates in the seventeenth century bitter and intricate: "the great value attached to personal merit, the emphasis on innovation as the creation of an individual talent, and the absence of a sociological theory of the growth of knowledge."[131] Beaugrand was nothing but a typical intellectual of the seventeenth century who had values which helped to entertain hostile feelings against 'independent or later discoveries,' to use today's terminology. The debates around Descartes's *Géométrie* were no exception. Beaugrand, who possessed the ordinary standards of value of his time, might not have admitted without reluctance independent discoveries in mathematics.

Did Descartes not read Viète at all before having drafted the *Géométrie*? We have seen that he does not consistently deny perusing Viète. In reality, it is at least certain that he looked at some works of Viète in the spring of 1632. To examine the plagiarism disputes between Beaugrand and Descartes more deeply and to place the *Géométrie* in a historically just place, let us see how Descartes responded to Viète before 1637.

B. Descartes's Acquaintance with Viète before the Composition of the *Géométrie*

It was especially due to his method of algebraic analysis that Descartes could pride himself on the novelty of his *Géométrie*. But the idea of algebraic analysis was not, of course, first thought out by him. As is well-known, it was François Viète who started the modern analytic school through his symbolic algebra. Descartes's rival, Fermat, received a great influence from Viète's analytic mathematics, probably through the aforementioned Beaugrand.[132]

Beaugrand was a mathematician who had a close relation to Viète's analytic program. In 1630 he became mathematician to Gaston d'Orléans and *Les cinq livres des zetetiques de François Viette* was dedicated to him by the

[130] *AT*, VI, pp. 471 & 473; Olscamp, pp. 249–250 for Cardano, *AT*, VI, p. 472; Olscamp, p. 249 for Ferro.

[131] A. Rupert Hall, *Philosophers at War: The Quarrel between Newton and Leibniz* (Cambridge, 1980), p. 6.

[132] Mahoney, *Fermat*, p. 49.

translator J. L. Vaulézard.[133] This work was a French translation of Viète's *Zeteticorum libri quinque*, first published in Tours in 1593, and constituted a companion volume of the same translator's *Introdvction en l'art analytic, ov novvelle algebre de François Viete*, published in Paris in the same year.

More importantly, in 1631 Beaugrand published with a commentary his edition of Viète's *In artem analyticem isagoge* and *Ad Logisticem speciosam notae priores*, the latter being bound with the former. It is noteworthy that parts of his commentary were incorporated into Frans van Schooten's 1646 version of the *Opera mathematica*.[134] The latter work *Ad Logisticem speciosam notae priores* had not seen the light of the day until the appearance of Beaugrand's 1631 edition, for Viète did not believe that it deserved publication. It contains elementary algebraic formulae that correspond to the propositions of Book II and IX of Euclid's *Elements*, and thereby provides essential rules on how to operate Viète's symbolic algebra, i. e. 'logistice speciosa'.[135] As we have already seen, it is supposed to have been Beaugrand who introduced the works of Viète to Fermat. Because of his apparent indebtness to the originator of symbolic algebra, Fermat was called 'a second Viète'.[136]

In contrast, even though Descartes was a prominent advocate of the analytic mathematics, it is quite difficult to establish how Viète influenced (or did not influence) him. As far as we can tell from extant documentary evidence, it was toward the end of April or at the beginning of May, 1632 that he surely read Viète's writings. At that time, he received a book by Viète which Mersenne had sent from Paris through a certain André Rivet in the Hague. That book was Beaugrand's work of 1631 which we referred to above. Mersenne, who at times communicated with Beaugrand, requested from the editor of the new Latin version of Viète two exemplars for Golius and Descartes.[137] The compiler

[133] See H. Nathan, *DSB*, I (n. 95), p. 541. An almost exhaustive catalog of Viète's works is now available: Warren Van Egmond, "A Catalog of François Viète's Printed and Manuscript Works," in Menso Folkerts und Uta Lindgren, hrsg., *Mathemata: Festschrift für Helmut Gericke, Boethius*, Bd. XII (Stuttgart, 1985), pp. 359–396. For Vaulézard's translations, see pp. 383-384. Recently a reprinted edition was published: Vaulézard, *La nouvelle algèbre de M. Viète* (Tours, 1986). For Vaulézard's dedication to Beaugrand, see p. 69.

[134] For example, a scholium added to Chapter IIII: "On the Rules of Specious Logistic" of the *Isagoge* in the 1646 edition was originally Beaugrand's. The scholium contains a concordance which compares the propositions of Diophantus's *Arithmetic* and those of Viète's *Zetetica*. *Francisci Vietæ in artem analyticem isagoge*, ed. I. de Beaugrand (Paris, 1631), pp. 69–71 correspnds to the *Opera mathematica* (1646), p. 4. Cf. J. Winfree Smith's translation in Jacob Klein's *Greek Mathematical Thought and the Origin of Algebra* (Ch. 4, n. 79), pp. 329–330. As to Beaugrand's 1631 edition, I refer to a copy in the Bibliothèque Nationale of Paris (Rés. V. 2075 (1)).

[135] *Vietæ Ad logisticem speciosam notae priores* (Paris, 1631): *Opera mathematica*, pp. 13–41. See also Witmer's English translation, "Preliminary Notes on Symbolic Logistic," pp. 33–82; F. Ritter's French translation, "Notes premières pour la logistice spécieuse," *Bulletino de bibliografia e di storia delle scienze matematiche e fisiche*, **1** (1868), pp. 245–276.

[136] Mahoney, *Fermat*, p. 26.

[137] Beaugrand wrote to Mersenne about geometry and astronomy on June 28 and July 31, 1630 (*CM*, II, pp. 504–507 & 509–525). On February 20, 1932 he also addressed to Mersenne

of Viète's two masterpieces seems to have been impressed by the new "analyse specieuse,"[138] and in his version the phrase "Nullum non problema solvere" (To leave no problem unsolved) appears several times, like a refrain.[139] "Nullum non problema solvere" was an expression used only once, at the end of Viète's original *Isagoge*.[140]

The book which fell into Descartes's hands was a small duo-decimo, which he referred to as "le livre d'Analyse" in his letter to Mersenne written on May 3, 1632. On examining this book of Viète's analytic art, he does not seem to have received a favorable impression from it, but he complained about the repeated words: "Nullum non problema solvere." He writes to Mersenne the following:

> I thank you for the book of Analysis which you sent to me. But, between ourselves, I do not think that it is very useful, nor that anyone can learn the method from reading it; I am not talking about "Nullum non problema solvere", but, about solving any problem as far as it is easy. It does not mean that I do not believe that the authors have a great knowledge, but I do not have sufficient ability to judge what is in this book nor what you wrote to me about the problem of Pappus. For one has to go far beyond the conic sections and the solid loci to solve it for any number of given lines, as a person must solve it who boasts of himself to "nullum non problema solvere" and as I think I have solved it.[141]

As far as the letter of May 3, 1632 is concerned, Descartes does not appear to have been impressed by Viète's art of algebraic analysis. Instead of speaking highly of Viète's book, he takes pride in having solved successfully Pappus's locus problem and grumbles at the repeated phrase "Nullum non problema solvere" in the *Isagoge* and its compiler's commentary about it. At least Descartes's mathematical talent, which solved the Pappus problem, was not affected by his reading Viète in the spring of 1632. His examination of Viète

a letter mentioning the problem of Pappus and his edition of Viète (*CM*, III, pp. 254–262). Mersenne asked Rivet to transmit exemplars of Viète's book to Golius and Descartes in his letter on March 14, 1632 (*CM*, III, pp. 268–272).

[138] Beaugrand called Viète's symbolic algebra "analyse specieuse" in his letter to Mersenne on February 20, 1632 (*CM*, III, 256).

[139] In Beaugrand's edition (n. 134), for example, pp. aij^{v}, 73–74 & 99.

[140] Viète, *In artem analyticem isagoge* (Tours, 1591), f. 9r; *Opera mathematica*, p. 12.

[141] "Descartes, à (Amsterdam), à Mersenne, à Paris. (3 mai 1632)," *AT*, I, p. 245; *AM*, I, p. 222; *CM*, III, pp. 296–297: "Ie vous remercie du liure d'Analyse que m'auez enuoyé; mais entre nous, ie ne vois pas qu'il soit de grande vtilité, ny que personne puisse apprendre en le lisant la façon, ie ne dis pas de *nullum non problema soluere*, mais de soudre aucun probleme, tant puisse-t-il estre facile. Ce n'est pas que ie ne ve'uille bien croire que les auteurs en sont fort sçauans, mais ie n'ay pas assez bon esprit pour iuger de ce qui est dans ce livre, non plus que de ce que vous me mandez du probleme de Pappus: car il faut bien aller au delà des sections coniques & des lieux solides, pour le resoudre en tout nombre de lignes données, ainsi que le doit resoudre vn homme qui se vante de *nullum non problema soluere*, et que ie pense l'auoir resolu."

at that time could have slightly influenced the notation and description of the *Géométrie*, but this took place after his having settled the ancient locus problem by means of the method of algebraic analysis in the winter of 1631–1632. So this alleged influence may be safely judged to have been inessential for Descartes's mathematical career.

Incidentally, Descartes doubtlessly heard something from Mersenne about the newly published Beaugrand version of Viète before having received it, because in his letter to Mersenne which was most probably written on April 5, 1632, he wrote about "someone who prides himself on solving all sorts of mathematical questions"[142] and advised the Parisian promoter of the mathematical sciences to propose him Pappus's problem to test whether or not he knew algebra comprehensively. "Someone" was apparently Beaugrand. This episode may corroborate the assertion that at least Beaugrand's 1631 edition did not influence Descartes's method of algebraic analysis enough to allow a solution of the problem of Pappus.

Can we conclude then that Descartes's mathematical thought was formed totally independently of Viète's? If available evidence is restricted to that which is directly ducumentary, the answer to this question might be affirmative. Nevertheless, there could be indirect and circumstantial evidence. In order to search for such evedence, let us retrace a few pivotal events in the formation process of modern analytic mathematics.

C. The Beginning of the Tradition of Algebraic Analysis in Arabic Mathematics and the Ramean Restart

By whom and when was the idea of algebraic analysis conceived for the first time? In other words, who was the first who understood algebra as the art of analysis? As we stated at the beginning of Subsection A, recent historians of mathematics have scrutinized the development of algebraic analysis in early modern period. In what follows we will attempt to survey the origins and development of the analytic tradition, insofar as they can be thought to have been some relation to Descartes's mathematical thought.

Incidentally, as recent historical studies have shown, the tradition of algebraic analysis is considered to have originated not necessarily in Western Europe, but in the Islamic world, as algebra, or more precisely *al-jabr* in Arabic. Here we don't try to describe Islamic or Arabic mathmatics comprehensively, but confine ourselves to the expalanation of it as far as it may be related to the mathematical thought of Descartes. Arabic mathematics flowered greatly with the almost entire translation and absorption of classics of ancient Greek mathematics. But in this jucture, we shouldn't forget that it received an impact from Indian mathematics. The influence from India relates to emphasizing

[142] "Descartes, à Amsterdam, à Mersenne, à Paris. (5 avril 1632)," *AT*, I, p. 244; *AM*, I, p. 221; *CM*, III, p. 291.

computation. To see this, for convenience we look at how Ibn Khaldūn, an outstanding historian of the Islamic world in the fourteenth century, classified mathematics in his *Muqaddimah*. According to Ibn Khaldūn, the mathematical disciplines begin with the sciences concerned with numbers. The first of these sciences is arithmetic (*arithmāṭīqī*) which originated from Greece. Its Arabic vocabulary 'arithmāṭīqī' itself is a transliteration from the Greek. Next comes the craft of calculation (*ṣinā'at al-ḥisāb*), and it in fact developed greatly in the Islamic world. Although Ibn Khaldūn doesn't mention it, the exemplar was the *Hindu Calculation* by Muḥammad ibn Mūsā al-Khwārizmī. The Arabic original of it has been lost, but its Latin manuscripts of the twelfth century are extant, well-known manuscripts beginning with "Dixit Algorizmi."[143] Through such writings, the craft of digital calculation beginning from 0 was transmitted from India, via the Islamic world, to the Western Christian world broadly.

Ibn Khaldūn writes a remarkable statement as an explanation of the craft of calculation.

> This craft is something newly created. It is needed for business calculations. Scholars have written many works on it. They are used in the cities for the instruction of children. The best method of instruction is to begin with (calculation), because it is concerned with lucid knowledge and systematic proofs. As a rule, it produces an enlightened intellect that is trained along correct lines. It has been said that whoever applies himself to the study of calculation early in his life will as a rule be truthful, because calculation has a sound basis and requires self-discipline. (Soundness and self-discipline) will, thus, become character qualities of such a person. He will get accustomed to truthfulness and adhere to it methodically.[144]

It cannot be said in the Greek world that the craft of calculation was respected. Ibn Khaldūn's statement seeing the craft of calculation at the basis of mathematics indicates a paticular character of Islamic mathematics.

Next comes 'Al-jabra wa'lmuqābalah.' Ibn Khaldūn defines this discipline as follows: "This is a craft that makes it possible to discover the unknown from the known data, if there exists a relationship between them requiring it."[145] This is a definition that should be noted. The oldest extant monograph on algebra is al-Khwārizmī's *Kitāb al-mukhtaṣar fī'l-ḥisāb al-jabr wa'l-muqābalah*

---

[143]For a Latin text and its French translation, see Muḥammad ibn Mūsā Al-Khwārizmī, *Le Calcul Indien (Algorismus): Versions latines du XIIᵉ siècle*, par André Allard, Préface de Roshdi Rashed (Paris, 1992). For a historical background, see Ahmad S. Saidan, "Numeration and Arithmetic," *Enclyclopedia of the History of Arabic Science*, edited by Roshdi Rashed in collaboraion with Régis Morelon, Vol. 2 (London/New York, 1996), pp. 331–348.

[144]*Prolégomèmes d'Ebn-Khaldoun*, texte arabe, par M. Quartermère, Tome premier, Troisième partie (Paris, 1858), p. 96; Ibn Khaldūn, *The Muqaddimah: An Introduction to History*, translated from the Arabic by Franz Rosenthal, Vol. 3 (London, 1958), p. 122.

[145]*Prolégomèmes*, p. 97; Rosenthal's translation, p. 124.

(Compendious Book on Calculation by Completion and Balancing).[146] Since the appearance of this book, the art of algebra developed exceedingly in Arabic mathematics. After referring to 'business arithmetic' and 'inheritance laws', Ibn Khaldūn proceeds to the next section on geometry.

Arabic mathematics became familiar with the pair conception of 'ἀνάλυσις' and 'σύνθεσις' of ancient Greece. In a sense, it may be said that it possessed a treasure of documents richer than the one on analysis and synthesis of classical Greece. The Arabic word 'taḥlīl' is correspondent to 'analysis', and 'tarkīb' to 'synthesis'. The two nouns have been derived from the verbs 'halla' and 'rakkaba' (< 'rakiba'), which mean 'resolve' and 'combine', respectively. One of the Muslim mathematicians who wrote very lucid writings on analysis and synthesis was Ibrāhīm ibn Sīnān ibn Thābit ibn Qurra, who was born from an illustrious family of scholars in Baghdad in the tenth century and who died a premature death. Roshdi Rashed and his disciple Hélène Bellosta have provided the entire shape of his mathematical achievements.[147] Ibn Sinān, in his *Treatise on the Method of Analysis and Synthesis, and the Other Ways of Solving Geometrical Problems*, discussing broadly the pair conception of analysis and synthesis, proceeded very closely to the point at which combining analysis with algebra. But he didn't succeed in having the two currents merge into each other. To quote Bellosta's observation, "Although Ibn Sinān forged the theoretical tools that the algebraists of the next centuries will use, his examples as well as his methods show that his work remained deeply rooted in the mathematical background of Greco-Hellenistic Antiquity. Ibn Sinān stands therefore at the point of junction of two chapters of mathematics: Greek geometry and Arabic algebra."[148] Although Ibn Sinān, a Blaise Pascal of Arabic mathematics, didn't become the first discoverer of algebraic analysis, he wrote just before geometrical analysis and algebra encountered each other.

Further, contributions by Abū 'Alī al-Ḥasan Ibn al-Ḥasan Ibn al-Haytham (Latin name, Alhazen) were also outstanding. Thanks to Rashed, Ibn al-Haytham's synthetic treatises *Analysis and Synthesis* and *The Known* have become available to us.[149] The latter treatise *The Known* is a work which de-

[146] Khwārizmī, *Al-Kitāb al-mukhtaṣar fī'l-ḥisāb al-jabr wa'l-muqābalah*, ed. by 'Alī Muṣṭafā Masharrafa and Muḥammad Mursī Aḥmad (Cairo, 1939); *The Algebra of Mohammed ben Musa*, edited and translated by Frederic Rosen (London, 1830). A critical Arabic text is a desideratum, and a version by Prof. Roshdi Rashed will be published in near future.

[147] Roshdi Rashed et Hélène Bellosta, *Ibrāhīm ibn Sinān: Logique et Géométrie au Xᵉ Siècle* (Leiden/Boston/Köln, 2000).

[148] Hélène Bellosta, "Ibrāhīm ibn Sinān: On Analysis and Synthesis," *Arabic Sciences and Philosophy*, Vol. 1 (1991), pp. 211–232, at p. 229.

[149] R. Rashed, "La philosophie des mathématiques d'Ibn al-Haytham, I: L'analyse et la synthèse," *Mélanges de l'Insitut Dominicain d'Études Orientales du Caire (MIDEO)*, **20** (1991), pp. 31–231, & "La philosophie des mathématiques d'Ibn al-Haytham, II: Les Connus," *MIDEO*, **21** (1993), pp. 87–273. Both the texts have been reproduced with elegant commentaries by Rashed in R. Rashed, *Les mathématiques infinitésimales du IXᵉ au XIᵉ siècle*, Vol. IV: *Ibn al-Haytham, Méthodes géométiques, Transformations ponctuelles et Philosophie des mathémaiques* (London, 2002). This volume is one of the most beautiful achievements which

veloped Euclid's *Data*, a writing Pappus referred to as necessary for analysis. Ibn al-Haytham's understanding of analysis and synthesis were quite broad and deep, but it didn't go beyond the framework of classical Greek mathematics. Rashed has thought that his art of analysis and synthesis, though interpreted as a kind of 'method', was essentially within the framework of geometry. According to Rashed, "sa généralité est l'éco de celle des objects de la discipline géométrique, et n'éxplime pas une universalité qui la caracteriserait comme methóde."[150] It may be said that the 'universality' of Cartesian method is presented here by way of comparison.

Rashed has also insisted that algebraic geometry began with Arabic mathematics, namely, by the renowned Persian mathematician of the eleven century Omar Khayyam (Arabic name, 'Umar al-Khayyāmī). Certainly his treatise on *Al-jabra wa'lmuqābalah* solved the cubic equation by the intersection of a parabola and a hyperpola, for example, and showed the merging the traditions of algebra and of the theory of conic sections. In this point, it has a similarity with Descartes's *Géométrie*. Khayyam thus must be admitted as a precursor of algebraic geometry. Nevertheless, it was exactly Descartes who started the tradition of the theory of algebraic equations of general dimensions and the general theory of algebraic curves.[151] Moreover, in his Commentary on the Difficulties of Certain Postulates of the Work of Euclid, Khayyam took a great step for the arithmetization of the theory of proportion on geometrical magnitudes as exemplifed by Book V of Euclid's *Elements*. That is to say, he interpreted the ratio of geometrical magnitues as an arithmetical division or fraction, and by intervening the concept of unit (*wāḥid*) proposed the new and ingenious idea that the ratio itself isn't a relationship as the ancient Greek thought, but can be interpreted as a kind of number.[152] This idea can be also evaluated highly as a consideration leading to the one of Descartes's *Regulae ad directionem ingenii*.

From these mathematical traditions as bases, there emerged a mathematician who can be said clearly to have cultivated the method of algebraic analysis. Insofar as we know, the Arabic mathematician of the twelfth century Ibn Yahyā al-Maghribi al-Samaw'al clearly understood the art of algebra to be a part of the analysis (*tahlīl*). In Book II, Chapter 1 "On That the Art of al-Jabr Is a Part of the Art of Analysis" of his *Al-bahir fī al-jabr* (The Dazzling in Algebra),

---

may be called an 'objet d'art' in recent scholarship of the history of Arabic mathematics. The dual concept of analysis and synthesis can be also seen in R. Rashed, *Les mathématiques infinitésimales du IXe au XIe siècle*, Vol. III: *Ibn al-Haytham, Théorie des coniques, Constructions géométriques et Géométrie pratique* (London, 2000).

[150] R. Rashed, "L'analyse et la syntèse selon Ibn al-Haytham," *Mathématiques et philosophie de l'Antiquité à l'âge classique*, Hommage à Jules Vuillemin, sous la direction de R. Rashed (Paris, 1991), pp. 131–162, at p. 147.

[151] R. Rashed et B. Vahabzadeh, *Al-Khayyām mathématicien* (Paris, 1999), pp. 12–29. Cf. *Idem*, "Fermat and Algebraic Geometry," *Historia Scientiarum*, Vol. 11, No. 1 (2001), pp. 1–23, at p. 23.

[152] *Al-Khayyām Mathématicien*, pp. 293–296, & 298; pp. 372–373; *Ibid.*, pp. 205–209, & 210–211; p. 250.

he defined the art of analysis as "the breaking of the composite into the simple elements that form its essence," and in another way as characterized as follows: "Whoever wants to find a certain requirement or to prove a certain issue, we will assume his requirement and see what results from that, until we reach simple results. If those simple results turn out to be true, one recomposes what he had analyzed and starts from the end of his analysis. But if those results turn out false he would have known that his requirement was in vain."[153] The latter definition of analysis reminds us of Pappus's characterization of the analytical method at the beginning of Book VII of his *Collection*. At the end of his description of the art of analysis, al-Samaw'al stated the relationship between analysis and algebra:

> And this practice is that which the art of 'al-jabr wa'l-muqabalah' requires and it is exacly that which the art of analysis requires.[154]

The term 'al-jabr wa'l-muqabalah' is, of course, the name given to the art of manipulating equations containing unknown quantities since al-Khwārizmī of the ninth century (see p. 74). Al-Samaw'al's conception of algebra as the art of analysis, however, took place in the world of medieval Islam. The *Al-bahir* is not presumed to have been transmitted to the European world.[155]

In any case, it is notable that there appeared the idea to undertand algebra as the analytic art. The art of algebra was assimilated in the Europan world of the twelfth century. As early as 1202, Leonardo da Pisa, or Fibonacci, wrote the *Liber abaci*, having been influenced by Islamic mathematics which had a different characteristic form classical Greek mathematics. 'Abacus' in the title of the book is thought to have been not 'abacus' in antiquity but a translation of the Arabic 'takhta', i. e. calculating board. In other words, in the Occidental

[153] *Al-Bahir en Algèbre d'As-Samaw'al*, édition, notes et introduction par Salah Ahmad et Roshdi Rashed (Damas, 1972), Arabic text, p. 73. The translation is taken from George A. Saliba, "The Meaning of al-jabr wa'l-muqabalah," *Centaurus*, **17** (1973), p. 195. For al-Samaw'al's career and works, see Adel Anbouba, "Al-Samaw'al, Ibn Yaḥyā Al-Maghribī," *DSB*, XII, pp. 91–95.

[154] *Op. cit.* (As-Samaw'al), p. 74. The translation has been aided by Prof. SUZUKI Takanori of Tokai University, but is essentially mine. See the following historical remark by S. Ahmad and R. Roshed in their French introduction (p. 74): "L'élément commun à l'algèbre et à la géométrie, ne subsiste que dans une démarche qui va en négligeant la synthèse pour ne garder que l'analyse. Algèbre et géométrie sont analogues, selon l'auteur, puisquue dans l'une et l'autre, le mathématicien part des prémisses connus, des conditions nécessaires, consécutives à ces prémisses et ainsi de suite jusqu'à la détermination des inconnuses. Or cet élément commun semble subordonner implicitement la géométrie à l'algèbre, dans la mesure où il ne retient que la voie analytique des géométres et où il néglige, dans la description la synthèse. As-Sawam'al saura plus tard tirer de cette concption la conséquence qui s'impose et identifiera précidément, et pour la première fois dans l'histoire de la philosophie mathématique, algèbre et analyse, géométrie et synthèse."

[155] Adel Anbouba, *Op. cit.* (n. 153), p. 93: "The book [*Al-bāhir*] was apparently altogether unknown in the west." For the problem of transmission of Arabic mathematics to Latin West in general, see André Allard, "The Influence of Arabic Mathematics in the Medieval West," *Enclyclopedia of the History of Arabic Science*, Vol. 2 (n. 143), pp. 539–580.

world there appeared a kind of mathematics influenced by the higher art of algebra in addition to the aforementioned art of calculation.[156] By this time, in both the Islamic and European worlds, symbols had started to be used for unknown quantities in equations and algebra tranformed gradually from the art of calculation closely connected with commercial activities to a theoretical subject constituting the mathematical disciplines.[157]

Then let's turn our eyes of historical consideration to Western Europe. The formation of the method of algebraic analysis in early modern Europe is understood today in relation to the thought of Peter Ramus (or Pierre de la Ramée) and the intellectual movement connected with it. This movement swept over Europe in the second half of the sixteenth and the early seventeenth centuries. Ramus's thought can be characterized first by his 'reform' or 'deconstruction' of Aristotelian logic or dialectic, which flourished in the Middle Ages and was called "the art of the arts and the science of the sciences" by Peter of Spain.[158] It has been at times pointed out that his understanding of Aristotle was superficial in important places and contained serious mistakes. But one thing which should not be forgotten is that his emphasis on pedagogical and utilitarian concerns in his reading and criticism of Aristotelian texts had a great effect in the transformation of the entire structure of learning, especially in the second half of the sixteenth century.[159] Frances A. Yates has very insightfully characterized his thought in general as "a kind of inner iconoclasm."[160]

This iconoclastic character of the Ramean thought worked very effectively not only in logic but also in mathematics. The target of Ramus's criticism in mathematics was Euclid, in particular the Euclid who put special emphasis upon synthetic demonstration. What Ramus searched for in his *Scholae math-*

---

[156]Kurt Vogel, "Fibonacci, Leonardo, or Leonardo of Pisa," *DSB*, IV, pp. 604–613, at p. 605.

[157]For the development of algebra from a practical art to a theoretical discipline, see, for example, Paul Benoit, "Calcul, algèbre et marchandise," *Éléments d'histoire des sciences*, sous la direction de Michel Serres (Paris, 1989), pp. 197–221, & Giovanna C. Cifoletti, "La *Question* de l'Algèbre: Mathématique et rhétorique des hommes de droit dans la France du 16e siècle," *Annales: Histoire, sciences sociales*, 50e Année, n° 6 (1995), pp. 1385–1416. This trend can be witnessed in the mathematical disciplines in general. See Mario Biagioli, "The Social Status of Italian Mathématicians, 1450–1600," *History of Science*, **27** (1989), pp. 41–95.

[158]Peter of Spain (Petrus Hispanus Portugalensis), *Tractatus* called afterwards *Summule Logicales*, First Critical Edition from the Manuscripts with an Introduction by L. M. De Rijk (Assen, 1972), p. 1: "ars artium et scientia scientiarum" (a variant).

[159]For the detail of Ramus's program for the reform of logic, consult, among others, Walter J. Ong, *Ramus, Method, and the Decay of Dialogue: From the Art of Discourse to the Art of Reason* (Cambridge, Mass., 1958); Wilbur Samuel Howell, *Logic and Rhetoric in England, 1500–1700* (Princeton, 1956). I have made an English translation of Book II, Chapter 1, "On the Definition of Logic," of Ramus's *Scholae dialecticae* (formerly called *Aristotelicae animadversiones*, which was published in Paris in 1543) in *Scholae in liberales artes* (Basel, 1569) with a historical introduction. See Chikara Sasaki, "Ramus' Struggle against Scholastic Logic: Historical Introduction and Translation," *The Journal of the Department of Liberal Arts*, The College of Arts and Sciences, The University of Tokyo, No. 15 (1982), pp. 119–136.

[160]F. A. Yates, *The Art of Memory* (Ch. 3, n. 35), pp. 234–235

ematicae of 1569 was a method for writing textbooks of mathematics easily understood "not only by learned people [...] but also by children, mechanics, architects and painters."[161] In Books VI–XXXI of the same work, he accused Euclid and the commentators of the Alexandrian compiler of the *Elements* for their sophistries (*elenchi*) and proposed alternatives for rewriting Euclid's *Elements* in a concrete manner. For example, in Book XIII "On the Definitions of Book V," he attempted to reinterpret Book V of the *Elements*, the theory of proportion for geometrical magnitudes, as a book on arithmetic.[162]

Ramus's primary interest in mathematics was to discover mathematical results. This attitude was reflected in his edition of Euclid, published anonymously in Paris in 1545.[163] A surprise to those who believe that mathematics consists of theorems and their deductive proofs is that Ramus's version belittled or totally neglected the latter aspect of mathematics, that is, synthetic demonstrations, simply by presenting propositions. It is reported that Ramus's teaching at the University of Paris attached importance to the discovery of proofs and the solutions to problems.[164] Ramus authored his own textbooks on mathematics, for example, *Arithmetica* (1555), *Algebra* (1560), and *Geometria* (1569). More significantly he was one of the first to have seen algebraic implications in some of the geometrical propositions of Euclid. Mahoney reports, for example, that in his *Geometria* of 1569 Ramus read in Euclid II, 4 the algebraic formula $(a+b)^2 = a^2 + b^2 + 2ab$, albeit without recognizing it symbolically.[165] There Ramus saw that the method of analysis is used: "For indeed it is the same way from Thebes to Athens as it is from Athens to Thebes. And the use of geometrical analysis remains here, as later with the cube, although in the entire *Elements* there is no other use of that at all."[166] Ramus's inclination to see

[161] Petrus Ramus, *Scholae mathematicae* (Basel, 1569): Lazarus Schöner's edition (Frankfurt am Main, 1627), p. 107: "[...] non solum a doctis hominibus [...] sed á pueris, á mechanicis, ab architectis, á pictoribus facilius ediscentur, facilius excercebuntur."

[162] See my paper "The Acceptance of the Theory of Proportion in the Sixteenth and Seventeenth Centuries — Barrow's Reaction to the Analytic Mathematics," *Historia Scientiarum*, No. 29 (1985), pp. 93–126, esp. VI. 2: "Ramus's Criticism of the Euclidean Tradition," pp. 117–123.

[163] [*Euclidis elementa mathematica*] (Paris, 1545). Cf. W. J. Ong, *Ramus and Talon Inventory* (Cambridge, Mass., 1958), p. 68, no. 34*; Charles Thomas-Stanford, *Early Editions of Euclid's Elements* (London, 1926), p. 35, no. 27.

[164] See Peter Scharratt, "La Ramée's Early Mathematical Teaching," *Bibliothèque d'Humanisme et Renaissance*, **28** (1966), pp. 605–614.

[165] Ramus, *Geometria* (1569): Lazarus Schöner's edition (Frankfurt am Main, 1599), p. 86. See Mahoney's explanation of this part in his *Fermat*, p. 32, note 16; "The Royal Road," p. 186.

[166] *Ibid.*: "Eadem siquidem via est Thebis Athenas, quæ Athenis Thebas: Et hic geometriae analyseos usus sperest, ut postea in cubo, cum alius in totis elementis nullus sit." Interestingly, in the English translation by William Bedwell, this part reads: "For it is the same way from Cambridge to London, that it is from London to Cambridge." *Via Regia ad Geometriam: The Way to Geometry, Being Necessary and Useful, from Astronomers, Geographers, Landmeaters, Sea-men, Engineres, Architecks, Carpenters, Paynters, Carvers, &c. ed. John Clerke* (London, 1636), p. 160. Note that the title is an exiting one: *The Royal Road to*

analysis algebraically seems to have been naturally accepted and consolidated in his mathematical followers.[167]

In 1560 Ramus published an anonymous work titled *Algebra*. How, then, did he view the discipline of algebra? In the introductory part of his *Algebra*, he defines algebra as follows:

> Algebra is the part of arithmetic which institutes a certain special numeration from continuously figurated proportions. For indeed various species of figures follow in that progression, of which a certain numeration is different from the numeration of the antecedent numbers.[168]

The "species of figures" can be written, if we set the unit of the ratio u = 1, the side l = 2, and so on, in the following manner:

1.	2.	4.	8.	16.	32.	64.	128.	256.	512.	1024.	2048.	4096.
u.	l.	q.	c.	bq.	ſ.	qc.	bſ.	tq.	cc.	ſq.	tſ.	bqc.

Note that the symbol 'ſ', , the long 's', stands for 'solidus'. Ramus's "species of figures" are taking the role of unknown numbers in ordinary algebra.

In his *Algebra* Ramus conceived algebra as a part of arithmetic. But unfortunately contrary to our expectation, the *Algebra* of 1560 said nothing about what kind of a part the algebra is.[169] One answer was given by an editor of the later editions of his *Algebra*. On compiling his *Ramus and Talon Inventory*, Walter J. Ong asked himself about the Ramist *Algebra*:

> In the Ramist methodization of the sciences, algebra is a source of permanent embarrasment. The first edition of this work is very loosely organized but suggests that algebra is a part of arithmetic.

Geometry! A description as the circumstances in which the *Via Regia ad Geometriam* was published can be seen in Howell, *Op. cit.* (n. 159), p. 246.

[167]This is pointed out persuasively in M. S. Mahoney's "The Royal Road," Chapter III: "The Formula Found: Algebraic Analysis in the Sixteenth and Seventeenth Centuries," pp. 145–225. As to Ramus's notion of analysis, especially non-mathematical, see Ong, *Op. cit.* (n. 159), Chapter XI, 13: "Analysis and Genesis (Synthesis)," pp. 263–267.

[168][Ramus], *Algebra* (Paris, 1560), f. 2r: "Algebra est pars Arithmeticæ, quæ é figuratis continué proportionalibus numerationem quandam propriam instituit. Variæ siquidem figuratorum species in eam progressionem succedunt, quarum numeratio quædam ab antecedentium numerorum numeratione diversa est." In his conmperhensive study of Ramean mathematics, J. J. Verdonk has pointed out that there are many parallel structures between Ramus's *Algebra* and Johannes Scheubel's *Algebrae compendiosa facilisque descriptio*, published in Paris in 1551. See Verdonk, *Petrus Ramus en de Wiskunde* (Assen, 1966), 411, "De Algebra van Ramus," pp. 209–224. For the life and works of Scheubel, see Mary S. Day, *Scheubel as an Algebraist* (New York, 1926). This book examines a manuscript of Scheubel's algebraic work preserved at the Columbia University Library.

[169]Cf. W. J. Ong, *Ramus and Talon Inventory* (n. 163), pp. 335–336: "There is nothing about the relation of analysis and synthesis to algebra or to arithmetic. [...] The fact that this first, tentative edition is anonymous suggests how Ramus' desire for fame as a mathematician was tempered by foreboding concerning his initial qualifications."

> But what sort of a part? The answer is provided after Ramus' death by Lazarus Schöner in his 1586 edition of Ramus' arithmetic and algebra. Here Schöner's commentary on the first chapter of the *Algebra* explains that algebra is what the Greeks call the "analytic" part of arithmetic, as against arithmetic proper, which they styled "synthetic"—this touting of analytic and synthetic is, of course, Ramist, not Greek.[170]

As Ong indicated, in Lazarus Schöner's version of *Algebra*, which first appeared in 1586, we find the following remarkable definition of algebra: "On the basis of this resolution, the Greeks called algebra analytic; they called absolute arithmetic synthetic."[171] Thus the Ramist mathematician Schöner explicitly defined algebra as the analytical part of arithmetic.

Insofar as we know, this statement by Schöner was the first in Western Europe to intepret algebra as the analytic part of arithmetic. In the Islamic world after al-Samaw'al there might have been a mathematician who understood algebra in the way of al-Samaw'al. And in the European world before Schöner there might perhaps have been a mathematics writer who had the same understanding as Schöner. The mathematical civilization of Islam was remote from the European intellectual world. To connect a missing link between al-Samaw'al and Schöner is a crucial problem for the history of mathematical thought. We should like to call this important problem, "the problem of the beginning of algebraic analysis, specifically connecting Islamic mathematics with European mathematics." To attack this problem, one has to equip oneself with the knowledge on the history of both Arabic and Latin mathematics, after having studied intellectual achievements such as Jakob Klein's "Die griechischen Logistik und die Entstehung der Algebra"[172] and Michael S. Mahoney's "The Royal Road: The Development of Algebraic Analysis from 1550 to 1650."[173] I believe that the time is ripe to resolve the problem.

In any case, the method of algebraic analysis in the Western Europe started with the Ramist intellectual milieu. But, in the history of mathematics as a whole, however, it was a new beginning or restart of the understanding

---

[170] Ong, *Ibid.*, p. 333.

[171] *Petri Rami Algebra a Lazaro Schonero emendata* (Francofvrti, 1599), p. 190. "A quare resoltione Algebra Græcis dicta fuit analytica, quibus absoluta Arithmetica dicebatur synthetica." Cf. Mahoney, "The Royal Road," p. 187, which mistakenly ascribes this statement to Ramus, not Schöner.

[172] Jacokob Klein, "Die griechische Logistik und die Entstehung der Algebra," *Quellen und Studien zur Geschichte der Mathemtik, Astronomie und Physik*, Abteilung B: Studien, Bd. 3-1 (1934), pp. 18–105 & Bd. 3-2 (1936), pp. 122–235. Its Eglish translation by Eva Brann is *Greek Mathematical Thought and the Origin of Algebra* (n. 134). Note that J. Klein was under the influence of Edmund Husserl's phenomenological philosophy, which can be seen in Jacob Klein, "Phenomenology and the History of Science," *Philosophical Essays in Memory of Edmund Husserl*, ed. by Marvin Farber (Cambridge, Mass, 1940), pp. 143–163.

[173] Mahoney, "The Royal Road" (n. 91), which was written under the supervision of Prof. Thomas S. Kuhn at Princeton University.

acquired in the Islamic world. At the same time, however, a new element was added in Europe. This addition was not by Ramus, but by Viète or Descartes.

Ramus's attempt to reform mathematics might not necessarily have led straight to Viète's creation of symbolic algebra as the art of analysis. The second generation of those who showed Ramist affinities did not conceal certain feelings of embarrassment about the hasty pedagogical motive of Ramus. Willebrord Snel, for example, is reported to have felt disappointed with Ramus's understanding of mathematics, especially when comparing Ramus's text of *Geometria* with those of the classical writers.[174] But interestingly, some Ramists or semi-Ramists had to innovate and in fact did reinterpret the master's "short-cut"[175] method in their own way.

It was Viète who, under some influence from Ramus, created a new form of analytic art, namely symbolic algebra, by providing symbols both for unknown and known quantities. Viète may possibly have met Ramus personally[176] and called the Huguenot reformer of mathematics in highest terms "homo *λογικώτατος*" (a most perspicacious man) in a work of 1579.[177] Moreover, his definition of analysis in the first chapter of his *In artem analyticem isagoge* (see p. 184) seems almost undoubtedly have been taken from Ramus's *Scholae mathematicae*.[178] And it is known that he wrote his *Isagoge* following the Ramean method stated in the *Scholae mathematicae*.[179]

---

[174] J. J. Verdonk, *Op. cit.* (n. 168), p. 370.

[175] Hardin Craig has named Ramus "the greatest master of the short cut the world has ever seen" in his *The Enchanted Glass: The Eizabethan Mind in Literature* (Oxford, 1936), p. 143. Ramus himself claimed that his method to reorganize the liberal arts is an "adrèsse et abrègement de chemin." See *Dialectique* (1555), Édition critique par Michel Dassonville (Gèneve, 1964), p. 145. Father Ong characterized Ramus's work in general as something similar to 'the calculus without pain' or 'topology without pain' in his lecture "Milton's Logic and the Pacification of Academia," delivered at Princeton University on May 4, 1979, at which I was in the audience.

[176] J. E. Hofmann, *Op. cit.* (Ch. 2, n. 105), p. XII*.

[177] Viète, *Universalium inspectionum ad canonum mathematicum liber singularis* (Paris, 1579), pl. 71.

[178] Ramus quoted and commented on the well-known interpolations to Propositions 1–5 of Euclid XIII (see Chapter 2, § 3, p. 72) on analysis and synthesis. In Ramus's text, analysis is defined as "assumptio quaesti tanquam concessi per consequentia ad verum concessu." *Scholae mathematicae* (n. 161), p. 300. It is totally identical with Viète's (see Ch. 4, n. 75: "Assumptio" in Ramus is "adsumptio" in Viète). Incidentally, one of the texts of Euclid that Ramus used for writing his *Scholae mathematicae* was doubtlessly the *editio princeps* of the Greek text of Theon of Alexandria, published by Simon Grynaeus the elder in Basel in 1533 (see Ch. 2, n. 46). See p. 239 of that edition. Cf. Commandino's definition of analysis (*resolutio*): "Resolutio est sumptio quaesti tanquam concessi per ea, quae consequuntur in aliquod uerum concessum." Commandino, *Op. cit.* (Ch. 2, n. 47), f. 229r. Note that Commandino's Latin translation appeared in 1572 after the publication of the *Scholae mathematicae* in 1569.

[179] For example, in Chapter VI, "On the Examination of Theorems by Poristics," of the *Isagoge*, Viète uses the Ramist three laws of method "*κατὰ πάντος*", "*κατ' αὐτὸ* (*sic*!)," and "*καθ' ὅλου πρῶτον*". *Isagoge* (1591), f. 8r; *Opera mathematica* (1646), p. 10. Ramus explained these three laws (in his own spelling) "*κατὰ πάντὸς*", "*κατ'αὐτο*", and "*κατ' ὅλου πρῶτον*" in the *Scholae mathematicae*, pp. 76–77 and 109 (see also his *Scholae dialecticae* (n. 159), p. 32.). These conceptions were originally used in Aristotle's *Posterior Analytics*

Viète, following Ramus, emphasized the heuristic aspect of algebra and geometrical analysis. So he called his analytic art, i. e. his new symbolic algebra, in a Ramistic way "the doctrine of finding well in mathematics."[180] If it is appropriate, and we believe it is, to name the development of algebraic analysis of the late sixteenth and early seventeenth centuries the "royal road,"[181] the first major builder of that road must have been Viète. Nevertheless, it is undeniable that Ramus, too, broke new ground for that road with his emphasizing the heuristic side of mathematics. He was a pathfinder to the "royal road." Certainly there might have been a great gulf between the way of the "short-cut" and that of the "royal road." Viète's and his followers' mathematical talents were indispensable for bridging this gulf. But Viète was influenced, as we have seen, to a considerable extent, by Ramus's methodology in the arts.

Now it should be noticed that Viète's works could not enjoy a large circulation. As Sarton has pointed out, "his many books were issued privately in Tours or Paris, in very small editions which he distributed to his friends."[182] Let us glance at how Viète's writings were distributed around 1603, the year in which he died. Viète's disciples were not many. He gave some of his manuscripts to Ghetaldi. Some of the other manuscripts were handed down to Viète's friend Pierre Alleaume by his heirs after his death. Then Pierre Alleaume left them to his son Jacques, a pupil of Viète's. Jacques in his turn entrusted Viète's Scottish

(Book A, Ch. IV, 73a21–74a3). Thus T. R. Witmer, for example, mistakenly ascribes them directly to Aristotle in his translation of Viète's works (p. 28, note 47). But, as has been at times pointed out, Ramus distorted Aristotle's original meanings and applied them to his reorganization of the liberal arts. For Ramus's misunderstanding of the Aristotelian conceptions, see Howell, *Op. cit.* (n. 159), pp. 149–153; Ong, *Op. cit.* (n. 152), pp. 258–262. Viète clearly follows Ramus's use and misspelling ("κατ'αὐτό" instead of "καθ'αὐτό")! For a discussion about Ramus's influence on Viète's thought, see Mahoney, "The Royal Road," esp. pp. 178–180 & 196–197.

180 Viète, *Isagoge* (1591), f. 4r; *Opera mathematica*, p. 1: "Doctrina bene inveniendi in mathematicis." In his *Scholae mathematicae* (p. 108), Ramus defined mathematics as follows: "What, then, is mathematics? I would say that it is the doctrine of quantity: As, in fact, arithmetic is defined as the doctrine of computing well; geometry, the doctrine of measuring well, so the art of mathematics might be defined by the general term and from its end be named, so to speak, the art of 'quantifying'." ("Quid igitur est mathematica? Est (inquam) doctrina quantitatis: ut veró definitur arithmetica bene nume-randi, geometria bene metiendi doctrina, sic generali verbo, modó ex suo fine definiretur mathematica ars, tamquam diceretur, quantitandi." Such a manner of defining the arts by Ramus reminds us of Cicero's *Topica*, one of Ramus's favorite books. Cicero called 'τοπική' as 'ars inveniendi'. *Topica*, § 6.

181 "The royal road" was, of course, used in Euclid's reply to the king Ptolemy the First. "It is also reported that Ptolemy once asked Euclid if there was not a shorter road to geometry than through the *Elements*, and Euclid replied that there was no royal road to geometry." See Proclus, *A Commentary on the First Book of Euclid's Elements*, Translated, with Introduction and Notes, by Glenn R. Morrow (Princeton, 1970; With a new foreword by Ian Mueller, 1992), pp. 56–57 (Friedlein, p. 68). I learned from Mahoney's Ph. D. thesis "The Royal Road" that this expression is applicable to algebraic analysis of early modern period. The construction of such a road, however, was consciously aimed at by contemporary mathematicians such as Galileo (see p. 124) and the English Ramist William Bedwell (see n. 166).

182 George Sarton, *Six Wings: Men of Science in the Renaissance* (Bloomington/London, 1957), p. 44. See also Van Egmond, *Op. cit.* (n. 133)

friend Alexander Anderson with the manuscripts *De Aequationum recognitione et emendatione tractatus duo*, *Ad angularium sectionum theoremata*, and *Notae ad logisticem speciosam posteriores*, the former two of which would be published in Paris in 1615.[183]

How was Viète's analytic art accepted by other mathematicians roughly before 1632? Ghetaldi's posthumous work *De resolutione et compositione mathematica* was published in Rome in 1630. Ghetaldi there used his teacher Viète's art of analysis to prove geometrical theorems.[184] In the same year, two French translations of the *Isagoge* and *Zetetica* appeared: one by Antoine Vasset (probably a pseudonym of Claude Hardy)[185] and the other by J. L. Vaulézard.[186] In 1631, Beaugrand's Latin version was published in Paris. We have seen (Subsection B) that Descartes took possession of a copy of it through Mersenne in the spring of 1632.

In London in 1631 the two monographs influenced by Viète's symbolic algebra appeared: Thomas Harriot's posthumous *Artis analyticae praxis*[187] and William Oughtred's influential *Arithmeticae in numeris et speciebus institutio* (later called *Clavis mathematicae*).[188] Such is a very rough description of the diffusion of Viète's analytic school (though invisible) before Descartes received the Beaugrand edition in 1632. The works listed above are certainly essential

---

[183] H. L. L. Busard, "Viète, François" (n. 118), pp. 23–24. It is known that Jacques Alleaume was residing in Holland and taught military sciences at the Military Academy there during the early seventeenth century. See Gustave Cohen, *Écrivans français en Hollande dans la première moitié du XVIIe siècle* (n. 3), pp. 341, 373 & 381; C. Laplatte, "Aleaume (Jacques) (1562–1627)," *Dictionnaire de Biographie Française*, t. 1 (Paris, 1933), cols. 1371–1372. For Anderson, see A. M. C[lerke], "Anderson, Alexander (1582–1619?)," *Dictionary of National Biography*, Vol. 1 (Oxford, 1921), pp. 371–372.

[184] Ghetaldi, *De Resolutione et compositione mathematica* (Rome, 1630): *Opera omnia* (n. 97), pp. 353–707. This book is examined in Mahoney, "The Royal Road", pp. 206–217; Ernest Stipanić, "L'Œuvre principale de Getaldić ⟪De resolutione et compositione mathematica⟫," *Actes* (Ch. 2, n. 107), pp. 91–104.

[185] See Van Egmond, *Op. cit.* (n. 133), p. 383.

[186] We have already mentioned this translation on p. 246.

[187] For Harriot as a mathematician, see J. A. Lohne, "Thomas Harriot als Mathematiker," *Centaurus*, **11** (1965), pp. 19–45, esp. pp. 39–42;"Dokumente zu Revalidierung von Thomas Harriot als Algebraiker," *Archive for History of Exact Sciencesre*, **3** (1966), pp. 185–205 ; Jacqueline A. Stedall, "Rob'd of Glories: The Posthumous Misfortunes of Thomas Harriot and His Algebra," *Archive for History of Exact Sciencesre*, **54** (2000), pp. 455–497. In the latter two papers, the authors have pointed out that Harriot's Praxis was published with many additions by the editor.

[188] An examination of Oughtred's *Clavis mathematicæ* can be seen in Mahoney, "The Royal Road," pp. 203-206. By examining the first edition of Oughtred's *Clavis*, H. Bosmans speculated about the possibility of its influence upon Descartes's Géométrie in his "La première édition de la *Clavis mathematica* (*sic*!) d'Oughtred. Son influence sur la *Géométrie* de Descartes," *Annales de la société scientifique de Bruxelles*, 35e année, 1910–1911 (Louvain, 1911), 1er partie, p. 121 and 2e partie, pp. 24–78. As Bosmans has admitted, there is no clear evidence to claim that Oughtred influenced Descartes. For Oughtred, see Florian Cajori, *William Oughtred: A Great Seventeenth Century Teacher of Mathematics* (Chicago/London, 1916).

ones, but our sketch is still a long way from an exhaustive understanding of how Viète's algebra was accepted in the early seventeenth century, of course.

Finally, let us pursue Descartes's mathematical career retrospectively up to the time he revealed his project of reforming mathematics on March 26, 1619. Before meeting Beeckman in 1618, he had already become familiar with, besides ordinary Euclidean geometry, the algebra of the moderns, especially Clavius's *Algebra*. It was the Descartes, having acquired this apparatus which emphasized a problem-solving approach to mathematics, who encountered Beeckman in Breda on November 10, 1618. The Dutch Calvinist scholar might not have known Viète's mathematics but certainly lived in the Ramist intellectual atmosphere. Willebrord Snel, an exponent of the Dutch Ramism, partly accepted, as we have seen (pp. 238–239), Viète's analytic art, though the basic character of his own mathematics was traditional. With the encounter with Beeckman, Descartes could have been influenced by Beeckman's Ramist temperament which particularly emphasized heuristic and practical aspects of mathematics. Such a catalytic agent might have been sufficient for Descartes, who had already studied Clavius's *Algebra*, to be spurred to conceive the idea for the unification of arithmetic and geometry, which was told as early as March 26, 1619 in his letter to Beeckman.

Accordingly, can we conclude that Descartes had no opportunity to come to know Viète's program of algebraic analysis before the spring of 1632? Is there not a person missing from our above list of mathematicians who was under the influence of Viète's mathematical thought or who possessed mathematical thought similar to Viète's? To answer this question in a critical manner, we have to examine the mathematical thought of Adriaan van Roomen, who was the main proponent of 'mathesis universalis'. It is he who is a prime candidate for the person who transmitted Viète's mathematical thought to Descartes.

D. The Supposed Intermediary Adriaan van Roomen

There remains the possibility that Descartes had indirect contact with the analytic mathematics which Viète initiated. One of the crucial figures who is conjectured to have furnished the connection between Viète and Descartes is Van Roomen. Historians of philosophy have examined his concept of 'mathesis universalis' in relation to Rule IV-B of the *Regulae ad directionem ingenii*, but rarely entered into his mathematics itself. Now it should be pointed out that Van Roomen was an advocate of 'mathesis universalis', and at the same time the possessor of a mathematical thought quite similar to Viète's.

Van Roomen's relation with Viète began when the ambassador from the Netherlands remarked to the French king Henri IV that in 1593 the Belgian mathematician Van Roomen issued a challenge to mathematicians "all over the world" to solve an equation of the 45th degree in his *Ideae mathematicae pars prima, sive methodus polygonorum.*[189] Viète almost immediately found one and shortly afterward twenty-two positive roots for it. He published his answer in *Ad problema, quod omnibus mathematicis totius orbis construendum proposuit Adrianus Romanus, responsum*, which appeared in Paris in 1595.[190] In turn, Viète, at the end of his booklet, proposed a problem which he took from Apollonius's *Tangencies*: To draw a circle tangent to three given circles. Van Roomen answered the question in his *Problema Apollonicum quo datis tribus circulis*, published in Würzburg in 1596.[191] He solved Viète's problem by means of the intersection of two hyperbolas. On the other hand, the proposer of the problem presented a better solution which required only the ruler and the compass. Viète's solution was contained in his *Apollonius Gallus* of 1600, which we referred to above (p. 239). Through these interrelationships, Van Roomen was so impressed by Viète's mathematical talent that he visited France to recover his health and personally met his French rival in the summer of 1601. On his arrival in Paris, he was told that Viète was in Poitou and traveled there to talk with him. It is said that they spent a month together and were very friendly.[192]

Van Roomen no doubt began to learn Viète's symbolic algebra from the latter's response to his friend, i. e. *Ad problema responsum* of 1595. In this work, there were some equations written in the symbolic manner, for example, "A3−A cubus/X quad. æquabitur D" for our "$3x - x^3/a^2 = d$."[193] As we have remarked in Chapter 2, § 4 (pp. 82–83), in 1597 Van Roomen informed Clavius of Viète's 'algebra speciosa'.

In the same year, 1597, Van Roomen published a book in three parts containing the *Apologia pro Archimede* as the second part, which is extremely important for our present study. It was in this work that he used his favorite concept 'mathesis universalis' for the first time. As has been stated briefly above (pp. 81–82), this work was composed to criticize Scaliger's book under the title *Cyclometrica elementa duo* (1594).[194] On attempting to attack Scaliger, Van

---

[189] Adrianus Romanus, *Ideæ mathematicæ pars prima* (Antwerpen, 1593), f. **iij^v. For Van Roomen's bibliography, see H. Bosmans, "Romain (Adrien)," *Biographie nationale*, publiée par l'Académie royale des sciences, des lettres et de beaux-arts de Belgique, t. XIX (Bruxelles, 1907), cols. 848–889; P. Bockstaele, *Nationaal Biografisch Woordenboek*, Vol. 2 (Brussels, 1966), cols. 751–765.

[190] Viète, *Opera mathematica* (Leiden, 1646), pp. 305–324.

[191] This work is examined in H. Bosmans, "Analyse de trois ouvrages célèbres d'Adrien Romain," *Annales de la société scientifique de Bruxelles*, 22e Année, 1904–1905 (Louvain, 1905), pp. 68–79, esp. pp. 68–74.

[192] G. Sarton, *Six Wings* (n. 182), p. 46. Sarton took this story from Jacques Auguste de Thou's *Historia sui temporis* (1604–1608). See notes 63 on pp. 251-252 and 76 on p. 253. See also P. Bockstaele, "The Correspondence of Adriaan van Roomen," *LIAS*, III (1976)/Separatum, p. 72, note 10; H. L. L. Busard, "Roomen, Adriaan van," *DSB*, XI, p. 532.

[193] Viète, *Opera mathematica*, p. 307. pp. 315–316 & 320–321

[194] Adrianus Romanus, *In Archimedis circuli dimensionem expositio et analysis. Apologia pro Archimede ad clarissimum virum Josephum Scaligerum. Exercitationes cycliae, contra Josephum Scaligerum, Orontium Finaeum et Raymarum Ursum* (Würzburg, 1597). The actual place of publication was Geneva instead of Würzburg. See Bockstaele, *Op. cit.* (n. 192),

Roomen first presents the Greek text of Archimedes's *On the Measurement of Circle* with its Latin translation and then in the *Apologia pro Archimede*, he defends Archimedes against Scaliger's criticism by justifying the use of arithmetical means to solve geometrical problems.

For this purpose, Van Roomen claims that there must be a science which treats properties common to both arithmetic and geometry. He was on the verge of introducing the concept of 'mathesis universalis'. In Chapter 6 under the title "There Is a Science Common to Geometry and Arithmetic which Considers Quantity Generally as Measurable," he asserts:

> Surely there is a certain mathematical science common to arithmetic and geometry to which properties common to all quantities pertain: since a proportion is common to all quantities, not only abstract as numbers and magnitudes, but also concrete as times, sounds, voices, places, motions, and forces (for all these and many others are said to have a proportion if their relation is considered from the viewpoint of quantity).[195]

The mathematical science common to arithmetic and geometry propounded by Van Roomen corresponds exactly to a discipline named 'mathesis universalis.' Arithmetic and geometry are regarded as unitary from the point of view of the theory of porportion, which, in Van Roomen's opinion, is concerned with general quantity. He elucidates this mathematical science in the next chapter. In the introductory part of Chapter 7 titled "An Idea of a Kind of Universal Mathematics (*universalis mathesis*), which We Shall Call First Mathematics (*prima mathesis*), Is Proposed," he insists that his 'mathesis universalis' is concerned with the subjects of all mathematical sciences, including arithmetic and geometry. In Chapter 7, § 2 below, we will quote Van Roomen's definition of 'mathesis universalis' *in toto* and discuss its mathematico-philosophical significance thoroughly (pp. 345–346).

On introducing a discipline which may be called 'mathesis universalis', Van Roomen appeals to the authority of Eutocius of Ascalon. According to

p. 44, note 3. J. J. Scaliger's *Cyclometria elementa duo* (Louvain, 1594) is examined briefly in Michel Chasles, *Aperçu historique sur l'origine et dévelopment des méthode en géométrie*, Seconde éd. (Paris, 1875), p. 443. The following article locates Van Roomen's work in the history of the quadrature of the circle in the sixteenth century: D. Bierens de Haan, "Notice sur quelques quadratures du cercle dans les Pays-Bas," *Bulletino di bibliografia e di storia delle scienze matematiche e fisiche*, **7** (1874), pp. 99–140, esp. pp. 127–138.

[195] Romanus, *Apologia pro Archimede*, "CAPVT SEXTVM. GEOMETRIAE, & Arithmeticæ communem esse scientiam, quæ quantitatem generaliter vti mensurabilem considerat," p. 23: "[...] Nimirum scientiam esse quandam Mathematicam communem Arithmeticæ & Geometriæ, ad quam spectarent affectiones omnibus quantitatibus: cùm autem proportio sit omnibus quantatibus communis, non abstractis tantum vt numeris et magnitudinibus, sed concretis etiam, vti temporibus, sonis, vocibus, locis, motibus, potentiis (Nam hæc omnia & plura alia proportionem dicuntur habere, si eorum habitudo consideretur secundum quantitatem) [...]."

Van Roomen's explanation in Chapter 6 of his *Apologia pro Archimede*, Eutocius ascribed some propositions concerning the theory of proportion to a discipline which is neither arithmetic nor geometry but is simply called by him "mathematics" (*mathematica*).[196] The place where Eutocius is supposed to have done so must be his commentary on Proposition 11 of Book One of Apollonius's *Conics*. There, discussing the concept of 'ratio of ratios',[197] or compound ratio, which we have argued about in Section 1, Eutocius stated that this notion is applicable not only to arithmetic but also to geometry, and he concluded his commentary with the following sentence: "For these mathematical disciplines seem to be sisters."[198] As can be easily seen, Van Roomen's interpretation is subtly different from Eutocius's actual assertion. In other words, he interprets Eutocius's sentence cited above conveniently in accordance with his concept of 'mathesis universalis'. This inclination of his becomes apparent again when he takes Euclid V, the theory of proportion on continuous magnitudes, to be common to both arithmetic and geometry (see our quotation in Chapter 7, § 2). Book V of Euclid's *Elements* was in reality about geometry.[199]

In any case, Van Roomen clearly believes that there exists a science which is so fundamental that it can be concerned with the principal part of the various mathematical sciences. This science is named 'universalis mathesis' or 'prima mathesis' on the analogy of 'prima philosophia', the queen of philosophy.

In what follows in Chapter 7, Van Roomen presents a specimen of his 'mathesis universalis'. First, definitions and axioms, and then theorems are stated. These are all considered to be applicable to both arithmetic and geometry and are mostly taken from Euclid's *Elements*, particularly Book V.

The part entitled "Methodus Arithmeticae Practicae Universalis" (A Universal Method of Practical Arithmetic), placed at the end of the *Apologia pro Archimede*, is notable because there Van Roomen provides simple symbols for the four arithmetical operations: A + B, A − B, A in B, and $\frac{\mathrm{A}}{\mathrm{B}}$, for addition, subtraction, multiplication and division, respectively. Let us examine only one example, i. e. Problem 7 in order to see how he manipulates these signs:

EXAMPLE

GIVEN. Let the fraction $\frac{\mathrm{A}}{\mathrm{B}}$ have been proposed, whose denominator is not lager than its numerator.

---

[196] *Ibid.*: "Eutochius (*sic*!) vocat absolutè Mathematicam."

[197] *Apollonii Pergaei Conicorvm Libri Qvattvor*, ed. F. Commandino (Bononiae, 1566), f. 14r: "Proportio ex proportionibus componi dicitur, quando proportionum quantitates inter se multiplicatæ aliquam producunt." Cf. Heiberg, *Apollonii Pergaei quae Graece exstant cum commentarijs antiquis*, Vol. 2 (Leipzig, 1893; Stuttgart, 1974), pp. 218–219.

[198] Commandino, f. 14v: "*ταῦτα γὰρ τά μαθήματα δοκοῦντι εἶμεν ἀδελφά*, hoc est, hæ enim mathematicæ disciplinæ germanæ esse uidentur." Cf. Heiberg, pp. 220–221.

[199] See a historically just interpretation. Ian Mueller, *Op. cit.* (n. 16), p. 136. As to Book V, Mueller asserts, "thus it seems quite certain that, for Euclid, magnitudes do not include numbers."

> REQUIRED. It is necessary to reduce the same to an integral or mixed number.
> PRAXIS. Let A be divided by B. Let the quotient be $C\frac{D}{B}$. I say that the number $C\frac{D}{B}$ is a mixed number equivalent to the fraction $\frac{A}{B}$. As you see, if the denominator B of the proposed fraction $\frac{A}{B}$ dividing the numerator A makes the quotient an integral number C, it will be equal to the fraction.[200]

He also exemplifies this by taking concrete numbers:

Proposed number	$\frac{A}{B}$	$\frac{14}{3}$
Equivalent number	$C\frac{D}{B}$	$4\frac{2}{3}$
Likewise		
Proposed number	$\frac{A}{B}$	$\frac{12}{3}$
Equivalent number	C	4.[201]

What Van Roomen is doing here may be a simple calculation of improper fractions by using algebraic symbols. But even if such an manipulation might seem to be very trivial, it would be undeniable that the step which he takes is a significant one. Quantities denoted by algebraic signes are not restricted to concrete arithmetical numbers. Operations applied to these quantities may be also applicable to geometrical magnitudes *mutatis mutandis*. Indeed, these quantities and operations can be those of 'mathesis universalis'. In this sense, the concept of 'mathesis universalis' and the idea of symbolic algebra (even Van Roomen's extremely primitive one) can be closely connected with each other.

Van Roomen's symbolism in his *Apologia pro Archimede* was thus in a sense close to Viète's. But unfortunately he does not seem to have developed his system of symbolism in that work to the extent to which it could be used to resolve geometrical problems. So with Michel Chasles[202] we highly esteem Van Roomen's use of primitive algebraic notation but conclude that a gap be-

[200]Romanus, *Apologia*, p. 43: "EXEMPLVM. DATA. Proposita sit fractio $\frac{A}{B}$, cuius denominator numeratore maior non est. QVAESITVM. Oportet eandem reducere ad numerum integrum vel mixtum. PRAXIS. Diuidatur A per B. quotiens fit $C\frac{D}{B}$. Dico numerum $C\frac{D}{B}$ esse numerum mixtum fracto $\frac{A}{B}$ æquiualentem. Sic si propositi fracti $\frac{A}{B}$ denominator B diuidens numeratorem A faciat quotientem numerum integrum C, is erit æqualis fractioni."

[201]*Ibid.*

[202]M. Chasles, "Histoire de l'algèbre—Note sur la nature des opérations algébriques (dont la connaissance a été attribuée, à tort, à Fibonacci).—Des driots de Viète méconnus)," *Comptes rendus des séances de l'Académie des sciences*, t. XII (Séance du Mercredi 5 mai 1841), p. 755: "Il semble donc que c'est Romanus qui a le plus approché de la conception de Viète, dans ce sens qu'il en a eu l'idée; mais il n'a pas su appliquer cette idée heureuse. Il faut donc reconnaître entre lui et Viète, dans l'histoire de l'algèbre, une distance immense, comme celle qui sépare Pythagore de Copernic et de Galilée dans la découverte et les preuves du mouvement de la Terre, ou Kepler de Newton dans l'histoire du principe de la gravitation universelle. Néanmoins la tentative de Romanus lui fait honneur, et rehausse le mérite et la gloire de Viète."

tween his and Viète's was immense. Van Roomen's algebraic symbolism in the *Apologia* seems to us to be justly placed in the history of algebra between those of Jordanus and Viète. There he surely manipulated algebraic symbols but did not proceed to apply them to geometry effectively.

Did Van Roomen write his *Apologia pro Archimede* knowing Viète's analytic art? We cannot provide a sure answer to this question, because Van Roomen may have possibly known Viète's algebra through the latter's *Ad problema responsum* of 1595, and the former could have finished composing the *Apologia* about the same year.[203]

But we can mention one work of Van Roomen which was indubitably influenced by Viète's new algebra. This work is his *Mathematicae analyseos triumphus*, published in Louvain in 1609. In this book, he calculates the ratios of the radius to the chord in the cases of nine-sided and eighteen-sided regular polygons to 108 decimal places. He employs the method of algebraic analysis to do so. As to how he approaches his problem, he states the following in his dedication to the Duke Julius of Franconia, Prince Bishop of Würzburg:

> And as certainly as all the efforts of the moderns as well as the ancients were of no effort, we also could have been frightened by this study if I had not noticed that most problems which were unsolvable by the powers of Geometry can be settled with the benefit of Analysis (commonly called Algebra), and therefore I have commenced to study whether [the problem of] polygons could be subject to the same doctrine.[204]

It is apparent that his approach to the polygonal problems was algebraic analysis. He regards analysis to be that which is commonly called algebra (*Analysis* (*Algebra vulgo dicta*)).

How, then, does he apply the analytic art to his problems? In order to calculate the sides of the regular polygons with 3, 6; 15 and 30 sides, for

---

[203] See Philippe Gilbert, "Notice sur le mathématicien louvaniste Adrianus Romanus, Professeur à l'ancienne Université de Louvain.—(XVIe siècle)," *Revue catholique*, t. 17 (Louvain/Brussels, 1859), pp. 277–286, 394–409 & 522–527, esp. p. 409: "Nous arrivons ainsi à cette conclusion, déjà justifiée d'ailleurs par le contenu du livre, que les idées de Romanus sur l'arithmétique universelle lui appartiennnent en propre, puisqu'il n'a pu connaître Viète et profiter de ses idées que vers 1597 ou 1598, tandis que la partie de son ouvrage que nous avons examinée plus haut doit évidement se rapporter, quant à la date de la composition, vers l'année 1595 ou peut-étre même à une époque antérieure."

[204] A. Romanus, *Mathematicæ analyseos triumphus* (Louvain, 1609), p. 3: "Et sanè conatus ij veterum tum recentionum omnes irriti, nos qvoque ab ea investigatione terrere potuissent, nisi pleraque Problemata, qvæ Geometriæ viribus solvi neqveunt, Analyticæ (Algebræ vulgo dictæ) beneficio expediri posse animadvertissem, ideoque num & Polygona eidem doctrinæ subessent investigare cepi." I refer to a copy of this very rare work at the Staat- und Stadtbibliothek Augusburg. See also A. Ruland, "Adrien Romain, premier professeur à la faculté de médicine de Wurzbourg," *Bibliophile belge* (Bulletin trimestriel publie par la Société des bibliophiles de Belgique), II (1867), pp. 56–100, 161–187 & 256–269, esp. pp. 86–87.

example, he sets up the following equation (to use H. Bosman's symbols):[205]

$$3 \text{ chord A} - \text{chord}^3\text{A}/\text{R}^2 = \text{chord 3A},$$

where A stands for the arc of the given polygon, chord A for the chord corresponding to the arc A, and R for the radius. In Van Roomen's own notation, chord 3A is represented as

$$3A(1) - 1\frac{A(3)}{B(2)}.$$ [206]

Similarly in the cases of the regular polygons with 5 and 10 sides he obtains the formula for chord 5A:

$$5A(1) - 5\frac{A(3)}{B(2)} + \frac{A(5)}{B(4)}.$$ [207]

It can be no longer be doubted that the influence of Viète's symbolic algebra appears in Van Roomen's work of 1609.

It is also reported that Van Roomen drafted a commentary on al-Khwārizmī's *Algebra*, *In Muhamedis algebram problegomena*, which was unpublished and now lost.[208] According to H. Bosmans, who examined the manuscript, Van Roomen's symbolism in this work is much more advanced than that of the *Apologia* of 1597. For example, we can see such algebraic symbols as "$A(2) + \frac{1}{4}B(2) + A$ in B" representing the square of $(A + 1\frac{B}{2})^2$.[209] This notation is identical with that of the *Mathematicae analyseos triumphus*. Furthermore, significantly, he explicitly identifies algebra with the analytic art and then connects it with his concept of 'first mathematics'. He puts it:

> Accordingly we have inclined to subject Algebra or the Analytical science to first Mathematics, which considers universal quantity.[210]

In Van Roomen's terminology, 'first mathematics' (*prima Mathesis*) is another name of 'mathesis universalis'.

In his treatise *Universae mathesis idea* (1602), which we have touched on in Chapter 4, §3, he places his concept of 'mathesis universalis' within the domain of the mathematical disciplines as a whole and discusses its significance

[205] H. Bosmans, *Op. cit.*(n. 191), pp. 77–79, where the author mathematically examines Van Roomen's present work.

[206] Romanus, *Mathematicæ analyseos triumphus*, p. 4.

[207] *Ibid.*, p. 6.

[208] H. L. L. Busard, *Op. cit.*(n. 192), p. 533; P. Bockstaele, *Op. cit.* (n. 189), cols. 759–760.

[209] H. Bosmans, "Le fragment du Commentaire d'Adrien Romain sur l'Algèbre de Mahumed ben Musa el-Chowârezmî," *Annales de la société scientifique de Bruxelles*, XXXe Année, 1905–1906 (Louvain, 1906), 2e partie, pp. 267–287, esp. p. 180.

[210] *Ibid.*, p. 271, note (***): "Nos itaque maluimus Algebram sive Analyticam scientiam revocare ad Mathesin primam, quae quantitatem universalem considerat." In the same manuscript, A. Romanus also refers to Viète's *Zetetica* of 1593. See Bosmans, p. 273.

rather philosophically. For this reason this work will be fully analyzed below in Chapter 7, § 2.

Incidentally, it has been made clear that Van Roomen, together with Viète, was in a Ramistic mathematical atmosphere. In fact, some editions of Ramus's *Geometria* were dedicated to him.[211]

But can we relate Van Roomen's mathematical thought to Descartes at all? From our investigation above, it is at least certain that Descartes was already familiar with the idea of 'mathesis universalis' when he composed the *Regulae ad directionem ingenii*. We do not know for sure that the notion which Descartes learnt was Van Roomen's. But it is quite likely from the content of Rule IV of the *Regulae* that Descartes's 'mathesis universalis' has a very close relationship with the method of algebraic analysis. Unless Van Roomen was a unique advocate of the concept of 'mathesis universalis' in some sense connected with symbolic algebra, Descartes could have been influenced by some mathematician(s) other than Van Roomen. Circumstantial evidence suggests, however, that Descartes most probably learned the idea of 'mathesis universalis' from the Belgian mathematician Van Roomen, a rival and then friend of Viète.

But did Descartes recognize the importance of the concept of 'mathesis universalis' only shortly before drafting the *Regulae*? Let us recall what he states in Rule IV. According to his statement, after having noticed that algebra contains the germ of true mathematics, that is, well-methodized mathematics, he turned from the particular studies of pure mathematics to "a general investigation of mathematics." (see pp. 92 and 195.) During the process of this general study, we are told, he encountered the very crucial idea of 'mathesis universalis.'

Thus we may ask: When did Descartes become familiar with 'mathesis universalis'? The date of his study of algebra and of the particular fields of pure mathematics must date from the time when he studied at La Flèche. If his autobiographical pronouncement is trustworthy, he began to learn that notion after he recommenced his mathematical studies, i. e. after the year 1618. And importantly, on March 26, 1619 he made known to Beeckman his program of "an entirely new science, by which all problems that can be posed concerning any kind of quantity, continuous or discrete, can be generally solved" (pp. 136–137). Reading through this statement, we immediately recognize that his plan of the spring of 1619 was quite similar to Van Roomen's concept of 'mathesis universalis'. An examination of Descartes's posthumous writing *Cogitationes privatae* shows us at least that we cannot detect a strong discernible influence fron Viète on that manuscript. The author of the *Cogitationes privatae* certainly represented a side of the rectangular tetrahedron with the base $\alpha\beta\gamma$ by the formula $\sqrt{.\frac{1}{2}\alpha q + \frac{1}{2}\beta q - \frac{1}{2}\gamma q}$ for our $\sqrt{\frac{1}{2}\alpha^2 + \frac{1}{2}\beta^2 - \frac{1}{2}\gamma^2}$ (p. 131). Such a formula might have been easily obtainable from Clavius's *Algebra*. That is to say, we

[211]See W. J. Ong, *Ramus and Talon Inventory* (n. 163), Nos. 615 & 616: *Geometriae libri septem et viginti, ad clarissimum Adrianum Romanum mathematicorum ocellum* (Hanover, 1604 & 1612).

cannot determine whether or not Descartes had known an analytic art of any sort when he wrote down the notes of *Cogitationes privatae* and his letter dated on March 26, 1619.

Now let us turn our attention in another direction. There is an anecdote told of Descartes's first meeting with Beeckman at Breda in the autumn of 1618. This account originated from Daniel Lipstorp's *Specimina philosophiae cartesianae*, published in 1653. According to Lipstorp,[212] a mathematical problem was posted on a street of Breda, which was written in Flemish. Looking at it, Descartes, Lipstorp tells us, asked someone who could translate it into Latin or French. The Dutchman who helped the young Frenchman was none other than Beeckman. Descartes succeeded in solving the problem. The problem which he is said to have tackled was one propounded in 1593 by Van Roomen on an algebraic equation of the 45th degree and was solved afterward by Viète.

Lipstorp's description of Descartes's first encounter with Beeckman was later reproduced with some additions in Baillet's *La Vie de Monsieur Des-Cartes* of 1591.[213] By comparing this legend with Beeckman's notes preserved in his *Journal*, Charles Adam suspected Lisptorp's story to be untrustworthy and uncertain,[214] for in the *Journal* Beeckman's version does not at all resemble Lipstorp's description in his biography of Descartes. We basically agree with Adam's suspicions about the historical authenticity of Lipstorp's story. Lipstorp might have desired to give the impression that the young Descartes's mathematical talent was equal to Viète's.

But the anecdote does not necessarily seem to have been sheer fiction. It is known that Lipstorp wrote the biography of Descartes based on what he heard from Van Schooten,[215] a mathematical disciple of Descartes and the editor of the two Latin versions of the *Géométrie*. Lipstorp's account as it stands must surely contain misinformation or exaggeration. But between Descartes and Beeckman, there could have been conversations not only about Ramus but also about Van Roomen, not say about Viète. Incidentally, Baillet described Beeckman as having a certain relationship with Jacques Alleaume, a friend of Viète and an one-time possessor of the latter's manuscripts. It is also plausible that in Holland in the years 1618–1619, Descartes, independently of Beeckman, might have found something connected with the idea of 'mathesis universalis'. He might have encountered and read some book(s) by Van Roomen, perhaps the *Apologia pro Archimede* and/or *Universae mathesis idea*, which mentioned 'mathesis universalis', probably not the *Mathematicae analyseos triumphus*, which made no mention of that significant concept. A trifling suggestion which may elicited to the idea of the unification of arithmetic and geometry might have been sufficient for Descartes to have attained the program

[212] D. Lipstorp, *Specimina philosophiæ cartesianæ* (Louvain, 1653), p. 77. Cf. *AT*, X, p. 48.
[213] Baillet, t. I, pp. 43–44. Cf. *AT*, X, pp. 49–50.
[214] *AT*, X, p. 51: "Le récit de Lipstorp et de Baillet nous laisse donc, quand même, dans la méfiance et l'incertitude."
[215] *AT*, X, p. 50; *AM*, I, p. 453.

which he stated explicitly in his letter to Beeckman on March 26, 1619.

Of course, this is only a possibility or mere conjecture. It must be admitted in candor that there is no definitive evidence. But examining Descartes's statements about his concept of 'mathesis universalis' in Rule IV-B and about his project of "an entirely new science," we believe that our conjecture that he knew the idea of 'mathesis universalis' during his sojourn in the Netherlands in 1618–1619 is substantially correct. What has been corroborated by our critical historical study is simply that in the spring of 1619 Descartes possessed a plan for the reorganization of the mathematical disciplines, and that before the composition of the *Regulae* he was familiar with the notion of 'mathesis universalis' and the method of algebraic analysis. It is also possible that he attained his program of unifying arithmetic and geometry and of combining geometrical analysis and algebra independently of Ramus, Van Roomen, and Viète, for example. It is also equally possible that he learnt some of these ideas from reading the books of others, for example of classical authors and Clavius.

There was also an opportunity for Descartes to acquire or consolidate his knowledge of algebraic analysis during his stay in Paris in the middle of the 1620s. For example, Mersenne dedicated an entire chapter of his *Les Verités des sciences, contre les septiques ou Pyrrhoniens*, published in Paris in 1625, on Viète's analytic art, quoting the last phrase "Nullum son problema solvere" from the latter's *Isagoge*.[216] It is sometimes pointed out that the mathematical desciplines including the new analytic art took an important role in Mersenne's plan for overcoming the Pyrrhonian thought which was rampant in France during the period of the 1620s.[217] It is probable that by that time at the latest Descartes possessed the method of algebraic analysis and shared a conception similar to Mersenne's.

In all events, in the winter of 1631–1632 Descartes, by means of his own reformed algebra, could succeed in solving the locus problem of Pappus. In this regard at least, he had already surpassed Viète as well as Apollonius and Pappus. So in 1632 when he received a package containing Beaugrand's 1631 edition of Viète's works, he quite understandably did not evaluate highly the Vietan art, simply complaining about the repeated words "Nullum non problema solvere" to the sender of the book, Mersenne, though it seems a little ironical that he himself expressed similar statements in 1619 when he revealed his program for "an entirely new science" (p. 102), around the same period when he proposed his project of composing a book with the title *Thesaurus mathematicus* (p. 110), and further in 1628 when he explained the content of his old *Algebra* to Beeckman (pp. 159–160).

---

[216]M. Mersenne, *Les Verités des sciences, contre les septiques ou Pyrrhoniens* (Paris, 1625), p. 452.

[217]J. A. Schuster, *Op. cit.* (Ch. 3, n. 7), p. 414; Peter Dear, "Marin Mersenne and the Probabilistic Roots of 'Mitigated Scepticism'," *Journal of the History of Philosophy*, **22** (1984), pp. 173–205, esp. p. 181. See P. Dear, *Mersenne and the Learning of the Schools* (Ithaca, 1988).

To speculate historically about the sources of Descartes's mathematical thought is not totally useless. Through such an investigation, at least we may be able to know something about how immediate predecessors and contemporaries of Descartes thought about mathematics in general and the extent of their understanding of algebraic analysis. Though we have suggested some possible or probable sources of his mathematics, it is undeniable that he had a general temperament to speculate and discover something by his own efforts as he stated in the *Cogitationes privatae*[218] and at the beginning of Rule X of the *Regulae*.[219] We have only attempted to show that even Descartes was not entirely independent of other mathematicians, despite his natural bent for self-instruction.

E. From Viète to Descartes

So far we have seen how Descartes formed his method of algebraic analysis and presented our conjecture that it was through studying Clavius's *Algebra* and then doubtlessly the concept of 'mathesis universalis' of Van Roomen, a disciple of Clavius and a friend of Viète, that he could have cultivated his idea for unifying and 'methodizing' pure mathematics while constructing a new algebraic discipline. It is not certain, however, whether he studied Viète's analytic art directly in his youth. The mere contact with 'mathesis universalis' might have been sufficient for him to establish the method of algebraic analysis. But he surely perused Viète's book in the spring of 1632 and made use of it at least slightly when composing the *Géométrie*.

On the other hand, what he insisted upon at the end of 1637, after the publication of the *Géométrie*, when he was accused of having plagiarized Viète's idea of analytic art, was not that he had not learnt from Viète but that he started where his predecessor had stopped. The present subsection will show that Descartes's claim that he went beyond Viète in his *Géométrie* is fundamentally sustainable.

To accomplish this end let us compare the main characteristics of Viète's symbolic algebra with those of Descartes's *Géométrie*.

First of all, Viète's analytic art could serve as a supplementary means for interpreting algebraically quantitative properties of geometrical propositions. By this new art, the quantitative properties in many propositions of Greek geometry began to be interpreted in a systematically algebraic way for the first time. Viète thus originated a new point of view for treating a certain number of propositions of Euclid, Apollonius, Pappus, and some other geometers as concerning 'geometrical algebra'. The label 'geometrical algebra' started to be used, as is well known, by Paul Tannery and H. G. Zeuthen in the late nineteenth century. Of course, algebraic interpretation of geometrical propositions had been attempted in the Arabic mathematical tradition beginning with

[218] *AT*, X, p. 214.
[219] *AT*, X, pp. 403–404.

al-Khwārizmī's treatise on algebra of the ninth century. But mathematical practitioners belonging to this tradition continued to represent algebraic relationships not in a symbolic manner but simply rhetorically at latest until about the twelfth century. Now the previous view is under revision—the view that algebraic symbolism of Arabic mathematics hadn't developped approximately until al-Qalāṣādī of the fifteenth century, who used some symbols for unknown numbers. We are beginning to know that in both the Islamic and European worlds algebraic symbolism started in the twelfth century.[220] However, when trying to prove algebraic relations, Arabic mathematicians in the end invoked geometrical theorems from Euclid's *Elements*, especially Book II.[221]

As we have seen, Ramus partly discerned an algebra in some geometrical propositions of Euclid. It was Viète who cultivated a powerful mathematical tool, symbolic algebra, to reinterpret them algebraically. But since his analytic art was closely linked with the purpose of restoring ancient geometrical works, it still left some geometrical vestiges, one of which was the "law of homogeneity"(*lex homogeneorum*). He, in Chapter III of his *Isagoge*, stated: "Homogeneous terms must be compared with homogeneous terms."[222] In geometry, heterogeneous magnitudes must not be conjoined and magnitudes of more than three dimensions must be excluded. Although his algebra violated the latter proscription in permitting magnitudes more than three dimensions, it accepted the former prohibition. Thus the world of quantities which were to be subjected to algebraic operations was ordered according to an intricate hierarchy of mutually incomparable species: the ladder of magnitudes (*scalaria magnitudinum*).[223]

With Descartes, the problem of the dimensionality of geometrical magnitudes was resolved in a deceptively simple manner that can be called the 'algebra of line segments'. This was a simple step but constituted a gestalt switch from the geometrical to the algebraic mode of thought in mathematics. As we have empha-sized in Section 1, it was through this conceptual change that Descartes could claim to have successfully presented the solution of the ancient locus problem. The line segment began to be viewed as consisting of various kinds of number. As Descartes himself claimed in his letter to Mersenne quoted above (pp. 241–242), in Book III of the *Géométrie* he commenced a new, general

[220] M. F. Woepcke, "Traduction du traité d'arithmétique d'Aboul Haçan Alī Ben Mohammad Alkalçadi," *Atti dell' Accademia Pontificia de' Nuovi Lincei*, t. XII (1858–1859), pp. 399–438; A. S. Saidan, "Al-Qalāṣādī (or Al-Qalṣādī), Abu'l-Ḥasan 'Alī Ibn Muḥammad Ibn 'Alī," *DSB*, XI, pp. 229–230. Acording to Prof. R. Rashed, mathematicians in Arabic began to use symbols for unknown numbers as early as the 12th century independently of European mathematicians.

[221] See the Arabic texts mentioned in n. 146 above. See also Martin Levy, ed., *The Algebra of Abū Kāmil: Kitāb fī al-jabr wa'l-muqābala in a Commentary by Mordecai Finzi* (Madison, 1966).

[222] Viète, *Isagoge* (1591), f. 4v; *Opera mathematica* (1646), p. 2: "Homogenea homogeneis comparari." See Witmer's translation, p. 15.

[223] *Isagoge*, f. 4v; *Opera mathematica*, p. 3.

non-intuitive way of thought in the theory of equations by employing this algebra of line segments thoroughly.[224] Mathematical relations which may appear in various fields of mathematics were to be written in the form of algebraic equations consisting of algebraically homogeneous quantities.

Second, Descartes's notation in the *Géométrie* was much simpler than Viète's. According to G. H. F. Nesselman's view of the history of algebra,[225] this discipline developed through three stages: (1) rhetorical algebra (*rhetorische Algebra*), (2) syncopated algebra (*synkopierte Algebra*), and finally, (3) symbolic algebra (*symbolische Algebra*). The algebra of the first stage is represented by al-Khwārizmī's algebra, which used no symbols at all. Syncopated algebra uses abreviated signs only for concepts and operations which are frequently repeated. Representative examples of this are Diophantus's *Arithmetic* and cossist algebra. In the third type, all formulae and operations are represented by well-developed symbols. Descartes's system belongs to this type. While Nesselmann did not state it explicitly, he apparently thought that Viète's algebra was a transitional sort between the second and the third stage, or a rudimentary form of symbolic algebra. "B in A − A quad. æquatur Z plano" in Viète's algebra, for example, is translated into the very simple formula "$ax - x^2 \propto b^2$" in Descartes's.

Third, since the synthetic procedure performed on algebraic equalities is in general simply the reverse of the analytic (in the case of theorematic analysis, to use Pappus's terminology), with the establishment of the modern analytic tradition the former process tended to be regarded as unnecessary. In most cases, the analysis alone became sufficient. As we have observed in Section 1, however, in problems on constructing algebraic curves, Descartes adhered to the classical manner and carried out the synthetic step (in this case, analysis was the problematic one). But, as H. J. M. Bos has shown in his paper titled "Arguments on Motivation in the Rise and Decline of a Mathematical Theory; the 'Construction of Equations', 1637–*ca.* 1750,"[226] within about one hundred years after the appearance of the *Géométrie*, even this synthetic procedure in the form of construction of equations gradually became forgotten among the followers of the Cartesian algebraic tradition.

Analysis thus gradually became a method for both invention and demonstration. This had been already recognized by Diophantus. At the end of the propositions in his Arithmetic he frequently repeated the sentence: "And

[224] See Mahoney, "The Beginnings of Algebraic Thought etc." (n. 21), and Abel Rey, "De Viète à Descartes," *Travaux du IXᵉ congrès international de philosophie: Congrès Descartes*, II. *Études Cartesiennes*, IIᵉ partie, pp. 27–32.

[225] G. H. F. Nesselmann, *Die Algebra der Griechen* (Berlin, 1842; Reprint: Frankfurt am Main, 1969), pp. 301–303.

[226] H. J. M. Bos, "Arguements on Motivation in the Rise and Decline fo a Mathematical Theory: the 'Construction of Equations', 1637–*ca.*1750," *Archive for History of Exact Sciences*, **30** (1984), pp. 331–380.

the proof is evident" (καὶ ἡ ἀπόδειξις φανερά),[227] because solutions of these propositions were obtained through arithmetical equalities. Ghetaldi in Book I of his *De resolutione et compositione mathematica* (1630) recognized that an algebraic analysis may make a synthetic proof useless.[228] Accordingly, it may be said that the algebraic proof system emerged with the formation of the tradition of algebraic analysis.

Descartes went further. We have seen in Chapter 4, § 2, B, that in his *Regulae* the method of algebraic analysis came to play a crucial role and subsequently became a model for the universal method which he supposed would serve to inquire about certain knowledge in general. This was not the sort of enterprise Viète had ever dreamed of. Bos's characterization that "Descartes' mathematics was a philosopher's mathematics"[229] can be completely justified. Further in the "Author's Reply to the Second Set of Objections" of his *Meditationes de prima philosophia*, published in 1641, Descartes, discerning two methods of demonstration, i. e. analysis and synthesis, stated: "It is analysis which is the best and truest method of instruction, and it was this method which I employed in my *Meditationes.*"[230] The method of algebraic analysis and then the more general philosophical method of analysis thus occupied a pivotal position in Descartes's thought in general, and was thought to be a kind of demonstration in mathematics and in philosophy, respectively.

But Descartes seems to have been satisfied simply with the recognition that algebraic analysis is for the most part adequate as a proof in mathematics. Needless to say, it was Leibniz who speculated on what a proof is in the algebraic mode of thought and tried to formulate an algebraic proof system, so to speak, formalistically,[231] by incorporating logic into a branch of his 'mathesis universalis'. His 'mathesis universalis' was much broader than Van Roomen's and Descartes's, although we will not discuss here whether Leibniz's attempt was in the end successful. (see Chapter 8, § 2). It is certain, however, that Descartes opened the way to this modern proof system of Leibniz. In this regard, Descartes by far surpassed not only his predecessors Viète and Van Roomen, but his mathematical rival Fermat as well.

Viète and Descartes can be counted as the two most important founders of the modern analytic tradition in mathematics. Viète, in the dedication to his

---

[227] See, for example, *Diophanti Alexandrini Opera omnia*, ed. P. Tannery (Leipzig, 1893; Stuttgart, 1974), Propositio 1, p. 16; Diophante d'Alexandrie, *Les six livres Arithmétiques et le livre des nombres polygones*, traduites par Paul Ver Eecke (Paris, Nouveau Tirage, 1959), p. 21. Cf. André Robert, "Descartes et l'analyse des ancients," *Archives de philosophie*, **13** (1937), pp. 301–325, esp. pp. 319–322.

[228] Mahoney, "The Royal Road", pp. 206–213.

[229] Henk J. M. Bos, *Redefining Geometrical Exactness: Descartes' Transformation of the Early Modern Concept of Construction* (New York/Berlin/Heidelberg, 2001), p. 228.

[230] Descartes, *Meditationes* (Paris, 1641): *AT*, VII, p. 156. See also Mahoney, *Op. cit.* (n. 21), p. 149;

[231] Ian Hacking, "Proof and Eternal Truths: Descartes and Leibniz," Chapter 7 of S. Gaukroger, ed., *Op. cit.* (n. 21), pp. 160–180.

In artem analyticem isagoge, declared a new vocabulary for the art of "Algebra or Almucabala" which is so spoiled and defiled "by the barbarians" (*à barbaris*) and gave it an elegant form.[232] In fact, he gave his "new algebra" (*algebra nova*) the vocabulary originated from the Greek "analysis." On the other hand, in his *Regulae ad directionem ingenii* Descartes stated that he had attached importance to the normative discipline of mathematics, not "by the outlandish name" (*barbaro nomine*) "Algebra" (pp. 92 & 194) but by the other name "mathesis universalis." Nevertheless, it is undeniable that the two mathematicians were successors of the heritage of Arabic mathematics which considered the art of calculation as important.

Moreover, it must be admitted at the same time that they succeeded in enhancing the art of algebraic analysis to a new theoretical horizon. On the one hand, Viète, having inspired by the Ramean insight, invented his symbolic algebra with very primitive symbolism simply as an ancillary means for achieving arithmetical and geometrical analyses. On the other hand, Descartes tried to extend the method of algebraic analysis to broader fields of knowledge. The former looked at ancient Greece as a mathematician of the late Renaissance whereas the latter turned toward the modern times as a thinker of the beginning of the Baroque period absolutism. Indeed, Descartes started where Viète stopped.

§ 4. Beyond Cartesian Mathematics

Near the beginning of Section 3, A, we stated that the *Géométrie* was not always warmly welcomed among contemporary French mathematicians, at least immediately after its appearance. But after the young Dutch mathematician Frans van Schooten (Figure 5.8) undertook its Latin translation and published it in 1649 with his extensive commentary along with Florimond de Beaune's *Notae breves*,[233] the situation concerning this mathematical work undoubtedly took a favorable turn, especially outside France.

The second Latin version, published in two parts in 1659 and 1661, surely enjoyed much greater success. This edition contained Van Schooten's enlarged commentary, De Beaune's notes which had already appeared in the first Latin edition, Jan Hudde's *De reductione aequationum* and *De maximis et minimis*, Hendrik van Heuraet's *Epistola de transmutatione curvarum linearum in rectas* (above, Part I published in 1659), Van Schooten's *Principia matheseos universalis*, de Beaune's *Duo tractatus posthumi, alter de natura & constitutione, alter de limitibus aequationum*, Jan de Witt's *Elementa curvarum linearum*, and finally

[232] Viète, *Isagoge* (1591), f. 2v; *Opera mathematica* (1646), p. [XI].

[233] *Geometria*, à Renato Des Cartes Anno 1637 Galliè edita; nunc autem Cum Notis Florimondi de Beavne, [...] Operâ atque studio Francisci à Schooten (Leiden, 1649). Cf. A. J. Guibert, *Bibliographie des œuvres de René Descartes publiées au XVIIe siècle* (Paris, 1976), pp. 27–29.

Van Schooten's *Tractatus de concinnandis demonstrationibus geometricis ex calculo algebraico.*[234] The fortune of the 1659–1661 version is also corroborated by the fact that it was undertaken by the distinguished Dutch publisher Daniel and Ludewijk Elsevier in Amsterdam.

It was Van Schooten's second Latin edition *Geometria à Renato Des Cartes* that the young Newton read and annotated as an undergraduate at the University of Cambridge as early as 1664[235] and that his rival Leibniz perused with a feeling of difficulty sometime in his undergraduate years 1661–1664 at Leipzig and Jena,[236] and studied seriously around 1673, the year of great discoveries during his sojourn in Paris.[237] Thus around the years of the publication of Van Schooten's second edition, the success of Cartesian analytic mathematics seems to have become undeniable.

Figure 5.8 Frans van Schooten (1615–1660)

In March of 1638, as we have stated, Descartes reaffirmed his resolution of fifteen years before that he would stop studying pure mathematics of his own will. But this did not mean that he totally lost interest in solving mathematical problems at the request of his friends. To mention a few examples, in fact, he contributed to the development of 'infinitesimal' geometry by giving the area[238] and the tangent[239] of a cycloid in 1638 and by solving an inverse problem of tangents

[234] *Geometria à Renato Des Cartes*, ed. F. à Schooten (Amstelædami, 1659); *Renati Des-Cartes Geometriae Pars Secunda* (Amstelædami, 1661). Cf. Guibert, *Op. cit.*, pp. 29–31. Jan de Witt's Elementa curvarum linearum, Book 1, in Part 2 is studied in *Jan de Witt's Elementa curvarum linearum, Liber primus*, Text, Translation, Introduction, and Commentary by Albert W. Grootendorst (New York/Berlin/Heidelberg, 2000). Incidentally, in *Tractatus de concinnandis demonstrationibus geometricis ex Calculo Algebaïco*, the name of Pieter Hartsinck, a Japanese student of Frans van Schooten, appears as "doctissimus juvenis D. Petrus Hartsingius, Iaponensis," on p. 414. He was the son of a Dutch merchant and a Japanese woman and was once a rival of Leibniz as a mining engineer.

[235] *The Mathematical Papers of Isaac Newton*, ed. by D. T. Whiteside, Vol. I (1664–1666), e. g. p. 31, notes (19) and (21). See also Richard Westfall, *Never at Rest: A Biography of Isaac Newton* (Cambridge, 1980), pp. 23–28 &100.

[236] G. W. Leibniz, *Mathematische Schriften*, hrsg. von G. I. Gerhardt, Bd. III/1 (Halle, 1855), p. 72. Cf. Joseph E. Hofmann, *Leibniz in Paris 1672–1676: His Growth to Mathematical Maturity* (Cambridge, 1974), p. 3.

[237] Hofmann, *Op. cit.*, p. 49, esp. note 18.

[238] See "Descartes, à (Santpoort), à Mersenne, à Paris. (17 et 27 mai 1638)," *AT*, II, pp. 135–137; *AM*, II, pp. 261–262; *CM*, VII, pp. 225–227, & "Descartes, à (Santpoort), à Mersenne, à Paris. 27 juillet 1638," *AT*, II, pp. 257–263; *AM*, II, pp. 355–360; *CM*, VII, pp. 407–412.

[239] See "Descartes, à (Santpoort), à Mersenne, à Paris. 23 août 1638," *AT*, II, pp. 308–312; *AM*, III, pp. 28–32; *CM*, VIII, pp. 34–39.

in 1639, proposed by Debeaune.[240] Accordingly, there is reason to claim that he did not necessarily lack the sense to pursue 'infinitesimal' mathematics.[241]

But this claim seems to be fully legitimate only with one reservation. In Section 2, we asserted that the mathematics of the *Géométrie* was basically the algebra of finite line segments. And in that monograph, Descartes stated that "the proportion (*proportion*) between straight and curved lines is unknown, and even, I believe, unknowable to men, so that we cannot thereby reach any exact and assured conclusions."[242] Here he followed the lead of the Greeks, believing in that there can be no ratio between the straight and the curved because of the two being different genera.[243] As to his attempts at solving 'infinitesimal' problems, the impression seems to us inevitable that they were basically ad hoc.

It was Leibniz who challenged intentionally Descartes's finitistic restriction and opened a new approach leading to the infinitesimal calculus. In his "Historia et origo calculi differentialis", drafted at the end of 1714, Leibniz extracted his new calculus from "ordinary or Apollonian geometry" (*Geometria communis seu Apolloniana*) by Viète and Descartes and legitimately called his own "Archimedean geometry" (*Geometria Archimedea*).[244] Descartes's mathematics stopped, in Leibniz's judgment, within the range of Apollonian geometry. By his calculus, Leibniz's asserted, "the whole of geometry is subjected to analytical calculation, and those transcendent lines that Descartes called mechanical are also reduced to equations chosen to suit them, by considering the differences dx, ddx, etc. and the sums that are the inverses of these differences, as functions (*functiones*) of the x's."[245] What Leibniz accomplished in his differential and integral calculus was in one sense a reinterpretation of Archimedian geometry by means of the method of algebraic analysis. Thus, Viète and Descartes were the founders of finite algebraic analysis, whereas Leibniz was the cultivator of infinitesimal algebraic analysis.

---

[240] "Descartes à M. de Beaune. 10 février 1639," *AT*, II, pp. 513–517; *AM*, III, pp. 187–193. For Descartes's solution of de Beaune's problem, see Jules Vuillemin, *Op. cit.* (Ch. 3, n. 154), Ch. II, §1, "La Théorie cartésienne de la courbe logarithmique," pp. 11–25; Christoph J. Scriba, "Zur Lösung des 2. Debeauneschen Problems durch Descartes: Ein Abschritt aus der Fröhgeschichte der inversen Tangentenaufgaben," *Archive for History of Exact Sciences*, **1** (1961), pp. 406–419.

[241] Christian Houzel, "Descartes et les courbes transcendantes," *Descartes et le Moyen Âge*, éd. par Joël Biard et R. Rashed (Paris, 1997), pp. 27–35. Cf. Pierre Costabel, "Descartes et la mathématique de l'infini," *Historia Scientiarum*, No. 29 (1985), pp. 37–49. Cf. P. Mancosu, *Philosophy of Mathematics and Mathematical Practice in the Seventeenth Century* (Ch. 2, n. 10), pp. 82–83

[242] *AT*, VI, p. 412; Olscamp, p. 206.

[243] For a mathematico-philosophical question on the rectification of curves, see Hofmann, *Op. cit.* (n. 236), pp. 101–102.

[244] G. W. Leibniz, *Mathematische Schriften*, hrsg. von C. I. Gerhardt, Bd. V (Halle, 1858), pp. 393–394; J. M. Child, *The Early Mathematical Manuscripts of Leibniz* (Chicago/London, 1920), p. 26. On the date of the composition, see K. Müller und G. Kronert, *Leben und Werke von G. W. Leibniz: Eine Chronik* (Frankfurt am Main, 1969), p. 250.

[245] Leibniz, *Ibid.*, p. 394; Child, *Loc. cit.*

Newton, on the other hand, began to tackle the field of infinitesimal geometry in an algebraic way inspired partly by Descartes's analytic method, and successfully attained a new calculus essentially equal to Leibniz's at latest in October 1666. The old Newton certainly became critical of Descartes's manner of algebraic analysis probably partly because he had noticed several errors in the *Geometria à Renato Des Cartes* (e. g. see above p. 223) and partly because he became critical of Descartes's strictly mechanical philosophy of nature[246] seeing in Descartes a shadow of Leibniz, with whom he engaged in caustic argument about the priority of the discovery of the calculus. In 1728 the year following Newton's death, Henry Pemberton reported about his teacher's liking of the style of mathematics:

> I have often heard from him [Newton] censure the handling geometrical subject by algebraic calculations; and his book of *Algebra* he called by the name of Universal Arithmetic, in opposition to the unjudicious title of Geometry, which Des Cartes had given to the treatise, wherein he shews, how the geometer may assist his invention by such kind of computations. He frequently praised Sluse, Barrow and Huygens for not being influenced by the false taste, which then has begun to prevail.[247]

David Gregory has also informed us of a rare utterance of the rival of Leibniz: "Algebra is the Analysis of the Bunglers in Mathematics."[248] It is now known, thanks to D. T. Whiteside's laborious editorial work, that between the years 1691 and 1695 Newton seriously studied the ancient geometrical analysis for composing his own authentic *Geometria*. In a draft for Book 1 of the *Geometria* we read the following observation by him: "Analysis (*resolutio*) is ill-suited to be taught to masses."[249] And in a draft for the revision of his *Principia mathematica philosophiae naturalis*, composed probably in the late 1710s, he stated:

> 'Specious' analysis is nothing other than arithmetic applied to geometrical matters. In calculations of this kind errors are easily introduced. But if the propositions so found out be demonstrated synthetically, then these acquire a dual certainty, both that which can be had from arithmetical calculation and that which is produced from geometrical demonstration; through such demonstration they

[246] Westfall, *Never at Rest* (n. 235), pp. 379–381.

[247] Henry Pemberton, *A View of Sir Isaac Newton's Philosophy* (London, 1728), Preface, [B2r].

[248] W. G. Hiscock, ed., *David Gregory, Isaac Newton and Their Circle: Extracts from David Gregory's Memoranda 1677–1708* (Oxford, 1937), p. 42.

[249] D. T. Whiteside, ed., *The Mathematical Papers of Isaac Newton*, Vol. VII (1691–1695) (Cambridge, 1976), pp. 306–307. See Mahoney's interesting review of Vol. VII: "Last Years at Cambridge," *Science*, Vol. 196, No. 4292 (20 May 1977), pp. 864–865.

> acquire geometrical certainty, and without geometrical ceraintty they ought not to be admitted into geometry. Here regard must be paid not only to the brevity of words and computations, but also to the evidence and certainty of things.[250]

In order to maintain mathematical rigor and his empirist epistemology, Newton continued to adhere to preserving the ancient intuitive style of geometry.

In spite of Leibniz's and Newton's criticisms, which had quite different characteristics,[251] we have to admit nevertheless that the two mathematicians of the late seventeenth century invented their new calculus on the horizon which Descartes cleared. That is to say, it was fundamentally by using Descartes's method of algebraic analysis to a considerable extent that they constructed their infinitesimal analysis. Remember that even Newton's method of fluxions had the character of an algebraic algorism.

Mathematicians of the generations after Descartes might have learned analytical techniques, for example, the method of tangents, from the appendices attached to Van Schooten's second Latin edition of the *Géométrie*. And it might have been mainly through Van Schooten's ***Principia matheseos universalis, seu Introductio ad Geometriae methodum Renati des Cartes***, which occupied the top of the second part of that version, not through Descartes's main text, that they studied even Cartesian symbolic algebra. But the facts which we have pointed out also can be interpreted not as showing the limitations of Descartes's mathematics itself, but as demonstrating the very fruitfulness of the Cartesian paradigm of algebraic analysis, although the main characteristics of the new calculus were different from the method of algebraic analysis for finite magnitudes and with the infinitesimal calculus the natural world was to be viewed in a new way.[252]

Michael J. Crowe, in his "Ten 'Laws' concerning Patterns of Change in the History of Mathematics," recently contended, intending to limit Thomas S. Kuhn's new historiography of the history of science to the sciences excluding mathematics, that "revolutions never occur in mathematics."[253] But he promptly had to accept a qualification to his tenth law, commenting that "the stress in law 10 on the preposition 'in' is crucial, for, as a number of the earlier laws make clear, revolutions may occur in mathematical nomenclature, symbolism, metamathematics (e. g. the metaphysics of mathematics), methodology (e. g. standards of rigor), and perhaps even in the historiography of mathematics."[254] It may certainly be doubted that there did, or does, exist a mathematics

---

[250] *Ibid.*, Vol. VIII (1697–1722) (Cambridge, 1981), pp. 450–453.

[251] See my paper, *Op. cit.* (n. 162), esp. pp. 123–125, where I discuss the tension between the two and its historical background.

[252] See M. S. Mahoney, "Changing Canons of Mathematical and Physical Intelligibility in the Later 17th Century," *Historia Mathematica*, **11** (1984), pp. 417–423.

[253] *Historia Mathematica*, **2** (1975), p. 165.

[254] *Ibid.*, pp. 165–166.

(*pace* Crowe) without its associated intellectual apparatuses and values, for example, its methodology, which Crowe has pointed out. In fact, fruitful debates by historians of mathematician after Crowe on 'revolutions in mathematics' seem to urge to revise Crowe's law 10 radically.[255] If so, we may be able to identify revolutions in mathematics.

René Descartes, among others, consciously revolutionized mathematics in the early seventeenth century. Of course, there were forerunners in the tradition of analytic mathematics, for example, Ramus, Viète, and Van Roomen. If we extend our view of history to Islamic mathematics on whom we have had much more knowledge than previously, the names of mathematicians such as Ibrāhīm Ibn Sinān, al-Haytham, Omar Khayyam, and al-Samaw'al occur immediately. There were contemporary comrades such as Fermat, who may have been more talented than Descartes in problem-solving techniques. There were also able successors, for example, Newton and Leibniz. But Descartes seems to have been different from his precursors and contemporary mathematicians at least in one thing, that is to say, in having established a coherent and methodical program of algebraic analysis. His revolutionary program, which may be called the paradigm of modern algebraically analytic mathematics, could have been conceived by only one able architect of thought who was a mathematician and philosopher at once, as he states at the beginning of Part II of his *Discours de la méthode*, an autobiographical and philosophical introduction to the *Géométrie*. Probably only Leibniz can be compared with Descartes for his comprehensive mathematical-philosophical program for constructing the 'chacteristica universalis'. Even for Leibniz Descartes was nothing other than "Descartes with a divine intelligence" (*divini ingenii Cartesius*).[256] It was a new paradigm[257] of mathematics that Descartes successfully constructed by replacing the classical one of Euclid, Apollonius, and Pappus. With him together with his comrade mathematicians like Fermat, a revolution certainly occured in mathematics, and with Newton and Leibniz the revolution was raised to the

---

[255]See various papers contained in Donald Gillies, ed., *Revolutions in Mathematics* (Oxford, 1992) and Elena Ausejo and Mariano Hormigón, eds., *Paradigms and Mathematics* (Madrid, 1996). The late Prof. Kuhn wrote to me "There must have been revolutions in mathematics" in his comment on my paper submitted to his course "Introduction to Philosophy of Science" of the academic year of 1976-1977. Previously, he had been skeptical of the possibility of applying his viewpoint of science to mathematics. It wouldn't be necessary to be pointed out that discussions should be made critically by defining the meaning of the word 'revolution' precisely. On the meaning of 'revolution', see Ilan Rachum, *"Revolution": The Entrance of a New Word into Western Political Discourse* (Lanham/New York/Oxford, 1999).

[256]"Leibniz to Hobbes, from Mainz, 13/23 Juy 1670," *The Correspondence of Thomas Hobbes*, edited by Noel Malcolm, Vol. II: 1660–1679 (Oxford, 1994), p. 716; *Die philosophische Schriften von G. W. Leibniz*, hrsg. von C. I. Gerhardt, Bd. 7 (Berlin, 1890), p. 572.

[257]The concept of 'paradigm' is taken, of course, from Thomas S. Kuhn's *Structure of Scientific Revolutions* (Chicago, 1962; $_2$1970; $_3$1996). Significantly, in his last years Kuhn rehabilitated and used the term 'paradigm' as 'hermeneutic bases'. See Thomas S. Kuhn, *The Road Since Structure: Philosophical Essays, 1970-1993, with an Autobiographical Interview*, edited by James Conant and John Haugeland (Chicago/London, 2000), p. 221.

level of Archimedean mathematics, the highest achievement of classical Greek mathematics and completed. Thus modern European mathematics took on a theoretically mature algebraically analytical form. With the mathematicians of the seventeenth century, the age of victorious analysis had begun.[258]

[258]We should keep in mind that algebraically analytic mathematics born in seventeenth century Europe was not accepted victoriously in European society in general. Its social acceptance must await the "analytical revolution" led mainly by Gaspard Monge and Sylvestre François Lacroix which tool place with the establishment of the École polytechnique soon after the French Revoution of 1789. This event was a mathematical version of the Second Scientitific Revolution. The appropriate naming of "anaytical revolution" was by Carl B. Boyer in his ***History of Analytic Geometry*** (n. 76), pp. 218, 256, and 268. This revolution in France was transmitted not only to England with the formation of 'Analytical Society' among students at the University of Cambridge, but to Germany and the United States of America. Its influence wasn't limited in the Western world. The mathematics introduced into China during the period of 'Westernization Movement' after the Opium War by the Protestant missionary Alexander Wylie can be fundamentally characterized as the Western mathematics after the 'analytical revolution'. In other words, Wylie published 数學啓蒙 *Shuxue qimeng* (Compendium of Arithmetic), an elementary treatise on Western arithmetic in written form, under his Chinese name 偉烈亞力 (Weilieyali) in Shanghai in 1853, 代数學 *Daishuxue* and 代微積拾級 *Daiweiji shiji*, both in collaboration with the Chinese mathematician 李善蘭 (Li Shanlan) in Shanghai in 1859. The second monograph was a Chinese translation of Augustus de Morgan's *Elements of Algebra* (London, 1935), and the third, the one of Elias Loomis's *Elements of Analytical Geometry and of the Differential and Integral Calculus* (New York, 1951). Accoding to the Japanese mathematician Teiji Takagi, Felix Klein "often spoke of three A, i. e., of Arithmetic, Algebra, and Analysis," as the main branches of nineteenth century mathematics in his lectures at the University of Gottingen around the year 1900 (see Shokichi Iyanaga, "On the life and work of Teiji Takagi," in Teiji Takagi, *Collected Papers*, Second enlarged edition, ed. by S. Iyanaga, K. Iwasawa, K. Kodaira, and K. Yosida, Tokyo/Berlin/Heidelberg, 1990, p. 356). Cf. Felix Klein, *Vorlesungen über die Entwicklung der Mathematik im 19. Jahrhundert*, Teil 1 (1926; Berlin/Heidelberg/New York, 1979), p. 25. In considering this episode, the three main disciplines of nineteenth-century mathematics, arithmetic, algebra, and analysis, were transmitted into East Asia by the effort of Wylie.

It should be mentioned that in his plan to arrange books in a libray Leibniz referred to "*Mathesis generalis* de quantitate in universum, seu calculo, ubi Arithmetica, Algebra, analytica infinitesimalis" next to "Mathematica communis" (Leibniz, *Opera omnia*, studio Ludovici Dutens, Vol. 5: *Opera philologica*, 1768; Hildesheim, 1989, p. 211). As "mathesis generalis" can be identified with 'mathesis universalis', Leibniz here seems to have thought arithmetic, algebra, and (infinitesimal) analysis as the essential constutients of 'mathesis universalis'. Thus, in this sense, we may regard Leibniz rather than Descartes as a pioneer of non-intuitive mathematics after the nineteenth century.

The three Chinese books by Wylie were introduced into Japan. *Shuxue qiment* was transformed into a Japanese textbook 筆算訓蒙 *Hitsuzan Kunmō* published in 1869 by Tsukamoto Akitake, who studied at the Nagasaki Kaigun Denshshujo Institute. Tsukamoto also translated the first three chapters of *Daishuxue* into Japanese and published it in 1872. Finally, *Daiweiji shiji* was copied and studied by Kanda Takahira, who become professor of mathematics of *Yōsho Shirabesho* (Institute for Investigating Western Books), whose predecessor was *Bansho Shirabesho* (Institute for Investigating Foreign [Barbarian] Books, established in 1856). Ōmura Kazuhide, a practitioner of traditional Japanese mathematics, translated it into Japanese, but remained unpublished. Kanda's and Ōmura's autograph copies are preserved at the Library of Tohoku University in Sendai. Fukuda Han translated it in part into Japanese and published it in Tokyo in 1872. These became bases for later publications of Western mathematics textbooks in modern Japan.

INTERIM CONSIDERATION

Descartes and the Beginnings of Mathematicism in Modern Thought

Descartes was historically important not only as a mathematician but also as a philosopher who thought exceptionally highly of mathematics. What we have delineated in Part I is how he constructed his method of algebraic analysis as it appears in its fully developed form in the *Géométrie* of 1637. He considered this novel manner of analysis to occupy a central place in his universal method for discovering certain knowledge. This method played a crucial role in his philosophy and in his program for reforming the traditional sciences.

Our intention in the interim consideration is to construct a bridge between Chapter 4 (especially § 3) and Part II, in other words, between mathematics and the philosophy of mathematics. Let us begin our discussion by remembering what Descartes contended in his *Regulae ad directionem ingenii*. In the medieval and Renaissance tradition of learning, dialectic or logic was regarded as the most important discipline (just reflect upon Peter of Spain, Peter Ramus, Alessandro Piccolomini, and Piccolomini's followers). Descartes disdained dialectic and esteemed mathematics. He praised the latter discipline for its certainty. In his *Regulae*, he insisted on not confining the objects of study to pure mathematics, but also on showing an interest to things which admit of as much certainty as the demonstrations of mathematics.

The establishment of the criterion for certainty implied to Descartes the expulsion of dialectic from its primary position in the general scheme of learning. In the *Regulae* he most mercilessly criticized dialectic as futile. In Rule II, he stated that "those chains with which dialecticians suppose they regulate human reason seem to be of little use" (see p. 178). In Rule IV he went so far as to say that mental operations of dialectic "are of no use here, or rather should be reckoned a positive hindrance, for nothing can be added to the clear light of reason which does not in some way dim it."[1] Moreover, he put it in Rule X as

[1] *AT*, X, pp. 372–373

follows:

> To make it even clearer that the aforementioned art of reasoning (*disserendi ars*) contributes nothing whatever to knowledge of the truth, we should realize that, on the basis of their method, dialecticians are unable to formulate a syllogism with a true conclusion unless they are already in possession of the substance of the conclusion, i. e. unless they have previous knowledge of the very truth deduced in the syllogism. It is obvious therefore that they themselves can learn nothing new from such forms of reasoning, and hence that ordinary dialectic is of no use whatever to those who wish to investigate the truth of things. Its sole advantage is that it sometimes enables us to explain to others arguments which are already known. It should therefore be transferred from philosophy to rhetoric.[2]

The decay of dialogue which Walter J. Ong believes began with Ramus found its completion in Descartes.

The title of Father Ong's monograph on Ramus published in 1958 was *Ramus, Method, and the Decay of Dialogue: From the Art of Discourse to the Art of Reason.*[3] The highest discipline for Ramus was still dialectic, whereas the art of reason for Descartes was not dialectic any longer, but pure mathematics *par excellence.* That is to say, pure mathematics occupied the central place in Descartes's early system of philosophy. As we have seen in Chapters 4 and 5, the method of algebraic analysis played a crucial role for him in the formation of true mathematics. In the latter half of Rule IV of his *Regulae*, he explained that 'mathesis universalis', a mathematical discipline, had functioned as the normative paradigm for developing and refining his general method. In this sense the Descartes of the *Regulae* placed mathematics in the center of his system of philosophy, and thus his thought has been at times characterized as a kind of 'mathematicism'. Étienne Gilson, for example, entitled a chapter on Descartes's philosophy in his book *The Unity of Philosophical Experience* (1937), "Cartesian Mathematicism," simply identifying 'method' with "Universal Mathematics."[4] Jean-Louis Allard followed Gilson's precedent by giving the

---

[2] *AT*, X, p. 406

[3] Walter J. Ong, published from Harvard University Press in 1958.

[4] E. Gilson, *The Unity of Philosophical Experience* (New York, 1937), Chapter V, "Cartesian Mathematicism," pp. 125–151. See especially pp. 142–143, where Gilson states: "Let us, with Descartes himself, call that method "Universal Mathematics"; it certainly was universal, but could it still be called mathematics? Descartes felt sure it could, because he was aiming at a complete liberation of knowledge from its objects." Gilson seems to have thought that Descartes had attained his idea of "Universal Mathematics" as early as 1618–1619. See Gilson, *Études sur la rôle de la pensée médiévale dans la formation du système cartésien* (Paris, $_3$1967), pp. 199–200: "Après avoir heureusement résolu quelques problèmes en 1618, il conçoit le projet d'une méthode universelle qui permettrait de résoudre tous les problèmes de géométrie quels qu'ils soient. C'est pour élaborer cette méthode qu'il s'arréte dans le poële d'Allemagne pendant l'hiver de 1619 et, par une nouvelle généralisation, il s'élève à l'i'dée de

title *Le Mathématisme de Descartes* to a monograph published in 1963.[5]

In his book, Gilson compared Clavius's mathematical thought with that of Descartes. According to him, Clavius argued as follows:

> Necessary knowledge is better than mere probability; mathematical knowledge alone is necessary; mathematical knowledge is better than all other knowledge. That was not original, but it was true. The young Descartes was following a much more risky way: true knowledge is necessary; mathematical knowledge alone is necessary; hence all knowledge has to be mathematical.[6]

Is this statement justifiable? Before answering to this, let us note Gilson's definition of 'mathematicism':

> Everything will be mathematically proved in your philosophy, save only this, that everything can, and must be, mathematically proved.
>
> There, at any rate, lies the deepest root of the Cartesian philosophy. If anything can be truly said to express its innermost spirit, it is what I venture to call "Mathematicism", for Descartes' philosophy was nothing else than a recklessly conducted experience to see what becomes of human knowledge when moulded into conformity with the pattern of mathematical evidence.[7]

'Mathematicism' ('mathématisme' in French) admittedly has a variety of connotations. The *Oxford English Dictionary* defines 'mathematicism' to be "the opinion that everything can be described ultimately in mathematical terms, or that the universe is fundamentally mathematical."[8] And according to *Le Grand Robert de la langue française*, 'mathèmatisme' means first of all a "tendance à faire prévaloir les méthodes mathématiques, dans un domaine quelconque."[9]

We believe that Descartes's philosophy may be called a kind of 'mathematicism' in some sense. In the *Regulae*, in fact he asserted that any reliable knowledge should obtain the certainty of mathematical demonstrations and stated that 'mathesis universalis' was employed to form his universal method. However, it seems to us quite doubtful that some of Gilson's implications are correct. For example, did Descartes actually think that "all knowledge has to be mathematical," as Gilson believed Descartes did? Descartes did not iden-

---

l'unité des sciences et de l'universalité absolue de méthode mathématique."

[5] J.-L. Allard, *Le Mathématisme de Descartes* (Ottawa, 1963). Allard clearly states that his work has succeeded Gilson's book *The Unity of Philosophical Experience*. See p. 7, note 2.

[6] Gilson, *The Unity of Philosophical Experience*, p. 140.

[7] *Ibid.*, p. 133.

[8] *A Supplement to the Oxford English Dictionary*, ed. by R. W. Burchfield (Oxford, 1976), p. 859.

[9] *Le Grand Robert de la langue française*, Deuxième éd. entièrement revue et enrichie par Alain Rey, t. VI (Paris, 1985), p. 303.

tify (*pace* Gilson) his universal method with 'mathesis universalis.' He simply regarded the concept of 'mathesis universalis' as a model for his method. That is to say, the Descartes described by Gilson is different from the historical Descartes.

More importantly, the youthful Descartes of the *Regulae* was different from the Descartes after the *Discours de la méthode*. It must be admitted that in his *Regulae* Descartes considered mathematical demonstration to be the model for certainty. But in Part Four of his *Discours de la méthode* of 1637, he revealed his new methodological position which was slightly different from that in the *Regulae*. There he asserted:

> [S]ince I now wished to devote myself solely to the search for truth, I thought it necessary to do the very opposite and reject as if absolutely false everything in which I could imagine the least doubt, in order to see if I was left believing anything that was entirely indubitable. Thus, because our senses sometimes deceive us, I decided to suppose that nothing was such as they led us to imagine. And since there are men who made mistakes in reasoning, committing logical fallacies concerning the simplest questions in geometry, and because I judged that I was prone to error as anyone else, I rejected as unsound all the arguments I had previously taken as demonstrative proofs. Lastly, considering that the very thoughts we have while awake may also occur while we sleep without any of them being at the that time true, I resolved to pretend that all the things that had ever entered my mind were no more true than the illusions of my dreams. But immediately I noticed that while I was trying thus to think everything false, it was necessary that I, who was thinking this, was something. And observing that this truth 'I am thinking, therefore I exist' (*ie pense, donc ie suis*) was so firm and sure all the most extravagant suppositions of the sceptics were incapable of shaking it, I decided that I could accept it without scruple as the first principle of the philosophy I was seeking.[10]

A significant change is stated here: the epistemological transformation from the viewpoint in which mathematical demonstration is considered to be the paradigm for certainty to one that regards the statement "Cogito, ergo sum" to be the first principle. Descartes did not necessarily deny the certainty of mathematical demonstrations. But at least once he "rejected as unsound all the arguments I had previously taken as demonstrative proofs." Instead of the older criterion of certainty, he adopted his new first principle "Cogito, ergo sum." This means that something important happened in the later Descartes. Can we still characterize Descartes's later philosophy as a kind of 'mathematicism'? We can at least answer to this question in the affirmative, because after all he

[10] *AT*, VI, pp. 31–32.

tried to restore the authority of the certainty of mathematical demonstrations through complicated metaphysical justification. In any case, what has been verified is that the problem of how Descartes thought about mathematics is very subtle. In any case, a historical and philosophical critical examination of Descartes's philosophy of mathematics is absolutely necessary. If it is just to criticize Descartes's view of mathematics, the target of our criticism must not be a fabricated Descartes, but the historical Descartes.

As we saw in an earlier part of our present considerations, Descartes's thought can be justly characterized as a kind of 'mathematicism'. To determine critically what kind of mathematicism Descartes supported, we will confine our strategy and tactics to a historical examination of 'mathesis universalis', a key notion in the latter part of Rule IV. Recent scholarship has shown that this idea can be traced back to some thinkers in the sixteenth century such as Adriaan van Roomen. Van Roomen in turn referred to Aristotle's concept of 'prima philosophia' so as to legitimize his notion of 'mathesis universalis', 'prima mathesis' (see Chapter 5, § 3, D and Chaper 7, § 2). So we will begin our historical examination with Aristotle.

There is another important reason why we should analyze Aristotle's texts. It was ultimately Aristotle's scheme of learning that Descartes aimed at replacing with his own. While the question of whether or not some notion of Aristotle influenced Descartes might be of minor imporance, undoubedly crucial is an examination of Descartes's 'mathesis universalis' against the background of traditional Aristotelian philosophy.

Following this we shall discuss some mathematical notions similar to 'mathesis universalis' which spread during the sixteenth century (Chaper 7). After that we shall examine how Descartes's philosophy of mathematics was transformed from the *Regulae* to the *Discours de la méthode*, *Meditationes*, and *Principia philosophiae* (Chapter 8, § 1). Next we will proceed to Leibniz, who considerably broadened the concept of 'mathesis universalis'. It was his systematized idea of 'mathesis universalis' which came to dominate a certain range of modern mathematical thought.

The concept of 'mathesis universalis' has been sometimes viewed by its proponents as the first discipline in the hierarchy of the mathematical sciences, so that it may be compared in a sense to today's foundations of mathematics. It should not be misunderstood as, for example, a universal method or science which is capable of covering any area of knowledge whatever. Just as the foundations of mathematics in the twentieth and twenty-first centuries is closely connected with both the mathematical disciplines proper and the philosophy of mathematics, so 'mathesis universalis' can be considered a significant notion which can reveal to us how its users thought of mathematics and its philosophy.

Our historical and philosophical study of 'mathesis universalis' in Part II will enable us to speculate upon the mathematico-philosophical significance of Descartes's mathematical thought more deeply than in Part I.

Part II

The Concept of 'Mathesis Universalis' in Historical Perspective

"[...] mesme que ces veritez qu'on nomme eternelles, comme que *totum est majus sua parte*, etc., ni seroient point veritez, si Dieu ne l'avoit ainsi etably, [...]."

—— Descartes à Mersenne, 17/27 mai 1638.

Chapter 6

'Universal Mathematics' in Aristotle

§ 1. Aristotle's *Metaphysics* and *Posterior Analytics*

In *Mathematics in Aristotle*, an anthology of Aristotle's passages on mathematics compiled with commentary by Thomas Little Heath and published posthumously in 1949, we encounter in the Chapter on *Metaphysics* a short section with the title "'Universal' Mathematics."[1] Jean-Luc Marion also refers to essentially the same passages as possible sources for Descartes's 'mathesis universalis' in the annotations to his French translation of the *Regulae ad directionem ingenii*.[2] Whether or not Descartes knew these passages of Aristotle, we should examine those which Heath thought to refer to 'universal mathematics.' The passages quoted by Heath can be classified into two groups: 'universal mathematics' as a science and 'universal propositions'. Let us begin with the first category.

According to Heath, two passages, Book VI(E), Chapter 1, 1026a23–27 and Book XI(K), Chapter 7, 1064b8–9 in the Bekker pagination, refer to 'universal mathematics'. In a paragraph of Book VI(E) of his *Metaphysics*, Aristotle discusses the three theoretical sciences, mathematics (μαθηματική), physics (φυσική) and theology (θεολογική), and presents his opinion as to which is the highest, or most universal science of the three. After asserting that the highest science deals with the highest genus and that theology is supreme among the theoretical sciences, he states in passage 1026a23–32 which includes 1026a23–27:

> One might indeed raise the question whether first philosophy is universal, or deals with one genus, i. e. some kind of being; for not even the mathematical sciences are all alike in this respect—geometry and astronomy deal with a certain particular nature, while the universal (ἡ καθόλου) applies alike to all. We answer that if there is no

[1] Thomas L. Heath, *Mathematics in Aristotle* (Oxford, 1949), pp. 222–224.
[2] René Descartes, *Règles utiles et claires etc.* (Ch. 4, n. 97), p. 163.

> substance other than those which are formed from nature, physics will be the first science; but if there is an immobile substance, the science of this must be prior and must be first philosophy, and universal in this way, because it is first. And it will belong to this to consider being *qua* being—both what it is and the attributes which belong to it *qua* being.[3]

Here Aristotle discusses whether first philosophy or metaphysics is universal in comparison to physics and the mathematical sciences. Concerning the mathematical sciences he observes that on the one hand both geometry and astronomy are special sciences which concern themselves with a particular nature; on the other hand "the universal" (*ἡ καθόλου*) is common to all. After arguing about mathematics, he proceeds to assert that the science of immobile substance, that is, metaphysics or theology, is first and universal. What this science considers is being *qua* being. Aristotle's sentences are in general, as is well known, very elliptical. And "the universal" in the quoted passage is today in most cases paraphrased as "universal mathematics." We will examine the problem of this translation below.

Heath also cites another passage on 'universal mathematics' from Book XI(K) which is very similar to the previous one. This similarity is usually considered to be natural because Book XI(K), Chapters 7 and 8, is a recapitulation of Book VI(E).[4] In 1064b6–9 Aristotle puts it:

> One might raise the quesion whether the science of being *qua* being is to be regarded as universal or not. Each of the mathematical sciences deals with some one class of things, but the universal (*ἡ καθόλου*) applies alike to all.[5]

Again Aristotle states that on the one hand each of the mathematical sciences is

---

[3]Aristotle, ***Metaphysica***: ***Opera***, ed. Immanuel Bekker, Academia regia Borussica, Vol. II (Berlin, 1931), 1026a23–32:"*ἀπορήσειε γὰρ ἄν τις πό-|τερόν ποθ'ἡ πρώτη φιλοσοφία καθόλου ἐστίν, ἤ περί τι γέ-|νος καὶ φύσιν τινὰ μίαν. οὐ γὰρ ὁ αὐτὸς τρόπος οὐδ' ἐν| τρόπος οὐδ' ἐν| ταῖς μαθηματικαῖς, ἀλλ' ἡ μὲν γεωμετρία καὶ ἀστρολογία| περί τινα φύσιν εἰσίν, ἡ δὲ καθόλου πασῶν κοινή. εἰ μὲν| οὖν μὴ ἐστί τις ἑτέρα οὐσία παρὰ τὰς φύσει συνεστηκυίας, ἡ| φυσικὴ ἂν εἴη πρώτη ἐπιστήμη· εἰ δ'ἐστί τις οὐσία ἀκίνητος, |αὕτη προτέρα καὶ φιλοσοφία πρώτη, καὶ καθόλου οὕτως| ὅτι πρώτη· καὶ περὶ τοῦ ὄντος ᾗ ὄν, ταύτης ἂν εἴη θεωρῆσαι,| καὶ τί ἐστι καὶ τὰ ὑπάρχοντα ᾗ ὄν.*" A variant of the line 27 reads "*ἐκείνη*" instead of "*ἡ*". The vertical line '|' stands for the demarcation between lines in the Bekker edition. For the texts of Aristotle, I basically refer to the English translation in Jonathan Barnes, ed., ***The Complete Works of Aristotle: The Revised Oxford Translation***, Vols. 1 and 2 (Princeton, 1984), but in many cases I have modified it according to the Greek original.

[4]See, for example, Giovanni Reale, ***The Concept of First Philosophy and the Unity of the Metaphysics of Aristotle***, edited and translated by John R. Catan (Albany, 1980), pp. 270–276.

[5]The original Greek is as follows (1064b6–9): "*ἀπορήσειε δ' ἄν τις πό-|τερόν ποτε τὴν τοῦ ὄντος ᾗ ὂν ἐπιστήμην καθόλου δεῖ θεῖναι ἢ| οὔ. τῶν μὲν γὰρ μαθηματικῶν ἑκάστη περὶ ἕν τι γένος ἀφω-|ρισμένον ἐστίν, ἡ δὲ καθόλου κοινὴ περὶ πάντων.*"

a special science; on the other hand "the universal" is common to all. Similarly today "the universal" is mostly interpreted as "universal mathematics."

Two passage in the *Metaphysics* belong to the second category on universal propositions in mathematics, 1077a9–10 and 1077b17–22 in Book XIII(M). We quote 1077a9–12 (including 1077a9–10) first:

> Again, there are certain propositions [proved] universally (*καθόλου*) by mathematicians, extending beyond these substances. Here then we shall have another substance intermediate between, and separate from, the Ideas and the intermediates—a substance which is neither number nor points nor spatial magnitude nor time.[6]

And passage 1077b17–22 reads as follows:

> Just as the universals (*τὰ καθόλου*) in mathematics deal not with objects which exist separately, apart from magnitudes and from numbers, but with magnitudes and numbers, not however *qua* such as to have magnitude or to be divisible, clearly it is possible that there should also be both statements and demonstrations about sensible magnitudes, not however *qua* sensible but *qua* possessed of certain definitive attributes.[7]

In the two passages quoted above, Aristotle makes his observation that there are certain mathematical propositions of a universal character which are not restricted to such substances as numbers (objects of arithmetic) and magnitudes (objects of geometry). So if it is correct to interpret "the universal" in passages 1026a23–32 and 1064b6–9 as "universal mathematics," the universal propositions in the second group can be related in a certain manner to the notion of 'universal mathematics'.

After presenting those passages we have shown above, the compiler of *Mathematics in Aristotle* writes an interesting comment on universal propositions:

> In speaking of universal propositions in mathematics applying to a wider class of entities than those of magnitudes, numbers, and the objects of particular mathematical sciences, Aristotle no doubt had in mind such propositions as that in which it is proved that, if four terms are in proportion, they are also proportional *alternando*. Aristotle pointed out that this proposition used to be proved separately

[6]1077a9–12: "*ἔτι γράφεται ἔνια καθόλου ὑπὸ τῶν μαθηματι-|κῶν παρὰ ταύτας τὰς οὐσίας. ἔσται οὖν καὶ αὕτη τις ἄλλη| οὐσία μεταξὺ κεχωρισμένη τῶν τ' ἰδεῶν καὶ τῶν μεταξύ, ἣ| ὄυτε ἀριθμός ἐστιν οὔτε στιγμαὶ οὔτε μέγεθος οὔτε χρόνος.*"

[7]1077b17–22: "*ὥσπερ γὰρ καὶ τὰ κα-|θόλου ἐν τοῖς μαθήμασιν οὐ περὶ κεχωρισμένων ἐστὶ παρὰ| τὰ μεγέθη καὶ τοὺς ἀριθμούς, ἀλλὰ περὶ τούτων μέν, οὐχ ᾗ| δὲ τοιαῦτα οἷα ἔχειν μέγεθος ἢ εἶναι διαιρετά, δῆλον ὅτι ἐνδέχεται καὶ τερὶ τῶν αἰσθητῶν μεγεθῶν εἶναι καὶ λόγους| καὶ ἀποδείξεις, μὴ ᾗ δὲ αἰσθητά, ἀλλ' ᾗ τοιαδί.*"

> for numbers, lines, solids, and times, 'but is now proved universally for all.' The reference is obviously to the proof forming part of Eudoxus' new theory of proportion. What Aristotle would have called the more general category of things to which the proof would apply does not appear, but it seems most likely that it would have been poso/n, quantity. What was the word used by Eudoxus in his proposition we do not know; Euclid, who reproduced Eudoxus' theory, speaks of 'magnitudes'.[8]

Heath contends that propositions proved universally are exemplified by Eudoxus's theory of proportion, which is embodied in Book V of Euclid's *Elements*. Euclid's Book V is, in Heath's opinion, not confined to geometrical magnitudes but concerned with the more general 'ποσόν' (quantity).

In interpreting Aristotle's statements on universal propositions in the *Metaphysics*, Heath refers the reader to a similar passage in Book I (A), Chapter 5 of the *Posterior Analytics*. In order to understand fully Aristotle's explanation in the *Metaphysics*, let us examine the passage in the *Posterior Analytics* in which the Peripatetic philosopher discusses 'alternate ratio.' According to Euclid's definition in Book V, " 'alternate ratio' (ἐναλλάξ λόγος) means taking the antecedent in relation to the antecedent and the consequent in relation to the consequent."[9] That is, if the two ratios (a, b) and (c, d) are given symbolically, the procedure of 'alternando' amounts to making the ratios (a, c) and (b, d). In 74a17–25, Aristotle observes that in proportion we can take terms alternately:

> And it might seem that proportion alternates for things as numbers and as lines and as solids and as times—as once it used to be proved separately, though it is possible for it to be proved of all cases by a single demonstration. But because all these things—numbers, lengths, times, solids—do not constitute a single named item and differ in sort from one another, it used to be taken separately. But now it is proved universally (καθόλου); for it did not belong to things as lines or as numbers, but as this which they suppose to belong universally.[10]

---

[8]Heath, *Op. cit.* (n. 1), p. 223.

[9]Heath, *The Thirteen Books of Euclid's Elements*, Vol. 2, pp. 114 & 134

[10]Aristotle, *Analytica posteriora*: *Opera*, Vol. I, ed. I. Bekker (Berlin, 1831), 74a17–25: *"καὶ| τὸ ἀνάλογον ὅτι ἐναλλάξ, ᾗ ἀριθμοὶ καὶ ᾗ γραμμαὶ καὶ| ᾗ στερεὰ καὶ ᾗ χρόνοι, ὥσπερ ἐδείκνυτό ποτε χωρίς. ἐνδε-|χόμενόν γε κατὰ πάντων, μιᾷ ἀποδείξει δειχθῆναι· ἀλλὰ| διὰ τὸ μὴ εἶναι ὠνομασμένον τι πάντα ταῦτα ἕν. ἀριθμοί| μέκη χρόνος στερεά, καὶ εἴδει διαφέρειν ἀλλήλων, χωρὶς| ἐλαμβάνετο. νῦν δὲ καθόλου δείκνυται· οὐ γὰρ ᾗ γραμμαὶ| ἢ ᾗ ἀριθμοὶ ὑπῆρχεν, ἀλλ' ᾗ τοδί, ὃ καθόλου ὑποτίθενται|ὑπάρχειν."* Cf. Heath, *Mathematics in Aristotle*, pp. 41–44.

Aristotle's topic here is the theory of proportion. The proposition that we can transform a proportion *alternando* was at one time proved separately for different substances such as numbers and lengths but is now proved universally for a particular character "which they suppose to belong universally." Noting this universal character of the theory of proportion, Heath emphasizes that that theory which is commonly ascribed to Eudoxus is a model for the Aristotelian statement on universal propositions in mathematics.[11]

Heath not only sees a model for universal propositions in the so-called general theory of proportion, which applies to all quantities alike. More noteworthy is that he views the best analogy to the above mentioned 'universal mathematics' to be the symbolic algebra of the moderns. He comments upon the passages related to 'universal mathematics' in the *Metaphysics*:

> Perhaps the best analogy for the sort of 'general' or 'universal' mathematics which Aristotle seems to have had in mind is our *algebra* which supplements arithmetical notation and operations by the use of general symbols for things known or unknown, and for operations. (Even so, the things denoted by the symbols are generally spoken of as 'quantities'.) But algebra in our sense was impossible for the Greeks in Aristotle's time, because no symbols had been invented, and such problems as are equivalent to the algebraical solution of quadratic or cubic equations the Greeks could only solve by geometry[12] [The emphasis is in the original].

Such a statement is appropriate for the eminent historian of Greek mathematics who, following Tannery and Zeuthen, systematically reinterpreted a number of geometrical propositions in terms of the so-called 'geometrical algebra'. Still, the legitimacy of Heath's explanation may be questioned from a historical point of view, for in the latter part of the cited passage, he asserts that symbolic algebra was impossible in classical Greece, while in the former part he looks at the mathematics in Aristotle from a standpoint which became historically possible for the first time only with the establishment of the algebraically analytic tradition in the late sixteenth and early seventeenth centuries. This seems to cause a certain conceptual confusion in his explanation.

But what did Aristotle really imply by 'universal mathematics'? There is an alternative interpretation which was first proposed by Hermann Bonitz. Bonitz thought that 'universal mathematics' (*ἡ καθόλου μαθηματική*) refers to arithmetic on the grounds of Aristotle's statement in 982a28 of the *Metaphysics*, Book I(A), Chapter 2 that arithmetic is more exact (*ἀκριβής*) than geometry.[13] According to Bonitz, Aristotle stated this because those disciplines

[11] Heath, *Mathematics in Aristotle*, p. 43.

[12] *Ibid.*, pp. 223–224.

[13] Hermann Bonitz, *Aristotelis Metaphysica: Commentarius* (Bonn, 1849), p. 285: "Arithmeticam didi ab Aristotele *τὴν καθόλου μαθηματικήν* probabile est conferenti A 2. 982a28."

which involve fewer principles are more exact (and universal) than those which involve additional principles.[14]

Giovanni Reale supported Bonitz's interpretation. To corroborate his explanation, Reale referred to 1004a2–9 of Books IV (Γ), Chapter 2. There Aristotle compared mathematics with philosophy, stating that " 'philosopher' is like 'mathematician'; for mathematics also has parts, and there is a first science and a second science and other successive ones within the sphere of mathematics."[15] Reale clearly identified arithmetic with 'first mathematics' and then 'universal mathematics'. Thus he puts it:

> The universal mathematical science, or as Γ.2 says, first mathematics, is "universal" in the sense that it has as its object of inquiry numbers, their laws, and their fundamental relations, and these are the *foundation* and condition of all the other mathematical sciences, which in spite of the diversity of their objects, are constructed only on the basis of numbers and numerical calculation.[16] [The emphasis is in the original.]

But there is a problem in Bonitz's and Reale's explanation. Reinterpreting 982a28, W. D. Ross contended that "arithmetic is a science alongside of geometry though more accurate" and he thought that as 'universal mathematics' "Aristotle contemplates a science wider than either arithmetic and geometry."[17] A specimen of that universal science is, according to him, Book V of Euclid's *Elements*, "of proportion as existing between any kind of magnitude."[18] It seems to us that Ross's interpretation of 982a28 is basically correct but his regarding Euclid V as an exemplar of 'universal mathematics' is doubtful, as we have suggested above in Chapter 5, § 3, D (p. 263) and when criticizing Heath (p. 293).

We have so far examined the translations which render "the universal" (*ἡ καθόλου*) as "universal mathematics." Nevertheless, the translators and commentators of the works of Aristotle have not all agreed with this rendering. For example, Charles B. Crowley has recently contended that "the universal" in 1026a27 and 1064b9 of the Aristotelian text should be translated not as "universal mathematics" but as "universal philosophy," that is to say, "first philosophy" (metaphysics). Further he has claimed support for his interpretation from works of Aristotle and Thomas Aquinas:

[14] Aristotle, *Metaphysica*, 982a26–28.

[15] 1004a6–9: "*ἔστι γὰρό φιλόσοφος| ὥσπερό μαθηματικὸς λεγόμενος· καὶ γὰρ αὕτη ἔχει| μέρη, καὶ πρώτη τις καὶ δευτέρα ἔστιν ἐπιστήμη καὶ ἄλλαι| ἐφεξῆς ἐν τοῖς μαθήμασιν.*"

[16] Giovanni Reale, *Op. cit.* (n. 4), p. 171.

[17] William David Ross, *Aristotle's Metaphysics*, A Revised Text with Introduction and Commentary, Vol. 1 (Oxford, 1924), pp. 356–357. Incidentally, Bertrand Dumoulin, who interprets "the universal" of 1026a27 as "Mathématique générale," generally agrees with Ross's explanation. See his *Analyse génétique de la Métaphysique d'Aristote* (Montréal/Paris, 1986), p. 142: "L'interpretation de Ross à propos de la 《Mathématique générale》 nous paraît convaincante."

[18] Ross, *Aristotle's Metaphysics*, p. 357.

> They [Aristotle and Thomas Aquinas] did not maintain that in mathematics there is prior to arithmetic and geometry a universal mathematical science of quantity in communi, and of the universal propositions or theorems of mathematics.[19]

In Crowley's opinion, "arithmetic and geometry are the two ultimate mathematical sciences of the ultimate mathematical measures in communi. There is no other universal mathematical science prior to them."[20] Crowley's solution might be a surprise to those who believe that there is a concept which can be labeled 'universal mathematics' in the Aristotelian corpus. An essential problem in his insightful interpretation is that he read the texts of Aristotle within the framework of medieval scholastic Aristotelianism, of which a representative philosopher was Thomas Aquinas.

Moreover, there have been scholars who doubted the genuineness, for instance, of passage 1026a23–32. Werner Jaeger for one regarded the whole passage as an annotation disconnected from the main context and thought it a later addition by an Aristotelian disciple.[21] It is well-known that the Jaegerian interpretation of Aristotle has generally emphasized the incoherency and changing variety of the *Metaphysics*. We basically agree with Reale in that the Aristotelian *Metaphysics* permits a unitary and consistent interpretation. Particularly, 1026a23–32 can be considered a genuine Aristotelian passage.[22] Even if Jaeger should be right, however, the passages have been assumed, historically at least, to be Aristotle's own sentences. So there is a certain meaning in continuing our consideration of those passages of the *Metaphysics* which we treat to be Aristotle's own. For the problem of translation and transmission of scientific texts in general, it is important to ask how the Aristotelian corpus as a whole was translated and developed. "The Aristotle we have today, the one that existed since the beginning, is a classroom assembly rather than a textbook. He is a loose commonwealth of thought and writing that has changed, irretrievably, over time."[23] Therefore, there is meaning for us to continue to consider the passages in the *Metaphysics* which have been treated as Aristotle's own.

What did Aristotle actually state in the passages we have presented? Certainly to answer this question is important. But in the present context it is more significant to ask how they had been interpreted up to Descartes's age. For this

---

[19] Charles B. Crowley, *Universal Mathematics in Aristotelian-Thomistic Philosophy: The Hermeneutics of Aristotelian Texts Relative to Universal Mathematics* (Washington, D. C., 1980), Introduction, p. xvii.

[20] *Ibid.*, p. 200.

[21] W. Jaeger, *Aristotle: Fundamentals of the History of His Development*, translated with the Author's Corrections and Additions by Richard Robinson (Oxford, $_2$1948), pp. 217–219. Jaeger states that this was also pointed out by H. Bonitz.

[22] Reale, pp. 170–171.

[23] Scott L. Montgomery, *Science in Translation: Movement of Knowledge through Cultures and Time* (Chicago/London, 2000), p. 9.

reason, we will delineate how ancient, medieval, and Renaissance translators and commentators interpreted "the universal" of 1026a27 and 1064b9. Such an examination will help us understand how the idea which Aristotle is thought to have conceived as 'universal mathematics' evolved till the Renaissance and in the end how it compares to Descartes's 'mathesis universalis'.

§ 2. Greek Commentators: Alexander of Aphrodisias and Asclepius of Tralles

What did Aristotle mean by "the universal" (*ἡ καθόλου*) in the sentences we are concerned with? To ask this question is important in order to understand what status mathematics occupies in the hierarchy of the Aristotelian sciences. In Section 1, we have shown that modern scholars have debated the interpretation of the Aristotelian concept of 'universal mathematics', which is thought to appear in 1026a27 and 1064b9. Among Greek commentators of Aristotle, at least two have left illuminating scholia specifically on 1026a27. One is Alexander of Aphrodisias and the other is Asclepius of Tralles. Their interpretations of the Aristotelian passage containing 1026a27 of the *Metaphysics* is found in the *Commentaria in Aristotelem Graeca*,[24] published by the Berlin Academy of Sciences between 1882 and 1907.

Alexander of Aphrodisias, who lived at the end of the second and beginning of the third century A.D., is known to have endeavored to interpret Aristotle genuinely, freeing the Peripatetic philosophy from the syncretism of the previous period. Because of the excellence of his commentaries on the writings of Aristotle, he was called the second Aristotle or "*ὁ ἐξηγητής*" (the Interpreter). Before Ibn Rushd (Latin name, Averroës) appeared about one thousand years later, he was considered to be the best authority for the interpretation of Aristotle. He wrote commentaries on the *Metaphysics*, of which those on Books I(A)–V(Δ) are credited to himself but those on the other books are thought to have been composed by someone else, most probably Michael of Ephesus of the eleventh century, although they may contain some genuine passages.[25] So the author of the commentary on Book VI(E), which is the most important to our present study, is usually referred to as Pseudo-Alexander. For the sake of

[24]See Richard Sorabji, "The Ancient Commentators on Aristotle" and Karl Praechter, "Review of the *Commentaria in Aristotelem Graeca*" in Richard Sorabji, ed., *Aristotle Transformed: The Ancient Commentators and Their Influence* (Ithaca, 1990), pp. 1–30 & 31–54.

[25]Paul Moraux, *Alexandre d'Aphrodise: Exégète de la Noétique d'Aristote* (Liège /Paris, 1942), pp. 14–19. Cf. Pierre Thillet, "Alexandre d'Aphrodise, IIe–IIIe siècle," *Dictionnaire des philosophes*, Directeur de la publication Denis Huisman, t. I (Paris, 1984),p. 46; K. Praechter, *Op. cit.* (n. 24), p. 51; Philip Merlan, "Alexander of Aphrodisias," *DSB*, I, pp. 117–120. We have an English translation of Alexander of Aphrodisias's commentary on Books I-V of Aristotle's *Metaphysics: On Aristotle's* Metaphysics 1, 2 & 3, 4, 5, translated by W. E. Dooney and Arthur Madigan, in 4 Volumes (Ithaca, 1989–93).

simplicity, however, we will call the commentator(s), Alexander and Pseudo-Alexander of all of the books of Aristotle's *Metaphysics* simply 'Alexander'.

The most authoritative ancient commentator on Aristotle writes about 1026a23–32 as follows:

> One might very well ask whether first philosophy is so universal as to contain everything, and these be under it in the same way as the parts are under the universal; or, it is not the universal, but like medicine and astronomy and the rest, it is also a part of being and nature partitioned off from the theoreticals. Since this is also found [to be the case] in the mathematical sciences—where geometry is about magnitudes, whereas astronomy is about the stars, and the others with other things, and mathematics simply is common to all [of them]—what would then keep physics from being about one part of being and mathematics about another part, and theology to be common to all together?[26]

First of all, our attention should be drawn to the fact that Alexander here clearly interprets "the universal" as "mathematics simply (or unqualifiedly)" (*ἡ ἁπλῶς μαθηματική*). "Mathematics simply" in turn can be rephrased "mathematics as a whole," embracing the various mathematical disciplines such as arithmetic, geometry, music (harmonics), and astronomy. Moreover, Alexander seems to take medicine and astronomy as a part of two more general theoretical sciences, i. e. physics and mathematics, respectively. He then compares first philosophy (metaphysics) with mathematics. In his explanation, as the particular mathematical sciences (e. g. geometry and astronomy) are to mathematics as a whole, so the two particular sciences, i. e. mathematics and physics are to first philosophy. Thus the Interpreter restates Aristotle's question as follows: Physics is about a part of being and mathematics is about another part of being; Is theology (another name of first philosophy in Book VI(E) of Aristotle's *Metaphysics*) common to all together?

It is noteworthy that Alexander distinguishes the universality (or generality) of mathematics as a whole from that of first philosophy, consciously using different wordings: that is to say, mathematics as a whole is common to all

[26] *Alexandrini Aphrodisiensis in Aristotelis Metaphysica Commentaria*, edidit Michael Hayduck: *Commentaia in Aristotelem Graeca,* Vol. 1 (Berlin, 1891), p. 447, *ll.* 9–19: "*ὅτι δὲ τῶν ἄλλων αἱρετωτάτη ἐπάγει ἀπορήσειε γὰρ ἄν τις πότερόν ποθ'ἡ πρώτη φιλοσοφία καθόλου ἐστίν, ὥστε περιέχειν ἁπάσας καὶ ὑπ'αὐτὴν εἶναι ὡς τὰ κατὰ μέρος ὑπὸ τὰ καθόλου, ἢ οὐκ ἔστι καθόλου, ἀλλ' ὥσπερ ἡ ἰατρικὴ κσὶ ἡ ἀστρονομία καὶ αἱ λοιπαὶ μέρος τι τοῦ ὄντος καὶ φύσιν τινὰ μίαν ἀποτεμόμεναι τοῦτο θεωροῦσιν, οὕτως καὶ αὕτη. ὥσπερ γὰρ καὶ ἐν τοῖς μαθηματικοῖς τοῦτο εὑρίσκεται, καὶ ἡ μὲν γεωμετρία περὶ μεγέθη, ἡ δὲ ἀστρονομία περὶ ἀστέρας, ἄλλη δὲ τερὶ ἄλλο, ἡ δὲ ἁπλῶς μαθηματικὴ κοινή ἐστι πασῶν, οὕτως τί κωλύει καὶ τὴν φυσικὴν μὲν περι'τι τοῦ ὄντος εἶναι καὶ τὴν μαθηματικὴν τερὴ ἄλλο, καὶ τὴν θεολογικὴν εἶναι κοινὴν ἁπασῶν;*" I have referred to Crowley's translation in *Op. cit.*(n. 19), pp. 11–12.

(κοινὴ πασῶν) whereas theology is common to all together (κοινὴ ἁπασῶν). "ἁπασῶν" is, of course, a strengthened form of "πασῶν" (< πᾶς). "Mathematics simply" is certainly more general than the particular mathematical sciences. In this sense, it is "*καθόλου καὶ πασῶν μαθηματικῶν κοινή*". But first philosophy in its turn in relation to physics and mathematics is "*καθόλου καὶ κοινὴ ἁπασῶν*." This is what Albert Schwegler especially took notice of.[27]

Nevertheless, Crowley has opposed Alexander's interpretation. According to him, Aristotle contrasts metaphysics with mathematics, emphasizing the difference between the two disciplines, while Alexander "*compares* metaphysics to mathematics."[28] (The emphasis is in the original.) On the other hand, Crowley seems to be right in understanding the Intrepreter as not supporting "a universal mathematical science prior to arithmetic and geometry which would treat of universal mathematical theorems, or of quantity as quantity."[29] Surely for Alexander, "the universal" (ἡ καθόλου) is simply the general name given to all the mathematical sciences. But is Alexander's explication of what Aristotle really means mistaken, as Crowley believes? Before answering this question, one thing should be noted here concerning our interpreting "mathematics simply" as "mathematics as a whole". After a brief note on this problem, let us turn to another commentator of ancient Greece.

As we have seen in Chapter 5, § 3, D (p. 263), Van Roomen interpreted a discipline which "Eutocius calls mathematics absolutely" in such a way as to give favorable support to his concept 'mathesis universalis'. We believe with Crowley that Alexander's "mathematics simply" implies "mathematics as a whole." But it should be noted that this interpretation itself is subtle and could be replaced with another.

The Neoplatonist Asclepius of Tralles, who lived about 500 A. D., also wrote a commentary on the first seven books of Aristotle's *Metaphysics*. He explicates the Aristotelian passage beginning with 1026a23 as follows:

> He [Aristotle] mentions the aporia by which he undermined his theory. For he asks whether first philosophy is concerned with all beings, or with one genus and some one nature. For not even the mathematical sciences are all alike in this respect: for, on the one hand, universal mathematics is concerned with the whole of mathematics; on the other hand, the particular is concerned with some one nature, as he himself states that geometry and astronomy deal with some particular nature, i. e. the former with magnitudes and the latter with the stars; but the universal is common of all, as is

[27]A. Schwegler, *Die Metaphysik des Aristoteles*, Grundtext, Übersetzung und Commentar, Bd. IV: Des Commentars Zweiter Hälfte (Tübingen, 1847–8), pp. 17–18. Reale supports Schwegler's interpretation. See Reale, *Op. cit.* (n. 4), pp. 187–190, footnote 31 of Chapter Four.

[28]Crowley, *Ibid.*, p. 19.

[29]Crowley, p. 18.

> said. In this manner, therefore, first philosophy argues about all unqualifiedly as it deals with the principles of beings.[30]

Asclepius takes "the universal" as "universal mathematics" (*ἡ καθόλου μαθηματική*) which is concerned with the whole of mathematics (*πάντα τὰ μαθηματικά*) and is common to all the particular mathematical sciences, e. g. geometry and astronomy.

Two interpretations may be possible for Asclepius's concept "*ἡ καθόλου μαθηματική*". One is that it means a universal mathematics common to arithmetic and geometry which may deal with general characteristics of the mathematical disciplines. Heath might support this explanation. The other is that it simply implies "mathematics as a whole," i. e. the whole realm of all the particular mathematical sciences.

Now what does the word '*καθόλου*' mean in Asclepius's phrase "*ἡ καθόλου μαθηματική*"? The Greek word '*καθόλου*' or '*καθ' ὅλου*' is derived from '*κατά*' plus '*ὅλος*', i. e. 'concerning whole', and denotes primarily as an adverb 'on the whole' or 'in general', and as an adjective 'general' or 'universal'. Notable here is that Asclepius paraphrases Aristotle's expression "*ἡ καθόλου*" as "*ἡ καθόλου μαθηματική*" where Alexander paraphrases it as "*ἡ ἁπλῶς μαθηματική*". We have rendered above Alexander's wording as "mathematics simply" or "mathematics unqualifiedly." Certainly we could be permitted to interpret "*ἡ καθόλου μαθηματική*" to be "mathematics as a whole", which is also the meaning of "*ἡ ἁπλῶς μαθηματική*" in the Alexandrian commentary. It is reported that in general Asclepius depended heavily on Alexander of Aphrodisias for his commentary on Aristotle's *Metaphysics*.[31] But if the commentary on Book VI(E) ascribed to Alexander, who was flourished around 200 A. D., was not totally composed by Alexander himself, but partly by Michael of Ephesus of the eleventh century, Asclepius was possibly unable to consult it. In any case, it is possible that by the words "*ἡ καθόλου μαθηματική*" Asclepius

---

[30] ***Asclepii in Aristotelis Metaphysicorum libros*** A–Z ***Commentaria***, Edidit Michael Hayduck: ***Commentaria in Aristotelem Graeca***, Vol. 6, Pars 2 (Berlin, 1888), p. 364, *ll.* 11–19: "*Τὴν ἀπορίαν δι' ἣν τὴν θεωρίαν ἐκίνησε ταύτην λέγει. φησὶ γὰρ ἆρά γε ἡ πρώτη φιλοσοφία περὶ πάντα τὰ ὄντα καταγίνεται ἢ τερί τι γένος καὶ φύσιν μίαν. οὔτε γὰρ ἐν ταῖς μαθηματικαῖς ἐπιστήμαις ὁ αὐτὸς τρόπος ἐστίν· ἡ μὲν γὰρ καθόλου μαθηματικὴ περὶ πάντα τὰ μαθηματικὰ καταγίνεται, ἡ δὲ μερικὴ περὶ μίαν φύσιν, οἷον, ὥς φησιν αὐτός, γεωμετρία καὶ ἀστρολογία περί τινα φύσιν μερικὴν ὑπάρχει, ἡ μὲν περὶ μεγέθη, ἡ δὲ περὶ ἄστρα, ἡ δὲ καθόλου κοινή ἐστι πασῶν, ὡς εἴρηται. οὕτως οὖν καὶ ἡ πρώτη φιλοσοφία ὡς ἔχουσα τὰς ἀρχὰς τῶν ὄντων περὶ πάντων ἁπλῶς διαλέγεται.*" The translation is mine. As to the works and life of Asclepius, consult A. Gercke, "Asklepios," ***Paulys Realencyclopädie der classischen Altertumswissenschaft***, Bd. IV (Stuttgart, 1896), col. 1697; [L. S.], "Asclepius," ***A Dictionary of Greek and Roman Biography and Mythology***, ed. William Smith, Vol. 1 (New York, 1967), p. 383.

[31] See Arthur Madigan, "Syrianus and Asclepius on Forms and Intermediates in Plato and Aristotle," ***Journal of the History of Philosophy***, **24** (1986), pp. 149–171, esp. pp. 170–171. Madigan's discussion is about Syrianus's and Asclepius's commentaries on Book III(B) of Aristotle's ***Metaphysics***.

means "mathematics as a whole."

Reading further on in the passage, however, the other possibility of interpretation once again comes to the fore. Asclepius contrasts "ἡ καθόλου μαθηματική" with the two particular mathematical sciences, geometry and astronomy. The latter two are concerned with some particular nature. But "the universal", he states, "is common to all." "The universal" (ἡ καθόλου) here clearly indicates "universal matheamatics". Then, on the analogy of "universal mathematics", he characterizes first philosophy with the following sentence: "In this manner, therefore, first philosophy argues about all unqualifiedly (ἁπλῶς) as it deals with the principles (ἀρχαί) of beings." Does not Asclepius's interpretation suggest that "universal mathematics" is a discipline which deals with the fundamental notions and propositions about mathematical beings? If it does, "universal mathematics" can be elucidated in the first sense, that is to say, as a mathematical science which is concerned with common characteristics of the mathematical sciences, not as "mathematics as a whole" in the second sense. But unfortunately Asclepius does not further elaborate here what precisely this "universal mathematics" means.

Here we will try to present our tentative solution concerning what a universal mathematical science is for Asclepius. Two answers seem to be possible. One is that it means a mathematical discipline which is common and applicable to and, moreover, different from both arithmetic and geometry. It was conceived as such by the Neoplatonists Iamblichus and Proclus, as we will show in Chapter 7, § 1. The other possibility is that that science signifies arithmetic, which Asclepius considered to occupy the first place among the mathematical disciplines. In his *Commentary to Nicomachus's Introduction to Arithmetic*, Asclepius, following Plato and Nicomachus of Gerasa, clearly stated that arithmetic is the most fundamental and universal mathematical discipline, to be placed prior to the other three mathematical sciences: geometry, music, and astronomy.[32] This explanation accords with Bonitz's and Reale's (see pp. 293–294). And a Newton might have adopted this interpretation (pp. 276, and 394).

For our present purposes, it is not necessary to decide what "ἡ καθόλου μαθηματική" signifies exactly, a universal mathematical discipline or mathematics as a whole, although the first interpretation appears to us more probable. Be that as it may, Asclepius's commentary may help, at least, to cast doubt upon Crowley's argument that Aristotle contrasts metaphysics with mathematics whereas Alexander compares the former to the latter. In our interpretation, both the Greek commentators compare first philosophy with something universal in mathematics. In doing so, they try to discern similarities and dis-

[32]See Asclepius of Tralles, "Commentary to Nicomachus' Introduction to Arithmetic," edited with an Introduction and Notes by Leonardo Tarán, *Transactions of the American Philosophical Society*, New Series, Vol. 59, Part 4 (1969), "I. λγ", pp. 30–32; see also Tarán's note on pp. 76–77. This part is a scholium to Book I, Chapter IV of Nicomachus's *Introduction to Arithmetic*, translated by Martin Luther D'Ooge (New York/London, 1926; Reprinted edition, New York, 1972), pp. 187–188.

similarities between the two general sciences. And there is agreement between the two that metaphysics is more universal than mathematics since the former treats all beings unqualifiedly. In this sense, they contrast first philosophy with mathematics.

Through our philological examination above, we have ascertained that "the universal" in 1026a27 and 1064b9 of Aristotle's *Metaphysics* was construed by the Greek commentators as the mathematical in any case, either a mathematical science common to the particular mathematical sciences or mathematics as a whole. But interestingly, the Aristotelian passages containing "the universal" were radically reinterpreted when they entered the medieval world of learning. Now let's see this.

§ 3. Medieval Commentators: Ibn Rushd, Albertus Magnus, Thomas Aquinas, and the Scotist Antonius Andreae

A. Albertus Magnus and Thomas Aquinas

How did medieval Schoolmen evaluate the importance of mathematics, especially in relation to metaphysics, the intellectual enterprise which they regarded most highly? To reply to this question at least in part, we will examine several prominent commentaries on the passages discussed in Section 1 of Aristotle's *Metaphysics*. The examination of the scholarly endeavor to assimilate, through commentaries, the philosophical heritage of ancient Aristotelianism is crucial for our understanding of scholasticism in the Middle Ages.[33] But to do this exhaustively is certainly an enormous task. So we will confine ourselves to examining fundamentally the commentaries on 1026a23–27 and 1064b6–9 of the *Metaphysics* by such representative interpreters Albertus Magnus, Thomas Aquinas, and Duns Scotus (actually his disciple Antonius Andreae). Together with these Latin authors, we will touch upon Ibn Rushd (Latin name, Averroës). Through the narrow channel of this textual analysis, we will be able to comprehend how the Schoolmen understood the value and the epistemological status of mathematics to the extent that we may compare their standpoint with Descartes's.

Before observing how medieval scholars interpreted the Aristotelian passages, it is helpful to know what kinds of Latin translation of the Aristotelian corpus were available to them. Aristotle's *Metaphysics* became known to Latin Schoolmen for the most part through the following translations[34] : (1) James

[33]For a general presentation of medieval scholasticism, see Josef Pieper, *Scholasticism: Perspectives and Problems of Medieval Philosophy*, tranlated from the German by Richard and Clara Winston (New York/Toronto, 1964).

[34]We refer to Lorenzo Minio-Paluello, "L' 《Aristoteles Latinus》," in *Opuscula: The Latin Aristotle* (Amsterdam, 1972), p. 472; *Idem*, "Aristotle: Tradition and Influence," *DSB*, I, pp. 267–281, esp. pp. 270–273; Gudrum Diem, "Les Traductions gréco-latines de la Métaphysique au moyen âge: Le problème de la Metaphysica Vetus," *Archiv für Geschichte des Philosophie*, **49** (1967), pp. 6–71.

(Jacobus) the Venetian-Greek: perhaps *Metaphysics* I–IV. 2, commonly called *Metaphysica vetustissima*, in the early twelfth century; (2) a systematic revision of Books I–III. 3 of the James version, made before 1236 and usually called *Composita* or *Vetus*; (3) Twelfth-century anonymous translation from the Greek: Books IV–X, XII–XIV and perhaps a revision of Books I–II and part of Book III of James's translation, known as *Metaphysica media*; (4) Michael Scot from the Arabic: most of the *Metaphysics* (parts of Books I and XII, and the whole of Books XI, XIII and XIV were not included), known as *Metaphysica nova*, with Averroës's *Great Commentary*, in the early thirteenth century; and (5) William Moerbeke: Book XI, and a revision of the other Books having recourse to the Greek text, in the thirteenth century.

Among these translations, important for us are the anonymous rendering ("Media") and Scot's translation ("Nova") of Book VI(E) and Moerbeke's Book XI(K). We may say that Moerbeke's version of 1026a23–27 is essentially the same as the *Media*'s with a slight syntactical deviation. The passage of 1026a23-27 in the *Metaphysica media* reads as follows:

> Now one will doubt whether first philosophy is universal or about some genus and one nature, for neither is there the same mode in mathematics, because geometry and astronomy are about some nature, that (*illa*), however, is universally common to all.[35]

Here the translator typically follows the medieval *verbum e verbo* method. For example, it may be said in a sense that the last sentence, "illa vero universaliter omnium est communis" corresponds to the Greek original, which the translator seems to have read as "*ἐκείνη δὲ καθόλου πασῶν κοινή*" rather than as "*ἡ δὲ καθόλου πασῶν κοινή*" among variants of the original Greek text. In other words, the Latin translation also leaves the interpretation of "*καθόλου*" ambiguous. There seems to be two important points which may be taken to transgress this principle of translation. The first point is that the Greek couplet "*μέν–δέ*," which means "on the one hand–on the other hand," is translated into a single word "vero." In the strict verbum e verbo method, it is usually rendered "quidem–vero" or "autem–vero." Of course, it is often necessary to leave "*μέν*" untranslated. The second point is that the Greek word "*καθόλου*" is translated unambiguously into the adverb "universaliter." As we saw above (p. 301), "*καθόλου*" can be used as both an adjective and an adverb. In any case, the Latin rendering clarifies what Aristotle meant no more than the Greek original.

And the passage of 1064b6–9 in the Moerbeke translation is as follows:

[35] *Aristoteles Latinus* XXV-2: *Metaphysica, Translatio Anonyma sive 'Media'*, edidit Gudrum Vuillemin-Diem (Leiden, 1976), p. 118, *ll.* 5–9: "Dubitabit autem aliquis utrum prima philosophia sit universalis aut circa aliquod genus et naturam unam (non enim idem modus nec in mathematicis, quia geometria et astrologia circa aliquam naturam sunt, illa vero universaliter omnium est communis)." The parentheses are the editor's. See "Praefatio", p. LXII. We refer to Crowley's literal translation on p. 5 of his book quoted above (n. 19).

> Now someone will doubt whether of being *qua* being it is necessary to posit a universal science or not. For each one of the mathematical sciences is, in fact, about some one determinate genus. Those which apply to the universal, however, are about all.[36]

In contrast with the previous rendering of 1026a23–27, here the *verbum e verbo* principle recedes. The Greek original "ἡ δὲ καθόλου κοινὴ περὶ πάντων" is rendered "Quae autem universali conveniunt de omnibus [sunt]." Again, the translator remains silent about what "καθόλου" ("the universal" or "those which apply to the universal" here) precisely indicates.

We will include Scot's translation of the passage in Book VI(E) when we discuss Ibn Rushd below.

How did influential commentators in the Middle Ages interpret these Aristotelian passages? To begin with, we will examine Albertus Magnus (Albert the Great), the first theologian-philosopher in the medieval Latin world who positively accepted Aristotelianism as a whole.[37] The German Dominican friar named 'Doctor universalis' composed a commentary on the *Metaphysics* probably during the period between 1264 and 1267. The text he used was the *Metaphysica media*, which lacked Book XI and was utilized only between 1250 and 1270.[38] Accordingly, only his paraphrase of Book VI is relevant to our topic.

Albertus Magnus commented on the passage 1026a23–27 in his *Metaphysica*, Book 6, Tract 1, Chapter 3, "That This Science Is Divine, Most Honorable, First and Universal."

> One will perhaps doubt whether it [first philosophy] is a universal science, just as we have already often said, or rather is one subject about some genus and about one divine nature, and thus is a particular science just as the other sciences, which consider one part of being. But this [problem] is solved as follows: The mode is not the same in mathematics and physics, which are about one determinate being, and in divine science, for geometry and astronomy are about some nature, of which principles are neither the principles of

[36]This is the Moerbeke text which the Marietti edition of Thomas Aquinas's *In Duodecim Libros Metaphysicorm Aristotelis Expositio* (Toronto, 1964) cites on p. 534: "Dubitabit autem quis, utrum entis inquantum ens scientiam universalem oportet poni, aut non. Mathematicarum quidem enim unaquæque circa unum aliquod genus determinatum est. Quæ autem universali conveniunt de omnibus." We will show another version of the Moerbeke text below (n. 82).

[37]Pieper, *Ibid.*(n. 33), p. 109.

[38]James A. Weisheipl, "Albert's Works on Natural Science (libri naturales) in Probable Chronological Order," Appendix 1 of J. A. Weisheipl, ed., *Albertus Magnus and the Sciences: Commemorative Essays 1980* (Toronto, 1980), p. 576. See also, *Idem*, "The Life and Works of St. Albert the Great," in *Ibid.*, p. 40. For a bibliography related to Albert's *Metaphysica*. See Charles H. Lohr, "Medieval Latin Aristotle Commentaries: Authors A-F," *Traditio*, **23** (1967), p. 341.

> all nor the principles of being insofar as it is being; and similarly for physics. On the other hand, the former (*illa*) which we call divine is common to all, for, even though it is about God and the divine [things], it is nevertheless about these insofar as they are the principles to the universal Being (*esse*) for the reason that they are truly the principles of being (*ens*) insofar as it is being, and this is substance composed simply of the principles of substance as far as it is substance.[39]

Amazingly to those who have seen the Greek commentators' interpretations, Albert here explicitly paraphrases "illa" into a term related not to mathematics in some sense or other but to first philosophy (metaphysics), which is identical with theology for the Schoolmen. He contrasts divine science (i. e. theology) with the other theoretical sciences, mathematics and physics. In his interpretation, mathematics deals with one determinate being and its principles are neither the principles of all nor the principles of being *qua* being. It can be concerned only with at most a partial aspect of being. What can deal with being *qua* being is first philosophy. It alone is common to all and, therefore, universal.

Thus, for the Dominican friar, "illa universaliter" in the *Media* text, which corresponds to "*ἐκείνη καθόλου*" or "*ἡ καθόλου*" in Aristotle's original, signifies nothing other than theology. In Albert's thought, the universality of theology overwhelms the qualified universality of mathematics, a loosely united body of all the particular mathematical sciences. He, like his disciple Thomas Aquinas, could not read Greek and had no means of referring to the Greek texts of Aristotle and his Greek commentators. What he relied on totally for his text was the *Media* translation which may syntactically allow one to interpret "illa universaliter" as first philosophy as well as the mathematical.

The paraphrase examined above seems to have been clearly influenced not only by the philological reason which we have just mentioned but also by Albert's general philosophical tendencies. First, he had, no doubt, a strong desire to harmonize Aristotle's doctrines with the teachings of the Christian Church. Thus he must have very naturally associated theology with an expression con-

[39] *Alberti Magni Opera omnia*, Tomus XVI, Pars II: *Metaphysica*, Libri VI–XIII (Münster, 1964), p. 305, *ll.* 64–76–p. 306, *ll.* 1–5: "*[D]ubitabit* fortasse *aliquis*, *utrum* ipsa *sit* scientia *universalis*, sicut saepe iam diximus, an forte sit *circa aliquod genus* subiectum unum *et* circa *naturam unam* divinam, et sic sit scientia particularis sicut et aliae scientiæ, quæ circa unam partem entis contemplantur. Sed ad hoc solvitur, quod *non* est *idem modus in mathematicis* et physica, quæ sunt circa unum ens determinatum, et in divina scientia, *quia geometria et astrologia circa aliquam naturam sunt*, cuius principia non sunt omnium principia neque sunt principia entis, secundum quod est ens; et similiter est de physica. *Illa vero* quam divinam vocamus, *est communis omnium,* quia licet sit de deo et divinis, est tamen de his, secundum quod illa sunt principia *universi* esse per hoc quod sunt principia entis vere, secundum quod est ens, et hoc est substantia non nisi ex principiis substantiæ, inquantum substantia est composita."

cerning what is universal, such as "illa universaliter."

Second, he was cautious about his contemporaries' opinions that rated the value of mathematics more highly than he thought necessary. As has been recently argued, he was concerned with criticizing the 'error' of the so-called 'Oxford Platonists', such as Robert Grosseteste and Roger Bacon.[40] In the introductory part of Book I of his *Metaphysics*, he openly attacks Plato's philosophy of mathematics:

> Here there is need to beware of the error of Plato, who said that naturals were founded in mathematicals and mathematicals in divines, just as the third cause is founded in the second, and the second is founded in the primary, and therefore he said that mathematicals were principles of naturals, which is completely false.[41]

It is certain that Albert evaluated the empirical method in the study of nature more highly than the mathematical approach. But he did not necessarily neglect the study of mathematics. Nothing is better evidence of this than the fact that he composed a commentary on Euclid's *Elements*.[42] In Albert's judgment, mathematics is abstract in a very strict sense and deals with only one aspect of substance, i. e. its quantitative side, so that it cannot in any way be universal. Physics, concerned with the sensible natural world, then cannot be explained sufficiently by mathematical tools. *Metaphysics* alone ultimately treats entire reality, i. e. being as such. It is incomparably more universal than mathematics.

In this regard, Thomas Aquinas basically adopted his teacher's exegesis. During the period between 1269 and 1272, Thomas wrote his commentary on the *Metaphysics*, referring to the *Metaphysica Moerbecana* insofar as Books VI and XI are concerned.[43] It is known that he consulted Albertus Magnus's

---

[40] For Albert's philosophy of mathematics especially in relation to the Oxford 'Platonists', see, among others, James A. Weisheipl, "Albertus Magnus and the Oxford Platonists," *Proceedings of the American Catholic Philosophical Association*, **32** (1958), pp. 124–139; A. G. Molland, "Mathematics in the Thought of Albertus Magnus," in Weisheipl's volume (n. 38), pp. 463–478; Benedict M. Ashley, "St. Albert and the Nature of Natural Science," in *Ibid.*, pp. 73–102.

[41] *Alberti Magni Opera Omnia*, Tomus XVI, Pars I: *Metaphysica Libri quinque priores*, Edidit B. Geyer (Münster, 1960), p. 2, ll. 31–35: "Cavendus autem hic est error Platonis, qui dixit naturalia fundari in mathematicis et mathematica in divinis, sicut tertia causa fundatur in secunda et secunda fundatur in primaria, et ideo dixit mathematica principia esse naturalium, quod omnino falsum est." We refer to Molland's translation quoted in *Op. cit.* (n. 40), p. 467.

[42] See Paul M. J. E. Tummers, "The Commentary of Albert on Euclid's *Elements* of Geometry," in Weisheipl's volume (n. 38), pp. 479–499; Anthony Lo Bello, "Albertus Magnus and Mathematics: A Translation with the Annotations of Those Portions of the Commentary on Euclid's *Elements* published by Bernhard Geyer," *Historia Mathematica*, **10** (1983), pp. 3–23. In the "Prologue" to his commentary, Albert compares mathematics with other theoretical sciences, physics and metaphysics, referring to Aristotle's *Metaphysics*, VI, 1. See Lo Bello, p. 56.

[43] See James C. Doig, *Aquinas on Metaphysics: A Historico-Doctrinal Study of the Com-*

commentary on the *Metaphysics* when composing his own.[44]

First we extract Thomas's paraphrase of 1026a23–17. It reads as follows:

> Someone may doubt whether first philosophy is universal, as though its consideration about being is universal, or its consideration is about some determinate genus and one nature. It seems that it is not of a determinate nature. For there is not one mode of this science and of the mathematical sciences, because geometry and astronomy, which are mathematical sciences, are about some determinate nature, whereas first philosophy is universally common to all.[45]

Clearly for Thomas "illa universaliter omnium est communis" is a statement applicable only to first philosophy. Geometry and astronomy are about some determinate nature not because they are particular mathematical sciences but because they are mathematics. Mathematics in any form, according to Thomas, cannot be a universal science.

The same point of view is further expounded in Thomas's interpretation of passage 1064b6–9. According to his commentary, Aristotle in this passage compared first philosophy with other sciences, e. g. physics and mathematics. Aristotle, states Thomas,

> compares this science to other sciences according to universality; and he says: it is doubtful whether this science which is about separable beings ought to be proposed as the universal science of being *qua* being, or not: and he shows that such is the case as it were by division.[46]

---

mentary on the Metaphysics (The Hague, 1972), Chapter I, "Literary and Chronological Aspects of the Commentary," pp. 3–22; James Weisheipl, *Friar Thomas d'Aquino: His Life, Thought and Works* (Washington, D. C., 1974), pp. 284 & 379. On editions of Thomas's exposition of the *Metaphysics*, consult Ch. H. Lohr, "Medieval Latin Aristotle Commentaries. Author: Robertus-Wilgelmus," *Traditio*, **29** (1973), pp. 164–165.

[44]See J. C. Doig, *Ibid.*, pp. 23, 46–54 & 84.

[45]*S. Thomæ Aquinatis Opera omnia*, 4: *Commentaria in Aristotelem et alios* (Stuttgart/Bad Cannstatt, 1980), "In Libro Metaphysicorum," p. 446a, n. 26; *S. Thomae Aquinatis In duodecim libros Metaphysicorum Aristotelis expositio* (Turin, 1964), p. 298, paragraph 1169: "[A]liquis potest dubitare, utrum prima philosophia sit universalis quasi considerans ens universaliter, aut eius consideratio sit circa aliquod genus determinatum et naturam unam. Et hoc non videtur. Non enim est unus modus huius scientiae et mathematicarum; quia geometria et astrologia, quæ sunt mathematicæ, sunt circa aliquam naturam determinatam; sed philosophia prima est universaliter communis omnium." The Marietti version above gives the following passage for Aristotle's text (p. 294, paragraph 541): "Dubitabit autem aliquis, utrum prima Philosophia sit universalis, aut circa aliquod genus, aut naturam unam. Non enim idem modus, nec in mathematicis; quia geometria et astrologia circa aliquam naturam sunt: illa vero universaliter omnium est communis."

[46]Thomas Aquinas, *Opera omnia*, 4 (*Op. cit.*), p. 491c, n. 20; Marietti ed., p. 536, paragraph 2266: "Comparat istam scientiam ad alias quantum ad universalitatem; et dicit: Dubitabile est, utrum istam scientiam, quae est circa entia separabilia, oporteat poni universalem scientiam entis, inquantum est ens, aut non: et ostendit, quod sic, quasi per divisionem." The Marietti Moerbeke text of paragraph 962 can be seen in our note 36.

Thomas affirms with Aristotle that there can be the universal science of being as such, which is identical with first philosophy. What of the universality of other sciences in comparison with that of this science, i. e. metaphysics? He answers:

> For it is manifest that the operative [practical] sciences are not universal. And so he passes over them.—But among the speculative sciences it is manifest that each one of the mathematical sciences is about one determinate genus, whereas a universal science is common to all beings.[47]

The Angelic Doctor affirms that a universal science ought to be common to all. But any mathematical science, in his judgment, is about some determinate genus. Then, he states that no sciences other than metaphysics are common to all beings. The latter science should be called, therefore, first and universal. Thomas concludes: "From the fact that it is first, it must be universal. For it is the same science which is of the first beings, and which is universal. For the first beings are the principles of the others."[48]

Thomas's assertion that metaphysics alone is a universal science does not imply that it is the most certain science. In the scholastic-Aristotelian tradition to praise the certainty of mathematics was quite common. Thomas for one, commenting on Aristotle's well-known passage 995a14–17, repeated his ancient master's claim that the most certain things may be acquired only in mathematics and may not be found for in physics which treats sensible matter in addition to mathematical forms.[49]

Thomas also stresses the certainty of mathematics in Question VI, concerning the method of speculative science, in his commentary on the *De Trinitate* of Boethius. In his opinion, "the method of mathematics is more certain (*certus*) than the method of divine science, because the objects of divine science are further removed from sensible things, from which our knowledge takes its origin."[50] He states: "It is clear," therefore, "that mathematical inquiry is easier and more certain than the natural and theological, and much more than that

---

[47] *Loc. cit.*: *Opera omnia*, 4, p. 491c, n. 21; Marietti ed., paragraph 2267: "Manifestum est enim, quod praedictæ scientiæ operativæ non sunt universales. Et ideo eas prætermittit.—Sed inter speculativas scientias, manifistum est, quod quælibet mathematicarum scientiarum est circa unam aliquod genus determinatum: universalis autem scientia communiter de omnibus est: unde nulla mathematicarum scientiarum est communis omnium entium."

[48] *Ibid.*: "[E]x eo quod est prima, oportet quod sit universalis. Eadem enim est scientia quæ est de primis entibus, et quæ est universalis. Nam prima entia sunt principia aliorum."

[49] Thomas, *Op. cit.*: *Opera omnia*, 4, p. 408c, n. 6–n. 7; Marietti ed., pp. 93–94, paragraphs 336–337.

[50] Thomas Aquinas, *Opera omnia*, 4 (*Op. cit.*), "In Librum Boethii de Trinitate," p. 536c, co 12; St. Thomas Aquinas, ***The Division and Methods of the Sciences: Questions V and VI of His Commentary on the De Trinitate of Boethius***, translated with Introduction and Notes by Armand Maurer (Toronto, ₃1963), p. 60.

of the other sciences that are practical."[51] Thomas reconfirms his statement, quoting the beginning part of Ptolemy's *Almagest*:

> Let us call the other two kinds of theoretical knowledge opinion rather than science: theology because of its obscurity and incomprehensibility, physics because of the instability and obscurity of matter. The mathematical type of investigation alone will give the inquirer firm and unshaken certainty through demonstrations carried out by unquestionable methods.[52]

This is exactly the same passage which was cited by Francesco Barozzi for consolidating his philosophical position in the 1560 controversy with Piccolomini on the certainty of mathematics (see above Chapter 2, § 2, p. 60). Notice that Thomas is here asserting that mathematics is the most certain science for us owing to the weakness of the intellects of us mortals, and that he does not necessarily intend to degrade the supreme status of theology.

But Thomas does not want us to be deluded by the certainty of the method of mathematics. For him as well as for his teacher Albert, mathematics deals simply with a defective quantitative aspect of being, not being as such. For both of them, the physical may be related to the mathematical as an addition so that the natural world cannot by any means be reduced to mathematical entities.[53] Physics tackles aspects of being in a richer way than mathematics. *Metaphysics* in turn deals with much richer and deeper structures of being than physics and mathematics. It treats being *qua* being without qualification. Thus it is the only science that can be called universal. If mathematics enjoys supreme certainty, it is because of the meagerness of the aspect of being with which it can deal.

As we have suggested above, Thomas was certainly influenced by Albertus Magnus's paraphrase which implies that "illa universaliter" in 1026a23–27 is first philosophy, not something related to mathematics. But such a paraphrase does not seem to have been unusual to the Schoolmen of the thirteenth century who took it for granted that metaphysics or theology alone was first and universal science. When Albert and Thomas were writing their commentaries on the *Metaphysics*, the thoughts of Ibn Sīnā (Avicenna) and of Ibn Rushd (Averroës) were in vogue in the Latin West. Both the Dominican doctors were familiar with how the two Islamic thinkers thought of metaphysics. All four philosophers agreed in that metaphysics as the unique universal science is the queen of the sciences, although they were of different views as to what this first

---

[51] *Ibid.*, p. 537a, co 12; Translation, pp. 60–61.

[52] *Ibid.*, p. 537a, co 12; Translation, p. 61. Cf. Claudius Ptolemaeus, *Syntaxis mathematica*, ed. J. L. Heiberg, Part I (Ch. 2, n. 39), p. 6; *Ptolemy's Almagest*, p. 36

[53] In regard to Thomistic ontology and epistemology concerning mathematics, see the discussion of Vincent Edward Smith, *St. Thomas on the Object of Geometry: The Aquinas Lecture, 1953* (Milwaukee, 1954).

philosophy ought to be.[54]

To examine closely how Albert and Thomas established their interpretations of passage 1026a23–27, let us broaden our historical perspective and analyze the text of the *Metaphysica Arabica* together with Ibn Rushd's commentary on it.

B. Ibn Rushd (Averroës) 's Commentary on the *Metaphysics*

Averroës (Ibn Rushd) wrote his renowned *Great Commentary* on Aristotle's *Metaphysics* in his last years after 1180.[55] And it along with the *Arabica* was translated into Latin by Michael Scot. Michael's translation of the Averroës *Commentary* was probably done during the period of between 1228 and 1235[56] and began to be read in Paris as early as the 1230s.[57]

We begin our examination with the *Arabica* text, and by using mainly Michael's Latin translation and referring to the original Arabic text when necessary we will see how Averroës paraphrased it. The *Arabica* translation of 1026a23–31 reads as follows:

> And it is worth asking whether first philosophy is universal, or of some genus, or of one nature. For neither is there one species in the mathematical sciences. But geometry and astronomy are of some nature and the universal science is common to all. If, therefore, there should not be some other substance than substances existing in nature, then first science is natural [science]. And if there should be some immovable substance, that will be first, and the science of it will be first philosophy and universal.[58]

[54]On the four philosophers' views on metaphysics, consult James C. Doig, *Op. cit.* (n. 43), Chapter II, "The Metaphysical Views of Avicenna, Averroës, and Albert," pp. 22–54; Chapter V, "The Relation of Metaphysics to the Other Sciences," esp. 1, "The Thought of Aquinas"; 2, "'Universal Science' and 'First Science' in the Predecessors of Aquinas," pp. 172–213.

[55]R. Arnaldez, "Ibn Rushd," *The Encyclopaedia of Islam*, New Edition, Vol. III (Leiden/London, 1971), p. 910.

[56]Henri Corbin, *Histoire de la philosophie islamique* (Paris, 1964), p. 337

[57]F. E. Peters, *Aristotle and the Arabs* (New York/London, 1968), pp. 228 & 231.

[58]Averroës, *Commentarium in Metaphysicorum,* in *Aristotelis Opera omnia*, Vol. VIII: *Metaphysicorum libri XIIII cum Averrois Cordubensis in eodem commentariis* (Venetiis, 1562), f. 146v: "Et dignum est quærere vtrum prima Philosophia sit vniuersalis, aut alicuius generis, aut vnius naturæ. Non enim est vna species, neque in scientijs Mathematicis etiam, sed Geometria, & Astrologia sunt alicuius naturæ, & scientia vniuersalis est communis omnibus. Si igitur non fuerit substantia aliqua alia à substantijs existentibus natura, tunc prima scientia est Naturalis. Et si aliqua substantia immobilis fuerit, illa erit primum, et scientia eius prima Philosophia, vniuersalis etiam." This version presents first Johannes Bessarion's Latin translation from the Greek, then the *Metaphysica Arabica*, and finally the Latin translation of Averroës's *Great Commentary.* An explanation for this edition is in "Editions of the Latin Aristotle Accompanied by the Commentaries of Averroës," in Edward P. Mahoney, *Philosophy and Humanism: Renaissance Essays in Honor of Paul Oskar Kristeller* (Leiden, 1976), pp. 116–128, esp. pp. 125–128.

Then Averroës comments on the passage above sentence by sentence. For our convenience, let us quote his commentary before examining the *Arabica* translation:

> That is: It is appropriate to ask, after he [Aristotle] explained that such a science is that which is said to be first philosophy, whether this science is universal to many genera, or of the same genus, or of the same nature. Then he said: For there is not one species; i. e. for it is evident that mathematics is not of one genus, but of different genera. For geometry is of a nature different from that of which astronomy is, and each of them is mathematics. Then he said: The universal science is common to all, i. e. but if this science should be universal, it is suitable that it is common to all beings. Then he said: If, therefore, there should not be etc.; i. e. it is manifest that if there should not be another substance here than the sensible, there would not be a science prior to natural [science]. Then he said: And if [there should be] some immovable substance etc.; i. e. it is manifest that if there is some immovable substance, this substance is first, and the science of it is the universal science and first philosophy. And he said: In Being (*esse*); for in the order of learning this science is posterior and therefore we become acquainted with it after physics.[59]

To begin with, we should make a few comments on the Latin translation from the Arabic ascribed to Michael Scot. It may be generally said that it is sufficiently accurate. "Scientiae mathematicae" in the *Metaphysica Arabica* corresponds to "al-ʿulūm al-tāʿlīmīya" (the mathematical sciences) in the Arabic text of Averroës's *Commentary on the Metaphysics* (*Tafsīr ma baʿd al-tabīʿa*), published by Maurice Bouyges in 1942.[60] "Mathematica" in the Latin

[59] *Op. cit.*, ff. 146v–147r: "Id est, Et rectum est quærere, postquam declarauit talem scientiam esse, quæ dicitur prima Philosophia, vtrum ista scientia sit vniuersalis pluribus generibus, aut eiusdem generis, aut eiusdem naturæ. Deinde d. Non est enim vna species. i. apparet enim, quod Mathematica non est vnius generis, sed diuersorum generum. Geometria enim est naturæ diuersæ à natura, cuius est Astrologia, & vtra que est Mathematica. Deinde d. & scientia vniuersalis communis est omnibus. i. sed, si scientia ista fuerit vniuersalis, dignius vt sit communis omnibus entibus. D. d. Si igitur non fuerit, &c. i. & manifestum est, quod si non esset hic alia substantia a sensibili, quod non esset scientia prior Naturali. D. d. Et si aliqua substantia mobilis [*sic!*], &c. i. & manifestum est quod, si aliqua substantia à immobilis est, quod ista substantia est prima, & quod scientia istius est scientia vniuersalis, & Philosophia prima. Et dixit in esse, quia ordo addiscendi istam scientiam est posterius, & ideo discimus eam post Physicam." We read "immobilis" for "mobilis". In the reproduced text, we transcribe all abbreviated symbols in the original into complete words except "D." (Deinde), "d." (dixit), and ".i." (id est).

[60] Averroës, *Tafisr Ma Baʿd At-Tabiʿat: Bibliotheca Arabica Scholasticorum, Série Arabe*, t. VI, Texte Arabe inédit etabli par Maurice Bouyges, deuxième volume de texte (Beyrouth, 1942), p. 713, *ll.* 3 & 14. (العلوم التعليمية). It is known that Averroës consulted two or three Arabic translations available to him. See Arnaldez, "Ibn Rushd," *The Encyclopaedia of Islam*,

rendering of the Averroës commentary corresponds to "al-taʿlīm" or "taʿlīmīya" (mathematics),[61] and "scientia universalis" to "al-ʿilm al-kullī" (the universal science).[62]

Now let us turn to the Latin text of the *Arabica* and Ibn Rushd (Averroës)'s commentary on it. First, in the *Metaphysica Arabica* text "ἡ καθόλου" in the Greek original is interpreted as "scientia universalis", which stands more or less independent of the discussion about mathematics. This translation may remind one of a science in its unqualified context, that is, in the context not directly related to the argument on mathematics. In fact, Ibn Rushd reinterprets the sentence referring to the universal science as "it is common to all beings" and connects it finally to first philosophy. The Commentator is known to have rejected the view of metaphysics as the universal science in the narrow sense that it gives all the other sciences their goals and principles.[63] One of the reasons which leads him to interpret "ἡ καθόλου" in the Greek original as "the universal science" in the unqualified sense can be surely attributed to a philological one, i. e. the Arabic translation of Aristotle which he consulted.

Second, Ibn Rushd's explanation that the universal science is not related to mathematics seems to have been determined not only by the philological reason pointed out above but also by his epistemological view on mathematics. It is reported that in general he did not value mathematics very highly. As a scholar studying his *Commentary on the Metaphysics* points out, in Ibn Rushd's opinion, "mathematics, instead of forming a bridge between physics and metaphysics, is degraded to a deficient (*nāqiṣ*) kind of being hardly worth studying at all."[64] Accordingly, it seems no accident that Ibn Rushd excluded mathematics from his notion of universal science.

Third, however, Ibn Rushd's exposition is different from Albert's and

---

Vol. III, p. 917. But the Bouyges edition ascribes the Arabic text of Book VI to Astāt, or Eustathius, who translated the *Metaphysics* for al-Kindī. See "Observations générales," by the editor on p. IX**; Tome, V, 1 (Notice) (Beyrouth, 1952), pp. CXVIII–CXIX and CXXIX. On Astāt see Peters, *Op. cit.* (n. 57), pp. 59, 159 & 160.

[61]The Bouyges edition, p. 713, *l.* 15 (التعاليم); p. 714, *l.* 2 (تعاليمية), respectively. 'taʿlīm' (تعليم; pl. 'taʿālīm, تعاليم) is derived from the verb 'ʿalima' (علم), i. e. "to know, to learn" (gerund, or *masdar*). 'taʿlīm' is different from 'riyāḍīyāt' (mathematics, رياضيات<راض, 'rāḍa', "to train") in current usage, and is close to the Greek noun 'μάθησις'. In fact, Thābit Ibn Qurra of the ninth century, for example, utilized 'taʿlīm' and 'al-ʿulūm al-taʿlīmīya' for 'μάθησις' and 'τὰ μαθήματα', respectively. See Wilhelm Kutsch, ed., *Ṯābit ibn Qurra's Arabische Übersetzung der* 'Αριθμητικὴ Εἰσαγωγή [*sic!*] *des Nikomachos von Gerasa zum Ersten Mal herausgegeben* (Beyrouth, 1959), "Wörterverzeichnis: Griechisch-Arabisch," p. 209; "Arabisch-Griechisch," p. 324.

[62]The Bouyges edition, p. 713, *l.* 4; . 714, *ll.* 3 & 11 (العلم الكلى).

[63]Roger Arnaldez and Albert Z. Iskandar, "Ibn Rushd," *DSB*, XII, p. 6. James C. Doig discusses Averroës's view of metaphysics by contrast with those of Albertus Magnus and Thomas Aquinas in *Op. cit.* (n. 43), pp. 196–199.

[64]Charles Genequand, *Ibn Rushd's Metaphysics: A Translation with Introduction of Ibn Rushd's Commentary on Aristotle's Metaphysics, Book Lam* (Leiden, 1984), pp. 13–14. This monograph provides very useful information on Averroës's general philosophical viewpoint, although its main subject is Book Lam (XII).

Thomas's. The Islamic Commentator takes Aristotle's original question to be one which asks whether first philosophy is universal to many genera or simply to one nature. And he makes Aristotle answer this question concerning first philosophy with comparison to mathematics. In his exegesis, geometry and astronomy, on the one hand, are of one genus; mathematics containing those two particular subjects, on the other hand, is of many different genera. In this sense, his paraphrase is similar to Alexander's in so far as the passage immediately before the sentences which discuss the universal science is concerned. He might have possibly studied the ancient commentary ascribed to Alexander of Aphrodisias, although we do not know whether he read the one which treats Aristotle's passage 1026a23–32.[65] But, as we have pointed out, he and the Arabic translator interpret "ἡ καθόλου" in the Greek text as "the universal science" and at last first philosophy.

It has been acertained that both Albert and Thomas studied Averroës (Ibn Rushd)'s *Commentary on the Metaphysics* while they were composing their own interpretations of the knotty book by the Prince of Greek philosophers.[66] From what we have considered above, we now know that the *Arabica* text and Ibn Rushd's elucidation of it could have influenced both of the Dominican friars who interpreted "illa universaliter" to be exclusively first philosophy as the only universal science.

C. Albert and Thomas against the Oxford 'Platonists'

From our textual examination above, it becomes clear that Albert and Thomas partially adopted Averroës's commentary. Their concern was to emphasize that metaphysics is the only universal science and that mathematics does not deserve to be called a universal science. They did not accept the Islamic Commentator's exegesis that mathematics as a whole is common to many genera.

As we mentioned in Subsection A, Albertus Magnus's commentary on the *Metaphysics* was conditioned by his keen awareness of a current problem, his contemporaries' committing the "error Platonis." It is today well-known that the Oxford 'Platonists' greatly valued the mathematical disciplines for studying the natural world. For example, in his *Opus tertium* Roger Bacon went so far as to claim that mathematics is the "prima scientiarum" without which the other sciences cannot be known.[67] In his *Opus maius* he praised the certainty of mathematics by stating that "mathematics alone remains certain

[65] A. Arnandez, "Ibn Rushd" (n. 55), p. 917. J. Freudenthal, "Die durch Averroes erhaltenen Fragmente Alexanders zur Metaphysik des Aristoteles: untersucht und übersetzt," *Abhandlungen der Berliner Akademie vom Jahre 1884* (Berlin, 1885) almost exclusively examines the commentary on Book XII(Λ), and remains silent about the one on Book VI(E).

[66] See Peters, *Op. cit.* (n. 57), p. 231; William A. Wallace, "Aristotle in the Middle Ages," *Dictionary of the Middle Ages*, **1** (New York, 1982), pp. 456–469, esp. p. 462.

[67] *Opus tertium*, in *Fr. Rogeri Baconi Opera quaedam hactenus inedita*, Vol. I, ed. J. S. Brewer (London, 1859), p. 105: "Atque propter hoc sequitur quod est prima scientiarum, sine quibus aliæ sciri non possunt."

and verified for us with the utmost certainty and verification. Therefore by means of this science all other sciences must be known and certified."[68] Thus in the writings of Bacon, the 'encomia' of mathematics are abundant.[69] For this reason, some historians of science have highly evaluated his methodology of natural philosophy asserting that "the difference between Bacon and ourselves is of degree, not one of kind."[70] This statement was formulated to corroborate Pierre Duhem's and Alistair Crombie's celebrated thesis which emphasizes an unbroken continuity between medieval and modern science.

In order to understand historically how Albert and Thomas responded to the so-called Oxford 'Platonists', whose principal figures were Robert Grosseteste and Roger Bacon, and not to commit an error of historical anachronism, let us briefly sketch what Grosseteste and Bacon achieved. First of all, Bacon did not claim by any means that the natural world can be exhaustively described by mathematical language. With his predecessor Grosseteste, he admitted that that mathematics is important for studying nature but that there exist certain physical princples which cannot be obtained from mathematics itself.[71]

Second, the mathematical approach to nature maintained at thirteenth century Oxford was based on the 'metaphysics of light' in the case of Grosseteste[72] and on the specific doctrine of the 'multiplication of species' (*multiplicatio specierum*) in the case of Bacon.[73] So the method of the Oxford 'Platonists' was specifically bound to the intellectual context of the thirteenth century and did not necessarily lead continuously to the mathematical physics

---

[68] *Opus maius*, ed. J. H. Bridges, Vol. I (Oxford, 1897), p. 107; *Opus Maius of Roger Bacon*: A Translation, by Robert Belle Burke, Vol. I (Philadelphia, 1928), p. 127.

[69] John E. Murdoch has pointed out that the essence of Bacon's philosophy of mathematics consisted in "les *encomia* des mathématiques" rather than creative mathematical activities. See his *"Rationes mathematice": Un Aspect du rapport des mathématiques et de la philosophie au moyen âge* (Paris, 1962), p. 7

[70] N. W. Ficher and Sebetai Unguru, "Experimental Science and Mathematics in Roger Bacon's Thought," *Traditio*, **27** (1971), p. 367.

[71] In this regard, consult David C. Lindberg, "On the Applicability of Mathematics to Nature: Roger Bacon and His Predecessors," *The British Journal of the History of Scienc*, **15** (1982), pp. 3–25, esp. p. 18. Lindberg in general provides a critical and sound view on Bacon's natural philosophy. Impressively, he begins this article with the following sentence on p. 3: "Roger Bacon has often been victimized by his friends, who have exaggerated and distorted his place in the history of mathematics." For the same point in Grosseteste, see James A. Weisheipl, *The Development of Physical Theory in the Middle Ages* (Ann Arbor, 1971; first published in 1959), p. 52.

[72] A. C. Crombie, *Robert Grosseteste and the Origins of Experimental Science 1100–1700* (Oxford, 1953), Chapter VI, "Metaphysics of Light," pp. 128–134. Crombie's general view on Grosseteste's philosophy, however, must be read critically. A criticism of Crombie can be found in James McEvoy, *The Philosophy of Robert Grosseteste* (Oxford, 1982), Part Three: "The Light of Nature," Chapter 2, "Grosseteste's Place in the History of Science," pp. 206–222.

[73] David C. Lindberg, *Roger Bacon's Philosophy of Nature, A Critical Edition, with English Translation, Introduction, and Notes, of De multiplicatione specierum and De speculis comburentibus* (Oxford, 1983), Lindberg's introduction, esp. "Bacon's Doctrine of the Multiplication of Species," pp. liii–lxxi.

advanced by Galileo and Descartes in the early seventeenth century.

Third, Bacon, following Aristotle (see Section 5), significantly held in his *Communia mathematica* that the truth of the principles of mathematics is ascertained not by mathematics itself but by metaphysics. According to him, it is a task of metaphysics to argue about their validity.[74] In respect to this, he admitted the superiority of metaphysics to mathematics.

Fourth and last, Bacon, who was sometimes called "doctissimus mathematicus," was historically significant not because he was a great working mathematician but because he was a promoter of mathematics who suggested on the value and importance of the mathematical sciences.[75] He wrote in praise of mathematics incomparably more abundantly than any other of his contemporaries. Grosseteste and Bacon may be characterized after all as Aristotelians, but their Aristotelianism, being still influenced by the theology of Augustinus and by twelfth-century translations of Arabic philosophers and scientists, was very syncretistic, or more precisely, heavily Neoplatonic. This is clear in Grosseteste, who tried to understand what the 'scientia demonstrativa' is by commenting on Aristotle's *Posterior Analytics* for the first time in the Western Latin world.[76]

At such a juncture, it was Albertus Magnus who attempted to reform the intellectual situation of the mid-thirteenth century, by introducing a 'genuine' full-fledged Aristotelianism by attempting to master the newly translated Aristotelian corpus. Albert's *Metaphysics* was composed in order to reform the Christian intellectual world of his time. To him the 'Platonic' philosophy of the Oxford Franciscans seemed to be harmful. Roger Bacon "did not emphasize the supreme role of metaphysics in human knowledge."[77] In fact, it is not a mere accident that Bacon, in Book VI(E) of his *Questions on the Metaphysics* took up Aristotle's question on the universal science, but there he did not praise the universality and superiority of metaphysics so highly as Albert.[78]

---

[74] *Communia mathematica*, in *Opera hactenus inedita Rogeri Baconi*, Fasc. XVI, ed. Robert Steele (Oxford, 1940), p. 66. Concerning mathematical principles such as "omne totum est majus sua parte," Bacon states "[...] quod requirat potestatem Methaphysice, cujus est omnium scientiarum principia declarare et a qua omnes sciencie accipiunt sua principia tamquam sibi probata est manifesta." I am indebted to Weisheipl's description in his, *Op. cit.* (n. 40), p. 138. Weisheipl has referred to p. 6 (instead of p. 66) of the *Communia mathematica* obviously by mistake.

[75] David Eugene Smith emphasizes this point in "The Place of Roger Bacon in the History of Mathematics," in A. G. Little, ed., *Roger Bacon: Essays Contributed by Various Writers on the Occasion of the Commemoration of the Seventh Centenary of His Birth* (New York, 1972; first published in 1914), pp. 153–183.

[76] See Eileen Serene, "Demonstrative Science," *The Cambridge History of Later Medieval Philosophy*, edited by N. Kretzmann, A. Kenny, & Jan Pinborg (Cambridge, 1982), pp. 496–517, esp. p. 498.

[77] J. A. Weisheipl, *The Development of Physical Theory in the Middle Ages* (n. 71), p. 56.

[78] *Questiones supra libros prime philosophie Aristotelis* (*Metaphysica*, I, II, V–X), in *Opera hactenus inedita Rogeri Baconi*, Fasc. X, ed. Robert Steele (Oxford, 1930), p. 184. See Charles H. Lohr, "Medieval Latin Aristotle Commentaries, Authors: Robertus-Wilgelmus," *Traditio*,

In contrast to the Oxford 'Platonists', Albert maintained that mathematics cannot demonstrate the physical proper reason. According to him, the doctrine that crucial aspects of nature are mathematically described is a serious mistake. In a discussion on the earth's sphericity he held that the astronomical (therefore mathematical) proof that the earth is a sphere does no more than establish the fact (*quia*), while the reason (*causa propter quid*), i. e. why the earth is round, cannot be established mathematically by the astronomer.[79] Grosseteste and Bacon maintained that the mathematical sciences such as astronomy and optics can provide the proper explanation of natural events. For Albert, however, such an opinion was an error derived from Plato's ontology.

The thirteenth century conflict between the Oxford 'Platonists' and Albertus Magnus on the status of mathematics[80] reminds us of the debates on the certainty of mathematics which occured in sixteenth century Italy, starting between Alessandro Piccolomini and Francesco Barozzi (see Chapter 2, § 2), though the scale and nature of the controversies were quite different. The school which became influential in the high and later Middle Ages was Albert's 'genuine' Aristotelianism. In the short term at least, 'the friends of Plato' (*amici Platonis*) were defeated. In the conflicts of the sixteenth century, the trend which emphasized the value of mathematics came to dominate the intellectual and social world. Descartes contributed greatly to the concluding of these controversies.

In any case, we can witness a difference in the intellectual scene in the conflicts of the thirteenth century and the debates of the early modern period. We have already seen that the Albertian view was accepted by his able disciple Thomas Aquinas. Next we will examine one more piece of evidence which confirms the fact that Albert's standpoint became dominant in the scholastic-Aristotelian tradition of the fourteenth century.

D. The Scotist Antonius Andreae

John Duns Scotus was one of the representative thinkers of the late thirteenth and early fourteenth centuries who commented extensively on the Aristotelian corpus. The Wadding-Vivès edition of Duns Scotus's *Opera omnia*, published in the late nineteenth century, includes a work entitled *In XII libros Metaphysicorum Aristotelis expositio*, a paragraph-for-paragraph commentary on the *Metaphysics*. This commentary, which is called *Metaphysica textualis*, was wrongly ascribed to Duns Scotus by Wadding-Vivès, actually having been written by An-

29 (1973), pp. 118–119.

[79]See Benedict M. Ashley, "St. Albert and the Nature of Natural Science," in Weisheipl, ed., *Op. cit.* (n. 71), pp. 96–97; Lindberg, *Op. cit.* (n. 71), p. 15.

[80]For the details of this debate, see among others, the works cited in note 40; and Weisheipl, *Op. cit.* (n. 71), "Physical Theory and Mathematics," pp. 48–60; *Idem*, "Classification of the Sciences in Medieval Thought," in J. A. Weisheipl (ed. by William E. Carroll), *Nature and Motion in the Middle Ages* (Washington, D. C., 1985), pp. 203–237, esp. pp. 216–237; Jeremiah M. G. Hackett, "The Attitude of Roger Bacon to the *Scientia* of Albertus Magnus," Weisheipl, ed., *Op. cit.* (n. 38), pp. 53–72.

thony Andreas (or Antonius Andreae), a disciple of Duns Scotus.[81] Andreas, called 'Doctor dulcifluus' (the 'Sweet-talking' Doctor), was a Franciscan, like his mentor 'Doctor subtilis.' The Franciscan Order of the late medieval period in general retained an Augustinian influence in its philosophy-theology. So it would be rather significant to observe how the Franciscan Friar commented on the Aristotelian passages which we have so far examined.

It was *Metaphysica Moerbecana* that the disciple of Duns Scotus referred to for his text. We read the following paraphrase of 1026a23–27 in Book VI(E) of his *Metaphysica textualis*:

> He [Aristotle] says that "someone will doubt indeed whether first philosophy", i. e. metaphysics, "is universal," as it certainly considers the common and universal being, "or is about some genus and" one determinate "nature"; and he mentions a reason for the negative part [of the question], stating that "for" there is "not the same mode" ⟨of this science⟩ "as in mathematics, because geometry and astronomy", which are mathematics, "are about some" determinate "nature"; "whereas the former," i. e. first philosophy is "universally common to all," and therefore it will consider being in common. He does not propose a reason for the opposite part [i. e. the affirmative part of the question], but it is possible to obtain it from what has been said previously; for if metaphysics is about immovable and separated being, it is seen that it would be of some determinate nature and not about being as a whole.[82]

[81] "Of the three *Metaphysicae* ascribed to Scotus by Wadding-Vivès, namely, *Metaphysica textualis, Conclusiones metaphysicae,* and *Quaestiones super libros metaphysicorum*, it can be said with certainty that only the *Quaestiones* belong to Scotus, and only as far as the first nine books are concerned. [...] The *Metaphysica textualis* was written by Anthony Andreas, a disciple who testified that he followed faithfully the works and statements of Scotus." Charles Balic, "The Life and Works of John Duns Scotus," in John K. Ryan and Bernardine M. Bonansea, eds., *John Duns Scotus, 1265–1965* (Washington, D. C., 1965), pp. 21–22. For Anthony Andreas, or Antonius Andreae, see Rudolf Eisler, *Philosophen-Lexikon: Leben, Werke und Lehren der Denker* (Vaduz/Liechtenstein, 1977), p. 16; C. H. Lohr, "Medieval Latin Aristotle Commentaries: Authors A-F," *Traditio*, **23** (1967), pp. 363–365.

[82] *In XII libros Metaphysicorum Aristotelis expositio: Joannis Duns Scoti Opera omnia,* t. VI: *Metaphysica textualis libri V–XII* (Paris, 1892), pp. 135–136: "Dicit ergo, quod *dubitabit utique aliquis utrum prima Philosophia*, id est, Metaphysica *sit universalis, ut* scilicet consideret ens commune et universale, *aut sit circa aliquod genus et naturam*, unam determinatam; et tangit rationem pro parte negativa, dicens, *non enim* est *idem modus*, supple hujus scientiæ, *ut in Mathematicis, quia Geometria et Astrologia*, quæ sunt Mathematicæ, *circa aliquam* determinatam *naturam sunt; illa vero*, id est, prima Philosophia est *universaliter omnibus communis*, et ideo considerabit de ente in communi. Rationem pro parte opposita non ponit, sed potest haberi ex prædictis; nam si Metaphysica est de ente immobili et separato, videtur quod sit alicujus naturæ determinatæ, et non de omni ente." In our translation, words within angle brackets ⟨ ⟩ indicate ones which the original author suggests to supply. The *Moerbecana* text which Wadding-Vivès has printed is slightly different from the *Marietti Moerbecana* quoted above (n. 45): "Dubitat enim utique aliquis, utrum prima Philosophia sit

As can be easily recognized, the Scotistic interpretation of 1026a23–27 closely resembles that of Thomas Aquinas's. First philosophy, in the commentator's opinion, is not comparable to mathematics. Geometry and astronomy, two fields of mathematics, are about some determinate nature and not universally common to all. The only science that is qualified to be universal is first philosophy.

In the same author's commentary on the passage 1064b6–14 of the *Metaphysics* XI(K), we witness a striking statement concerning the status of mathematics in the scholastic hierarchy of the sciences of the late Middle Ages. Here the disciple of Scotus, following Aristotle, contrasts sharply rather than compares metaphysics with the other sciences:

> He [Aristotle] compares this science [first philosophy] to the other [sciences] with respect to universality, saying that "someone will doubt whether" ⟨this science⟩, which is about divine beings, "it is necessary to posit a universal science of being *qua* being, or not"; and he proves so by division, saying that "each one of the mathematical sciences is about some one determinate genus," [and that] what is the "universal" ⟨science⟩ is "about all in common," and in consequence no mathematical science is universal to all beings. And he puts it: "if in fact, therefore, natural substance", namely those which are movable and sensible, "are the first of beings," it follows that "natural science will be indeed the first of sciences," because the order of sciences is with respect to the order of subjects. "If, however," [there is another] "nature and substance, separable" ⟨from the natural [beings]⟩ "and immovable, it is necessary that there be another science of it prior to natural [science] and also univeral because it is prior", because the same science is what explores first beings, namely immaterial substances; and is the universal [science] considering the entire being *qua* being, because those separated substances are the causes and principles of being *qua* being, as was already often stated. Accordingly neither natural nor mathematical [science] is a universal science. As to the practical sciences also, it is evident that none of them are universal; therefore it remains by division that that should be the universal science considering the entire being.[83]

universalis, aut circa aliquod genus, et naturam unam, non enim idem modus, nec in Mathematicis, quia Geometria, et Astrologia circa aliquam naturam sunt, illa vero universaliter est omnibus communis."

[83] *Ibid.*, p. 472: "Comparat hanc scientiam ad alias penes universalitatem, dicens quod ***dubitabit aliquis utrum*** supple hanc scientiam, quæ est de entibus divinis, ***oportet ponere scientiam universalem entis inquantum ens, aut non***; et probat quod sic per divisionem, dicens quod ***unaquæque scientiarum Mathematicarum, est circa aliquod unum genus determinatum***, quæ est supple scientia ***universalis communiter de omnibus***, et per consequens nulla scientia Mathematica est universalis omnium entium. Et subdit: ***siquidem igitur substantiæ naturales***, scilicet quæ sunt mobiles set sensibiles ***sunt primæ entium***, sequitur quod ***et scientia***

What the Scotist commentator emphasizes in his commentary is that metaphysics alone is universal. According to him, "no mathematical science is universal to all beings" and "neither natural nor mathematical [science] is a universal science." We have witnessed similar statements in Albert's and Thomas's writings. But nothing is comparable to the above quoted declarations which deny straightforwardly that mathematics is universal. It is worth remembering that the intellectuals of the fourteenth century took for granted the overwhelming universality of metaphysics or theology.

We will see, of course, a quite different intellectual scene in the late sixteenth and early seventeenth centuries. Before looking at the scene of the early modern period, let us see how Renaissance commentators read the Aristotelian passages that we are concerned with.

§ 4. Renaissance Commentators: Agostino Nifo and Pedro da Fonseca

A. Agostino Nifo

In the preceding section we have seen that Aristotle's passages which are supposed to contain a discussion of 'universal mathematics' were very specifically reinterpreted to such a degree that they indicate the particularity of mathematics by comparison with the universality of first philosophy. The present section discusses how they were explained by Renaissance commentators of Aristotle. In what follows we examine only two exegetes: Agostino Nifo and Pedro da Fonseca. The former seems important as he was clearly one of the first who presented commentaries on Aristotle based on the Greek text or texts and thus freed himself from medieval interpretations. And the latter Jesuit commentator attracts our special interest because he was an influential commentator

---

naturalis erit utique prima scientarum, quia penes ordinem subjectorum est ordo scientiarum. ***Si autem est natura, et substantia separabilis***, supple a naturalibus, ***et immobilis, necesse est ipsius alteram esse scientiam et priorem naturali, et etiam universalem eo quod priorem***, quia eadem est scientia, quæ speculatur prima entia, puta substantias immateriales; et quæ est universalis considerans de omni ente inquantum ens, quia istæ substantiæ separatæ sunt causæ, et principia entis inquntum ens, ut jam dictum fuit sæpe. Quare igitur nec naturalis, neque Mathematica est scientia universalis. De practicis etiam scientiis patet quod nulla earum est universalis; relinquitur ergo a divisione, quod ista sit universalis considerans de omni ente." The Wadding-Vives version has supplied the following moerbeke text: "Dubitabit autem utique quis, utrum entis inquantum ens scientiam universalem oportet poni, aut non? Mathematicarum quidem unaquæque circa unam aliquod genus determinatum est. Quæ autem universali conveniunt de omnibus. Siquidem igitur naturales substantiæ primæ entium sunt, et naturalis utique prima scientiarum erit. Si autem est altera natura et substantia separabilis et immobilis; alteram et necesse ecientiam ipsius esse, et priorem naturali, et universalem eo quod priorem."

of Aristotle in the Society of Jesus of the late sixteenth and early seventeenth centuries, well within the intellectual milieu in which Descartes was educated.

One of the significant differences between medieval and Renaissance Aristotelianism was that the latter brand of Aristotelianism had the advantage of the large number of Greek texts and commentaries of the *Corpus Aristotelicum* printed in both Greek and Latin.[84] As to the *Metaphysics*, in the Middle Ages the commentary by Alexander of Aphrodisias, for example, was only fragmentarily and indirectly known through Averroës's *Great Commentary.* But in the early sixteenth century, it came to be read both indirectly through the Latin translation from the Greek first printed in 1527 and directly through codices.[85] In that Latin translation by the Spanish theologian and historian Juan Ginés de Sepúlveda, "ἡ καθόλου"or "ἐκείνη καθόλου" in line 1026a27 of Aristotle's text and "ἡ ἁπλῶς μαθηματική" in the Alexandrian commentary (p. 297) were rendered into "illa universalis" and "mathematica simpliciter," respectively.[86] We should note the fact that through this exposition, for example, the Latin learned world of the sixteenth century discovered that "ἡ καθόλου" possibly signified philologically not something related to the metaphysical but the mathematical.

Alexander's influence was clearly visible, for example, in an exposition of the *Metaphysics*, first published in Venice in 1518. The author of this commentary was Agostino Nifo or Augustinus Niphus. He is reported to have learned Greek and come to prefer the Greek commentators of Aristotle over Averroës.[87] In this book he presents his exposition of the Latin translation of Aristotle's text after each paragraph. We read the passage corresponding to 1026a23–27 of the Bekker text:

---

[84] On the main characteristics of Renaissance Aristotelianism, consult Charles B. Schmitt, *Aristotle and the Renaissance* (Cambridge, Mass., 1983).

[85] [W. A. G.], "Alexander Aphrodisiensis," *A Dictionary of Greek and Roman Biography and Mythology*, ed. W. Smith, Vol. I (New York, 1967), p. 113: "*Comment. in Metaphysicorum XII libros*; ex versione J. G. Sepulvedae, Rome 1527, Paris, 1536, Ven. 1544 & 1561."

[86] *Alex▶Aphrodisiei Commentaria in duodecim Aristotelis libros de prima philosophia*, Interprete Ioanne Genesio Sepulueda Cordubensi (Venezia, 1561), p. 174. We present below the translations of Aristotle's text (1026a23–27) and Alexander's commentary on it: (Aristotle's text) "Dubitauerit autem, quispiam, vtrum vniuersalis sit ipsa prima philosophia, an circa genus quippiam vnamque naturam versetur. Neque enim vt ipsis mathematicis scientiis, idem est modus, sed Geometria quidem atque Astrologia circa naturam quandam versantur, illa vero est vniuersalis omnium sanè communis." (Alexander's commentary) "Dubitauerit enim quispiam vtrum vniuersalis sit ipsa prima philosophia prorsus vt omnes complectatur, eique caeterae vt particulares vniuersalibus subiiciantur, an non? Sed quemadmodum medicina, vel astronomia, & doctrinae caeterae in vna entis parte versantur, & vnam quandam naturam velut ex vniuersitate abcissam contemplantur, sic item theologia partem quampiam entis considerat: quoniam vt in disciplinis mathematicis notatur, Geometria disputat de magnitudinibus, Astronomia de stellis, & alia de alio. Mathematica vero simpliciter omnium est communis. Sic quid obstat quo minus physica aliquam entis partem consideret, aliam mathematica, thologia vero omnium communis habeatur?"

[87] Edward P. Mahoney, "Nifo, Agostino," *DSB*, X, pp. 122–124, esp. p. 122; Charles H. Lohr, "Niphus (Nifo), (Eutychus) Augustinus, Suessanus," in *Latin Aristotle Commentaries*: II. *Renaissance Authors* (Firenze, 1988), pp. 282–287, esp. p. 284.

> Someone, however, might doubt whether first philosophy is universal or about one genus and nature. For neither is there the same mode even in the mathematical sciences, on the one hand, geometry and astronomy are concerned with some proper nature. On the other hand, that (*illa*) which is universal is common to all.[88]

Nifo appears to have made the above translation from the Greek, for the Greek couplet "μέν–δέ" is rendered word for word into "quidem–vero." Before him, Johannes Bessarion, an eminent Greek scholar of the fifteenth century, also preserved the same word for word correspondence when he translated the *Metaphysics* directly from the Greek original.[89]

In the commentary which follows the above text, Nifo proposes his paraphrase as to what "that" (*illa*) means:

> He [Aristotle] says, "for neither is there the same mode even in the mathematical sciences," namely because it is of particular sciences and of 'universalis mathematica', "but, on the one hand, geometry and astronomy are concerned with some proper nature, on the other hand, that (*illa*)," namely, 'communis mathematica', which holds these, is "common to all."[90]

What does "universalis mathematica" or "communis mathematica" in this passage mean? Let us examine the next passage without answering this question immediately.

An examination of the passage shows that Nifo is familiar with Averroës's and Alexander's commentary on Aristotle's statement under discussion here:

> What then prevents physics from being considered some part of it [being] and mathematics another [part], whereas by theology what is common to all may be handled? The Expositor relates the word "that" (*illa*) to theology itself, but Alexander to 'communis mathematica' itself, for in this way the argument can be taken to be similar in the mathematical [sciences].[91]

---

[88] *Augustini Niphi Medicis Philosophi Suessani Expositiones, in Aristetelis libros Metaphysices* (Venezia, 1559), p. 350: "Dubitauerit autem quispiam sit ne Prima Philosophia uniuersalis, an circa genus, & naturam unam. neque enim in ipsis mathematicis scientijs idem modus est, sed geometria quidem, atque astrologia circa propriam quuandam naturam uersantur, illa uero uniuersalis omnium communis."

[89] *Aristotelis Opera omnia*, Vol. VIII: *Metaphysicorum libri XIIII, Op. cit.* (n. 58), f. 146v: "Non enim idem modus est atque in Mathematicis, sed Geometria quidem & Astrologia circa quandum naturam sunt, illa vero vniuersaliter omnibus communis est."

[90] Augustinus Niphus, *Loc. cit.*: "[D]icit {neque enim in ipsis mathematicis scientijs est idem modus} quod scilicet sit particularium scientiarum, & vniuersalis mathematicae {sed geometria quidem, atque astrologia circa propriam quandam naturam versantur, illa vero} scilicet communis mathematica, quae has continet, est {omnium communis}."

[91] *Ibid.*: "[S]ic quid obstat, vt Physica aliquam eius partem consideret, aliam mathematica, theologia vero omnium communis habeatur? verbum {illa} Expositor ad ipsam theologiam

The Renaissance commentator thus knew both Averroës's and Alexander's interpretations. And he at last adopted the Greek commentator's exposition. He was influenced by Renaissance humanism which in general placed high value on Greek texts and commentaries.

Now we can examine what "universalis mathematica" or "communis mathematica" signifies in Nifo's exposition. Alexander paraphrased "ἡ καθόλου" into "mathematics simply," that is "mathematics as a whole." If we interpret the phrase "communis mathematica, quae has continet" more lucidly than the way presented above, for example, as "'communis mathematica' which contains these", our exegesis will maintain that "communis mathematica" (or "universalis mathematica") means "mathematics as a whole." This interpretation certainly accords with Alexander's, but unfortunately comes to contain the difficulty that Nifo's "communis mathematica" contradicts the alternative usage of the sixteenth century: 'communis mathematica' as a mathematical discipline common and applicable to the particular mathematical sciences (see Chapter 7, § 1).

But is it impossible to interpret the above phrase "communis mathematica, quae has continet" in another way, for example, "'communis mathematica' which embraces these [from inside]," or more explicitly as "'communis mathematica' which makes a connection with these"? Once we have adopted this rendering, Nifo's "communis mathematica" or "universalis mathematica" comes quite naturally to mean "a discipline common and applicable to the particular mathematical disciplines," a notion which surely has been used since antiquity. To translate "contineo" into "make a connection with" is at least possible but not common. Nevertheless, if we interpret it as "embrace from inside", the problem will be eliminated. Of course, this explanation contradicts Alexander's paraphrase. It seems to me more reasonable, thus, that here we interpret "communis mathematica" as "mathematics as a whole." However, we do not need to determine which is the better interpretation for the present argument and are satisfied simply with having recognized at least "η καθόλου" indicates something mathematical in Nifo's exegesis.

B. The Jesuit Commentator Pedro da Fonseca

Next we will examine another Renaissance commentary on the *Metaphysics* which could have been influenced by Alexander's interpretation. This exposition is historically more significant for us than Nifo's, because it was fully utilized in the Jesuit Order of the early seventeenth century. It was Pedro da Fonseca's *Commentarii in libros Metaphysicorum Aristotelis Stagiritae*, in four volumes published in 1577, 1589, 1604, and 1612. Fonseca is known to have been one of the representative scholars who planned the famous *Commentarii Collegii*

refert, Alex. autem ad ipsam communem mathematicam, nam sic argumentum poterit accipi per simile in mathematicis."

Conimbricensis of the works of Aristotle. And because of his commentary on the *Metaphysics* of Aristotle, he received the title of 'the Portuguese Aristotle'.[92]

In *Tomus tertius commentariorum in Metaphysicorum Aristotelis Stagiritae libros*, first printed posthumously in Évora in 1604, Fonseca provides his explanation of passage 1026a23–27. In his *Commentaries*, Fonseca first establishes the authentic Greek text, translates it into Latin and then presents his own interpretation, while referring to commentaries of his predecessors such as Alexander, Averroës, Albertus Magnus, and Thomas Aquinas.

What first attracts our attention in Fonseca's Greek text[93] is that he has chosen as Aristotle's text of 1026a27 "*ἐκείνη δὲ καθόλου πασῶν κοινή*" [*sic*!] instead of "*ἡ δὲ καθόλου πασῶν κοινή*" which is found in the main Bekker edition. In Bekker's version, the former is shown as a variant. "*ἐκείνη*" means "that", "the more remote" or "the former", for the Latin word "illa".

On examining his Latin translation of passage 1026a23–27,[94] we notice that its syntactic structure is fundamentally the same as Nifo's and Bessarion's Latin rendering which we have shown above. For example, the couplet "*μέν–δέ*" is translated verbatim into "quidem–vero." He renders the crucial sentence "*ἐκείνη δὲ καθόλου πασῶν κοινή*" into "illa vero omnibus universe communis est" (that, on the other hand, is universally common to all).

Does Fonseca interpret "that" (*ἐκείνη*, *illa*), as first philosophy or something mathematical? Before attempting to solve this problem, let us examine his explanation which follows immediately after the Greek text and its Latin translation. The Jesuit commentator amplifies Aristotle's passage as follows:

> From what has been said, the following doubt of not little importance arises: whether [on the one hand] first philosophy is universal, that is, it deals with being as it is being, and with its principles and attributes, hence with all beings; or on the other hand, it is par-

[92] For the life and works of Fonseca, see Joaquin Ferreira Gomes, "Pedro da Fonseca: Sixteenth Century Portuguese Philosopher," *International Philosophical Quarterly*, **6** (1966), pp. 632–644.

[93] *Comentariorum Petri Fonsecae in Metaphysicorum Aristotelis Stagiriae Libros tomus III* (Köln, 1615), p. 6: "*ἀπορήδειε γὰρ ἄν τις, πότερὸν ποθ'ἡ πρώτη φιλοσοφία καθόλου ἐστὶν, ἢ περί γένος καὶ φύσιν τινὰ μίαν, οὐ γὰρ ὁ αὐτὸς τρόπος, οὐδ' ἐν ταῖς μαθηματικαῖς: ἀλλ'ἡ μὲν γεωμετρία, καὶ ἀστρολογία περί τινα φύσιν ἰδίαν εἰσὶν, ἐκείνη δὲ καθόλου τασων κοινὴ.*" Note that Fonseca's procedure for editing the Greek text was critical. Gomes states: "The philological rigor of his methods of establishing the Greek text is still admired today by critics." *Op. cit.* (n. 92), pp. 639–640. Cf. Ch. H. Lohr, "Fonseca, Petrus de, S.J.," in *Latin Aristotle Commentaries*: II. *Renaissance Authors* (n. 87), pp. 150–151."

[94] *Loc. cit.*: "At enim dubitauerit aliquis, vtrum prima philosophia sit vniuersalis, an potius in genere vno, & natura vna versetur? Neque enim idem est modus etiam in Mathematicis; sed Geometria quidem, & Astrologia, propriam quandam naturam considerant: illa vero omnibus vniuerse communis est." Incidentally, Francisco Suarez highly praises Fonseca's Latin translation in his *Disputationes Metaphysicae* (1597) by stating that it is "so elegant and lucid that it can be understood by anyone almost without exposition." Suarez, *Disputationes metaphysicae*, Vol. 1 (Paris, 1866), "Index Locupletissimus in Metaphysicam Aristotelis," Lib. I, Ch. VII, p. IV. Quoted by Gomes, *Op. cit.* (n. 92), p. 639.

> ticular, and does not go beyond the boundary of some determinate genus.[95]

To remove this doubt, Fonseca first inquires into the former case in which first philosophy is thought to be universal. He states directly after the sentence above:

> If indeed what has been said in Book IV, Chapter 1 and the beginning of this [Book VI, Chapter 1] are true, first philosophy will be universal and extend to the genera of all things. For it has been said in those places that it does not deal with things in the manner in which geometry, astronomy, and the other particular sciences do (assuredly so because the particular science partitions off some particular genus from the total universality of things and discusses it alone), but deals with being in common, or insofar as it is being, which the other [sciences] do not treat.[96]

In fact in Book IV, Chapter 1 and the introductory part of Book VI, Chapter 1, Aristotle asserts that none of the particular sciences such as mathematics deal generally with being *qua* being because they divide off a part of being and investigate the attributes of this part. So far our commentator faithfully follows the tradition of his medieval predecessors. He emphasizes that the particular mathematical sciences like geometry and astronomy are not comparable to first philosophy with regard to the universality of their subjects.

Fonseca then proceeds to the latter case in which first philosophy is viewed as a particular science. Concerning this case, the Coimbra professor of philosophy states:

> As it [first philosophy] is distinguished from physics and mathematics for the reason that all these are concerned with things which are not separated from matter (***although it is distinguished from mathematics to a lesser degree, which considers the immaterial in a certain manner***), the former [first philosophy] surely gives attention to disputing totally about that which is separated from matter, a certain peculiar genus of beings, not the total being.[97] [The emphasis is in

[95] Fonseca, *Op. cit.* (n. 93), p. 6: "Ex his, quæ dicta sunt, oritur dubitatio hæc non parui momenti: num prima Philosophia sit vniuersalis, hoc est, anne agat de ente, vt *ens*, deque eius principijs & affectionibus, ac proinde de omnibus entibus; an vero sit particularis, quæ non egrediatur fines ceti alicuius generis."

[96] *Ibid.*: "Si enim vera sunt ea, quæ lib. 4. cap. 1. & initio huius dicta sunt; erit prima Philosophia vniuersalis, quæ ad omnium rerum genera se extendat. Dictum est enim ijs locis non agere eam de rebus eo pacto, quo agit Geometria, & Astrologia, & aliæ particulares, ita nempe, vt ex tota rerum vniuersitate particulare aliquod genus abscindat, deque eo solo disserat, sed agere de ente in commune, siue quatenus ens est, quod aliæ non faciunt."

[97] *Ibid.*: "[C]um ea ratione distincta sit à Physica, & Mathematicis, quod hæ omnes versentur in rebus non separatis à materia (***tametsi minus à Mathematicis, quæ modo aliquo immaterialia considerant***) illa vero suscipiat curam disputandi de omnino separatis à materia, quod

the original.]

So according to Fonseca's interpretation, Aristotle (or rather someone) is asking the question whether first philosophy is concerned with all beings (and therefore universal), or deals simply with things separated from matter (in this case it is particular).

Then Fonseca tries to propose a solution of this problem:

> Consequently for two reasons this science is distinguished from other [sciences] just as the universal is from the particulars: one [reason] is because it deals with the first principles and the causes of all beings, which are immaterial things; the other [reason] is because it deals with those which are common to the material and the immaterial: anything other than this has not been admitted by anybody.[98]

In this statement Fonseca seems to emphasize that in any case first philosophy is distinguished from other particular sciences. Thus he first of all ascertains that the only universal science in the unqualified sense is first philosophy.

In agreement on this point, at the end of his commentary about 1026a23–32, the Jesuit philosopher presents his exposition of the following sentence, which is our present concern: "For neither is there the same mode in the mathematical sciences, etc." His paraphrase is as follows:

> For neither is its mode of investigation the same as in other sciences, even in the mathematical sciences, which are closer to first philosophy in the mode of considering their subjects, although not in the subjects considered.[99]

Fonseca recognizes here that the method of studying the mathematical sciences is similar to that of metaphysics, but not the objects of study. Taking a broad view, however, the mathematical sciences, which are particular, are contrasted with first philosophy, the unique universal science. What Fonseca wants to claim in the commentary examined above is the last point, that first philosophy is uniquely universal.

One thing which may perplex us in his exposition is that Fonseca gives no explicit interpretation of what the sentence 1026a27 "that is universally common to all" actually signifies. In editing the Greek text, he is conjectured to have noticed that "that" (ἐκείνη, *illa*), on balance, means something mathematical

---

est peculiare quoddam genus entium, non totum ens."

[98] *Ibid.*, pp. 6–7: "Itaque duabus de causis distinguitur hæc scientia à cæteris tanquam vniuersalis à particularibus: vna, quia agit de primis principijs, & causis omnium entium, quæ sunt res immateriales; altera, quia agit de ijs, quæ communia sunt materialibus, & immaterialibus: quorum neutrum vlli cæterarum concessum est."

[99] *Ibid.*, p. 7: "Neque enim eius tractandi modus idem est, qui in alijs scientijs, etiam Mathematicis, quæ in modo considerandi res suas magis ad primam philosophiam accedunt tametsi non in rebus consideratis."

rather than first philosophy. Moreover, he could have recognized that Alexander had interpreted "that is universally" to be "mathematics simply" (ἡ ἁπλῶς μαθηματική, *mathematica simpliciter*). Certainly, the phrase "in the mathematical [sciences]" (ἐν ταῖς μαθηματικαῖς, *in Mathematicis*) appears, but a substantial word (e. g. ἡ μαθηματική, *mathematica*) does not. In relatively remote places, we find that Aristotle frequently uses the word "ἡ μαθηματική" (*mathematica* in Fonseca's translation).[100]

Despite these circumstances which might have allowed the Portuguese commentator to adopt the exposition in which "that" means the mathematical, Fonseca does not give a clear explanation of what "that" implies. But in a sense such an attitude may be expected from a scholar who had "the right to be ranked among the outstanding representatives of the 'Second Scholasticism',"[101] the Scholasticism which inherited the great medieval tradition and which was at the same time influenced by Renaissance humanism. Fonseca, in any case, does not state explicitly what "that is universally common to all" means, simply emphasizing the incomparable superiority of metaphysics. This points again to Fonseca's 'syncretism', a typical intellectual predisposition for scholastics of the Society of Jesus.

We may imagine that the above interpretation was intended by Fonseca. The project of publishing his *Commentarii in libros Metaphysicorum* was conceived as a work conforming with the *Commentarii Collegii Conimbricensis* of which he was one of the planners. And its first volume appeared in 1577, one year after the publication of Pereira's monograph, which intended to confront the Neoplatonic proclamation of mathematics by Barozzi.[102] Fonseca might have entertained at least implicitly an idea to support Perira's thought. This teachs us that even an attempt at interpreting classical texts may reflect an intellectual background or a 'hermeneutical basis' of its interpreter.

Unfortunately Fonseca did not live long enough to compose a detailed commentary on passage 1064b6–9 comparable to that on 1026a23–27. He left only its Latin translation along with a very brief explanation, which was printed in the fourth volume of his commentary on the *Metaphysics*. What is noteworthy for us is only that he renders the sentence "ἡ δὲ καθόλου κοινὴ περὶ πάντων" of 1064b9 into "quae autem universalis est, communiter de omnibus disserit" (what, on the other hand, is universal treats in common about all).[103]

Incidentally, in Book VI, Chapter 1 of his *Scholae metaphysicae*, which was first published in 1566 and can be considered to be his commentary "on—

[100]For example, see 1026a7–8, 12–13, 14 & 19.

[101]Gomes, *Op. cit.* (n. 92), p. 637.

[102]Charles H. Lohr, "Metaphysics," in Charles B. Schmitt & Quentin Skinner, eds., *The Cambridge History of Renaissance Philosophy* (Cambridge, 1988), p. 607.

[103]*Commentariorum Petri Fonsecae in Metaphysicorum Aristotelis Stagiritae decimum, undecimum, & duodecimum tomus IV* (Coloniae, 1629), p. 63. he corresponding Greek text by Fonseca is "ἡ καθόλου κοινὴ περὶ τάντων" lacking "δέ".

or rather, against—the fourteen books of Aristotle's *Metaphysics*,"[104] Ramus critically reinterpreted Aristotle's statement that "metaphysics is exceedingly universal" (*Metaphysica est maximè universalis*). In his opinion metaphysics in Aristotle consists of some substantial component disciplines, for example, dialectic and theology. It is because of the superiority of these two sciences that metaphysics is supposed to be universal. Thus he states:

> The doctrine of matter, form, and the other, tenets of the dialectical art on the one hand are exceedingly universal and common for knowing and explicating all beings, but theology about God is said in another sense to be extremely estimable, desirable, and universal.[105]

It should be especially noticed that he praised the preeminence of the art of dialectic. In this one sees a 'Platonic' standpoint of Ramus. In any case, Ramus's metaphysics was a representation of humanist logic of sixteenth-century Protestantism, which regarded the role of human spirit to be important.[106] That is to say, Ramus was presented with an interpretation as to what a universal science should be. His answer was that it is in a sense dialectic and in another sense theology.

§ 5. The Status of Mathematics in the Aristotelian Scheme of Learning

How, then, did Aristotle himself actually think of "ἡ καθόλου" in 1026a27 and 1064b9? This problem is certainly related to how mathematics is viewed in the Aristotelian scheme of learning. As may be easily agreed, in Aristotle's general judgment, mathematics in any form cannot be viewed as a universal science if the latter means a knowledge concerned with being as such, or being *qua* being. In this sense, metaphysics alone deserves to be called a universal science. This is what the medieval commentators emphasized.

But how does Aristotle define metaphysics? According to Giovanni Reale, who has stressed the unitary character of Aristotle's conception of first philosophy, metaphysics is defined in various ways as follows: sometimes as (1) an aetiology or a theory of the causes and principles, sometimes as (2) an ontology

[104] Ong, *Ramus and Talon Inventory* (Cambridge, Mass., 1958), p. 355.

[105] Ramus, *Scholae metaphysicae*, in *Scholae in liberales artes* (Basileae, 1569), col. 918: "Doctrina materiae, formae, & reliquorum dialecticae artis dogmatum, maximé quidem est universalis & communis ad omnia entia, cognoscendum & explicandum: at theologia de Deo, longé alio sensu honorabilis, optabilis, universalis appellatur."

[106] For instance, see the following paper: Riccardo Pozzo, "Ramus' Metaphysics and its Criticism by the Helmstedt Aristoterians," in Mordechai Feingold, Joseph S. Freedman, and Wolfgang Rother, eds., *The Influence of Petrus Ramus: Studies in Sixteenth and Seventeenth Century Philosophy and Sciences* (Basel, 2001), pp. 92–106, esp. pp. 95–97.

or a doctrine of being *qua* being, sometimes as (3) an ousiology or a theory of substance, and finally as (4) theology or a doctrine of the immobile, eternal, and transcendent substance.[107]

From the Aristotelian point of view, mathematics cannot apparently satisfy these characteristics of first philosophy. First, mathematical knowledge is thought to be obtainable only 'through abstraction' (ἐξ ἀφαιρέσεως or δι' ἀφαιρέσεως), in other words 'by taking away' something from being as such, which is a subject of metaphysics.[108] Ian Mueller has thought it to be appropriate to give the name of 'abstractionism' to the epistemological standpoint of Philoponus and Ammonius among the commentators of Aristotle in late antiquity who emphasized the concept of 'abstract' originated from Aristotle.[109] In fact, there are several points in Aristotle's epistemological thought on mathematics conformative with the name of 'abstractionism'. Representative discussions can be seen in Chapters 1–3 of Book XIII (M) of the *Metaphysics*. For this reason mathematics is ontologically and epistemologically 'incomplete' particularly in comparison with first philosophy. It can simply provide us knowledge about the quantitative aspect of substance.

Second, according to Aristotle's understanding, mathematical knowledge cannot prove all of its own principles. In attempting to do so, the mathematician would commit the error of *petitio principii*. To study the validity of the mathematician's principles themselves is at least partly a task of first philosophy (not mathematics), as Aristotle clearly states in his *Metaphysics* XI (K), Chapter 4.[110] This is a further reason why mathematical knowledge is necessarily obliged to be 'incomplete'. Mathematical studies should always be supplemented, both ontologically and logically, by a metaphysical inquiry which makes being *qua* being the object of investigation. The ontological trueness of mathematical propositions will be guaranteed only after thorough metaphysical consideration about being.

---

[107] G. Reale, *Op. cit.* (n. 4), especially Conclusions: "The Horizons of the Aristotelian Metaphysics," pp. 349–363. See also G. Reale, *A History of Ancient Philosophy*, II. *Plato and Aristotle*, edited and translated by John R. Catan (Albany, 1990), pp. 265–292.

[108] For a detailed discussion on Aristotle's concept of 'abstraction', see Ian Mueller, "Aristotle on Geometrical Objects," *Archiv für Geschichte der Philosophie*, **52** (1970), pp. 156–171; also in J. Barnes, M. Schofield and R. Sorabji, eds., *Articles on Aristotle*, 3. *Metaphysics* (London, 1979), pp. 96–107. Cf. John J. Clearly, *Aristotle & Mathematics: Aporetic Method in Cosmology & Metaphysics* (Leiden/New York/Koln, 1995), 'abstraction' in General Index.

[109] Ian Mueller, "Aristotle's Doctrine of Abstraction in the Commentators," in Sorabji, ed., *Aristotle Transformed* (n. 24), pp. 463–480, esp. p. 467.

[110] *Metaphysica*, 1061b17–27: "Since even the mathematician uses the common axioms in a special application, it must be the business of first philosophy to examine the principles of mathematics also. That when equals are taken from equals the remainders are equal, is common to all quantities, but mathematics marks off a part of its proper matter and studies it separately, e. g. lines or angles or numbers or some other kind of quantity—not, however, *qua* being but in so far as each of them is continuous in one or two or three dimensions; but philosophy does not inquire about particular subjects in so far as each of them has such and such attributes, but considers each subject in relation to being *qua* being."

It is true that Aristotle argued what appropriate mathematical princples should be as starting points of mathematical discussions, but in such occasions he warned not to take a mistaken viewpoint such as Plato's theory of idea.[111]

Now, let us return to a more cocrete problem. A problem to be taken up here is what Aristotle means by the sentence "ἡ καθόλου" in 1026a27 and 1064b9. Is it more natural syntactically to take "ἡ καθόλου" to be related to something mathematical than to first philosophy? In fact, as we have sometimes pointed out previously, in 1026a25–27 Aristotle seems to contrast the particulars which follow the particle "μέν" with the universal which comes after "δέ" *within* the domain of the mathematical sciences. In the overall scheme of discussion, however, he tries to contrast mathematics and physics, which are both particular sciences with respect to metaphysics, with first philosophy. This interpretation is the one the Greek commentators take.

Thus it might seem to become more convincing that "ἡ καθόλου" signifies after all something mathematical. Let us continue our discussion for the present under the supposition that "ἡ καθόλου" means something mathematical. Does "ἡ καθόλου" mean in this case 'universal mathematics' in Heath's sense? Heath, after presenting some passages of Aristotle's *Metaphysics*, argued that "ἡ καθόλου" is related to universal propositions in mathematics which are exemplified by Eudoxus's general theory of proportion, now known through Euclid V.

But Heath's argument seems to us dubious with respect to the latter claim. That is to say, Book V of Euclid's *Elements* has been usually called the general theory of proportion which is commonly applicable to both numbers and magnitudes. This is notwithstanding, as we have pointed out in connection with Ian Mueller, Book V in on magnitude (μέγεθος), the object of geometry. As Mueller asserts, "magnitudes are geometric objects only and do not include numbers."[112] It is thus a historical misunderstanding to interpret Book V as the 'general theory' of arithmetical numbers and of geometrical magnitudes.

Heath, furthermore, was inclined to connect 'universal mathematics' to the algebra of the modern era. This interpretation certainly originated from his general position supporting the so-called 'geometrical algebra' first proposed by Tannery and Zeuthen. According to an interpretation of, for example, Euclid II by means of the concept of geometrical algebra, the main body of Book II

[111] Ian Mueller, "Aristotle's Approach to the Problem of Principles in Metaphysics M and N," in Andreas Graeser, ed., *Mathematik und Metaphysik bei Aristoteles: Akten des X. Symposium Aristotelicum, Sigriswil, 6.-12. September 1984* (Berne, 1987), pp. 241–259. For a very general discussion of Aristotle's criticism of Plato's concept of ideas, see Harold Cherniss, *Aristotle's Criticism of Plato and the Academy* (New York, 1962; first published in 1944).

[112] Ian Mueller, *Philosophy of Mathematics and Deductive Structure in Euclid's Elements* (Cambridge, Mass., 1981), Introduction, p. xi. Mueller also argues that "[T]here is no reason to connect Euclid's theory of proportionality with calculation, an argument which also weighs against algebraic readings of the *Elements*" (p. xi). See also the detailed discussion in Chapter 3, "Magnitudes in Proportion," pp. 118–151.

is simply an algebra formulated geometrically.[113] This point of view should be carefully reexamined and replaced by a more scrutinized historical theory.[114] Thus, there are serious problems in relating 'universal mathematics' to the notion of algebra.

Then, is it legitimate to identify a system of universal propositions in mathematics with 'universal mathematics'? No serious contradictions seem to follow from this identification. In fact, not only Heath but also Hippocrates G. Apostle supported this explanation. Apostle has devoted the first chapter of his book *Aristotle's Philosophy of Mathematics* to 'universal mathematics' which he considers to be a general theory of quantity placed prior to arithmetic and geometry.[115] But what was in reality a general theory of quantity in the time of Aristotle? Euclid V itself can be safely excluded from this body of mathematical knowledge, as we have just argued. Axioms and common notions in Euclid's *Elements* may have been judged to belong to 'universal mathematics' in Apostle's sense. For example, "things which are equal to the same thing are also equal to one another" is common to both arithmetic and geometry. So 'universal mathematics' would be such mathematical principles plus universal propositions which Aristotle refers to in several places in the *Metaphysics* and the *Posterior Analytics* (see Section 1).

Now, let us examine how Aristotle looks at these universal propositions in mathematics. In his *Posterior Analytics* he warns against a misunderstanding about universal demonstration in mathematics:

> If the universal is not a thing apart from the particulars, and demonstration instils an opinion that that in virtue of which it demonstrates is some thing, and that this belongs as a sort of natural object among the things there are (e. g. a triangle apart from the individual triangles, and a figure apart from the individual figures and a number apart from the individual numbers), and a demonstration about something there is is better than one about something that is not, and one by which we will not be led into error is better than one by which we will be, and universal demonstration is of this

---

[113] A standard view of 'geometrical algebra' may be seen in B. L. van der Waerden, *Science Awakening*, I, English Translation by Arnold Dresden (New York, $_3$1971), pp. 118–126.

[114] See, among others, Árpád Szabá, "Zum Problem der so. geometrischen Algebra in Euklids Elementen," ΠΡΙΣΜΑΤΑ : *Naturwissenschaftsgeschichtliche Studien, Festschrift für Willy Harter*, hrsg. von Y. Maeyama und W. G. Saltzen (Wiesbaden, 1977), pp. 373–393; I. Mueller, *Op. cit.* (n. 110), pp. 41–52; Ken Saito, "Book II of Euclid's *Elements* in the Light of the Theory of Conic Sections," *Historia Scientiarum*, No. 28 (1985), pp. 31–60.

[115] Apostle states at the end of Chapter 1, entitled "Universal Mathematics": "Since quantity is prior in existence and in definition to number and to magnitude, and since we arrive at the unit (which is the principle of number) by abstracting continuity itself, the order of scientific learning and of knowledge in mathematics is at once evident. Universal mathematics comes first, and it is concerned with quantities qua quantities and their properties. Next follows arithmetic and geometry." Hippocrates George Apostle, *Aristotle's Philosophy of Mathematics* (Chicago, 1952), p. 80.

> type (for as they go on they prove as in the case of proportion, e.g. that whatever is of such a type—neither line nor number nor solid nor plane but something apart from these—will be proportional)—so, if this is more universal and is less about something there is than the particular demonstration, and instils a false opinion, then the universal will be worse than the particular.[116]

Aristotle asserts here that the universal demonstration is worse than the particular under certain conditions, at least. But this statement may be deduced easily from Aristotle's ontology, which asserts that the universal demonstration in mathematics is concerned with more remote and abstract entities from being as such or substance than the particular. Accordingly we doubt that Aristotle ontologically and epistemologically attached importance to such a body of knowledge which Apostle has named 'universal mathematics'.[117] But from the viewpoint of the history of Greek mathematics, there could have been a general mathematical theory common and applicable to arithmetic and geometry (see in detail Chapter 7, § 1). Perhaps Asclepius may be regarded to have admitted this kind of theory. Nevertheless, to discuss whether the Greeks thought of such a system as a substantial mathematical discipline is one thing while to argue about what epistemological value Aristotle attached to this system of knowledge is quite another.

From our above consideration, we may draw the conclusion that it is not impossible to paraphrase "ἡ καθόλου" as "universal mathematics." But to exposit "universal mathematics" as Heath did may be anachronistic. It also appears to us that Alexander of Aphrodisias's explanation is not so unnatural as Crowley thinks and that it is an exposition that demands our further careful examination.

Our discussion above is under the presupposition that "ἡ καθόλου" signifies the mathematical. We are inclined to adopt this reading as did most of the commentators who read the Greek original. But we may admit, at the same time, that the possibility of interpretating "ἡ καθόλου" as metaphysics cannot be excluded altogether both syntactically and contexually. This paraphrase is the one which the medieval commentators almost unanimously chose, as far as we know.

To determine what Aristotle meant by "ἡ καθόλου" is not crucial for our present study. What has been ascertained is that in the Aristotelian tradition the concept of 'universal mathematics' in the sense of a system of mathematical knowledge commonly applicable to arithmetic and geometry was not epistemologically highly valued. Rather it was emphasized in particular during the

---

[116] *Analytica Posterior*, 85a31–85b3.

[117] We have been inspired by Crowley's criticism of Heath's and Apostle's interpretation of 'universal mathematics'. See *Op. cit.* (n. 19). But Crowley in general exposits Aristotle's texts as Thomas Aquinas read them. We have to reconstruct in a historically critical manner what Aristotle actually meant.

Middle Ages that mathematics is not universal in comparison with metaphysics. We will observe a different scene with the rise of a new intellectual tradition in the sixteenth century. The intellectual milieu will constitute the direct background in which Descartes conceived his 'mathesis universalis'.

Chapter 7

'Mathesis Universalis' in the Sixteenth Century

§ 1. Proclus Diadochus and Francesco Barozzi

It was through Proclus's *Commentary on the First Book of Euclid's Elements*, among others, that a novel mathematical idea entered the intellectual world of the sixteenth century. It is not too much to say that there has been no impact greater than Proclus regarding the status of mathematical sciences among the philosophies of mathematics in early modern Europe. We have seen in Chapter 2, § 2 that Clavius was greatly inspired by Proclus's philosophy of mathematics. By attempting to graft, as it were, the mathematical thought of Proclus on the trunk of the scholastic-Aristotelian scheme of learning, the Jesuit mathematician promoted the study of mathematics. Descartes is conjectured to have acquired the thought of Proclus through Clavius or others. Proclus also played a crucial role in spreading the idea that there is a substantial mathematical discipline prior and superior to arithmetic and geometry. This idea was useful to convince thinkers of the sixteenth century that there existed a mathematical discipline of 'mathesis universalis'.

Proclus's mathematical thought became known in modern Europe, as was noted in Chapter 2, § 2, through the Greek text of Simon Grynaeus the elder, published in 1533, and through the Latin translation of Francesco Barozzi, which appeared in 1560. We will examine Proclus's idea of a science which anticipates the later concept of 'mathesis universalis' by referring to the these two versions together with the critical Greek edition published by Gottfried Friedlein in 1873.

It is in *A Commentary*'s Book I (Prologue: Part One) where Proclus discusses comprehensively things commonly applicable to the mathematical sciences. Barozzi, the translator into Latin, has divided Book I into fifteen chapters. Since this division has been reproduced in Glenn R. Morrow's English translation, we will be able to follow Barozzi's partitioning of the chapters.

Proclus recognizes in Book I, Chapter III that there exists "the single science (ἡ μία ἐπιστήμη; *una scientia* in Barozzi's Latin translation) that embraces alike all forms of mathematical knowledge."[1] Barozzi called it "a divine science"[2] (*Diuina scientia*) in the marginal notes printed in his Latin translation of 1560. In the beginning of Chapter VII, the same notion appears as "the single science which we showed to be prior to the several mathematical sciences."[3] We will name this single mathematical discipline which is prior to the other particular mathematical sciences 'common mathematics' (*communis mathematica*). The legitimacy of this terminology will become clear as our discussion develops.

Proclus also points out in Chapter XIV that there is a unifying bond between the mathematical sciences and calls it "the single and entire mathematics"[4] (ἡ μία καὶ ὅλη μαθηματική; *una, & tota Mathematica*). He characterizes this notion as follows:

> As for the unifying bond (ὁ σύνδεσμος; *coniunctio*) of the mathematical sciences, we should not suppose it to be proportion (ἀναλογία; *proportio*), as Eratosthenes says. For though proportion is said to be, and is, one of the features common to all mathematics, there are many other characteristics that are all-pervading, so to speak, and intrinsic to the common nature of mathematics. We should prefer to say that the immediate bond of union between them is the single and entire mathematics which contains in itself in simpler form the principles of all the particular sciences, that science which teaches their common nature as well as their differences and what traits are the same in them all and what belong to more or fewer of them. Those who study mathematics in the proper way advance to this science from the particular ones.[5]

Certainly "the single and entire mathematics" is supposed to be similar to "the single science" in Chapters III and VII. There is a room of questioning whether

[1] *Procli Diadochi in primum Euclidis Elementorum librum commentarii*, ed. G. Friedlein (Leipzig, 1873), p. 7, *ll.* 17–18; *Procli Diadochi Lycii in primum Euclidis Elementorum commentariorum libri IV*, a Francisco Barocio (Padua, 1560), p. 4. Note that in the Greek original the word "commentary" was not used: Εἰς τὸ πρῶτον τῶν Εὐκλείδου στοιχείων βιβλίον. As for the English translation, we basically follow G. R. Morrow's translation: Proclus, *A Commentary on the First Book of Euclid's Elements* (Ch. 5, n. 180). His translation of the words for 'common mathematics' is not necessarily careful and consistent. Therefore, our translation sometimes deviates from it without notice. Morrow has included Friedlein's pagination, hence we do not indicate the pages of his translation. We also refer to Paul Ver Eecke's French translation: Proclus de Lycie, *Les Commentaires sur le premier livre des Éléments d'Euclide*, traduits par Paul Ver Eecke (Bruges, 1948).

[2] Barozzi, p. 4.

[3] Friedlein, p. 18, *l.* 16; Barozzi, p. 10.

[4] Friedlein, p. 44, *ll.* 2–3; Barozzi, p. 25.

[5] Friedlein, pp. 43–44.

the two are entirely identical, but it is certain that Proclus possesses a notion which may be seen as a unifying bond between all the particular mathematical sciences and which can justifiably be called 'common mathematics'.

How, then, does Proclus characterize 'common mathematics'? The Neoplatonic philosopher believes, first of all, that there exist "all-pervading principles that generate everything from themselves," that is to say, the Limit (*πέρας*; *finis*) and the Unlimited (*ἄπειρον*, *ἀπειρία*; *infinitum*).[6] Not only the primary principles of mathematical being but also the secondary principles that proceed from the former are the offspring of the Limit and the Unlimited. Both number and magnitude are to be generated by this pair of notions. Important here is that the Limit and the Unlimited are considered to precede the objects of arithmetic and geometry. In Chapter III Proclus argues that there are common principles and common theorems that pervade all classes of mathematical objects. It is here that he mentions the concept which we call 'common mathematics' for the first time.

> Let us enumerate the simple theorems that are common to them all, that is, the theorems generated by the single science that embraces alike all forms of mathematical knowledge; and let us see how they fit into all these sciences and can be observed alike in numbers, magnitudes, and motions. Such are the theorems governing proportion, namely, the rules of compounding, dividing, converting, and alternating; likewise the theorems concerning ratios of all kinds, multiple, superparticular, superpartient, and their counterparts; and the theorems about equality and inequality in their most general and universal aspects, not equality or inequality of figures, numbers, or motions, but each of the two by itself as having a nature common to all its forms and capable of more simple apprehension.[7]

What the head of the Platonic Academy includes in common mathematics are the general theorems concerning proportion, ratio, and equality, i. e. theorems about the concepts prescribing quantitative characteristics indifferently to numbers, magnitudes, and motions. Moreover, he adds to his catalogue of 'common mathematics', beauty (*κάλλος*; *pulchritudo*), order (*τάξις*; *ordo*) and the renowned methods of analysis and synthesis.[8]

Even the Aristotelians familiar with the mathematical sciences would probably admit the existence of such a body of mathematical knowledge as has been characterized above. But the following characterization of 'common mathematics' by Proclus differs from that of the Peripatetics ontologically and epistemologically. The Neoplatonic commentator of Euclid puts it this way in Chapter IV:

---

[6] Friedlein, p. 5; Barozzi, p. 2.

[7] Friedlein, pp. 7–8; Barozzi, p. 4.

[8] Friedlein, p. 8; Barozzi, p. 4.

> Consequently we must not regard these common theorems as subsisting in these many separate forms of being, nor as later born and deriving their origin from them, but as prior to their instances and superior in simplicity and exactness. For this reason, knowledge of them takes precedence over the particular sciences and furnishes to them their principles; that is, these several sciences are based upon this prior science and refer back to it.[9]

According to Proclus 'common mathematics' is prior and superior to all the particular mathematical sciences. This Platonic point of view is, as we have seen in Chapter 6, § 5, the reverse of Aristotle's observation on universal demonstration.

It is ascertained in Book XIII (M) of the *Metaphysics*, Book I, Chapter 10 of the *Posterior Analytics* and so on, as we have already seen in Chapter 6, § 5, that Aristotle and his followers admitted a collection of universal propositions in some sense or other. Then, how are differences between the Aristotelian understanding of such a collection and Proclus's 'common mathematics'?

For Aristotle, this collection of propositions are nothing other than a universal generalization of particular examples. Prior are beings which we conceive through our perception. Mathematical propositions and propositions common to them are grasped through abstraction. Ian Mueller has called such an epistemological point of view 'abstractionism' (p. 467). Proclus understands 'common mathematics' in a different manner. According to him, the certainty and eternity of mathematical arguments cannot be explained if we understand mathematical beings through 'abstraction' from sensible beings as Aristotle. To overcome this difficulty, Proclus understands mathematical beings as projections of the Forms (*προβαλαί τῶν εἰδῶν*; *formarum emissiones*)[10] onto a screen of imagination of the human intellect. Mueller calls such an epistemological recaption 'projectionism'.[11] If we conceive mathematics in these ways, the understanding of 'common mathematics' becomes different between Aristotle and Proclus. That is to say, since for Proclus 'common mathematics' is closer to the human intellect, they are given a higher status than the particular mathematical disciplines. Consequently, it is no accident that he sees this discipline substantial and calls it "the single and entire mathematics."

But Proclus does not view 'common mathematics' as the ultimate and highest science. Let us examine in details what kind of epistemological status it has. For Proclus, as for Plato, mathematical being occupies the middle ground between the subject of physics and that of philosophy per se. Our knowledge

---

[9]Friedlein, pp. 8–9; Barozzi, p. 4.

[10]Friedlein, p. 13; Barozzi, p. 7.

[11]For the concept of 'projectionism' as opposed to 'abstractionism', see Ian Mueller, "Forward to the 1992 Edition," Morrow's English translation (n. 1), p. xxvi. See also *Idem*, "Mathematics and Philosophy in Proclus' Commentary on Book I of Euclid's *Elements*," in *Proclus: Lecteur et interprète des ancients*, publiées par Jean Pépin et H. D. Saffrey (Paris, 1987), pp. 305–318, esp. p. 317; *Idem*, "Aristotle's Doctrine of Abstractionism in the Commentators," in *Aristotle Transformed* (Ch. 6, n. 24), pp. 463–480, esp. p. 473.

in general "ascends from the more partial to the more general until at last we reach the science of being *qua* being (ἡ τοῦ ὄντος, ᾗ ὄν ἐστιν ἐπιστήμη; *ipsa eius, quod est, quatenus est scientia*)."[12] The last science which Proclus mentions is what Aristotle called 'first philosophy' in Book VI(E), Chapter 1 of his *Metaphysics*. This science alone can provide the ultimate principles for the other inferior sciences, including the various areas of the mathematical sciences. Proclus does not argue in Chapter IV clearly how this ultimate science precedes the other sciences but answers this question in Chapter XIV when he refers to "the single and entire mathematics" (see p. 334).

The topic of Chapter XIV concerns the way in which dialectic is the capstone of mathematical knowledge. Proclus's argument is adopted faithfully from Book VII of Plato's *Republic*. The Neoplatonic philosopher discusses the preeminence of philosophy per se over the mathematical sciences in the sense that dialectic is superior to mathematical understanding. Proclus reconfirms Plato's thesis that dialectic is the capstone of the mathematical sciences, for "dialectic (ἡ διαλεκτική; *Dialectica*), the purest part of philosophy, hovers attentively over mathematics, encompasses its whole development, and of itself contributes to the special sciences their various perfecting, critical, and intellectual powers—the procedures, I mean, of analysis (ἀναλυτική; *resolvens*), division (διαιρετική; *dividens*), definition (ὁριστική; *definiens*), and demonstration (ἀποδεικτιλεή; *demonstrans*)."[13] Mathematical sciences can be made complete only through the agency of dialectic. Even "the single and entire mathematics" which he calls the unifying bond between mathematical sciences cannot take the role of dialectic. Dialectic is higher than "the single and entire mathematics" and is affirmed to be the true bond of union among the mathematical sciences. And it is Nous (νοῦς; *mens*) that governs the dialectical methods and constitutes the perfect terminus of knowledge.[14] The concept of Nous, distinguished from the faculty of mathematical reasoning 'intelligence' (διάνοια; *cogitatio*), is the highest faculty of the intelligence. Dianoetic thinking characterizes the function of the entire mathematics (ἡ ὅλη μαθηματική; *tota Mathematica*)[15] and is rightly placed between opinion, which is full of errors, and Nous.

In order to understand thoroughly, however, what Proclus means in Chapter XIV, we have to examine further how Plato understood the relationship between mathematics and dialectic in a locus classicus of the *Republic* VII. Plato there discussed the 'incompleteness' of mathematical knowledge. Let us quote what Socrates taught in 533B–C (as usual, in pages and paragraphs of the Stephanus edition):

> As for the rest, geometry and the like, though they have some hold

[12] Friedlein, p. 9; Barozzi, p. 5.
[13] Friedlein, pp. 42–43; Barozzi, p. 25.
[14] Friedlein, p. 44; Barozzi, p. 26.
[15] Friedlein, p. 18; Barozzi, p. 10.

> on reality, we can see that they are only dreaming about it; they can never wake and look at it as it is so long as they leave the hypotheses (ὑποθέσεις) they use undisturbed and cannot account for them. For if one's starting point (ἀρχή) is something unknown, and one's conclusion and intermediate steps are made up of unknowns also, how can the resulting consistency ever by any manner of means become knowledge (ἐπιστήμη).[16]

Then Socrates continues his argument by praising the supremacy of dialectic over geometry:

> Dialectic, in fact, is the only procedure which proceeds by the destruction of hypotheses to the very first principle (ἀρχή), so as to give itself a firm base. [533C]

According to Plato, mathematical reasoning cannot be a sufficient knowledge and must always be supplemented by dialectical thinking. Mathematics is certainly better than opinion (δόξα), but always leaves the hypotheses unproved. To examine the appropriateness of these hypotheses is a task of dialectic, or noetic thought (νόησις), i. e. the function of Nous. Thus Socrates concludes:

> Then you agree that dialectic is the capstone (θριγκός) that tops our educational system; it completes the course of studies and there is no other study that can rightly be placed above it. [534E–535A]

It is known that Plato praised mathematics greatly, but at the same time he did not neglect to point out that mathematical knowledge becomes perfect only when complemented by dialectical thinking.[17]

Seeing thus, Proclus formulated an ascending structure for the mathematical disciplines: 'Common mathematics' combines the mathematical disciplines within them and is superior to the particular mathematical sciences; dialectic flies over and controls it; and moreover Nous stands above it and reign over the whole.[18]

Proclus, maintaining that there exists 'common mathematics' prior to arithmetic and geometry, follows Plato's lead, reporting that dialectic is the capstone (θριγκός; *vertex*) of the mathematical sciences. Aristotle also seems

[16]Plato, *The Republic*, translated with an introduction by Desmond Lee (London, $_2$1974). We indicate the Stephanus pagination only.

[17]As for the relationship between mathematics and dialectic in Plato's thought, see F. M. Cornford, "Mathematics and Dialectic in the *Republic* VI.–VII.," *Mind*, XLI (1932), pp. 37–52 & 173–190; Julia Annas, *An Introduction to Plato's Republic* (Oxford, 1981), Chapter II, "Philosophy and Mathematics," pp. 272–293. On Plato's mathematical thought in general, see Harold Cherniss, "Plato as Mathematician," *The Review of Metaphysics*, **4** (1951), pp. 395–425.

[18]See I. Mueller, "Forward to the 1992 Edition" (n. 1), p. xxix. Mueller uses "universal mathematics" deliberately for our "common mathematics." This use is, I think, legitimate.

to have inherited the Platonic insight on the demonstrative science.[19] But, although the disciple degraded the role of dialectic more than his master of the Academy did, Aristotle still continued to regard dialectic as an important function of philosophy and to consider philosophy to be superior to mathematics.[20] It might be argued that great ancient thinkers fundamentally preserved the understanding that mathematics was based on the method of dialectic, originally the questioning and answering between two people. This is the very point that Árpád Szabó has insisted upon in his study on the history of early Greek mathematics.[21]

It was such a conception of mathematics that was introduced into sixteenth-century Europe through Grynaeus's Greek original and Barozzi's Latin translation and that played a crucial role in enhancing the status of mathematics among various branches of science. It is clear from its Latin preface that Grynaeus, a humanist friend of Philip Melanchton, attached the Proclus commentary to his edition of Euclid's *Elements* and made efforts to introduce Proclus's mathematical thought to the world of the Renaissance.[22] And, as we observed in Chapter 2, § 2, it was by adducing this thought that in 1560 the translator of Proclus's work, Barozzi, began to criticize the Aristotelian epistemology of Alessandro Piccolomini and to emphasize the preeminence and certainty of mathematics. Barozzi fully utilized at that time the Platonic epistemology and ontology to consolidate his own position. Important for our concern here is that he handed down Proclus's significant idea of 'common mathematics' to philosophers and mathematicians of the Renaissance through his translation. On the other hand, Proclus, following Plato—Proclus claimed himself to be the 'Succesor' (διάδοχος) of Plato—maintained that dialectic was superior to even

[19]Aristotle states in Book I of *Topics* (101a37–101b4): "It [the treatise] is a further use in relation to the principles used in the several sciences. For it is impossible to discuss them at all from the principles proper to the particular science in hand, seeing that the principles are primitive in relation to everything else: it is through reputable opinions about them that these have to be discussed, and this task belongs properly, or most appropriately, to dialectic; for dialectic is a process of criticism wherein lies the path to the principles of all inquiries." For Plato's influence on Aristotle, see H. D. P. Lee, "Geometrical Method and Aristotle's Account of First Principles," *The Classical Quarterly*, XXIX (1935), pp. 113–124, esp. pp. 119–124.

[20]See J. D. G. Evans, *Aristotle's Concept of Dialectic* (Cambridge, 1977), esp. "Summary: Aristotle and Plato on Dialectic and Science," pp. 49–52. "Aristotle preserves the Platonic idea of dialectic as unrestricted in its scope. On the other hand this lack of restriction is for Aristotle an indication of the unscientific character of dialectic, whereas for Plato it had been an indication that dialectic was the only true science." (p. 50.) Similarities and dissimilarities in general between Plato's and Aristotle's philosophy are discussed by G. E. L. Owen in his excellent paper "The Platonism of Aristotle," in *Logic, Science and Dialectic: Collected Papers in Greek Philosophy*, ed. Martha Nussbaum (Ithaca, 1986), pp. 200–220. On mathematics and dialectic, consult especially pp. 211–216.

[21]Árpád Szabó, *Anfänge der Griechischen Mathematik* (München/Wien, 1969).

[22]See Simon Grynaeus's "Praefatio" to his Greek *editio princeps* of Euclid's *Elements* (n. Ch. 2, n. 51). An interesting discussion can be seen in E. Kessler, "Clavius entre Proclus et Descartes" (Ch. 2, n. 16), pp. 290–294. Cf. H. Schüling, *Die Geschichte der axiomatischen Methode im 16. und beginnenden 17. Jahrhundert* (Hildesheim, 1969), p. 37.

that 'common mathematics' and the *sine qua non* for mathematical reasoning. Nor did Barozzi go beyong his Neoplatonic master. Even if we may be allowed to ascribe 'mathesis universalis' to Proclus as some historians of philosophy have done,[23] it would be completely in error to assume that the ancient Neoplatonic philosopher possessed the same concept that the Descartes of the *Regulae* did, as will be discussed in Chapter 8, § 1. He asserted that dialectic took precedence even over the noblest discipline among the mathematical sciences.

A historical note should be added here as to why we have assigned the term 'common mathematics' to Proclus's notion of a mathematical science prior and superior to the particular mathematical disciplines. Proclus had a forerunner who used the terminology 'common mathematical science' for the principles and theorems common to all the mathematical disciplines. The Syrian Neoplatonist Iamblichus of Chalcis composed a work *On the Common Mathematical Science* (Περὶ τῆς κοινῆς μαθηματικῆς ἐπιστήμης) as Book III of *On Pythagoreanism*, which had a certain similarity with Proclus's commentary on Euclid's *Elements*.[24] Iamblichus wrote that book more than a century before Proclus's time. It has been debated whether or not there had been an influence of it upon Proclus's commentary on the *Elements* of Euclid and upon Proclus's epistemological viewpoint of 'projectionism'. It may be certain that Proclus had a forerunner in his Neoplatonic epistemology. Iamblichus and Syrianus, who was his teacher at the Academy, are candidates for this role. But Proclus got out of Pythagorean mysticism, and tried to provide philosophical foundations coherently concerning ordinary mathematics. This was a major difference between Iamblichus and him.[25] Before Proclus, the Neoplatonic intellectual currents possibly shared as a common doctrine the idea that there existed a mathematical discipline which is expected to be the 'common mathematics'. In fact, they identified a substantial doctrine which Iamblichus called 'common mathematical science'. This may be a sufficient reason for us to call "the single science prior to the several mathematical sciences" by the name 'common

---

[23]See Stanislas Breton, *Philosophie et mathématique chez Proclus* (Paris, 1969), p. 89; Annick Charles-Saget, *L'Architecture du divin: Mathématique et philosophie chez Plotin et Proclus* (Paris, 1982), pp. 195, 205, & 295. On the other hand, Nicolai Hartmann doesn's seem to see the concept of 'mathesis universalis' in Proclus's *Commentary*, for, according to him even the theory of proportion is not common to both geometry and arithmetic. Nicolai Hartmann, *Des Proklus Diadochus philosophische Anfangsgründe der Mathematik nach den ersten zwei Büchern des Euklidkommentars* (Gießen, 1909), p. 15: "Die Geometrie stellt den Grundsatz der Proportion für dei Größen, die Arithmetik für die Zahlen auf, jede aus ihren eigenen Mitteln und für ihre eigenen Zwecke." Cf. Geneviève de Pesloüan's French translation in Breton, *Ibid.*, p. 198.

[24]Iamblichus, *De communi mathematica scientia liber*, ed. Nicolaus Festa (Stuttgart, 1975). The content of this treatise is analyzed in Dominic J. O'Meara, *Pythagoras Revived: Mathematics and Philosophy in Late Antiquity* (Oxford, 1989), 4. "Pythagorean Mathematical Science (Book III: *On General Mathematical Science*)," pp. 44–51. Cf. Descartes, *Règles utiles et claires etc.*, éd. par J.-L. Marion et P. Costabel (La Haye, 1977), p. 162.

[25]See Ian Mueller, "Iamblichus and Proclus' Euclid Commentary," *Hermes*, Bd. 115 (1987), pp. 334–348; O'Meara, *Op. cit.* (n. 24), pp. 157–166; Morrow, *Op. cit.* (n. 1), p. 345.

mathematics'.

A 'hermeneutical basis' on which Asclepius interpreted "ἡ καθόλου" in 1026a27 of Book VI (E) of Aristotle's *Metaphysics* as "ἡ καθόλου μαθηματική" (see p. 299) seems to have been exactly the exsitence of a Neoplatonic concept like 'common mathematics'. A disciple of Proclus was Ammonius, who taught in Alexandria. Asclepius was a disciple of his and is known to have taken notes of his teacher's lectures on metaphysics. Asclepius's commentary on the *Metaphysics* referred to Syrianus, Proclus's teacher.[26] For Asclepius, "ἡ καθόλου" in a context of mathematics naturally may have invoked an interpretation based on "ἡ καθόλου μαθηματική".

Another significant point is that during the Middle Ages there was a proponent of a concept similar to the 'common mathematics' of the ancient Neoplatonists. One of the so-called Oxford 'Platonists' Roger Bacon wrote a work with the title *Communia mathematica* (literally, Mathematical Commons), which we referred to in the previous chapter (p. 314), probably toward the end of the 1260s. According to Bacon, mathematics has two principal parts, of which one is about "commons," the other is about "particulars." He thus defines 'mathematical commons' as follows: "And the former part of mathematics should be called neither Geometry nor Arithmetic nor Astrology nor Astronomy nor Music, but is about the elements and the foundations of the entire mathematics which should be premised before special parts."[27] What, then, constitutes 'mathematical commons'? He counts, for example, "the principles of Book V" in Euclid's *Elements* as a component of that discipline.[28] In sum, 'mathematical commons' according to his understanding may be said to be a mathematical science which considers the principles and theorems common and prior to the particular branches of mathematics as a whole. Further noticeably in Part 4 of his most comprehensive work *Opus majus* Bacon pointed out "the power of logic as a whole is dependent upon mathematics,"[29] and "only in mathematics there exists a most powerful demonstration (*demonstratio potissima*) according to necessary causes."[30] Such an understanding about mathematical demonstration was 'novel' and 'bold'. The 'most powerful demonstration' is defined as a

---

[26] O'Meara, *Op. cit.* (n. 24), p. 121.

[27] Roger Bacon, *Communia mathematica: Opera hactenus inedita Rogeri Baconi*, Fasc. XVI, ed. Robert Steele (Oxford, 1940), p. 38: "Et hec pars Mathematice non debet vocari Geometria nec Arithmetica nec Astrologia nec Astronomia nec Musica, sed de elementis et de radicibus totius Mathematice que debent premitti ante partes speciales." On the date of the composition, see A. C. Crombie and J. D. North, "Bacon, Roger," *DSB*, I, p. 378.

[28] R. Bacon, *Op. cit.*, p. 84.

[29] *The 'Opus Majus' of Roger Bacon*, ed. by John Henry Briges (Frankfurt/Main, 1964), p. 103: "[...] virtus tota logicae dependet ex mathematica." Cf. *The Opus Majus of Roger Bacon*, A Translation by Robert Belle Burke, Vol. 1 (New York, 1962), p. 120

[30] *Ibid.*, p. 106: "[...] in sola mathematica sunt demonstrationes potissimae propter per causam necessariam." Cf. Burke's translation, p. 124. See also *Communia mathematica*, p. 15: "[...] cum sola mathematica sit vere demonstrativa secundum demonstrationem potissimam [...]."

demonstration which combines 'demonstration of reason' and 'demonstration of fact' (Chapter 2, § 2, pp. 52–53). This point may lead us to label Bacon's insistence as the brand name of a kind of 'mathematicism'. Nevertheless, on the other hand, the same Bacon did not neglect to notice, as we have seen in Chapter 6, § 3, C (p. 314), the superiority of metaphysics over mathematics, stating that "metaphysics is prior to mathematics and more abstract."[31] Moreover, we cannot identify Bacon's 'mathematical commons' with the 'common mathematics'. The former concept can be naturally imagined particularly from Book I, Chapter 10 of Aristotle's *Posterior Analytics*, whereas the former concept of 'common mathematics' implies 'projectionism' in its epistemological understanding, although the terminology itself may be admitted by Aristotelians.

Consequently, throughout late antiquity and the Middle Ages there was an intellectual tradition which recognized a mathematical discipline, at times called 'common mathematics', distinguished from the other particular mathematical sciences. Proclus was one of the main proponents who belonged to this tradition. And it was the Greek original text of Proclus's *Commentary on the First Book of Euclid's Elements* and its Latin translation by Barozzi that introduced this idea to the intellectual world in the sixteenth century. This work by Proclus not only contributed to the development of the concept of 'mathesis universalis' in the sixteenth century, which will be discussed in the next section, but also encouraged exposing the mathematical structures of the natural world. This is exemplified by Kepler's *Harmonice mundi*, among others, which referred frequently to Proclus's *Commentary on the First Book of Euclid's Elements* when the author developed his Neoplatonic-Neopythagorean thought.[32]

We have to keep in mind that there is a great divide between the concept of 'common mathematics', which was introduced into the Renaissance world, and Descartes's 'mathesis universalis'. We see a unifying bond of the mathematical disciplines different from that of antiquity and also a metaphysical 'boldness' to negate the role of dialectic. The latter point we will be discussed in Chapter 8. In the next section we will see a mathematician—not a philosopher—who reintroduced the mathematical unifying bond into the sixteenth century.

§ 2. Adriaan van Roomen

Proclus's *Commentary on the First Book of Euclid's Elements* stimulated some philosophers and mathematicians of the Renaissance to restore and construct a novel philosophy of mathematics different from that of the Aristotelians. When

[31] *Communia mathematica*, p. 63: "Methaphysica est prior mathematica et magis abstracta [...]."

[32] Johannes Kepler, *Gesammelte Werke*, Bd. VI: *Harmonice mundi* (München, 1940). In "Personenregister" "Proklus" appears 34 times.

this commentary on Euclid was brought into the Renaissance, the idea of 'common mathematics' advocated in it was naturally connected with mathematics in the ancient world. That mathematical concept, however, would find a new expression in the mathematical avant-garde of the sixteenth century. Adriaan van Roomen (Latin name, Adrianus Romanus) played a crucial role in providing a paradigm for the new combination and in transmitting it to theoreticians of the early seventeenth century.

Van Roomen was one of the mathematicians who called the notion of 'common mathematics' 'mathesis universalis'. It is with sufficient reason, therefore, that Giovanni Crapulli, in the conclusion of his *Mathesis universalis: Genesi di una idea nel XVI secolo* states that "we find the definition or rather the more complete description [of 'mathesis universalis'] in Van Roomen, that is at the end of the [sixteenth] century" and that he quotes, by way of recapitulation, the Belgian mathematician's definition of 'mathesis universalis' from Chapter Six of the *Apologia pro Archimede* (1597): "There is a science common to arithmetic and geometry which considers quantity generally as mensurable [...], to which properties common to all quantities pertain, [...] not only abstract as numbers and magnitudes, but also concrete as times, sounds, voices, motions, and forces [...], it should be very justly judged that those ratios which pertain to proportions are related to a certain universal science."[33]

As we have suggested in the previous section, not a few predecessors advanced a notion almost equivalent to Van Roomen's 'mathesis universalis'. Crapulli, for instance, has brought together several authors who used terminologies similar to 'mathesis universalis': Alessandro Piccolomini, Cunradus Dasypodius, Benito Pereira, Johann Heinrich Alsted, and so forth.[34]

Leaving detailed discussion to the monograph by Crapulli, here we mention only a few memorable expressions akin to '*mathesis universalis*'. In his tract *Commentarius de certitudine mathematicarum disciplinarium* (1547), which we referred to in Chapter 2, § 2, Piccolomini admitted that in the human mind there exists "a certain faculty common to arithmetic and geometry,"[35] following Proclus's discussion of the common principles of mathematics being as a whole in Chapter II of the latter's commentary on Euclid. Of course, pointing out such a faculty of mind and holding an explicit concept corresponding to 'mathesis universalis' are distinct points.

---

[33]A. Crapulli, *Mathesis universalis* (Roma, 1969), p. 146. The quoted Latin passage is as follows (A. Romanus, *Apologia pro Archimede*, pp. 22–23): "Geometriae & Arithmeticae communem esse scientiam, quæ quantitatem generaliter vti mensurabilem considerat [...] ad quam spectarent affectiones communes omnibus quantitatibus [...] non abstractis tantum vt numeris & magnitudinibus, sed concretis etiam, vti temporibus, sonis, vocibus, locis, motibus, potentiis [...] propositiones eas quæ spectant ad analogias [...] ad scientiam aliquam vniuersalem iure merito pertinere existimandum est." We have referred to Van Roomen's original rather than what Crapulli cited.

[34]See in detail Crapulli, p. 150.

[35]Piccolomini, *Op. cit.* (Ch. 2, n. 18), f. LXXXXVIIv: "cuiusdam facultas communis ad Geometriam, & Arithmeticam."

Cunradus Dasypodius used a more significant terminology. In his book *Analyseis geometricae sex librorum Euclidis*, written with his teacher Christian Herlinus and published in Strasbourg in 1566, Dasypodius ascribed some theorems to a discipline which he named "illa universalis *μαθηματική*." In Chapter II of the *Commentarium* of that book he wrote:

> There are other [theorems] which are transferred, since they are common, from geometry to arithmetic and from the latter to the former, or hold on both sides, and are considered equally in geometry and arithmetic. Assuredly they can be applied in other disciplines as well, because they are taken from 'that universal mathematics', as are *alternando* and conversion of ratios [and] synthesis and division, which are of this genus.[36]

Dasypodius's concept of "universalis *μαθηματική*" may be almost identical with the 'common mathematics' which we discussed in Section 1. As can be imagined, he attributed Book V of Euclid's *Elements* to this discipline which this time he called "universalis illa disciplina mathematica." Commenting on Book V, he states:

> This Book V provides a peculiar doctrine of proportions which are extended through the universal nature of things and pertain not to geometry or arithmetic alone but to the other sciences, and rather it should, therefore, be ascribed to 'that universal mathematical discipline', which grasps all other species under it.[37]

It is now clear that Dasypodius believed that there exists a substantial mathematical discipline common to the particular mathematical sciences.

It is known, moreover, that even the strict Aristotelian Benito Pereira, who was at one time a professor of philosophy at the Collegio Romano and an adversary of Clavius in the controversy on the 'Quaestio de certitudine mathematicarum', assented to the notion of 'common mathematics'. Pereira, in his widely read book on natural philosophy *De Communibus omnium rerum naturalium principijs & affectionibus* (1576), asserted the following:

> Thus there is no doubt that there is a certain common mathematical science, which should explore properties common to magnitude and

[36] Dasypodius, *Analyseis Geometricæ sex librorum Euclidis* (Argentinensi, 1566), f. LXXXVIv: "[S]unt etiam quæ cum sint communia ex geometria in arithmeticam: et ex hac in illam transferuntur: uel etiam utrinque se habent: et peræque in geometria, atque in arithmetica considerantur: imo in cæteris quoque disciplinis adhiberi possunt: quia ex illa uniuersali *μαθηματικὴ* desumuntur: ut sunt *ἐναλλαγαὶ, καὶ ἀναστροφαὶ τῶν λογῶν: συνθέσεις, διαιρέσεις*: et quæ huius sunt generis."

[37] *Ibid.*, f. XCVIIr: "His quintus liber peculiarem dectrinam instituit de proportionibus: quæ diffusæ sunt per uniuersam rerum naturam, non ad geometriam tantum aut arithmeticam pertinent, sed ad alias scientias atque idcirco potius referri debet, ad uniuersalem illam disciplinam mathematicam, quæ reliquas omnes tanquam species sub se comprehendit."

> number, and which, notwithstanding, is not regarded as a science separated from the mathematical [disciplines], that is, from geometry and arithmetic.[38]

This statement was one which Van Roomen cited together with Eutocius's commentary on Apollonius's *Conics* I in Chapter Six of his *Apologia pro Archimede* for justifying his introduction of "a certain universal mathematics".[39] We know therefore this that in the seventeenth century the term 'mathematica universalis' was commonly conceived to be identical with 'mathematica communis'. (Cf. Nifo's commentary on Aristotle's *Metaphysics*, which we examined in Chapter 6, § 4, A.) Attention should be paid to the fact that in Pereira's epistemological point of view the 'common mathematical science' was not regarded as being distinguished from geometry and arithmetic, which were usually viewed to be more concrete than the former general science.

It is also reported that Alsted employed a notion similar to Van Roomen's, 'mathematica generalis', in his *Methodus admirandorum mathematicarum*, published in 1623.[40]

The concepts which we have so far mentioned might be said, more or less, to have been derived from ancient passages which discuss 'universal demonstrations' or 'common mathematics'.

Van Roomen hit upon his idea of 'mathesis universalis', which he first developed in his *Apologia pro Archimede* of 1597, having studied writers like Pereira and then having contested Scaliger's accusation that arithmetic should not be used to solve geometrical problems. In Chapter 5, § 3, D, we saw that Van Roomen's concept of 'mathesis universalis' signifies a common mathematical discipline prior to arithmetic and geometry, and that it can be identified with a kind of algebra. He gives a detailed definition of this concept in Chapter Seven of his *Apologia pro Archimede.* Its introductory part, printed beautifully in a reverse triangular form, is worth quoting *in toto*. Indeed it contains an idea which leads to a clear—perhaps, too clear—exposition of the Aristotelian passages which we have examined in Chapter 6:

> CHAPTER SEVEN
>
> AN IDEA of a kind of universal Mathematics, which we shall call first Mathematics, is proposed.

---

[38]Benedictus Pererius, *De Communibus omnium rerum naturalium principijs & affectionibus, libri qvindecim* (Rome, 1576), p. 34: "Quaemadmodum non est dubium quin sit aliqua scientia Mathematica communis, quae debeat speculari affectiones communes magnitudini & numero, quae tamen scientia, à Mathematicis non numeratur distincta à Geometria & Arithmetica." Crapulli mistakenly refers to this statement on p. 37 instead of p. 34. See Crapulli, p. 95. Notice that Pereira admitted 'scientia Mathematica communis' while still claiming that "the mathematical doctrine is not properly a science" (doctrina Mathematica non est propriè scientia) (Pereira, p. 24).

[39]A. Romanus, *Apologia*, p. 23. See our discussion on pp. 354–356.

[40]Crapulli, pp. 125–143.

> Now although by reason as well as by the authority of Eutocius we have declared that there is a kind of Universal Mathematics, in order that every ambiguity be removed, we will propose a certain outline or idea of it so that anyone may understand clearly that propositions and their demonstrations which should be ascribed to Universal Mathematics are not purely Arithmetical, since almost no mention is made of numbers and in other cases, in fact, in addition to numbers quantities of another kind are also taken, nor Geometrical, since no mention is made of magnitudes, that is, length, width or depth. And now we will give to this science the name of ***First Mathematics (Mathematica)***, or ***First Mathesis***, by the analogy of First Philosophy. For, for instance, it is called 'First' because it grasps the subjects of all the other sciences under it and further it proves the principles of the others if they need demonstration. Thus this First Mathematics is concerned with the subjects of all mathematical sciences, pure and mixed. It also proves the principles of the other sciences. For all conclusions of this science can be taken for the principles in the other sciences. So much for the name. The method of proceeding will be as follows: Principles, that is, definitions and axioms are premised. Then several theorems follow. But I have not wanted to bring out many because our mind can depend upon very few which we have adduced. But if anyone should have wanted, he would add to ours the entire fifth book of Euclid's *Elements*. For all propositions which are propounded about magnitudes they can be accomodated to any quantity whatever, with the formula remaining the same for every demonstration. For principles which are assumed for their demonstration are common to every quantity. But it is better that the aforementioned thing itself should be begun.[41]

This is a remarkable pronouncement. Van Roomen's 'mathesis universalis', which is also called 'prima mathesis', is compared with first philosophy or metaphysics, because it is placed prior to any other mathematical science, pure and mixed, and "proves the principles of the other sciences." It is concerned with 'general quantity', which is thought to be the object of Euclid V. It is no exaggeration to say that what Heath desired to designate by the 'universal mathematics' of Aristotle is contained in this very passage. Neither Alexander nor Asclepius nor the Renaissance commentator Nifo, as we saw in Chapter 6, so clearly formulated the 'universality' of 'something universal in mathematics', which each of the exegetes chose as the interpretation of Aristotle's "ἡ καθόλου."

[41] A. Romanus, ***Apologia***, p. 23. Cf. Philippe Gilbert, "Les sciences exactes dans l'ancienne université de Louvain," ***Revue des questions scientifiques***, XVI (1884), pp. 438–453, esp. p. 449. Figure 7.1 shows the Latin original. The Komaba Library of the University of Tokyo possesses a copy of this rare book.

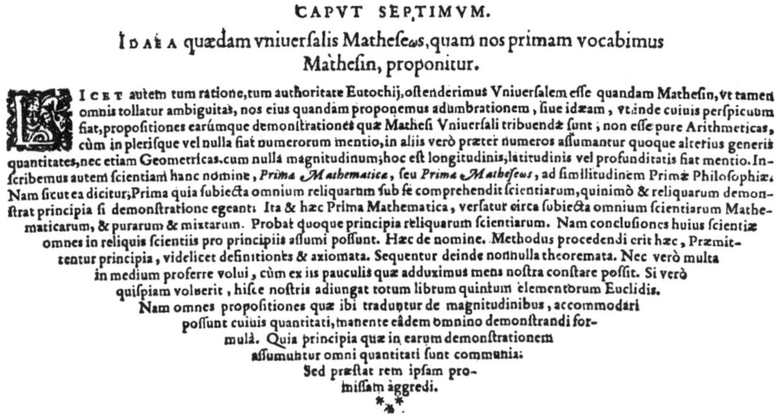

CAPVT SEPTIMVM.

IDAEA quædam vniuerſalis Matheſeως, quam nos primam vocabimus Matheſin, proponitur.

LICET autem tum ratione, tum authoritate Eutochij, oſtenderimus Vniuerſalem eſſe quandam Matheſin, vt tamen omnis tollatur ambiguitas, nos eius quandam proponemus adumbrationem, ſiue idæam, vt inde cuiuis perſpicuum fiat, propoſitiones earúmque demonſtrationes quæ Matheſi Vniuerſali tribuendæ ſunt; non eſſe pure Arithmeticas, cùm in pleriſque vel nulla fiat numerorum mentio, in aliis verò præter numeros aſſumantur quoque alterius generis quantitates, nec etiam Geometricas, cum nullâ magnitudinum; hoc eſt longitudinis, latitudinis vel profunditatis fiat mentio. Inſcribemus autem ſcientiam hanc nomine, *Prima Mathematica*, ſeu *Prima Matheſeως*, ad ſimilitudinem Primæ Philoſophiæ. Nam ſicut ea dicitur; Prima quia ſubiecta omnium reliquarum ſub ſe comprehendit ſcientiarum, quinimò & reliquarum demonſtrat principia ſi demonſtratione egeant: Ita & hæc Prima Mathematica, verſatur circa ſubiecta omnium ſcientiarum Mathematicarum, & purarum & mixtarum. Probat quoque principia reliquarum ſcientiarum. Nam concluſiones huius ſcientiæ omnes in reliquis ſcientiis pro principiis aſſumi poſſunt. Hæc de nomine. Methodus procedendi erit hæc, Præmittentur principia, videlicet definitiones & axiomata. Sequentur deinde nonnulla theoremata. Nec verò multa in medium proferre volui, cùm ex iis pauculis quæ adduximus mens noſtra conſtare poſſit. Si verò quiſpiam voluerit, hiſce noſtris adiungat totum librum quintum elementorum Euclidis. Nam omnes propoſitiones quæ ibi traduntur de magnitudinibus, accommodari poſſunt cuiuis quantitati, manente eâdem omnino demonſtrandi formulâ. Quia principia quæ in earum demonſtrationem aſſumuntur omni quantitati ſunt communia: Sed præſtat rem ipſam promiſſam aggredi.

Figure 7.1 Chater 7 of Van Roomen's *Apologia pro Archimede* (1597)

When we proceed, furthermore, to examine Van Roomen's *Universæ mathesis idea*, it becomes clear that he studied the theories of mathematics of the sixteenth century very extensively. To develop his own theory, he quoted, for example, Plato, Aristotle, Pappus, and Proclus among ancient authors; and Ramus, Dasypodius, Pereira, and Clavius among his contemporaries. It might not be too much to say that the concept of 'mathesis universalis' was most explicitly and schematically formulated in this work insofar as we restrict our discussion to the period before Descartes. Let us analyze the content of the *Universae mathesis idea* in detail.

The full title of that small volume is *Universæ mathesis idea, qua mathematicae universim sumptae natura, praestantia, usus et distributio brevissimo proponuntur* (The Idea of Mathematics as a Whole, in which the Nature, Preeminence, Use and Division of Mathematics Taken Generally Are Displayed Very Briefly) (see Fig. 7.2). First of all, as we have touched upon in Chapter 4, § 3, it should be noted that in the sixteenth and seventeenth centuries 'universa mathesis' was in general distinguished from 'universalis mathesis.'[42] The former term meant 'mathematics as a whole' which embraces all the mathematical disciplines, whereas the latter usually denoted 'a mathematical discipline common to

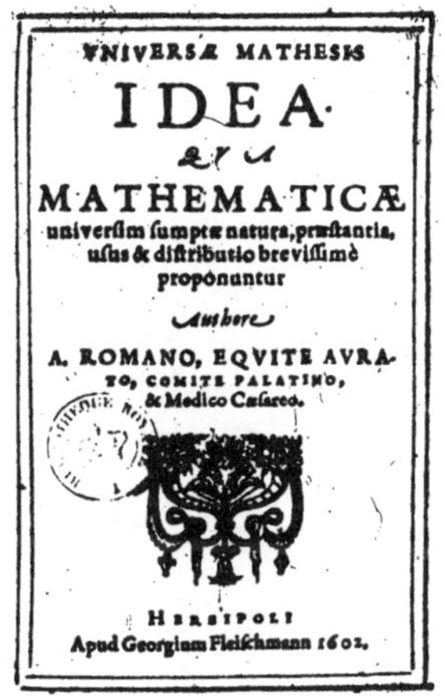

Figure 7.2 The Title Page of *Universæ mathesis idea* (1602)

[42]This is the point Crapulli emphasizes. See Crapulli, p. 8.

arithmetic and geometry'. Van Roomen, of course, strictly maintained this custom. This book was published in 1602 in Herbipolis, that is Würzburg, where he had been professor of medicine since 1593 and also 'mathematician' from 1596 to 1603.[43]

This work is essentially concerned with classification of the mathematical sciences. In Chapter I, the author reveals the core of his theory of mathematics; in Chapter II, he argues generally about the division of mathematics; and in the successive chapters, about various branches of the mathematical sciences. The concept of 'mathesis universalis' appears in Chapters III and IV.

Chapter I is titled "On the Nature, Preeminence and Use of Mathematics Taken Generally." The author at the beginning draws attention to the fact that the Latin word 'mathematica' has been derived from the Greek word 'μάθημα' or 'μάθησις' and denoted originally 'disciplina' or 'doctrina' (p. 3). He, therefore, does not distinguish in essence between 'mathesis' and 'mathematica'. And he begins his substantial explanation with the following sentence: "With Proclus we derive the method of description from the schools of Platonists and Pythagoreans as follows."[44] This is a sentence which tells us how great Proclus's influence upon Van Roomen was.

The mathematician at Würzburg does not ignore the well-known debates on the certainty of mathematics which began with Barozzi's attack on Piccolomini in 1560. While adducing Pereira's statement that mathematical demonstrations are not 'most powerful', he emphasizes the certainty of mathematics:

> It has been disputed among several people whether mathematical demonstrations are most powerful. Benedictus Pereira believes that they are not, because they may not advance beyond principles which are causes of a conclusion as he has demonstrated by many examples taken from Euclid's *Geometrical Elements*. Nevertheless, whatever it may be (that is, whether they are most powerful or not), the following one thing is indubitable among all: Mathematical demonstrations are the most certain among all the demonstrations of whatever capacity. Truly, we are forced to say almost always in natural and sensible things (Aristotle being a witness) that this or that happens ὡς ἐπὶ τὸ πολύ (for the most part) and also ἐὰν μή τι ἐμποδίζει (unless anything impedes). But on the other side, (Book II, Chapter 3 of the *Metaphysics*) the certainty of mathematical sciences is the highest, exceeding any certainty of the other capacities.[45]

[43] H. L. L. Busard, "Roomen, Adriaan van," *DSB*, XI, p. 532.

[44] Romanus, *Vniversæ mathesis idea* (Würzburg, 1602), p. 3.

[45] *Ibid.*, pp. 6–7: "Demonstrationes Mathematicæ an sint potissimæ, apud nonnullos controversum est; Benedictus Pererius sentit potissimas non esse, cùm non progrediantur ex principijs, qvæ sint causæ conclusionis, uti pluribus ostendit exemplis ex elementis Geometricis Euclidis desumptis. Verùm qvicqvid sit, (hoc est sive potissimæ sint, sive non) hoc unum indubitatum est apud omnes, demonstrationes mathematicas esse certissimas inter omnes

Here Van Roomen's philosophical standpoint is equivocal. He admits, on the one hand, following Pereira, that mathematical demonstrations are not powerful enough to demonstrate the presumed principles, and following Aristotle concedes that they may be permitted only insofar as the treated objects do not contain matter. On the other hand, the final aim of his argument is to insist that the certainty of mathematics exceeds that of any other science.

How, then, does Van Roomen position mathematics with respect to the other two theoretical sciences, i. e. metaphysics and physics? His point of view in this regard is clearly Platonic. He puts it concisely: "It [mathematics] is surpassed by metaphysics whereas it surpasses physics."[46] Moreover, Van Roomen invokes Book VII of Plato's *Republic* and asserts that "it [mathematics] seems to be not only useful but also necessary for metaphysics."[47] In the same place Plato maintained that mathematical studies raise the human soul to divine things and occupy the place between physics and metaphysics. On this point at least Van Roomen sides with his Jesuit mentor Clavius and the latter's friend Barozzi.

In Chapter II, Van Roomen presents his classification of the mathematical sciences. He first divides them into two broad categories: "true mathematics"

---

demonstrationes qvarumcumqve facultatum. In rebus sanè naturalibus & sensibilibus dicere ferè semper cogimur (Aristotele teste) hoc vel illud fieri *ὣς ἐπὶ τὸ πολύ*, vel etiam *ἐὰν μὴ τὶ ἐμποδίζει*; at contrà (2. Metaphys. 3) Mathematicarum scientiarum summa est certitudo excedens omnem aliarum facultatum certitudinem." Pererius, *Op. cit.* (n. 38), Lib. I, Cap. XII, "Scientia speculatiuam non dici vniuovè de Mathematicis disciplinis & alijs, quoniam doctrina Mathematica non est propriè scientia," pp. 24–26. Aristotle's statement in the *Metaphysics*, II, 3 (995a14–17) can be seen in Ch. 2, n. 19.

[46] *Ibid.*, p. 8: "Vnde à Metaphysica superatur, Physicam verò superat."

[47] *Ibid.*, p. 12: "Metaphysicae non modo vtilis sed & necessaria videtur." Van Roomen continues to state the reason: "For the entrance to metaphysics opens to nobody but through the mathematical disciplines. Certainly if we should try to raise the power and sharpenness of our intellect without any mediation from sensible things, which the physicist considers, to things divided and separated from every sensible matter, which the metaphysician contemplates, we ourselves will be made blind just as it happens to one who is sent out of a certain dark cave in which he has been hidden long into the clearest light of the sun. And for this reason, before the intellect ascends from physical things, to which matter subjected to senses is joined, to metaphysical things, which are separated greatly from matter, in order not to be filled with the brightness of the latter it is necessary, earlier than that, to be accustomed with less abstract things as are considered by mathematicians in order that one can comprehend them easier. And therefore the Divine Plato rightly maintains that the mathematical disciplines raise the soul to the contemplation of divine things and sharpen the mind." (p. 12.) "Nulli etenim ad Metaphysicam pater aditus, nisi per mathematicas disciplinas. Nam si à rebus sensibilibus, qvas Physicus considerat, ad res ab omni materia sensibili secretas, seiunctasqve qvas contemplatur Metaphysicus, vires, aciemqve nostri intellectus attollere absque ullo medio tentemus: nos metipsos excæcabimus, non secus, ac ei contingit, qvi è carcere aliqvo tenebricoso, in qvo diu latuit, in lucem Solis clarissimam emittitur. Qvamobrem, anteqvam à rebus physicis, qvæ materiæ sensibus obnoxiæ, sunt conjunctæ, ad res metaphysicas, qvæ sunt ab eadem maxime avulsæ, intellectus ascendat; necesse est, ne harum claritate offundatur, prius eum assvefieri rebus minùs abstractis, qvales à Mathematicis considerantur, ut faciliùs illas possit comprehendere. Qvocirca rectè Divinus Plato mathematicas disciplinas erigere animum, ad divinarum rerum contemplationum exacuere mentis aciem affirmat."

(*vera mathematica*) and "quasi-mathematics" (*quasi mathematica*): The former category is further divided into two classes: "pure mathematics" (*pura mathematica*) and "mixed mathematics" (*mixta mathematica*). "Pure mathematics" is then partitioned into two genera: "universal mathematics" (*universalis mathematica*) and "special mathematics" (*specialis mathematica*). "Universal mathematics" can be viewed in two ways: "logistic" (*logistice*) as an organon of science and "first mathematics" (*prima mathesis*) as an appellation of a science. Two species, arithmetic and geometry, belong to special mathematics. Stereometry is placed under geometry. This summarizes the content of pure mathematics.

The class of mixed mathematics is, then, attended by many disciplines. Among others, mechanics belongs to this class. It is further divided into "universal or elementary mechanics" (*universalis sive elementaris mechanica*) and "particular mechanics" (*particularis mechanica*). "Universal [mechanics] contains the elements of violent motion and is said to be mechanics in the strict sense and *Manganaria* by Pappus."[48] It is known from this usage that Van Roomen employs the adjective 'universalis' to signify a 'general', 'introductory', and 'elementary' part of a science in contrast with special or particular branches of that science. "Mathesis universalis," by the same token, implies the elementary and general part of pure mathematics.

The division of mathematics according to Van Roomen may be diagramed as below (see the next page). The Roman numerals in parentheses after the names of sciences indicate the chapters of the *Universae mathesis idea* in which the author explains the contents of the disciplines concerned (math.=mathematica).

Terms that may be unfamiliar to us appear in the diagram: 'Euthymetria' in Chapter XIV and 'Taumato-poetica' in Chapter XIX. 'Euthymetria' is derived from the Greek 'εὐθυμετρία', which means literally 'the measurement of straight lines', but Van Roomen himself defines this art as follows: "'Euthymetria' is a science which measures finite straight lines by exactly straight visual rays."[49] That is, this discipline is the art of measuring rectilinear figures by the aid of appropriate optical instruments. 'Thaumato-poetica' is also Greek in origin 'θαυματοποιϊκή', the art of making wonderful machines. Hero of Alexandria expounded this art.[50]

We will be able to find English equivalents easily for the following terms: 'astronomia', 'ouranographia' (celestial geography), 'chronologia', 'chronome-

[48]Romanus, *Ibid.*, p. 16: "Vniuersalis continet elementa motus violenti; diciturqve propriè *Mechanica*, Pappo *Manganaria*." The term 'Manganaria' is used in Pappus's *Collection*, VIII: "ἡ τέχνη τῶν μαγγαναρίων" (Hultsch, III, p. 1024, *l*. 14); "ars manganorum" (Commandino, f. 305v); "l'art des artificiers" (Ver Eecke, II, p. 810). Incidentally, Van Roomen wrote the title of Chapter XVI as "De Manganaria sive mechanaria proprie dicta," probably by mistake. "Mechanaria" should be corrected to "mechanica", of course.

[49]P. 56: "Evthymetria est scientia quæ per radios visuales aptè directos, lineas rectas finitas metitur."

[50]See Heath, *A History of Greek Mathematics*, Vol. 1, p. 18.

- Universa mathesis
  - Vera math.
    - Pura. math.
      - Universalis math. = Supputatrix sive Logistice (III) sive Prima mathesis (IV)
      - Specialis math.
        - Arithmetica (V)
        - Geometria(VI)–Streometrica
    - Mixta math.
      - Astronomia et Ouranographia (VII)
      - Chronologia sive Chronometria (VIII)
      - Cosmographia, Geographia etc.(IX)
      - Astrologia sive Astronomia (X)
      - Geodesia (XI)
      - Musica (XII)
      - Optica sive Perspectiva (XIII)
      - Euthymetria (XIV)
      - Sphaeropoeia (XV)
      - Manganaria sive Mechanica (XVI)
      - Mechano-poetica (XVII)
      - Organo-poetica (XVIII)
      - Thaumato-poetica (XIX)
  - Quasi mathematica (XX)
    - Architectura
    - Ars militaris
    - etc.

tria', 'cosmographia', 'geographia', 'astrologia', 'geogedia', 'musica', 'optica', 'perspectiva', 'sphaeropoeia' (art of making the sphere), 'mechano-poetica', 'organo-poetica' (art of making the organ), 'architectura', and 'ars militaris' (military art).

A notable point in Van Roomen's manner of classification of the mathematical sciences is that he explicitly introduces the discipline of 'mathesis universalis'. His understanding of 'mathesis universalis' is set forth in Chapters III "De supputatrice" and IV "De prima mathesi." Let us examine Chapter III first.

Van Roomen defines the discipline of 'supputatrix' as follows: "'Supputatrix', called λογιστική in Greek, is that which brings out the unknown from given numbers in suitable circumstances by means of general canons."[51] This definition invokes necessarily the art of *al-jabr* in Arabic mathematics. He elaborates on this definition as follows:

> It might have been called 'Arithmopraxia' if the expression had been accepted. It has been agreed that this 'supputatrix' is distinguished from other mathematical sciences because it does not treat properties of a particular object, yet considers every concrete number, that is, any numerable thing whatsoever without any property of numbers so that it is able to serve all parts of mathematics, even every civic use. 'Supputatrix' can be placed in mathematics just as logic is in philosophy as a whole.[52]

[51] P. 17: "Svpputatrix Græcis λογιστική dicta, est quæ beneficio canonum universalium, ex datis numeris rebus accommodatis quæsitum elicit."

[52] P. 17: "Posset non incommodè si vox recepta esset Arithmopraxia vocari. Supputatricem hanc à reliquis Mathematicis scientijs separare placuit, quod ea non peculiaris obiecti tradat proprietates, verùm omnem numerum respiciat concretum; hoc est, rem quam vis nu-

'Supputatrix' is the art of computation, which is identified by Van Roomen with the logistic of ancient Greece. It is the instrumental side of the 'mathesis universalis' just as logic is an organon of philosophy in general. He, then, describes the difference between 'supputatrix' and arithmetic:

> Furthermore, 'supputatrix' differs from arithmetic, because the latter shows certain properties of numbers by demonstration, whereas the former shows, without any demonstration and any property of numbers, very few canons of finding and judging and very general canons which are different by far from theorems and problems of arithmetic.[53]

What Van Roomen stresses here is that 'supputatrix' is quite general and requires little preliminary knowledge. "It hardly has other principles than the explanation of appropriate terms of the science; nevertheless, the use of 'supputatrix' demands the knowledge of properties of those things which numbers are applied to."[54] It is for this reason that 'supputatrix' is placed ahead of all the mathematical sciences:

> 'Supputatrix' can be justly placed in advance of all the mathematical sciences. For 'supputatrix' can be taught without any preexisting knowledge of the other mathematical sciences.[55]

Van Roomen notes that 'supputatrix' may be called practical arithmetic. But in this chapter he states nothing about algebra. Nor does he mention it at any other place in the same book. Algebra was of Arabic origin and was not contained in the catalog of the names of mathematical sciences in classical Greece. This might be conjectured to be a reason why the author of the *Universae mathesis idea* does not refer to the word 'algebra' at all. We have no definitive evidence to determine whether or not Van Roomen intended to include algebra in the discipline called 'supputatrix'. But, as he considers 'supputatrix' to be synonymous with 'λογιστική' in Greek, we may think that it contains algebra. It may be regarded that there is at least a large overlap between the two terms, for in 1602 Van Roomen surely knew Viète's mathematical works as a whole, and Viète called the ordinary art of computation 'logistice numerosa'

---

merabilem sine ulla numerorum proprietate, ita ut omnibus servire queat Matheseos partibus, imò omni ciuili usui. Quodque Logica est in universa Philosophia, idipsum Supputatrix censeri potest in Mathesi."

[53] *Ibid.*: "Differt porrò Supputatrix ab Arithmetica, quod hæc numerorum exhibeat proprietates demonstratione confirmatas, illa verò sine ulla demonstratione, sine ulla numerorum proprietate, canones paucissimos exhibeat inveniendi & iudicandi, eosque generalissimos longè à Theorematibus & Problematibus Arithmeticæ diversos."

[54] P. 18: "Principia præter terminorum scientiæ propriorum explicationem vis habet alia; usus tamen supputatricis requirit cognitionem proprietatum earum rerum quibus numeri applicantur."

[55] *Ibid.*: "Supputatrix ergo meritò omnibus mathematicis præmitti potest. Nam Supputatrix doceri sine ulla reliquarum Mathematicarum præexistenti cognitione."

and his own symbolic algebra 'logistice speciosa' in his *In artem analyticem isagoge*. In fact, Van Roomen personally met Viète in 1601, and he defines 'supputatrix' or 'λογιστική' as "that which brings out the unknown from given numbers in suitable circumstances by means of general canons," which contains the definition of algebra.

As we have already pointed out, Van Roomen certainly ascribed 'algebra' or 'analytical science' to his 'first mathematics,' the discipline of general quantity (see p. 267) and Biancani, a mathematical writer in the early seventeenth century and a disciple of Clavius, understood that 'supputatrix' contains algebra (p. 196). Hence, we may safely assert that Van Roomen considered the 'supputatrix' to be a broad art of computation including the newly invented symbolic algebra in his book of 1602 as well as in his later writings.

Van Roomen, then, attempts to corroborate his opinion that logistic is crucial and preliminary to the remaining mathematical sciences by quoting a long passage from Ramus's *Scholae mathematicae*, where the French 'usuarius' has claims about the importance of the art of computation.[56] Van Roomen further cites a *locus classicus* of Book VII of Plato's *Republic* in which the ancient philosopher makes his master Socrates state that logistic and arithmetic are preliminary disciplines for the philosopher ruler. But the paragraph Van Roomen cites is in reality nothing other than a faithful reproduction of a part of, again, Ramus's *Scholae mathematicae*.[57]

Van Roomen proceeds to the next chapter in which 'prima mathesis' is argued about rather briefly. He begins this chapter with the definition:

> First mathematics is that which is concerned with quantity taken absolutely.
>
> Its object is quantity absolutely taken.
>
> On the other hand, [its] end is to show properties common to all quantities.
>
> It has appropriate principles alone.
>
> It occupies the first place in mathematics for the same reason that first philosophy is the first among the remaining philosophical sciences.[58]

According to Van Roomen's definition of 'mathematica universalis' in Chapter II, titled "De mathematicae divisione," 'first mathematics' is the name of that discipline when placing it within various mathematical sciences, whereas 'logistic' is the name from the viewpoint of its use as an instrument for the other mathematical sciences. We can sense Pereira's influence in the above statement

[56] Ramus, *Scholae mathematicae* (Ch. 5, n. 161), Lib. 2, p. 52.

[57] *Ibid.*, p. 53.

[58] Pp. 20–21: "Prima Mathesis est quæ versatur circa quantitatem absolutè sumptam. Objectum eius est Quantitas absolute sumpta. Finis verò, affectiones quantitatibus omnibus communes exhibere. Principia habet tantum propria. Locum in Mathesi obtinet primum. Eadem ratione qua Prima Philosophia inter reliquas philosophicas scientias prima est."

about the purpose of 'first mathematics': "to show properties common to all quantities." Pereira ascribed the same role to the 'scientia mathematica communis'. What is especially notable is that Van Roomen consciously compares the structure of mathematical sciences with that of philosophy. If it is right that first philosophy treats the principles of the entirety of philosophy and is thus placed first among various branches of philosophy, 'first mathematics' should play a similar role in the mathematical sciences. This is the very reason why he called 'mathesis universalis' 'first mathematics' in Chapter Seven of the *Apologia pro Archimede* of 1597.

Through this definition Van Roomen became the first modern, as far as we know, to conceive the notion of 'mathesis universalis' in a form consciously connected with the concept of 'prima mathesis'. Pereira might have recognized that Aristotle's passage of the *Metaphysics* 1026a23–32 should be interpreted from the viewpoint of comparing metaphysics with mathematics. He frequently referred to Alexander of Aphrodisias's commentary on Aristotle's *Metaphysics* in his aforementioned work of 1576. So he could have reasonably perceived that "ἡ καθόλου" in 1026a27 implies something mathematical. Nevertheless, it seems to have been Van Roomen who grasped 'mathesis universalis' as being identical with 'first mathematics' and related to 'first philosophy' for the first time. Thus, he established the conceptual framework interpreting the Aristotelian passages 1026a23–32 and 1064b6–9 as Heath did in the twentieth century. Of course, it can still be disputed as to how Aristotle himself thought about these passages. But the Belgian mathematician seems no doubt to have been the first in modern Europe.

The author of the *Universae mathesis idea* did not develop further the manner in which his 'mathesis universalis' should be concretely constituted. He concludes Chapter IV, "De prima mathesi" with the following words:

> Someday in a sufficient volume we will display this science which nobody hitherto has described.
>
> We have proposed a specimen of this in our *Apologia*, but commentaries of various people furnish more veracious material, which we will take pains to put in order.[59]

That is to say, Van Roomen does not display mathematically in his present *Universae mathesis idea* what he conceives by the notion of 'mathesis universalis' or 'prima mathesis' and more substantially than the description in his *Apologia pro Archimede*. The *Universae mathesis idea* was reproduced in a revised and augmented form in his *Mathesis polemica* published in Frankfurt am Main in 1605. But it does not seem that the author added any new material to the

[59] P. 21: "Hanc scientiam à nemine adhuc descriptam iusto volumine aliquando proponemus. Huius specimen in Apologia nostra proposuimus, at veriorem materiam suppeditant variorum commentarij, quam in ordinem redigere studebimus."

chapters on 'mathesis universalis' in this republication.[60]

After Chapter IV of the *Universae mathesis idea* follow Chapter V "De arithmetica" and Chapter VI "De geometria." In the author's opinion, it is necessary to learn first mathematics before arithmetic,[61] and first mathematics and arithmetic before geometry.[62] In any case, first mathematics is placed prior to any other mathematical discipline.

As we saw above, Van Roomen in Chapter Seven of his *Apologia* asserted that the principles of the mathematical sciences should be treated in first mathematics. But it is doubtful that he considered the principles treated in first mathematics to be all the principles of mathematics, i. e. that he viewed mathematics as an absolutely autonomous discipline independent of metaphysics, for he surely referred to Pereira's statement that mathematics cannot necessarily prove its own first principles in the affirmative and asserted that metaphysics is superior to mathematics (see p. 346).

With few exceptions such an epistemological belief seems to have been widely held among most philosophically trained theoreticians of mathematics in the sixteenth century, even among those who advanced a notion equivalent to Van Roomen's 'mathesis universalis.' As Crapulli has pointed out, Dasypodius clearly viewed his notion representing 'mathesis universalis' as dependent on metaphysical considerations.[63] Pereira categorically maintained that mathematics is not properly a science and that the only science is metaphysics.[64] For him mathematics was epistemologically and ontologically 'incomplete'. And for Ramus, too, as we saw in Chapter 6, § 4 (p. 326), the universal science was theology as science and dialectic as organon. His point of view esteeming dialectic was the same as Plato and Proclus, who claimed that dialectic is the capstone of the sciences. André Robinet recently tried to find one of the sources of Descartes's *Regulae ad directionem ingenii* in Ramus. We can learn greatly from his study which also mentioned Van Roomen.[65] We must not forget, at the same time, that there was a sharp difference between Ramus and Descartes on the evaluation of dialectic. The situation does not seem to have been different in other

---

[60]The first twenty chapters constituting the first part of *De principiis ex mathesi desumendis* may correspond to the previous *Vniversæ mathesis idea.* See Henri Bosmans, "Romain (Adrien)" (Ch. 5, n. 189), cols. 879–880.

[61]Romanus, *Ibid.*, p. 22: "Principia habet tum propria, tum ex Prima Mathesi desumpta. Locum in Mathesi post primam, obtinet primum."

[62]See "Principia præter propria, etiam desumit, tum ex prima Mathesi, tum ex Arithmetica." (p. 23) & "Proximum in Mathesi locum post Arithmeticam obtinet." (p. 24)

[63]Crapulli, p. 146.

[64]B. Pererius, *Op. cit.* (n. 38), pp. 24–27. See also Crapulli, IV. "La 'scientia mathematica communis' in analogia alla 'prima philosophia' secondo B. Pereira," pp. 93–99; G. C. Giacobbe, *Op. cit.* (Ch. 2, n. 29).

[65]André Robinet, *Aux Sources de l'esprit cartésien: L'Axe la Ramée-Descartes de la Dialectique de 1555 aux Regulae* (Paris, 1996) & *Idem*, "L'Axe La Ramée-Descartes: Position dans la Règle IV de la «Mathesis universalis»," *Descartes et la Renaissance: Actes du Colloque internationale de Tours des 22–24 mars 1996*, Réunis par Emmanuel Faye (Paris, 1999), pp. 67–76.

thinkers of the sixteenth century.

Importantly, it can be stated in general, with Crapulli,[66] that in the sixteenth century 'mathesis universalis' was never advocated in a form clearly independent of metaphysics or dialectic. Certainly Van Roomen attempted to present a mathematical specimen of his 'mathesis universalis' in the *Apologia* and in his *Universae mathesis idea*. And the contemporary development of algebra contributed to the crystallization of such a specimen. But even he was not totally free from the aforesaid philosophical belief of sixteenth-century thinkers. In other words, his 'mathesis universalis' may be said to be 'first' and 'universal' relatively within the domain of mathematical sciences, but not 'first' and 'universal' in the field of human knowledge as a whole.

Let us consider in a comprehensive manner of history of mathematics how Van Roomen's mathematical program for displaying 'mathesis universalis' became possible in the sixteenth century. First of all, it was a product of the Renaissance during which Proclus's Neoplatonic mathematical thought was restored. At the same time we must not forget that he was a mathematician. In the European world of his time, the Indian art of computation through the Islamic world and the art of algebra, which had been totally symbolized for unknown numbers, was sinking firm roots within the mathematical disciples. Let us remember the definition of 'mathesis universalis' in his *Apologia pro Archimede*: "A science common to arithmetic and geometry which considers quantity generally as mensurable [...], to which properties common to all quantities pertain, [...], it should be very justly judged that those propositions which pertain to proportions are related to a certain universal science." Here we should pay special attention to the expression "considers quantity generally as mensurable (*mensurabilis*)." In the European world of the sixteenth century, "quantity" had become not simply an object of philosophical speculation but a "measurable" entity. Europe after the thirteenth century is characterized as a 'pantometrical' society which in thought and social activities emphasized *measure* in general, including pictorial and musical sensibility as well as a theoretical science of mathematics.[67] This kind of society was formed not simply by the power of Europe. As the transmission of Hindu numerals and the mathematical discipline called 'al-jabr' in Arabic indicates, it was a product of the entire Eurasian Continent which recreated the mathematical framework of ancient Greece as a whole. True, the mathematical terminology which Van Roomen used was unmistakably classical. But, mathematical contents associated with Hindu numerals and algebra had already penetrated into the meaning of the Latin word 'supputatrix' in Van Roomen.

[66]Crapulli, *Op. cit.*, VII. "Conclusioni sulla *mathesis universalis* nel sec. XVI," pp. 145–154.

[67]Alfred W. Crosby, *The Measure of Reality: Quantification and Western Society, 1250–1600* (Cambridge, 1997). On the technical side, see John J. Roche, *The Mathematics of Measurement: A Critical History* (London, 1998), esp. Chaps. 3–5.

In a chapter of his *Scholae mathematicae* discussing Book V the *Elements*, Ramus, whom Van Roomen referred to, accused its author Euclid of sophistry (*elenchus*) in that it had Book VII on the arithmetical theory of proportion in addition to Book V. According to Ramus, "Proclus considers the entire book (Book V) to be common to arithmetic and geometry,"[68] but in fact that book is concerning logic and arithmetic. Ramus's criticism was hasty, but not totally misdirected. The reason for this is that a kind of arithmetization (*Arithmetisierung*) of mathematics as a whole had been already practically developed for several centuries at the time of Ramus with the merging of the arithmetical and geometrical theories of proportion in practical computations which would culminate at the German mathematicians Georg Cantor and Richard Dedekind of the nineteenth century. Such a theoretical tendency existed in the Islamic world, for example, in Omar Khayyam, as we have seen in Chapter 5, § 3, C. Section 12 of Jacob Klein's "Die griechische Logistik und die Entstehung der Algebra" has argued the formation of the new concept of numbers in Simon Stevin, Descartes, and John Wallis in early modern Europe.[69] Klein's discussion has an incomparable philosophical depth, but at the same time it is still narrowly Eurocentric. If he had extended the subject of his compehensive consideration not simply to Viète but also to Van Roomen, the vision of his monograph would have been much broader and more persuasive. Let us remember what Ibn Khaldūn stated in his *Muqaddimah*: He expected that the new art of calculation which had been transmitted from India "produces an enlightened intellect that is trained along correct lines" (p. 249). The art of calculation at the time of Ibn Khaldūn was on the low level of theoretical refinement. But such was not the case at the period of Van Roomen. In his mathematics, the barrier between arithmetic and geometry was actually removed and the unified concept of numbers was about to grow. This was a practical implication of the 'mathesis universalis' originating from Proclus.

To conclude, Van Roomen's concept of 'mathesis universalis' was nothing other than 'Eurasian mathematics', so to speak, containing speculations on the philosophy of mathematics in ancient Greece, the art of numerical calculations in India, algebra in the Islamic civilization, and the intention to 'pantometry' in the Latin Middle Ages.

It shall be a task of historians in a new age to construct a new historiography by liberating Klein's history of mathematical thought from the European world to include the Indian and Islamic worlds. This is what we, historians of mathematics in an age of unprecedented globalization in the twenty-first century, have to undertake.

---

[68]Ramus, *Scholae mathematicae* (Ch. 5, n. 161), Book XIII, p. 212. See my article, "The Acceptance of the Theory of Proportion in the Sixteenth and Seventeenth Centuries," *Historia Scientiarum*, No. 29 (1985), pp. 107–108.

[69]Klein, "Die griechische Logistik etc." *Quellen und Studien*, Bd. 3–2 (1936) (Ch. 5, n. 172), pp. 195–235; English translation, pp. 186–224.

The mathematical realization of Van Roomen's concept of 'mathesis universalis' will be accomplished substantially in the seventeenth century, by Frans van Schooten, John Wallis, Newton, and Leibniz. The philosophical implication of that notion, will be elaborated in one sense but discarded in another sense by Descartes in the *Regulae ad directionem ingenii*. We will examine this novel intellectual scene in the following chapter.

Chapter 8

'Mathesis Universalis' in the Seventeenth Century

§ 1. Reviewing Descartes's Concept of 'Mathesis Universalis' from His Philosophy of Mathematics

A. The Philosophy of Mathematics in the *Regulae ad Directionem Ingenii*

How did Descartes grasp 'mathesis universalis'? We have already discussed its mathematical nature in Chapter 4, § 3. Now we turn to his philosophy of mathematics and attempt to support an argument about how 'mathesis universalis' in his youthful work *Regulae ad directionem ingenii* should be understood philosophically. For this purpose, we consider retrospectively his theory of mathematics there, turning our attention to its philosophical aspect.

It should be noted that in Descartes's corpus the term 'mathesis universalis' appears only in the *Regulae*. He never used it in his writings after the *Regulae*. To understand the reasons why this concept does not appear in his later writings, we examine in detail his later philosophy of mathematics, which seems to be closely tied with the philosophy of "Cogito, ergo sum." Our speculations about Descartes's use or disuse of the concept of 'mathesis universalis' will deepen our understanding of the development of his philosophy of mathematics. Furthermore, our examinations of his philosophy of mathematics may shed light on his entire system of philosophy, especially his later philosophy based on "Cogito, ergo sum."

Let us begin with the *Regulae ad directionem ingenii*. In his explanation of Rule II Descartes asserts "we ought to concern ourselves only with objects which admit of as much certainty as the demonstrations of arithmetic and geometry" (see p. 179). The two branches of pure mathematics are especially extolled, for "they alone," he states, "are concerned with an object so pure and so simple that they make no assumptions that experience might render uncertain; they consist in deducing conclusions by means of rational arguments" (see p. 178). Then the author of the *Regulae* continues and compares arithmetic and geometry with other disciplines, especially philosophy, with respect to certainty:

> They are therefore the easiest and clearest of all the sciences and have just the sort of object we are looking for. Where these sciences are concerned it scarcely seems humanly possible to err, except through inadvertence. Yet we should not be surprised if many prefer of their own accord to apply their minds to other arts, or to philosophy. The reason for this is that everyone feels free to make more confident guesses about matters which are obscure than about matters which are clear. It is much easier to hazard some conjecture on this or that question than to arrive at the exact truth about one particular question, however straightforward it may be.[1]

It is due to a method which works most effectively in pure mathematics, i. e. deduction (*deductio*), that these two disciplines can enjoy their incomparable certainty. In this regard philosophy cannot rival them.

But the mode of knowing in these disciplines does not necessarily consist merely in deduction. In Rule III Descartes sets forth the need for intuition (*intuitus*) in addition to deduction. "Intuition" is defined by him as "the indubitable conception of a clear and attentive mind which proceeds solely from the light of reason" (see p. 179). In his opinion, intuition and deduction, therefore, are the two essential modes of knowing which guarantee the certainty of pure mathematics. According to him, the first principles of mathematics are known only through intuition (see p. 179). The author of the *Regulae* bestows a pivotal position upon intuition.

What Descartes has argued in Rules II and III is the synthetic side of pure mathematics. It is in Rule IV that he sheds light on its analytic and heuristic aspects. According to his explanation, true mathematics can be obtained from a reformed and 'well-methodized' algebra which functions as an organon for discovering mathematical truths. This algebra functions as mathematical analysis, in particular as problematic analysis.

From another point of view in the classification of mathematical sciences, there is a discipline, Descartes tells us, which can be called "mathesis universalis," "not a far-fetched designation but a venerable term with a well established meaning" (see p. 195). This science is concerned simply with "order" and "measure" and contains "everything on account of which the other sciences are called parts of mathematics." These particular sciences are, for example, arithmetic, geometry, astronomy, music, optics, and mechanics. 'Mathesis universalis' for Descartes is to be placed prior to arithmetic, geometry and other

[1] *AT*, X, pp. 365–366.

mathematical sciences. How this characterization of 'mathesis universalis' resembles that in Van Roomen's *Universæ mathesis idea* which we examined in Chapter 7, § 2 !

Descartes writes concerning this specific discipline of mathematics as follows:

> How superior it [mathesis universalis] is to these subordinate sciences both in utility and simplicity is clear from the fact that it covers all they deal with, and more besides; and any difficulties it involves apply to these as well, whereas their particular subject matter involves difficulties which it lacks.[2]

The author does not indicate explicitly what difficulties 'mathesis universalis' could involve. At least, in his judgment, it involves fewer difficulties than the other particular mathematical sciences. That is to say, it is the simplest and easiest among all the mathematical disciplines and the starting point of the search for certain knowledge. So it constitutes a paradigm without rival for his general method of attaining certain knowledge.

We can find very few words on the philosophical character of 'mathesis universalis' in Rule IV. And, as pointed out previously, remarkably Descartes did not utilize this concept in any other place in his all writings. In any case, in the *Regulae* 'mathesis universalis' is presented as a paradigm discipline for the mathematical sciences and then for certain knowledge in general.

In evaluating the theory of mathematics Descartes presented in his *Regulae*, we may be permitted to ask the following question. Was not this philosophy of mathematics 'naive', not in comparison to the critical standard of today's philosophy of mathematics, but as compared with certain arguments which were pervasive in the sixteenth century and which we have examined in the previous chapter? Here we can point out two problems: One is concerned with the logical and epistemological status of the first principles in mathematics, and the other the philosophical nature of 'mathesis universalis'.

Descartes claimed that the mathematical principles were to be known through intuition, "the indubitable conception of a clear and attentive mind," whereas, he contended, dialectic was useless. What follows from this standpoint? As we have seen in Chapter 7, § 2, Van Roomen, in contrast to this, argued, adducing Pereira, that mathematical demonstrations cannot prove the principles themselves and that they are not "potissima" (see p. 348). The mathematical principles must be presupposed without mathematical demonstrations. According to the Platonists, these must be chosen in the course of a dialectical discussion, while according to the Aristotelians, these may be determined through sufficient metaphysical considerations.[3] In any case, the first principles

---

[2] *AT*, X, p. 378.

[3] On Aristotle's discussions on the first principles, see T. H. Irwin, *Aristotle's First Pinciples* (Oxford, 1988). The relation of metaphysical principles to mathematical priciples are

themselves cannot be proved within the domain of mathematics. In this sense, mathematics cannot stand autonomously and has no choice but to remain 'incomplete'. Can the 'naive' point of view of Descartes in his *Regulae* be adequate against this argument? He praised the certainty of mathematical demonstrations and, at the same time, was in contempt of the function of dialectic in intellectual activites. We may say at least that his understanding of dialectic was extremely onesided.

The second difficulty concerns 'mathesis universalis'. Descartes asserted that this mathematical discipline is superior to the particular subordinate sciences in utility and simplicity. This discipline is supposed to have been similar to Van Roomen's and may be characterized as a science of general quantity common to magnitudes and numbers. The author of the *Regulae* certainly developed an argument that this science can work effectively as a tool for mathematical discoveries, that is, as an analytical art. We willingly grant this. But there is a problem in that he did not describe how his 'mathesis universalis' could be constructed. In what way can the rules of this science be obtained? How is this science related to arithmetic, geometry, and other subordinate mathematical sciences? Can it be autonomous without its 'subordinate' disciplines or more concrete experiences?

We doubt that 'mathesis universalis' can answer these epistemological and ontological questions. We have to admit, that there was some naivete or difficulty in the young Descartes's philosophy of mathematics. But did this conception in the *Regulae* remain unchanged in the later Descartes? As we have suggested in the "Interim Consideration," he came to change his theory of mathematics when he began to tackle metaphysical considerations seriously. Next let us pursue how this transformation took place.

B. The Later Philosophy of Mathematics: Mathematical Truths as Creatures of God

Descartes did not continue to maintain the 'naive' philosophy of mathematics developed in his *Regulae ad directionem ingenii*. In Part IV of his *Discours de la méthode* of 1637, we find that he comes to involve a criterion on the certainty of mathematics different from the one in the *Regulae*. What occurred between the *Regulae* and the *Discours*?

According to his autobiographical description in Part IV of the *Discours*, a transformation in his philosophical meditations after he moved to the Netherlands. Part IV of the *Discours* begins with the following sentence: "I do not know whether I should tell you of the first meditations that I had there, for they

---

more subtle. See, among others, Ian Mueller, "Sur les principes des mathématiques chez Aristote et Euclide," *Mathématiques et philosphie de l'Antiquité à l'âge classique*, éd. par R. Rashed (Paris, 1991), pp. 101–113, and Richard D. McKirahan, Jr., *Principles and Proofs: Aristotle's Theory of Demonstrative Science* (Princeton, 1992).

are perhaps too metaphysical and uncommon for everyone's taste."[4] Yet, the author of the *Discours* ventures to tell the reader how he meditated, believing that he is obliged to speak of the speculations in order to make it possible to judge whether the foundations of his new philosophy are firm enough.

As is known from the passage we have already quoted in the "Interim Consideration" (p. 284), "I thought it necessary to [...] reject as if absolutely false everything in which I could imagine the least doubt, in order to see if I was left believing anything that was entirely indubitable." Noteworthy here is that examples of "the least doubt" have been taken from mathematical demonstrations, among others. Men sometimes make, he tells us, "mistakes in reasoning, committing logical fallacies (*paralogismes*) concerning the simplest questions in geometry" so that he has determined to reject "as unsound all the arguments I had previously taken as demonstrative proofs." What he rejects here is his previous 'naive' point of view concerning mathematical truth. Instead he adopts the novel criterion "Je pense, donc je suis," or "Cogito, ergo sum." This is, we are told, the very first principle of the philosophy which he was seeking, which even "the skeptics were incapable of shaking."

It may be no exaggeration to say that all the essential factors in Descartes's philosophy subsequent to the *Regulae* are contained in this description. He inaugurated the so-called methodological or hyperbolic doubt and at last, he claims, he attained the absolute truth "Cogito, ergo sum," which according to him even the most fanatic skeptic cannot refute. In his letter to Mersenne in January 1630, Descartes characterized such a hyperbole by a metaphor that "Just as a curved stick is to be set right, it is not only to be set back but to be bent to the opposite direction."[5] The truth thus obtained is, according to him, the one which even the most stubborn skeptic cannot reject. Important for us is not only that he claims to have discovered the irrefutability of "Cogito, ergo sum", but also that in the process of his metaphysical meditations he at one time rejects as unsound all the geometrical arguments that he had previously taken as demonstrative proofs. Something significant occurred in his philosophy of mathematics, in other words, the mathematical certainty dependent on the procedure of intuition-deduction stated in the *Regulae* became doubtful.

In order to understand the details of his metaphysical speculations which he speaks of in the *Discours*, we need to examine other documents. It is, as far as we know, in his letter from Amsterdam to Mersenne of April 15, 1630 that Descartes first told of a turn or deepening of his philosophical standpoint. He writes to his friend in Paris:

> At least I think that I have found how to prove metaphysical truths in a manner which is more evident than the proofs of geometry—in my own opinion, that is: I do not know if I shall be able to convince

[4] *AT*, VI, p. 31.

[5] "Descartes, à (Amsterdam), à Mersenne, à Paris. (mi-janvier 1630)," *AT*, I, p. 110; *AM*, I, p. 112; *CM*, II, p. 373.

> others of it. During my first nine months in this country I worked on nothing else. [...] [I]n my treatise on Physics I shall discuss a number of metaphysical topics and especially the following. The mathematical truths which you call eternal have been laid down by God and depend on Him entirely no less than the rest of this creatures. Indeed to say that these truths are independent of God, to talk of Him as if He were Jupiter or Saturn, and to subject Him to the Styx and the Fates. Please do not hesitate to assert and proclaim everywhere that it is God who has laid down these laws in nature just as a king lays down laws in his kingdom.[6]

This letter marks the beginning of Descartes's recording the fruit of his first metaphysical thought in the Netherlands. He meditated on metaphysical problems, he writes, "during my first months in this country," that is, during his sojourn in Franecker, Frisia, from the end of 1628 to the summer of 1629. These speculations crystallized into the beginning part of a treatise on metaphysics which would grow into the *Meditationes de prima philosophia*, published in 1641.[7]

What does Descartes claim in the letter just quoted? Even mathematical truths, in his opinion, are created by God and thus subordinate to Him. This thesis must surely have bewildered Mersenne[8] and much later would be rejected categorically by Leibniz. For brevity we shall call this thesis 'the creation doctrine', following some predecessors.[9] This doctrine might appear *prima facie* curious. Alexandre Koyré for one was embarassed with this theory and wanted to believe that it is incompatible with the overall Cartesian philosophy and was held by Descartes only temporarily.[10] It has been confirmed, however, that the

---

[6]"Descartes, à Amsterdam, à Mersenne, à Paris. 15 avril 1630," *AT*, I, pp. 144–145; *AM*, I, pp. 135–136; *CM*, II, pp. 430–431.

[7]See the following statements: "[P]erhaps I may some day complete a little treatise of Metaphysics, which I began when in Frisia." "Descartes, à (Leiden ou Amsterdam), à Mersenne, à Paris. (25 novembre 1630)," *AT*, I, p. 182; *AM*, I, p. 173; *CM*, II, p. 564. "Eight years ago, however, I wrote in Latin the beginnings of a treatise of metaphysics [...]." "Descartes, à (Leyde), à Mersenne, à Paris. (vers le 20 avril 1637)," *AT*, I, p. 350; *AM*, I, p. 329; *CM*, VI, p. 234.

[8]See "Descartes, à Amsterdam, à Mersenne, en voyage. (6 mai 1630)," *AT*, I, p. 149; *AM*, I, p. 139; *CM*, II, p. 481: "As for the eternal truths, I say once more that *they are true or possible only because God knows them as true or possible. They are not known as true by God in any way which would imply that they are true independently of Him.*" This statement seems to have been written in reply to Mersenne's question concerning Descartes's creation theory of 'eternal truths' stated in the letter on April 15, 1630.

[9]Margaret Dauler Wilson, *Descartes* (London/Henley/Boston, 1978), p. 121. Cf. Harry Frankfurt, "Descartes on the Creation of the Eternal Truth," *The Philosophical Review*, **86** (1977), pp. 36–57; *Eternal Truths and the Cartesian Circle: A Collection of Studies*, ed. by Willis Doney (New York/London, 1987) and Geneviève Rodis-Lewis, *Idées et vérités éternelles chez Descartes et ses successeurs* (Paris, 1985).

[10]A. Koyré, *Essai sur l'idée de Dieu et les preuves de son existence chez Descartes* (Paris, 1922), pp. 19–22.

philosopher continued to retain his creation doctrine from 1630 on throughout his life. Indeed his philosophy of mathematics maintained in the *Meditationes de prima philosophia* of 1641 and the *Principia philosophiae* of 1644 is certainly derived from this creation theory.

To scrutinize what Descartes implied in his creation doctrine, let us examine a few more documents.[11] In a letter of May 17 (or 27), 1638, he repeats to Mersenne again his thesis that God created mathematical propositions:

> I believe that the intellect can reach the truth of the matter, which is, in my opinion, that not only would there not be any space, but even those truths which are called eternal—as that the whole is greater than its part (*totum est majus sua parte*)—would not be truths if God had not so established, as I think I wrote you once before.[12]

This letter is notable in that a mathematical principle (or axiom), not mathematical demonstrations, is considered to be dependent on the will of God.

A similar statement can be seen in his letter addressed probably to Father Mesland on May 2, 1644. The topic is about the truth of mathematics and of logic. Descartes here answers the questions the correspondent has raised concerning his *Meditationes*:

> I turn to the difficulty of conceiving how it would be free and indifferent for God to make it not be true that the three angles of a triangle were equal to two right angles, or in general that contradictories could not be true together. It is easy to dispel this difficulty by considering that the power of God cannot have any limits, and that our mind is finite and so created as to be able to conceive as possible things which God has wished to be in fact possible, but not to be able to conceive as possible things which God could have made possible, but which he has in fact wished to make impossible. The first consideration shows us that God cannot have determined to make it true that contradictories cannot be true together, and therefore that he could have done the opposite. The second consideration shows us that even if this be true, we should not try to comprehend it since our nature is incapable of doing so. And even if God has willed that some truths should be necessary, this does not mean that he willed them necessarily; for it is one thing to will that they be necessary, and quite another to will them necessarily, or to be necessitated to will them.[13]

[11]For almost exhaustive documents on this doctrine, consult Émile Boutroux, *Des Vérités éternelles chez Descartes*, tr. par M. Canguilhem (Paris, 1985), Introduction, "Textes de Descartes relatifs aux vérités éternelles," pp. 41–48.

[12]"Descartes, à (Santpoort), à Mersenne, à Paris. (17 et 27 mai 1638)," *AT*, II, p. 138; *AM*, II, p. 263; *CM*, VII, pp. 227–228.

[13]"Descartes au P. [Mesland] [Leyde, 2 mai 1644?]," *AT*, IV, pp. 118–119; *AM*, VI, pp. 145–

Descartes considers mathematical and logical truths in terms of theology and metaphysics. God in his understanding far and away transcends us mortal and finite beings. God is omnipotent, so He can make any mathematical and logical propositions or their opposites to be true. On the other hand, we, human beings, cannot judge how He has willed them. Certainly we do not arbitrarily fix as true certain statements in mathematics and logic, and these can be judged evidently to be necessarily true. But both we and statements in mathematics and logic are, in any case, finite creatures. God alone is infinite and totally free to make it be true or not that, for example, the three angles of a triangle were equal to two right angles. Here again Descartes takes an example from the mathematical propositions which are commonly regarded as trivially apparent. In any case, Descartes's conception of God is 'voluntaristic God'. This can be distinguished from Leibniz's later notion of 'intellectualistic God'. But whether a concept is 'voluntaristic' or 'intellectualistic' may be relative, and it has many variations. Since Descartes regards mathematical truths to be 'eternal' and does not adopt a skeptical and empirical viewpoint as Gassendi does, his conception of God may be said to be 'intellectualistic' in comparison with Gassendi's.[14]

Let us examine, next, how Descartes conceives mathematical truths metaphysically in the *Meditationes de prima philosophia* and the *Principia philosphiae*. We begin with the *Meditationes*. By a brief sketch of Descartes's systematic argument on mathematical knowledge, we will see how Descartes reformulates his philosophy of mathematics and recovers mathematical truths. In the first Meditation the author is determined to reject all things for which he could find the least reason for doubt. His methodological doubt ranges over not only those things which are perceived through our senses but also mathematical statements such as "two and three added together are five." To justify this hyperbolic doubt he supposes that "not God, who is supremely good and the source of truth, but rather some malicious demon (*genius malignus, malin génie*) of the utmost power and cunning, has employed all his energies in order to deceive me."[15] This evil genius might lead us, for example, to be constantly mistaken in calculating the sum of two and three.

But in the Second Meditation Descartes notices that "Archimedes used to demand just one firm and immovable point in order to shift the entire earth"[16] and then looks for the Archimedean point that is absolutely certain and unshakable for his entire philosophy. Fortunately he has discovered, we are told, that "this proposition, I am, I exist (*Ego sum, ego existo*), is necessarily true whenever it is put forward by me or conceived in my mind.[17] That is to say, he had

146.

[14]See Margaret J. Osler, "Divine Will and Mathematical Truth: Gassendi and Descartes on the Status of the Eternal Truths," in Roger Ariew and Marjorie Grene, eds., *Descartes and His Contemporaries: Meditations, Objections, and Replies* (Chicago/London, 1995), pp. 145–158.

[15]*AT*, VII, p. 22.

[16]*Ibid.*, p. 24

[17]P. 25.

found his Archimedean point, the necessary truth of "Cogito, ergo sum." This proposition can outwit, in his opinion, the deceiver of supreme power whom he has supposed.

From this first principle, in the Third Meditation Descartes asks whether God exists or not. He formulates his well-known proof for the existence of God as follows: It is established that, so long as I think, I exist. Then, he raises a question: "From whom would I derive my existence?"[18] As the answer to this question, he concludes that God exists as the *ultimate* efficient cause of my existence, for only God can exist through His own power. If God exists, it is clear that He cannot be deceiver. For otherwise He would not be perfect. The imperfection of God contradicts the definition of Him. Thus finally he claims to have proved under what conditions mathematical truths are warranted. According to Descartes, only those people who believe in God can acknowledge mathematical truths with sufficient grounds.

Now we can understand the reason why Descartes states the following in the "Author's Replies to the Second Set of Objections":

> The fact that an atheist can be 'clearly aware that the three angles of a triangle are equal to two right angles' is something I do not dispute. But I maintain that this awareness of his is not true knowledge, since no act of awareness that can be rendered doubtful seems fit to be called knowledge. Now since we are supposing that this individual is an atheist, he cannot be certain that he is not being deceived on matters which seem to him to be very evident (as I fully explained). And although this doubt may not occur to him, it can still crop up if someone else raises the point or if he looks into the matter himself. So he will never be free of this doubt until he acknowledges that God exists.[19]

In Descartes's opinion, only the Christian can possess the true knowledge of mathematics, for one is incapable of having perfect knowledge about anything until he knows God. This is the very essential core of his later philosophy of mathematics and his creation doctrine of mathematical knowledge.

In Part I, "The Principles of Human Knowledge," Paragraph 5, of the *Principia philosophiae*, Descartes states "the reasons for doubting even mathematical demonstrations." Here we observe again his creation doctrine:

> Our doubt will also apply to other matters which we previously regarded as most certain—even the demonstrations of mathematics and even the principles which we hitherto considered to be self-evident. One reason for this is that we have sometimes seen people make mistakes in such matters and accept as most certain and self-

[18] P. 48

[19] P. 141.

> evident things which seemed false to us. Secondly, and most importantly, we have been told that there is an omnipotent God who created us. Now we do not know whether he may have wished to make us beings of the sort who are always deceived even in those matters which seem to us supremely evident; for such constant deception seems no less a possibility than the occasional deception which, as we have noticed on previous occasions, does occur. We may of course suppose that our existence derives not from a supremely powerful God but either from ourselves or from some other source; but in that case, the less powerful we make the author of our coming into being, the more likely it will be that we are so imperfect as to be deceived all the time.[20]

Noteworthy in this statement is that, as we have seen above in Descartes's letter to Mersenne of May 17 (or 27), 1638, not merely mathematical demonstrations but in addition the mathematical principles which are usually considered to be self-evident are subject to doubt. But again God, whose existence is proved through a similar argument developed in the *Meditationes* from the first truth of "Cogito, ergo sum," guarantees the following: "The light of nature or faculty of knowledge which God gave us can never encompass any object which is not true in so far as it is clearly and distinctly perceived."[21] Thus Descartes declares: "Mathematical truths should no longer be suspect, since they are utterly clear to us."[22]

Now we ask: Does Descartes think that mathematical propositions which are judged to be true by us are necessarily and absolutely true? If the answer is affirmative, in what sense are they true and why does he need to invoke the unqualified truth of *cogito* and the existence and veracity of God? What he means in his formulation of the creation doctrine seems to be that mathematical truths to some degree come through our mental powers—although they are extremely manifest.

In order to see the Cartesian creation doctrine more clearly, we have to choose weapons to attack the problem as to what it is and how it was so formulated.

This doctrine is surely related to Descartes's conception of God. So some people may find a key to solve this problem from the viewpoint of the history of theology. It is today known that it was, among others, Francisco Suárez's doctrine on eternal truths presented in his *Disputationes metaphysicae*, that Descartes seriously objected to.[23] Suárez is known to have insisted

[20] *AT*, VIII-1, p. 6 (Part I, Art. 5).
[21] *AT*, VIII-1, p. 16 (Part I, Art. 30).
[22] *Ibid.*, p. 17 (Part I, Art. 30).
[23] T. J. Cronin, "Eternal Truths in the Thought of Descartes and of His Adversary," *Journal of the History of Ideas*, **21** (1960), pp. 553–559; *Idem*, "Eternal Truths in the Thought of Suárez and of Descartes," *The Modern Schoolman*, **38** (1961), pp. 269–288, & **39** (1961),

that mathematical and logical truths are independent of God's will. He, like Thomas Aquinas, believed in an intellectualistic concept of God in contrast with Descartes's voluntaristic notion of God. And it has sometimes been pointed out that Descartes's religious point of view was to a considerable degree influenced by the Augustinian theology of Oratorians such as the Cardinal Bérulle, the founder of the Society in France and mentor of Descartes, and Guillaume Gibieuf, a correspondent and friend of Descartes.[24] The two Oratorians maintained a voluntaristic concept of God and asserted that God's will is absolutely free. We judge these two theological interpretations of Descartes's creation doctrine to be probable and persuasive.

Other people turn their attention to the facts that Descartes conceived his creation doctrine during the period in which he was looking for the "foundations of physics" (*les fondemans de la Physique*)[25] in the words of his letter to Mersenne on April 15, 1630, and that Descartes's physics tended to identify matter with mathematical extension. From these facts they believe that the doctrine was thought out as a part of his attempt at finding the metaphysical and theological foundations of natural philosophy. For example, Rodis-Lewis compares the theory by Descartes with the fate of Giordano Bruno one generation before: "In 1600, Giordano Bruno was burnt at the stake in Rome for saying that the world was infinite, thus allowing it to be confused with God. In contrast, Descartes was going to accentuate God's transcendence, the only thing truly 'infinite,' in contrast to this world for which [...] he uses the term 'indefinite,' or sometimes, 'towards the infinite' [*à l'infinit*], end*less*. He was the first to detail the difference between the divine, and infinite in a positive sense, and the absence of determinate limits."[26] (The emphasis is in the original.) In *La Philosophie naturelle de Descartes*, Michio Kobayashi discusses on the relation of this doctrine to Descartes's physics, insisting that the creation theory of eternal truths was necessary for metaphysically founding his specific mathematical physics.[27] These interpretations, which pay attention to the relation of Descartes's natural philosophy to his metaphysics, are insightful and, subesquently, can be fundamentally supported.

These interpretaions will become more persuasive if we read the concluding part of the *Principia philosophiae*. That is to say, in Part IV, Paragraph 205

pp. 23–38; *Idem, Objective Being in Descartes and in Suarez, Analecta Gregoriana*, **154** (Roma: Gregorian University Press, 1966).

[24]See É. Gilson, *La Liberté chez Descartes et la théologie* (Paris, 1987), Première Partie, Ch. IV: "La doctrine cartésienne de la liberté divine et la théologie de l'Oratoire," pp. 157–210. A more comprehensive study can be seen in Henri Gouhier, *Cartésianisme et augustinisme au XVIIe siècle* (Paris, 1978).

[25]*Ibid.* (n. 6), *AT*, I, p. 144; *AM*, I, p. 135; *CM*, II, p. 430.

[26]G. Rodis-Lewis, *Descartes* (Introduction, n. 4), pp. 146–147; The English translation, p. 101. See also the same author's "From Metaphysics to Physics," in Stephen Voss, ed., *Essays on the Philosophy and Science of René Descartes* (New York/Oxford, 1993), pp. 242–258, at p. 247.

[27]Michio Kobayashi, *La Philosophie naturelle de Descartes* (Paris, 1993), pp. 27–42.

of the book, Descartes stated that "Those things which I have explained here do seem at least morally (*moraliter*) certain," and in Paragraph 206, "They seem more than morally certain" on the ground that his physics is mathematical, but, in the final Paragraph 207, he concluded that "I submit all my opinions to the authority of the Church."[28] From this argument, we know that he finally entrusted the theological authority as to the final judgment of his natural philosophy.

Other people may use another weapon in arguing about the logical structure of Descartes's doctrine on mathematical and logical truths. One expositor, for example, has rather ingeniously reformulated Descartes's theory in terms of today's modal logic.[29] Another has elaborated it further by making a distinction between absolute and hypothetical (conditional) necessity.[30] Moreover, it may be observed that the structure of Cartesian epistemology and philosophy of mathematics can be interpreted to permit the possibility of non-Euclidean geometry[31] and other competitive mathematical theories. These logical or philosophical approaches are quite interesting. As they are taking notice of the consequences rather than the reasons Descartes conceived his doctrine, they easily lead us into the danger of committing ourselves to an anachronistic interpretation.

We do not deny that the aforesaid interpretations shed light on and deepen our understanding of Descartes's creation doctrine. Particularly the explanations from the viewponts of theology and of natural philosophy are historically plausible. Nevertheless, they do not completely satisfy us. They may be sufficient to explain the origins of Descartes's concept of a voluntaristic God, but seem to us inadequate, at least partly, to interpret his creation theory of *mathematical truths*. We need an approarch which clarifies why and how Descartes's philosophical point of view was transformed from the philosophy of the *Regulae ad directionem ingenii* to that of "Cogito, ergo sum." Certainly there is a continuity between the two. The former monograph was, as its title may indicate, a writing which argues the Rules for the direction of the intelligence (*ingenium*)

[28] *AT*, VIII-1, pp. 327–329 (Part IV, Arts. 205–207).

[29] See E. M. Curley, "Descartes on the Creation of the Eternal Truths," *The Philosophical Review*, **43** (1984), pp. 569–597.

[30] See, for example, Martial Gueroult, *Descartes' Philosophy Interpreted according to the Order of Reasons*, Vol. II: *The Soul and the Body*, translated by Roger Ariew (Minneapolis, 1985; The original French edition, 1953), p. 28; Jacques Bouveresse, "La Théorie du possible chez Descartes," *Revue internationale de philosophie*, 37e année, N° 146 (1983), pp. 293–318; Hidé Ishiguro, "Reply to Jacques Bouveresse," *Ibid.*, pp. 311–318; *Eadem*, "The Status of Necessity and Impossibility in Descartes," in Amélio O. Rorty, ed., *Essays on Descartes'* Meditations (Berkeley/Los Angeles/London, 1986), pp. 459–471. And A. Boyce Gibson invokes Émile Boutroux's distinction between absolute and relative contradiction in his "The Eternal Verities and the Will of God in the Philosophy of Descartes," *Proceedings of the Aristotelian Society*, N. S., **30** (1929–30), pp. 31–54.

[31] M. Wilson, *Descartes* (n. 9), pp. 125–126; H. Ishiguro, "Reply to Jacques Bouveresse" (n. 30), p. 317; *Eadem*, "The Structure of Necessity and Impossibility in Descartes" (n. 30), p. 468.

among vaious functions of the mind. Thus, we can easily understand that the author placed 'mathesis universalis' of which an axis is algebraic analysis in its central position. It may be quite natural to think that the philosophical analysis in the *Regulae* led ultimately to the discovery of "Cogito, ergo sum," the first indubitable principle. This is why the *Meditationes*, as its "Second Set of Replies" typically show, was 'analytic' par excellence.[32] Nevertheless, an important discontituity separates between the epistemology of the *Regulae* and the metaphysics of the *Meditationes*. This is the creation doctrine of eternal truths.

We will take a different and historically sound approach to the present problem about the manner in which Descartes formed his creation doctrine. Our approach takes the viewpoint of the philosophy of mathematics, or more exactly the philosophy of mathematics considered in its historical context.

A remarkable fact about the formation of his creation doctrine is that Descartes almost always characterized his conception of a specifically voluntaristic God in connection with mathematical truths. And it is concerning the status of these truths that we can sense a discontinuity between his philosophy of mathematics in the *Regulae* and that maintained from 1630. In the *Regulae*, the principles of mathematics were regarded as being grasped by indubitable intuition. It may be said that in a sense this view was retained even in the later Descartes. But the philosophical status of the mathematical principles seems to have begun to shake after his first serious metaphysical meditations which took place during the years 1628–1629 in Franeker. Since then he appears to have abandoned his previous theory which viewed the principles of mathematics 'naively', it seems, as self-evident through intuition.

Before considering how and why this transition occurred, let us reformulate briefly the epistemological nature of Descartes's philosophy of mathematics after 1630. Both in the *Regulae* and in the later writings, mathematical statements were viewed generally as products interposed more or less by the human mind. We should not interpret the Cartesian epistemology as a naive psychologism, as Martial Gueroult has warned.[33] But it seems certain at any rate that Descartes represented a tendency opposed to Leibniz's logicism. Consequently,

[32]E. M. Curley, "Analysis in the *Meditations*: The Quest from Clear and Distinct Ideas," in Stanley Tweyman, ed., *René Descartes: Meditaions on First Philosophy in Focus* (London/New York, 1993), pp. 159–184. Cf. Daniel Garber and Lesley Cohen, "A Point of Order: Analysis, Synthesis, and Descartes' *Principles*," and S. Tweyman, "Professor Cottingham and Descartes' Methods of Anaysis and Syntheis," in *Ibid.*, pp. 135–147 & 148–158, respectively. See also Stephen Gaukloger, *Cartesian Logic: An Essay on Descartes's Conception of Inference* (Oxford, 1989).

[33]M. Gueroult, *Descartes' Philosophy Interpreted according to the Order of Reasons*, Vol. I: *The Soul and God*, tr. by Roger Ariew (Minneapolis, 1984; The French edition, 1952), p. 47. For the use of the terminology 'psychologism', Gueroult refers to Edmund Husserl's *Logische Untersuchungen*, Bd. I (Halle, 1900), 7. Kap. "Der Psychologismus als skeptischer Relativismus." Cf. Yvon Bélaval, "Intuitinnisme cartésien et psychologisme," *Revue internationale de philosophie*, 37ᵉ année, N° 146 (1983), pp. 319–325.

it is suspect to interpret his introduction of the malicious demon, who causes us to make mistakes in mathematical reasoning, simply to be equivalent to the following situation: The demon intervenes in our mental functions and leads us to make errors in geometrical demonstrations and arithmetical calculations whereas mathematical truths stand as true independent of our minds which may err. This simplified interpretation is one which Leibniz took later in criticizing Descartes's standpoint. We venture to speculate that Descartes devised the deceiving demon hypothesis in order to cope with a more complicated problem in the philosophy of mathematics.

To confirm that our speculation is not without reason, let us recall what Descartes states about the principles of mathematics in his later philosophical writings. We have already seen that in his letter to Mersenne of May 17 (or 27), 1638 and in Part I, Paragraph 5 of his *Principia philosophiae*, the French philosopher applied his methodological doubt even to the mathematical principles which had been hitherto regarded to be self-evident.

Such discussions about principles may have been common knowledge in the age of Descartes. But what is important is that he came to be more attentive to principles in the later writings than he was in the *Regulae*. In the "Second Set of Replies" of his *Meditationes*, he pays attention to how dialecticians handle principles: "Awareness of [first] principles is not normally called 'knowledge' by dialecticians."[34] He then proceeds to refute these dialecticians by the self-evident intuition of the cogito. Note that it is by means of the intuition of his first principle "Cogito, ergo sum," not by means of the one of the mathematical principles, that he tries to overthrow the dialecticians. Furthermore, in the letter addressed to Clerselier in June of 1646, Descartes advocated a careful observation on the first principles.[35]

Clearly we witness a change in Descartes's attitude toward the intuition of mathematical principles. In his epistemology of mathematics, which has a certain tendency toward psychologism, simple mistakes in geometrical demonstrations and in arithmetical calculations, on the one hand, and difficultiess in judging the mathematical principles to be self-evident and true, on the other, are inseparably mixed. And indeed it might be arduous for us finite beings to distinguish the two categories clearly and *a priori*. Insofar as mathematical

---

[34] *AT*, VII, p. 140: "Principiorum enim notitia non solet a dialecticis scientia appellari."

[35] "Descartes á Clerselier. [Egmond, juin ou juillet 1646]," *AT*, IV, pp. 443–444; *AM*, VII, pp. 84–85. Descartes maintained that the logical principle of contradiction is a principle which is valid in general and that our soul exists is the first principle. The latter case seems to be closely related to the *cogito*. And he stated generally about the first principle: "I will also add that one should not require the first principle to be such that all other propositions can be reduced to it and proved by it. It is enough if it is useful for the discovery of many, and if there is no other proposition on which it depends, and none which is easier to discover. It may be that there is no principle at all to which alone all things can be reduced. They do indeed reduce other propositions to the principle that *the same thing cannot both be and not be at the same time*, but their procedure is superfluous and useless." This discussion is certainly related to mathematical principles.

statements are not independent of human minds, at least potentially we could always be deceived even in our mathematical activities. And in reality we finite beings cannot distinguish the two categories evidently and *a priori.* Because mathematical statements are not independent of the human mind, it is always possible for us to be deceived at least implicitly in our mathematical activities. In his later philosophy of mathematics, Descartes came to suspend his judgment at one time about whether the mathematical principles which were commonly taken to be true are clear and distinct. We may use the above considerations to solve the problem of how and why he formed his creation theory around the years 1628–1629.

C. Descartes against the Skeptical Arguments on Mathematics: How Was the Philosophy of "Cogito, Ergo Sum" Discovered?

Now we turn to a historically more important question: How and why did the transformation which we have outlined above take place? Recently it has been recognized that his skeptical speculations during the years 1628–1629 played a crucial role in Descartes's philosophical development away from the point of view expressed in the *Regulae* to that of the *Meditationes.* Against this view one might contend that we can sense the influence of skepticism upon the concept of intuition as presented in the *Regulae*, especially through Mersenne's *La Verité des sciences contre les septiques ou Pyrroniens*, published in Paris in 1625.[36] This criticism can be easily answered by our consideration of the later Descartes's attitude toward the principles of mathematics and above all by the fact that Mersenne's philosophy in this work of 1625 can be characterized as a kind of 'probabilism' or 'mitigated skepticism', one of whose means of refuting skepticism is "a presentation of mathematical truths as an antidote to Pyrrhonism,"[37] which is radically different from Descartes's later standpoint. Consequently we may safely conclude that there is a visible discontinuity between the *Regulae* and the *Meditationes.*[38] We claim, furthermore, that Descartes's later philosophical point of view was formed through the intellectual struggle to confront and overcome, above all, Pyrrhonian skepticism during the period 1628–1629.

It was the historian of philosophy Richard H. Popkin who has tried to advance the above thesis most comprehensively. In the most illuminating chapter "Descartes: Conqueror of Scepticism" of his monumental *History of Scepticism*

[36] O. A. Kubitz, "Scepticism and Intuition in the Philosophy of Descartes," *The Philosophical Review*, **48** (1939), pp. 472–491.

[37] Peter Dear, "Marin Mersenne and the Probabilistic Roots of 'Mitigated Scepticism'," *Journal of the History of Philosophy*, **22** (1984), p. 202. See also *Idem, Mersenne and the Learning of the Schools* (Ithaca/London, 1988) and "Mersenne's Suggestion: Cartesian Meditaion and the Mathematical Model of Knowledge in the Seventeenth Century" in *Descartes and His Contemporaries: Meditations, Objections, and Replies* (n. 14), pp. 44–62.

[38] For a discussion on this discontinuity from the point of view of general philosophy, see M. Gueroult, *Op. cit.* (n. 33), pp. 30–32.

from Erasmus to Descartes, he wrote about Descartes's introducing methodological doubt into our mathematical knowledge:

> The simple starting place of the *Regulae*, that reason, in intuiting and deducing, was infallible, and hence that mathematics was indubitably true, was now challenged by a scepticism wich regard to our faculties, and a scepticism with regard to our ability to use them. As long as we might be the victims of some force or agent who purposely misleads us, what we consider most certain, what we are unable to doubt (psychologically), may actually be false or dubious. In introducing this level of doubt, creating the possibility of the *malin génie*, Descartes overthrew the mathematical intuitionism of the *Regulae* as the foundation of all certainty. The *crise pyrrhonienne* had been pressed to its farthest limit. Not only had all the opinions and theories of all previous thinkers been cast in doubt, but also those of the young René Descartes. But from this voyage into the depths of complete scepticism, Descartes was to find a new metaphysical and theological justification for the world of human rationality.[39]

It must be admitted, as Popkin has argued, that skeptical thought constituted an enormously important factor, if not the only factor, in the making of modern philosophy, particularly in the development of Descartes's metaphysical thought. Incidentally, it is noticeable that this excellent historian of philosophy began his academic career from the philosophy of mathematics.[40]

More recently E. M. Curley in his *Descartes against the Skeptics* (1978) developed and deepened Popkin's view, and contrasted Descartes's theses stated in the *Meditationes* with the thoughts of ancient and modern skeptics.[41] A little later than Curley, but independently of him, in his *La Philosophie première de Descartes* Jean-Marie Beyssade tried to relate Descartes's arguments on mathematical truths to those of Montaigne.[42]

The importance of skeptical thought in the Scientific Revolution has also begun to be understood. Popkin himself wrote an article entitled "Scepticism, Theology and the Scientific Revolution in the Seventeenth Century"[43] and em-

[39]Richard H. Popkin, *The History of Scepticism from Erasmus to Descartes* (Assen, 1964; The First Edition, 1960), p. 184. The revised and expanded edition of the former was published under the new title of *The History of Scepticism from Erasmus to Spinoza* (Berkeley/Los Angeles/London, 1979). The cited passage is on p. 180.

[40]R. H. Popkin, "Intellectual Autobiography: wars and all," in Richard A. Watson and James E. Force, eds., *The Sceptical Mode in Modern Philosophy: Essays in Honor of Richard H. Popkin* (Dordrecht/Boston/Lancaster, 1988), pp. 103–149.

[41]E. M. Curley, *Descartes against the Skeptics* (Cambridge, Mass., 1978).

[42]Jean-Marie Beyssade, *La Philosophie première de Descartes: Le temps et la cohérence de la métaphysique* (Paris, 1979), pp. 105–106.

[43]Popkin, "Secpticism, Theology and the Scientific Revolution," in Imre Lakatos and Alan Musgrave, eds., *Problems in the Philosophy of Science* (Amsterdam, 1968), pp. 1–28; See also the discussion which follows Popkin's paper, pp. 29–39.

phasized in it that the skepticism fashionable during the seventeenth century played a role in the construction of faith and the mechanical world-view. In the field of the history of biology, Phillip R. Sloan's excellent paper "Descartes, the Sceptic, and the Rejection of Vitalism in Seventeenth-Century Physiology"[44] contributed to promoting our understanding of the fact that Descartes's mechanist physiology was motivated by his confrontation with Pyrrhonian skepticism.

Despite these contributions, however, it seems to us that Descartes's turn in his philosophy of mathematics has not been sufficiently explored from the viewpoint sketched above. Certainly Popkin's book on the history of skepticism is cautious at times about the relationship between mathematics and skepticism. Unfortunately, he does not discuss it as a main subject. However, Sloan's article has left an insightful suggestion. He put it:

> By Spring of 1630, Descartes claims to have attained metaphysical foundations for his science more certain and more readily demonstrable than the propositions of geometry, which rendered his science now incapable of doubt. Such an assertion, moving significantly beyond the intuitive 'geometrical' certitude claimed in the *Regulae*, only seems explicable if Descartes had taken full cognizance of pyrrhonist arguments which could readily undermine the appeal to the self evidence of axioms.[45]

Sloan wrote this comment in a footnote to his paper. Moreover the main topic of his discussion was the history of physiology. This suggests only that we should scrutinize more minutely how Descartes changed his view on mathematical truths in connection with skepticism. Rodis-Lewis, whom we referred to as an author attempting to connect the creation doctrine to Descartes's natural philosophy, began to point out the relation of that theory to mathematical truths,[46] and thus has encouraged us. Such an intellectual situation seems to require a nouvel synthetic discussion.

Let us study, then, how Descartes is supposed to have been influenced by the skeptics, concentrating on their mathematical thought. In the "Second Set of Replies" of his *Meditationes*, Descartes stated that he "had some time ago seen many books written by the Academics and Skeptics."[47] Unfortunately, we

[44]Phillip R. Sloan, "Descartes, the Sceptics, and the Rejection of Vitalism in Seventeenth Century Physiology," *Studies in History and Philosophy of Science*, **8** (1977), pp. 1–28.

[45]*Ibid.*, pp. 11–12, footnote 29.

[46]G. Rodis-Lewis, "Création des vérités éternelles, doute suprême et limites de l'impossible chez Descartes," *Idées et vérités éternelles chez Descartes et ses successeurs* (n. 9), p. 120.

[47]*AT*, VII, p. 130: "[...] libros eâ de re complures ab Academicis & Scepticis scriptos dudum videssem [...]. The translation of this sentence by J. Cottingham et al. reads: "I had seen many ancient writings by the Academics and Sceptics on this subject." This is a serious mistranslation. "Dudum" should be traslated into "sometime ago." In this point, the translation by Elizabeth S. Haldane and G. R. T. Ross in *The Philosophical Works of Descartes*, Vol. II (Cambridge, ${}_2$1931) is basically correct: "I had long ago seen several books written by the Academics and Sceptics about this subject." (p. 31.)

do not know what he read exactly. Nevertheless, thanks to recent historical research, especially that of Richard H. Popkin[48] and Charles B. Schmitt,[49] we now know what among the writings of ancient skeptical thinkers became available during the sixteenth and early seventeenth centuries. Here we examine the mathematical thoughts of those skeptics who were influential in Descartes's time. We will examine Sextus Empiricus and Cicero, and then study Michel de Montaigne, who became one of the most important thinkers among the 'nouveaux pyrrhoniens', being influenced by the ancient skeptics just mentioned. Popkin pointed out that there were a number of followers of Montaigne in late sixteenth and early seventeenth century France. And more recently, it has been conjectured that the writings by Franscico Sanches, Pierre Charron, François de la Mothe le Vayer, and so on could have directly stimulated Descartes to compose the *Meditationes*.[50] Admittedly our examination will be limited, but will make clear the essential features as to how the skeptics thought of or rather attacked the authority of mathematics.

Two preliminary remarks should be made before examining the mathematical thought of the skeptics. One is that they did not allow any exceptions to their suspension of judgment (ἐποχή). That is to say, they cast doubt on everything, including mathematics, which was usually thought to be the most certain science, except on their tenets that the appearance (φαινόμενον) should basically be trusted[51] and that they suspend judgment generally and continue their activity in investigation and inquiry.[52] It was quite natural for Sextus Empiricus to challenge the trustworthiness of the strongpoint of Greek geometry exemplified by Euclid's *Elements* in Book III ("Against the Geometers") of his *Against the Professors* (*Adversus mathematicos*). The second remark is that they attacked mathematics in order to destroy it rather than to build something. Accordingly even if there is a case in which the skeptics' criticism is legitimate, it is doubtful that it will be used for the purposes of constructing a science of mathematics straightforwardly and positively.

Criticisms of mathematics by Sextus Empiricus, the spokesman of Pyrrhonian skepticism, may be said to consist mainly of two parts: one logical and

[48]Popkin, *The History of Scepticism from Erasmus to Spinoza* (n. 39), Chapter II "The Revival of Greek Scepticism in the Sixteenth Century," pp. 18–41.

[49]C. B. Schmitt, "The Rediscovery of Ancient Skepticism in Modern Times," in Myles Burnyeat, ed., *The Skeptical Tradition* (Berkeley/Los Angeles/London, 1983), pp. 225–251. An earlier version with the title "The Recovery and Assimilation of Ancient Scepticism in the Renaissance" appeared in *Rivista critica di storia della fiosofia*, **27** (1972), pp. 363–384.

[50]On the writings which are supposed to have influenced Descartes, see Roger Ariew, John Cottingham, and Tom Sorell, eds., *Descartes'* Meditations: *Background Source Materials* (Cambridge/New York, 1998).

[51]Sextus Empiricus, *Outlines of Pyrrhonism*, translated by R. G. Bury (London / Cambridge, Mass., 1933), Book I, Chaps., X–XI, pp. 14–17. I have also referred to Descartes's contemporary Greek and Latin version *Sexti Empirici Opera quae extant* (Paris, 1621), which has the same chapter divisions as Bury's edition.

[52]*Ibid.*, Book I, Chap. III, pp. 4–7.

the other epistemological. The most convincing and interesting point that Sextus utilizes in his attacking any theoretical science whatsoever is related to the modes of demonstration. In Book I, Chapter XV of his *Outlines of Pyrrhonism*, he notes the five modes used by the later skeptics (modes ascribed to the ancient Pyrrhonist Agrippa) which lead to suspension of judgment: "The later Sceptics hand down Five Modes leading to suspension, namely these: the first based on discrepancy, the second on regress *ad infinitum*, the third on relativity, the fourth on hypothesis, the fifth on circular reasoning."[53]

All these modes are likely to appear in mathematical proofs. There may be a discrepancy among mathematicians as to whether or not some axioms are acceptable, for example, the fifth postulate on parallel lines, which was a subject of discussion in antiquity (Mode I).[54] These axioms become objects of discussion, relative to mathematicians (Mode III). The other three modes are stated as follows:

> [Mode II] The mode based upon regress *ad infinitum* is that whereby we assert that the thing adduced as a proof of the matter proposed needs a further proof, and this again another, and so on *ad infinitum*, so that the consequence is suspension, as we possess no starting-point for our argument. [§ 166]
> [Mode IV] We have the mode based on hypothesis when the Dogmatists, being forced to recede *ad infinitum*, take as their starting-point something which they do not establish by argument but claim to assume as granted simply and without demonstration. [§ 168]
> [Mode V] The mode of circular reasoning is the form used when the proof itself which ought to establish the matter of inquiry requires confirmation derived from that matter, in this case, being unable to assume either in order to establish the other, we suspend judgment about both. [§ 169][55]

The mode V on circular reasoning is sometimes called the mode of mutual dependence. It would be clear enough that the mere intuitive affirmation of mathematical principles and demonstrations is not sufficient to refute these logical assaults of the Pyrrhonists.

These five modes of the logical incompleteness of demonstrations are also documented in Diogenes Laertius's "Life of Pyrrho" contained in Book IX of his *Lives of Eminent Philosophers*.[56]

Sextus's charge against geometers in Book III of his *Against the Professors*

---

[53]*Ibid.*, Book I, Chap. XV, §164, pp. 94–95. See also Julia Annas and Jonathan Barnes, *The Modes of Scepticism: Ancient Texts and Modern Interpretations* (Cambridge, 1985), p. 182, and J. Barnes, *The Toils of Scepticism* (Cambridge, 1990).

[54]Heath, *The Thirteenth Books of Euclid's Elements*, Vol. 1 (New York, 1956), pp. 202–220.

[55]Sextus, *Ibid.*, pp. 94–95.

[56]Diogenes Laertius, *Lives of Eminent Philosophers*, Vol. II, translated by R. D. Hicks (London/Cambridge, Mass., 1925), pp. 500–501 (§§88–89).

is grounded on Mode IV, on hypotheses, in the first place, and secondly on his strictly empiricist epistemology. According to Sextus, the science of geometry has to begin with a certain hypothesis. Supposing a hypothesis is used as the starting-point of proof. Sextus continues his discussion:

> For this is the hypothesis which the Geometers adopt when they wish to prove anything geometrically. Consequently, we must state at once that since those who assume a thing by hypothesis are satisfied with mere assertion, without proof, for its confirmation, one will interrogate them, employing some such reasoning as this:—Assuming a thing by hypothesis is either a strong and firm confirmation or unreliable and weak.[57]

The argument will proceed *ad infinitum*. After all, Sextus asserts: "Nothing, then, must be assumed by hypothesis."[58]

Furthermore Sextus criticizes basic notions such as point, line, surface and so forth which appear in geometrical definitions and principles. Remember that Sextus's nickname is 'Empiricus' (empiricist). From his empirical point of view, the concept of line in the Euclidean definition is absurd. According to Euclid's definition: "A line is breadthless length."[59] But the empiricist Sextus attacks such a definition:

> For the line, as one may learn from the Geometers themselves, is "length without breadth," but when we have examined the matter closely, we shall not find either amongst intelligibles or amongst sensibles anything that is capable of being perceived as length without breadth.[60]

"Length without breadth" is an expression permitted in the ideal world and inconceivable to the empiricist. Then, in his concluding remark Sextus claims categorically:

> Thus, as regards the principles of geometry, the result is that they are unfounded; and as these are abolished no other geometrical theorem can subsist. For the theorem, of whatever sort it be, must be proved by a diagram, but we have shown that the generic line is nothing, and from this it follows that none of the specific lines exist, whenever one assumes a straight one, or a curved one, or one of some other form.[61]

[57]Sextus Empiricus, *Against the Professors*, Vol. IV, translated by R. G. Bury (London/Cambridge, Mass., 1949), pp. 246–247 (§§7–8).

[58]*Ibid.*, pp. 248–249 (§ 8).

[59]Heath, *Op. cit.* (n. 54), p. 153.

[60]Sextus, *Ibid.*, pp. 262–263 (§ 37).

[61]*Ibid.*, pp. 290–291 (§ 92).

Such a criticism appears *prima facie* simply ridiculous. But this statement is in a form similar to that which Pierre Gassendi utilized when he criticized Descartes's program of physics as the subject matter of pure mathematics in his "Fifth Set of Objections" to the latter's Sixth Meditation.[62] The similarity is no accident. Sextus's epistemology, as Ian Mueller contends, may well have been influenced by Epicurus's criticizing of geometrical physics in order to advocate qualitative physics.[63]

Cicero's *Academica*, which contains important doctrines of academic skepticism, also discusses mathematical subjects, although his argument seems to us much less exciting than Sextus's. Cicero seems to have known Epicurus's attack on geometry. He presents a short story about Polyaenus, a friend of Epicurus: "Polyaenus is said to have been a great mathematician: after he had accepted the view of Epicurus and come to believe that all geometry is false, surely he did not forget even the knowledge that he possessed?"[64] Cicero does not state the reason why the friend of Epicurus came to discredit geometry. But the only probable explanation of the event seems to have been his becoming acquainted with Epicurus's empiricist epistemology which denies the describability of the natural world by geometry. In his *Essais*, Montaigne cited Cicero's story about Polyaenus mentioned above.[65] At least Cicero does not seem to have thought that the geometrical method is the only approach to natural philosophy, for he insists that there are various approaches to it, all of which appear equally plausible.

The three works on Pyrrhonian skepticism which we have referred to, Sextus's *Outlines of Pyrrhonism*, *Against the Professors*, and Diogenes Laertius's

---

[62] Gassendi criticizes Descartes as follows: "I have no objection to what you say at the beginning of the Sixth Meditation, namely that material things are 'capable of existing in so far as they are the subject matter of pure mathematics.' In fact, however, material things are the subject matter of mixed, not pure, mathematics (objectum mixtae, non purae, matheseos), and the subject matter of pure mathematics—including the point, the line, the surface, and the indivisible figures which are composed of these elements and yet remain invisible—cannot exist in reality." *AT*, VII, pp. 328–329; Gassendi, *Disquistio metaphysica: seu Dubitationes et instantiae adversus Renati Cartesii metaphysicam et responsa*, Texte établi, traduit et annoté par Bernard Rochot (Paris, 1962), pp. 518–519. For Gassendi's view of qualitative physics in contrast with the Cartesian mathematical physics, see B. Rochot, "Gassendi et les mathématiques," *Revue d'histoire des sciences*, X (1957), pp. 69–78.

[63] For a discussion about the epistemological aspect of Sextus's attack on geometry, see Ian Mueller, "Geometry and Scepticism," in Jonathan Barnes, *et al.*, eds., *Science and Speculation: Studies in Hellenistic Theory and Practice* (Cambridge/Paris, 1982), pp. 69–95. On the Epicurian's view of geometry, see, especially, "Appendix: Epicureans and Geometers," pp. 92–95. Study I "Indivisible Magnitudes" in David J. Furley's *Two Studies in the Greek Atomists* (Princeton, 1967), pp. 1–158 also contains some interesting arguments on the Atomists' criticism of geometry.

[64] Cicero, *Academica*, Vol. XIX, With an English Translation by H. Rackham (London/Cambridge, Mass., 1933), Book II, § 106, pp. 602–603.

[65] Michel de Montaigne, *An Apology of Raymond Sebond*, translated and edited with an introduction and notes by M. A. Screech (London, 1987), p. 109; *Œuvres complètes*, textes établis par Albert Thiboudet et Maurice Rat (Paris, 1962), p. 516.

"Life of Pyrrho," were all widely known through Latin translations before the decade of the 1560s.[66] Cicero's *Academica* was read during the Renaissance and became influential enough to have caused the formation of the 'nouveaux académiciens', to which Omer Talon, a follower of Ramus, belonged.[67] Henri Gouhier has suggested that there are signs that Descartes read Cicero's *Academica.*[68]

It was, above all, Michel de Montaigne's *Essais*, first published in 1580 in two books and republished in 1588 in three books, that played the most significant role in widely diffusing ancient skeptical thought in France in the milieu of the Counter Reformation. He was clearly familiar with ancient skepticism in the forms of both the Pyrrhonists and Academics. Historians of philosophy have pointed out that there are many parallelisms between the arguments of Montaigne and those of Descartes.[69] Regrettably, however, only a few have examined closely Montaigne's skeptical theory of mathematics, in striking contrast to the fact that many have discussed Descartes's philosophy of mathematics. It must be our task here to amend this defect.

Montaigne, the champion of the 'nouveaux pyrrhoniens' of the sixteenth century, advanced his criticism of mathematics in Book II, Chapter 12 "Apologie de Raymond Sebond" of his *Essais.* Notably in the letter which is supposed to have been addressed to the Marquess of Newcastle on November 23, 1646, Descartes explicitly referred to Montaigne. There Descartes mentioned Montaigne's opinion which emphasized a similarity between human beings and animals.[70] Montaigne's rather detailed discussion on the closeness of animals to human beings appears near the beginning part of the chapter "Apologie de Raymond Sebond." It seems highly probable that Descartes read the entire chapter. In Chapter 12, Montaigne indicates that he was not necessarily blind to mathematical theories. There are clear signs showing that he was interested in discussions about mathematics. He invokes an episode from geometry in order to support his Pyrrhonian attitude toward the dogmatists. By referring to an example in mathematics, he argues that effects and reason are very often incompatible:

> I then retorted that I would rather be guided by results than by reason—for they are always clashing! I have even been told that

[66]See Popkin' and Schmitt's writings quoted in notes 48 and 49.

[67]Consult Charles B. Schmitt, *Cicero Scepticus: A Study of the Influence of the Academica in the Renaissance* (The Hague, 1972), esp. pp. 78–91.

[68]See Henri Gouhier, *La Pensée métaphysique de Descartes* (Paris, 1978; First edition, 1962), pp. 34–35.

[69]E. g. Léon Brunschvicq, *Descartes et Pascal lecteurs de Montaigne* (Paris, 1995; First edition., 1942); Popkin, *Op. cit.* (n. 48), p. 172; Michael G. Paulson, *The Possible Influence of Montaigne's* Essais *on Descartes'* Treatise on the Passions (Lanham/New York/London, 1988).

[70]"Descartes au [Marquis du Newcastle] [Egmond, 23 Novembre 1646]," *AT*, IV, pp. 573 & 575; *AM*, VII, pp. 225–226.

> in geometry (which claims to have reached the highest degree of certainty among the scientists) there are irrefutable demonstrations which overturn truth based on experience. Jacques Peletier, for example, in my own home, told me how he had discovered two lines drawing ever closer together but which, as he could prove, would meet only in infinity. And the sole use Pyrrhonists have for their arguments and their reason is to undermine whatever experience shows to be probable.[71]

Jacques Peletier du Mans was a poet, medical doctor, and mathematician of sixteenth-century France. The two lines about which he talked to Montaigne seem to be interpreted, for example, as a hyperbola and its asymptote.[72] This simple anecdote would have been enough to overturn Montaigne's initial belief that two infinite lines which become close to one another necessarily touch. Quoting this story, he warns us not to believe too readily those people who assure us that now all is found, all is seen.

The following remarks are much more significant than the previous one and essential for our argument. Montaigne seems to have been familiar with Agrippa's five modes which lead to suspension of judgment. We will discuss Montaigne's discourses corresponding to Modes II, IV, and V, to which we paid attention above (p. 377). What follows indicates that Montaigne's theory of mathematics, while not always original, reached a considerable depth. The first example is related to the later skeptics' Mode IV based on hypothesis. The main target of criticism is the axiomatic method as used by geometers:

> Base yourself on admitted postulates and you can build up any case you like; from the rules which order the original principles the remainder of your construction will follow on easily without self-contradiction.
>
> This method allows us to bowl our arguments with the jack in view (and so be satisfied that our foundations are rational ones); before they even begin, our professors (like geometricians with their postulated axioms) establish such a hold over our beliefs that they can subsequently reach any conclusion they want. We give them our agreement and consent: they can then pull us this way and that way, spinning us about at will. Once we accept anyone's postulates he becomes our professor and our god: for his foundations he will grab territory so ample and so easy that, if he so wishes, he will drag us up to the clouds. In the practice and business of scholarship we have accepted Phythagoras's contention as legal tender: every expert, he says, must be believed in his own speciality. So, for the meaning of words the dialectician turns to the grammarians; for the matter

[71] *Apology* (n. 65), p. 277; *Œuvres*, p. 555.

[72] See footnote 359 of *Apology* and footnote 439 of *Œuvres*, p. 1574, n. 1.

> of his arguments the rhetorician borrows from the dialectician; the poet takes his rhythms from the musician; the geometer takes his proportions from the arithmeticians; the metaphysicians make their foundations out of the conjectures of physics. For their principles, all branches of learning take admitted postulates, which restrain human judgment on every side. If you come up against the barrier behind which their error of principle is sheltering, they have an axiom ready on their lips: Never argue with those who deny first principles.[73]

Montaigne here appears to detect the secret and Achilles's heel, so to speak, of the certainty of geometrical reasoning. If the axiomatic method of geometers is assumed to be infallibly certain, the reason for that is that they have presupposed sufficiently ample principles to deduce what they want. All propositions contained in conclusions have been transferred, in reality, to the principles from which they can be inferred. Thus what of the principles? The voyant Montaigne proceeds to this problem:

> But there can be no first principles unless God has revealed them; all the rest—beginning, middle and end—is dream and vapor.
>
> Whenever a case is fought from preliminary assumptions, to oppose it take the very axiom which is in dispute, reverse it and make that into your preliminary assumption. For any human assumption, any rhetorical proposition, has just as much authority as any other, unless a difference can be established by reason. So they must all be weighed in the balance—starting with general principles and any tyrannous ones. To be convinced of certainty is certain evidence of madness and of extreme unsureness.[74]

Human beings are necessarily at a loss for principles. But remarkable for our discussion in connection with Descartes's creation doctrine is the following: Montaigne states that only God warrants their truth whereas all human presuppositions have equal authority "unless a difference can be established by reason." In other words, the truth of the principles is surely guaranteed when God wills them to be so and reason makes the difference. What Montaigne

[73] *Apology*, p. 115; *Œuvres*, pp. 521–522. For a possible source of this passage, see Henry Cornerius Agrippa, *De incertitudo et vanitate scientiarum atque artium* (Antwerp, 1530); *Opera*, t. II (Hildesheim, 1970), pp. 6–7; *On the Vanitie and Vncertaintie of Artes and Sciences*, ed. by Catherine M. Dunn (Northridge, CA, 1975), p. 17. In his "Introduction" to the 1970 *Opera* edition, Popkin writes on this work: "*De Vanitate* has often been regarded as one of the early sceptical treatises of the Renaissance, leading to the "nouveau pyrrhonisme" of Montaigne and Charron, and contributing to the sceptical crisis that was to engulf the late 16th century and early 17th cenury intellectuals. Aucually *De Vanitate* is more anti-intellectual than sceptical." (Introduction to *Opera*, I (Hildesheim, 1970), p. XVI.) "*De Vanitate* at least probably played a significant role in preparing the climate of opinion in whcich the revival of Greek scepticism could take place as a serious movement." (*Ibid.*, p. XX.)

[74] *Apology*, pp. 115–116; *Œuvres*, p. 522

really intends here is a degrading of the method which mathematicians have traditionally taken pride in.

Montaigne furthermore gives attention to Mode II, infinite regress; and Mode V, circular reasoning:

> We register the appearance of objects; to judge them we need an instrument of judgment; to test the veracity of that instrument we need practical proof; to test that proof we need an instrument. We are going round in circles.
>
> The senses themselves being full of uncertainty cannot decide the issue of our dispute. It will have to be Reason, then. But no Reason can be established except by another Reason. We retreat into infinity.[75]

The quoted sentences as a whole pertain to Mode II. In addition to this, by the words "going round in circles" they partly suggest a danger of Mode II. Therewith Montaigne restored the ancient Pyrrhonist's methods for limiting human knowledge.[76]

It is notable that Montaigne's theory of mathematics is closely related to his theological view. We have already observed that the truth of mathematical principles, according to him, is first warranted by the divine power (p. 550). Human reason is incomparably powerless. Montaigne repeatedly emphasizes this contrast. For example, he states:

> We confidently use words like might, truth, justice. They are words signifying something great. But what that 'something' is we cannot see or conceive. We say that God 'fears', that God 'is angry', that God 'loves':
>
> > *Immortalia mortali sermone notantes.* [Denoting immortal things in mortal speech.
> > —Lucretius, *De rerum natura*, V. 122.]
>
> But they are disturbances and emotions which in any form known to us find no place in God. Nor can we imagine them in forms known to him. God alone can know himself; God alone can interpret his works.[77]

How Montaigne's God resembles Descartes's! The former's notion of God is formed so as to make us know the weakness of human beings. It is in order to justify his skeptical attitude that Montaigne maintains God's voluntarism. He writes with a feeling of powerlessness: "I think Man will confess, if he speaks

---

[75] *Apology*, p. 185; *Œuvres*, p. 585

[76] This does not necessarily mean that Montaigne was a strict Pyrrhonist. The following article has pointed put that Montaigne was closer to Socrates, not Pyrrho or Sextus. See Craig Walton, "Montaigne on the Art of Judgment: The Trial of Montaigne," in *The Sceptical Mode in Modern Philosophy* (n. 40), pp. 87–102.

[77] *Apology*, p. 66; *Œuvres*, p. 479.

honestly, that all he has gained from so long a chase is knowledge of his own weakness. By long study we have confirmed and verified that ignorance does lie naturally within us."[78]

Whether Descartes read Montaigne's discussions about mathematics or not, he resisted that line of argumentaition. He had to vacate his previous fortress in which mathematics is the paradigm for certain knowledge. It can now be safely said that it was because he had to grant, even though against his will, the persuasiveness of the skeptics' attack on mathematical principles and demonstrations. Here we present definitive evidence to show that Descartes surely knew the way in which the skeptics criticized and attempted to destroy mathematical knowledge. In the "Fifth Set of Replies" of his *Meditationes*, he relates to Gassendi the following crucial words:

> I cite the case of the skeptics who did have doubts about these very geometrical demonstrations. And I insist that they could not have done so had they known the true nature of God.[79]

We do not know exactly whether or not Descartes was familiar with Montaigne's conception of the voluntaristic God. What is certain, however, is that he knew the skeptics' doubt about mathematical demonstrations. He seems to have conceived or confirmed his idea of the transcendent God not only from his theological consideration *per se* but also in relation with such a skeptical discussion about mathematical demonstrations.

The supporter of mathematical certainty at one time gave up. Descartes was again challenged to build another fortress of certainty. He believed that he after all had found it. The new strong fortress was titled the philosophy of "Cogito, ergo sum." We contend that this philosophy was thought out at least partly in order that he could defend the most severe attack of the skeptics in which they invoke, among others, their five Modes which necessarily lead to suspension of judgment.

Since in mathematical demonstrations their first principles cannot be generally proved, they may contain a 'vicious circle' (*circulus vitiosus*), or question-begging (*petitio principii*) if certain mathematical propositions, for example, the Pythagorean theorem, are supposed to be true by themselves without any presuppositions (Mode V). But the proposition "Cogito, ergo sum" is not so. It is recognized to be true not by means of the deduction of a syllogism, but by "a simple intuition of the mind" (*simplici mentis intelectu*),[80] in Descartes's own words in the "Second Set of Replies" of his *Meditationes*. And, as E. M. Curley puts it in his *Descartes against the Skeptics*, notably "the *cogito* is not question-begging."[81] It is infallibly evident that this statement is ab-

[78] *Apology*, p. 67; *Œuvres*, p. 480.
[79] *AT*, VII, p. 384.
[80] *AT*, VII, p. 140.
[81] E. M. Curley, *Descartes against the Skeptics* (n. 41), p. 86.

solutely true. As Spinoza wrote in his commentary on Descartes's *Principia philosophiae*, "*Cogito ergo sum* is one proposition equivalent to the statement *ego sum cogitans*."[82] Just reflect on whether I exist or not while I am thinking. The answer must always be in the affirmative. The *cogito* does not bring about an infinite regress, for, in Martial Gueroult's skillful formulation, "within the Cogito consciousness and consciousness of consciousness are identical."[83] It is now quite understandable why in the "Seventh Set of Objections and Replies" of the *Meditationes* Descartes declared with confidence to Father Bourdin that "I became the first philosopher ever to overturn the doubt of the skeptics."[84]

Certainly, it may be admitted that Descartes discovered his first truth "Cogito, ergo sum", which no skeptic can refute. We claim that it is, at least partly, in order to rebut the skeptics' criticism of mathematical truths that he found it. Therewith he seems to have abandoned his 'naive' point of view in the philosophy of mathematics in his *Regulae* and acquired the new philosophy based on "Cogito, ergo sum," which was fully developed in his *Meditationes* of 1641. Any mathematical proposition is 'petitio principii', strictly speaking, whereas "Cogito, ergo sum" is not 'petitio principii' and can resist the skeptics' attack, according to Descartes. By his later philosophy, it may be said that he first attained a depth in the philosophy of mathematics which had been achieved by some of his predecessors from antiquity through the Renaissance, whom we have discussed in Chapters 6 and 7.

We do not entirely deny, however, another possible route leading to the philosophy of the cogito. For example, theological considerations per se could have played an important role. Knowledge of a statement similar to "Cogito, ergo sum" as well might have guided Descartes to his later philosophy. In fact, in his "Fourth Set of Objections" to Descartes's *Meditationes*, Antoine Arnauld pointed out that the statement "unless you exist, you never make mistakes" of St. Augustine can function as the first truth as well as can Descartes's *cogito*.[85] As Leibniz would ingeniously recognize, all of the primitive truths that cannot be proved by anything more certain can substitute for "Cogito, ergo sum" (see § 2, B, p. 416).

What we do claim is that Descartes's deepened thought on mathematics in addition to his considerations on theology and natural philosophy was a crucial factor along with other factors for the establishment of his later philosophy.

---

[82]Benedictus de Spinoza, *The Principles of Descartes' Philosophy*, translated by Halbert Hains Britan (La Salle, Illinois, 1905), p. 14.

[83]Gueroult, *Op. cit.* (n. 33), p. 60.

[84]*AT*, VII, p. 550.

[85]St. Augustine's statement in Book II, Chapter 3 of *De Libero arbitrio* which Arnauld cited is as follows: "First, if we are to take as our starting point what is most evident, I ask you to tell me whether you yourself exist. Or are you perhaps afraid of making a mistake in your answer, given that, if you did not exist, it would be quite impossible for you to make a mistake?" (*AT*, VII, p. 198.) Cf. Gareth B. Matthews, *Thought's Ego in Augustine and Descartes* (Ithaca/London, 1992), p. 11; Léon Blanchet, *Les Antécedents historiques du ⟪Je pense, donc je suis⟫* (Paris, 1920), pp. 27–33.

This recognition was nothing other than a mark of the greatness of Descartes's unparalleled intellectual project for reconstructing the entire structure of learning from "Cogito, ergo sum" to theology, natural philosophy, and mathematics.

D. Difficulties of the Later Philosophy of Mathematics: The 'Cartesian Spiral'

Through metaphysical meditations Descartes thought that at least one proposition which could be asserted to be absolutely certain had been attained. But is it true, as he himself claimed, that this new philosophical point of view was totally successful? What did he actually obtain? He surely acquired the truth of "Cogito, ergo sum." But what does this mean for the new philosophy of mathematics?

From his newly conceived first truth Descartes intends to recapture the mathematical certainty which was once doubted and abandoned. To do so, as we have seen (p. 367), he tries to prove the existence of God as the cause of the existence of the *ego* (or that of the notion that I have an idea of God), and lets this God, who is considered veracious, guarantee the truth of mathematics, which is supposed to be clear and distinct.

Is this argument sustainable? Descartes's contemporaries, Marin Mersenne, Antoine Arnauld, and Pierre Gassendi, for example, saw a logical 'circle' (*circulus*) in this reasoning of the *Meditationes*. We quote Gassendi's formulation: "A circular reasoning is committed in proving the existence of God and His veracity because the knowledge of Him is clear and distinct; and that the knowledge of God is clear and distinct because of the existence of God and His veracity."[86] Here appears the problem of the 'Cartesian circle', which may have distressed perhaps Descartes himself and his faithful advocates. Take only one example. To attempt a definitive refutation of the skeptics, Descartes set himself to composing *La Recherche de la vérité par la lumière naturelle*, which remained unpublished until 1701.[87] Interestingly this treatise was abandoned soon after the truth of the *cogito* was proved. Was this a mere accident? This problem concerning the 'Cartesian circle' appears to have vexed Cartesian scholars even up until the present time.[88]

Now let us formulate briefly in terms of the philosophy of mathematics a difficulty in Descartes's argument for restoring the certainty of mathematical knowledge. He let his veracious God guarantee the truth of determinate mathe-

---

[86]Gassendi, ***Disquisitio metaphysica*** (n. 62), "In Meditationem IV, Dubitatio IV, Instantia 2," p. 463: "Circulum commissum, probando, esse Deum, insumque veracem, quia clara, distinctaque est ejus notitia; et, Notitiam Dei esse claram, distinctamque, quia Deus est, ipseque verax." For Mersenne's objection, see *AT*, VII, pp. 124–125; and for Arnault's see *AT*, VII, p. 214.

[87]*AT*, X, pp. 489–527.

[88]For a description of this problem, see, for example, John Etchemendy, "The Cartesian Circle: Circulus ex tempore," ***Studia Cartesiana***, **2** (1981), pp. 5–42. See also papers contained in ***Eternal Truths and the Cartesian Circle: A Collection of Studies*** (n. 9)

matical propositions. There is a certain mistake in this procedure. It was some *determinate* mathematical statements, for example, the proposition that the three angles of a triangle are equal to two right angles, which is known today to be derivable from the parallel postulate in the Euclidean geometry, that he presented as ones which he thought God should warrant to be true. Though from hindsight, we know today that alternatives to the presented mathematical propositions, for example, a negation of the parallel postulate, may produce fruitful mathematical results. In this sense, mathematical truth is characterized at most by having a conditional validity, not an ontological or absolute truth. It was through some determinate mathematical propositions that he, a finite being, 'chose' that Descartes let his transcendent God warrant to be true. There was an unreasonable jump in this reasoning. Descartes's methodological doubt thus was not totally in vain, but he was too hasty in restoring the certainty of mathematical demonstrations and the truth of determinate mathematical statements.

The problem about Descartes's proof of the existence of God and about his way of restoring the certainty of mathematics could be explained, we insist, at least in part, if we suppose that he constructed his reasoning for securing mathematical truths so as to be sufficient to defend himself from a particular kind of skeptic. Suppose that that skeptic was Montaigne. As we have pointed out above (pp. 380–384), one might argue that Descartes certainly can protect his position from the skeptical criticism of mathematics presented in the *Essais*. There Montaigne maintained that finite creatures can have no principles unless they are revealed by God. Now even the skeptic, if he is a Christian, will approve of any reason whatever that God exists and is veracious. Strictly speaking, Descartes's logic warranting mathematical truths by invoking God's goodness has a certain difficulty. But he might have been satisfied, it seems, with his reasoning as it stood if and only if it could be defended from some skeptics' attack on mathematical certainty. Our reconstruction seems to help us deepen our understanding as to how and why Descartes reached his creation doctrine which appeared to his contemporaries to be very idiosyncratic.

Certainly, there have existed advocates of Descartes asserting that the so-called 'Cartesian circle' is not a 'circle' as Descartes himself avoided it. For example, in his *La Pensée métaphysique de Descartes* Henri Gouhier observed that as the veracity of God intervenes between the direct truth of the *cogito* and mathematical truths, the two conceptions of truth are distinguished despite his recognizing both to be evident truths.[89] For this reason, the the so-called 'Cartesian circle' may be rephrased the 'Cartesian spiral' in Anthony Kenny's ingenious words.[90] However, one would not able to insist obstinately

[89]See Gouhier, *La Pensée métaphysique de Descartes* (n. 68), pp. 293–319.

[90]Anthony Kenny, "The Cartesian Spiral," *Revue internationale de philosophie*, 37e année, N° 146 (1983), pp. 247–256. On Kenny's precedent understanding of the 'Cartesian circle', see A. Kenny, *Descartes: A Study of His Philosophy* (New York, 1968), pp. 188–196.

that Descartes clearly and sucessfully avoided the circle. It is no doubt that the two concepts of truth overlap in Descartes's terminology itself.

More importantly, Descartes's motive for reestablishing the certainty of mathematics is indubitable. He conceived his creation doctrine while he was attempting to discover the foundations of physics, as was stated clearly in his letter to Mersenne of April 15, 1630 (see pp. 363–364). This new physics for which he was seeking firm grounds was *par excellence mathematical* physics. According to his view in the "Fifth and Sixth Meditations", what God warrants is, first of all, that mathematics is certain, and then that corporeal nature exists insofar as it is the "subject matter of pure mathematics." At the end of of the Fifth Meditation he states the following concerning the result of his metaphysical speculations:

> Thus I see plainly that the certainty and truth of all knowledge depends uniquely on my knowledge of the true God, to such an extent that I was incapable of perfect knowledge about anything else until I knew him. And now it is possible for me to achieve full and certain knowledge of countless matters, both concerning God himself and other things whose nature is intellectual, and also concerning the whole of that corporeal nature which is the subject matter of pure mathematics (*purae matheseos objectum*).[91]

Descartes's expression "that corporeal nature which is the subject matter of pure mathematics" may sound curious. But his conception of nature as the subject matter of pure mathematics was indispensable for providing the same degree of certainty to mathematical physics as to arithmetic and geometry, the two traditional disciplines of pure mathematics. In the later *Principia philosophiae*, the same theme will be formulated concisely as follows: "The only principles which I accept, or require, in physics are those of geometry and abstract mathematics (*Mathesis abstracta*); those principles explain all natural phenomena, and enable us to provide quite certain demonstrations regarding them."[92] The young Descartes's program of a "physico-mathematician" (*physico-mathematicus*) which he reaffirmed with Beeckman in 1618–1619 (see Chapter 3, § 1, p. 99) was thus accomplished or rather, strictly speaking, replanned based on the newly formulated metaphysical foundations.

We have stated previously that Descartes's motive for providing certainty to mathematics again was absolutely clear. It was intended to justify his project for mathematical physics. This aim is more certain than the metaphysical grounds of Descartes's mathematical physics itself. He confided to Mersenne in

[91] *AT*, VII, p. 71. See also similar passages in the Sixth Meditation, pp. 71, 74 and 80. For the expression "objectum purae matheseos," see David R. Lichterman, "*Objectum Purae Matheseos*: Mathematical Construction and the Passage from Essence to Existence," in A. O. Rorty, ed., *Essays on Descartes' Meditations* (n. 30), pp. 435–458.

[92] *AT*, VIII-1, p. 78 (Part II, Art. 64).

a letter of January 28, 1641 on his discussions in the *Meditationes*:

> These are the things that I want people mainly to notice. But I think I included many other things besides; and I may tell you, between ourselves, that these six Meditations contain all the foundations of my physics. But please do not tell people, for that might make it harder for supporters of Aristotle to approve them. I hope that readers will gradually get used to my principles, and recognize their truth, before they notice that they destroy the principles of Aristotle.[93]

Descartes's *Meditationes* was, then, "a Trojan horse" that he sent behind the lines of the scholastic-Aristotelian scheme of learning, to borrow the skillful words of the historian of philosophy Daniel Garber.[94] The main enemies of his metaphysics were undoubtedly the Aristotelians. Recall Aristotle's assertion on the certainty of mathematics. We read in the *Metaphysics*: "The accuracy [or certainty] of mathematics is not to be demanded in all cases, but only in the case of things which have no matter."[95] The problem of the applicability of mathematical method to physics was disputed by medieval philosophers (e. g. Albertus Magnus vs. Roger Bacon) and some Renaissance thinkers (e. g. Alessandro Piccolomini vs. Francesco Barozzi). The Philosopher for the Scholastics continued directly after the sentence which we quoted: "Therefore its method is not that of physics; for presumably all nature has matter." Descartes simply negated this thesis. With him "matter" became the "extended thing" (*res extensa*) which was supposed to be the "subject matter of pure mathematics." This new insight delivered a heavy blow to the Aristotelians. Certainly the main target of Descartes was Aristotelian physics and its metaphysical basis. But we should add to this that he thought out the philosophical foundations for his intended mathematical physics in such a way as to sidestep attacks of the skeptics.

Here we present a significant contemporary observation which may confirm that Descartes's *cogito* was very closely related to his program of mathematical physics (and naturally to mathematics itself as well). In his *De l'esprit géométrique et de l'art de persuader* (composed about 1657–58), Blaise Pascal argued:

> I would like to ask reasonable persons if this principle: "Matter has a natural, invincible incapacity to think," and this one, "I think, therefore I am," are actually the same thing in the mind of Descartes

[93] "Descartes, à Leyde, à Mersenne, à Paris. (28 janvier 1641)," *AT*, III, pp. 297–298; *AM*, IV, p. 269; *CM*, X, p. 440.

[94] Daniel Garber, "*Semel in vita*: The Scientific Background to Descartes' *Meditations*," in A. O. Rorty, *Op. cit.* (n. 30), p. 82; *Idem, Descartes Embodied: Reading Cartesian Philosophy through Cartesian Science* (Cambridge, 2001), p. 223.

[95] *Metaphysica*, 995a14–16.

> and in the mind of St. Augustine, who said the same thing twelve hundred years earlier.
>
> In actuality, I am very far from saying that Descartes is not the real author of these words, even if he only learned them by reading that great saint; for I know how much of a difference there is between writing a word by chance without making a very long and extended reflection about it and perceiving in this word an admirable series of consequences that proves the distinction of material and spiritual natures, and making of it a firm and sustained princple of a whole system of physics, as Descartes claimed to do. For without examining if he succeeded effectively in his contention, I shall suppose that he has done so, and it is in terms of this supposition that I say that this word is as different in his writings from the same word in others who have stated it in passing, as a dead man from a man full of life and strength.[96]

According to Pascal, Descartes held as his first principle a statement resembling one of St. Augustine. But Descartes was quite different from his alleged predecessor in that only he thought of the *cogito* as the starting point for justifying his project of mathematical physics. Indeed, in a certain sense, Pascal's observation supports our attempt to construct how Descartes discovered his philosophy of "Cogito, ergo sum."

The age of mathematical physics and the mechanical world-view along with that of algebraic analysis opened with Descartes's metaphysics. The practical success that followed was overwhelming. At the same time, however, we should bear in mind the historical background and metaphysical foundations of the manner in which Descartes became the first full-fledged thinker of mathematical physics.[97] Furthermore, we should not forget that in reality Descartes's physics vigorously held to metaphysical elements greatly, contrary to its mathematical ideal and object.[98]

---

[96]Blaise Pascal, "The Geometrial Mind and the Art of Persuation," translated by R. H. Popkin, in *Pascal: Selections*, ed. by Popkin (New York/London, 1989), p. 192; *Œuvres complètes*, Édition établie et annotée par Jacques Chevalier (Paris, 1954), pp. 599–600.

[97]For example, see Élie Denisoff, *Descartes, premier théoreticien de la physique mathématique: Trois essais sur le 《Discours de la méthode》* (Louvain/Paris, 1970).

[98]See, among others, Garber, *Descartes' Metaphysical Physics* (Ch. 4, n. 132). On various aspects of Cartesian physics, see S. Gaukroger, J. Schuster and J. Sutton, eds., *Descartes' Natural Philosophy* (London/New York, 2000), and on the development of Cartesian natural philosophy which culminated in his *Principia philosophiae* of 1644, see Stephen Gaukroger, *Descartes' System of Natural Philosophy* (Cambridge, 2002).

E. The Disuse of the Concept of 'Mathesis Universalis' in the Later Writings

The preparations appear sufficient to tackle the original problem taken up in Part II. How do we evaluate Descartes's 'mathesis universalis' in the light of his later metaphysical framework? We have noted that he used the notion of 'mathesis universalis' only in Rule IV of the *Regulae ad directionem ingenii*. Did the shift of the philosophical point of view from the *Regulae* to the later standpoint influence his understanding of 'mathesis universalis'?

In his insightful book, *L'Imagination et les mathématiques selon Descartes*, published in 1900, Pierre Boutroux conjectured that "the idea of 'mathesis universalis' was in Descartes only a dream of youth."[99] He redefines the Cartesian concept of 'mathesis universalis' as a science about the "quantités absolument nues, c'est–à–dire sur des quantités qui n'aient aucune réalité concrète."[100] According to him, Viète attempted to form a discipline of general algebraic quantities. It was his idea that Descartes followed. Boutroux continues his observation and states that there is a difficulty in this newly established discipline:

> But both the two [Viète and Descartes], hoping to establish algebra as an autonomous science, came up against great difficulties when they sought to found the first propositions. Since begging the question (*pétition de principe*) is involved in hoping to demonstrate everything (because every science presupposes a point of departure obtained by a way other than by reasoning) is the intellect obliged, from the beginning, to appeal to another faculty for aid?[101]

Boutroux further suggests that some preparatory, for example geometrical, exercises are needed to commence algebra. This preliminary is called in general "imagination." "Imagination," in his opinion, "will still play a non-negligible role in algebra, although more modest."[102] He goes on to conclude that "Descartes, a grown man, never dreamed of creating a universal mathematics and did not even dare to write a treatise on pure algebra.[...] The science of naked quantities, the absolutely universal science, is not possible for man. It would be possible for an understanding which could be independent completely of imagination and senses."[103]

What Boutroux wants to claim concerning Descartes's 'mathesis universalis' can be summarized as follows: For 'mathesis universalis' as a science of

[99]Pierre Boutroux, *L'Imagination et les mathématiques selon Descartes* (Paris, 1900), p. 42. Boutroux composed the work when he was barely nineteen years old! Ronald S. Calinger, "Boutroux, Pierre Léon," *DSB*, II, p. 357.

[100]Boutroux, *L'Imagination* (n. 99), p. 25.

[101]Pp. 25–26.

[102]P. 31.

[103]P. 34.

abstract quantities, there must be a starting point from which reasoning commences. This starting point may be called "imagination." "Imagination" can be fostered only in a domain more concrete than that of 'pure quantities'. For this reason, the mature Descartes never used 'mathesis universalis'.

Boutroux's emphasis upon imagination and a critical attitude toward the logicist or formalist point of view in the foundations of mathematics which was becoming influential around 1900 could have been affected by his uncle Henri Poincaré's intuitionist philosophy of mathematics.[104] Boutroux's conjecture is admittedly attractive, but one problem about it consists in the fact that it was based mainly on his own philosophical speculations. Though they were ingenious, his conjecture itself is not based on any certain historical evidence, and he simply pays attention to Descartes's later disuse of 'mathesis universalis'.

Martial Gueroult, on the other hand, has attempted to locate afresh the notion of 'mathesis universalis' in Descartes's later philosophy of the *Meditationes*. With the new metaphysical framework, some mathematical conceptions presented in the *Regulae* may be discussed in a new perspective. Gueroult suggests that 'mathesis universalis' has the same philosophical status as 'universalis', which had been a subject of the metaphysical debates since the Middle Ages. Concerning 'universal mathematics' he writes:

> The algebraist who abstracts from figures and numbers by increasing the abstraction further can perceive with the greatest distinction "the nature of the difficulties," the universal mathematics that is detached from mathematics by abstraction from its *integumentum* [outer garment], carrying the abstraction to the maximum, and allowing one to grasp in its purity the *system that renders all combinations of essences possible.* These disciplines, in spite of their extreme generality, by assuring their complete possession to the understanding, instead of weakening the objective validity of science, strengthen it; therefore if science can be universalized to the maximum, it is because simple natures are no more than "the general terms of the difficulty."[105] [The emphases are in the original.]

Thus, in Gueroult's opinion, the Cartesian 'mathesis universalis' has a certain meaning even in the framework of the *Meditationes.* He attempts to prevent a possible misinterpretation of that notion and to recover the sense in which Descartes might have conceived it with traditional mathematical terminology. 'Mathesis universalis' is, according to Gueroult's interpretation, nothing other than a discipline produced by carrying abstraction to the maximum.

---

[104]See, for example, H. Poincaré, "Sur la nature du raisonnement mathématique," *Revue de métaphysique et de morale*, **2** (1894): Reprinted as Chap. I of *La Science et l'hypothèse* (Paris, 1903); English translation, "On the Nature of Mathematical Reasoning," in Paul Benacerraf and Hilary Putnum, eds, *Philosophy of Mathematics: Selected Readings* (Cambridge, $_2$1983), pp. 394–402.

[105]Gueroult, *Op. cit.* (n. 33), p. 273.

After having presented the previous explanation, Gueroult makes a remark which is important for interpreting Descartes correctly: "Many basic errors of pre-Cartesian science are due, according to Descartes, to the confusion between the concepts issued from the needs of our knowledge and the real articulations of things."[106] He further elaborates this point: "Many aberrant interpretations relative to Descartes's doctrine arise from an often unconquerable confusion between what he has carefully distinguished, namely, the *veritas rationum* and the *veritas rei*, the necessary requirements in me in order that I reach clear and distinct ideas, and the necessary requirements in themselves for the real and actual essence outside of me."[107] If we follow Gueroult's comment, 'mathesis universalis' will be the *veritas rationum*, that is to say, the most abstract mathematical discipline.

But how can one verify the validity of that discipline as the *veritas rationum*? More importantly, we must admit at least that neither Descartes nor his advocate Gueroult has stated almost anything about the other aspect of truth, i. e. the *veritas rei*, as it relates to 'mathesis universalis'. The concept of 'mathesis universalis' used in the *Regulae* thus can mean only *relatively* 'universal' mathematics or, in other words, a mathematics 'as universal as possible', strictly speaking, and not 'universal mathematics' in the most rigorous sense, in the framework of the *Meditationes*, because the first principle there is not some mathematical truth obtained through intuition but "Cogito, ergo sum."

After all, we cannot judge for certain whether or not the later Descartes continued to retain 'mathesis universalis' within his mathematico-philosophical terminology. There remain the historical facts that he became silent about the idea after the *Regulae* and that the *Regulae* was left unfinished and published only posthumously in 1701.

It is certain, in addition, that Descartes's and his forerunners' concepts of 'mathesis universalis' were all primarily mathematical, not strictly philosophical. The Cartesian idea discussed in Rule IV may be characterized by a general theory of quantities viewed with a special emphasis on its methodological aspect. In Chapter 4, § 3, we understood it to be identical with the reformed algebra applicable to both arithmetic and geometry. It is also almost certain that Descartes inherited this concept from certain predecessors, of whom the most important figure was Adriaan van Roomen. In the entire *Regulae*, 'mathesis universalis' was considered to function as the central paradigm for the universal method, which was to attain, in turn, certain knowledge as a whole.

Thus, by invoking the traditional notion of 'mathesis universalis', Descartes did *not* necessarily claim that this discipline is absolutely autonomous or a universal science straightforwardly applicable to any field of certain knowledge which might be acquired. We know that he began to use this idea implicitly in the form of a science useful for solving both arithmetical and geometrical

[106] P. 380.
[107] *Ibid.*

problems in his letter of March 26, 1619 to Beeckman. 'Mathesis universalis' was discussed further in Rule IV as the paradigmatic method for certain knowledge. It was a discipline within the domain of mathematics, as Jürgen Mittelstrass has emphasized, in comparison with Leibniz's quite general counterpart.[108] If it is legitimate to characterize Descartes's philosophy as a 'mathematicism', it will be because and only because in his youthful work *Regulae* he required of all branches of knowledge the same certainty as that acquired in pure mathematics. He was simply attracted to the idea of 'mathesis universalis' which was current among mathematicians of the sixteenth century and used it as his own terminology in the *Regulae*. But for some reason which we simply do not know he ceased to employ it. This is what we can assert from historical and philosophical considerations.

F. 'Mathesis Universalis' as the Discipline of Symbolic Algebra among Other Mathematicians

It may be corroborated by examining some works of seventeenth-century mathematicians that 'mathesis universalis' was primarily not a philosophical concept but a mathematical one.

We will observe that 'mathesis universalis' was a widely used conception among mathematicians of the mid-seventeenth century. During Descartes's lifetime, his mathematical disciple Frans van Schooten delivered a general introductory course of mathematics at the University of Leiden. The content of Van Schooten's lecture was published in 1651 in Leiden with the significant title *Principia matheseos universalis, seu Introductio ad Geometriae methodum Renati des Cartes* under the editorship of Erasmus Bartholin, the author's student from Denmark.[109] The *Principia matheseos universalis* was later incorporated into the second part of the second Latin edition of Descartes's *Géométrie* (Fig. 8.1), which came into wide use (see p. 275). Since as early as around 1639 Van Schooten composed a treatise titled *Recueil de Calcul pour l'intelligence de la Geometrie de Monsr. des Cartes*, accoding to a letter of his to Christiaan Huygens,[110] it may be said that the Dutch mathematical disciple of Descartes had entertained for a long time a project to write a suitable introductory work to Cartesian algebra, and which bore fruit as the *Principia matheseos universalis* in 1651.

---

[108] Jürgen Mittelstraß, "Die Idee einer Mathesis Universalis bei Descartes," *Perspektiven der Philosophie*, **4** (1978), pp. 177–192; *Idem*, "The Philosopher's Conception of *Mathesis Universalis* from Descartes to Leibniz," *Annals of Science*, **33** (1979), pp. 593–610.

[109] J. E. Hofmann, *Frans van Schooten der Jüngere, Boethius*, II (Wiesbaden, 1962), p. 33, note 52.

[110] "Fr. van Schooen à Christiaan Huygens. 3 Juin 1648," *Œuvres complètes de Christiaan Huygens*, t. I (La Haye, 1888), p. 98. We do not know exactly whether this treatise is identical with "Calcul de Mons. de Cartes. [Introduction a sa Geometrie]," a short monograph on elementary symbolic algebra, contained in *AT*, X, pp. 659–680. But the *Principia* has characteristics similar to that French tract.

In the *Principia*, quantity is conceived quite generally and abstractedly by alphabetical letters. For example, combinations of letters, $a + b$, $a - b$, ab, and a/b, stand for addition, subtraction, multiplication, and division, respectively, of the two quantities a and b, where a and b represent any quantity whatever, continuous, discrete or 'mixed'. As the subtitle of this booklet indicates, it was composed as an introduction primarily to Descartes's *Géométrie* and secondarily to any field of the mathematical sciences. The latter intention is reflected in the main title: *Principia matheseos universalis*.

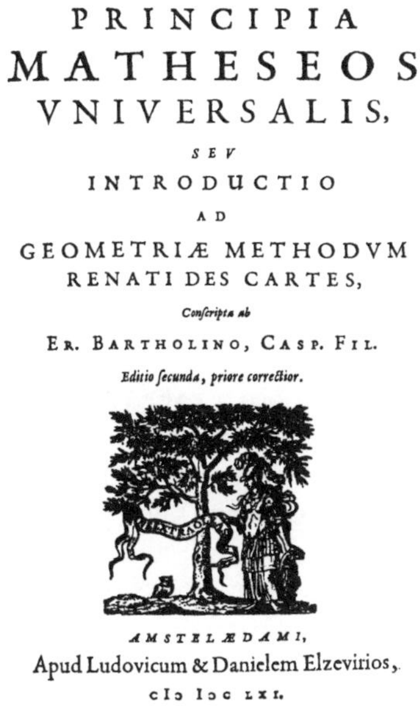
PRINCIPIA
MATHESEOS
VNIVERSALIS,
SEV
INTRODUCTIO
AD
GEOMETRIÆ METHODVM
RENATI DES CARTES,
Conscripta ab
ER. BARTHOLINO, CASP. FIL.
Editio secunda, priore correctior.
AMSTELÆDAMI,
Apud Ludovicum & Danielem Elzevirios,
CIↃ IↃC LXI.

Figure 8.1 The Title Page of Van Schooten's *Principia matheseos universalis* in the Second Part of the Second Latin Version of Descartes's *Geometria* (1661)

John Wallis's *Mathesis universalis: sive, Arithmeticum opus integrum* (Universal Mathematics: or Pure Arithmetical Work) has almost the same characteristics as Van Schooten's. Wallis delivered elementary lectures as Savilian professor of geometry from 1649 at Oxford at the time of the Puritan Revolution.[111] The outcome of these lectures was the *Mathesis universalis*, later published in 1657. He expressed in the "Dedicatio" of his book his indebtedness to Viète, Oughtred, Descartes and Van Schooten. According to him, the new discipline formed by them was also called by various names, "Arithmetica Speciosa, sive Symbolica," "Calculus Geometricus," or "Mathesis Universalis."[112] Wallis seems to have borrowed the title of Van Schooten's book for his own.

One thing which can be distinguished from the previous uses of the concept of 'mathesis universalis' is that Wallis took his treatise to be a pure arithmetical work, as was shown in its subtitle. In his opinion, of the two disciplines of pure mathematics arithmetic is purer and more abstract than geometry.[113] And algebra is a kind of arithmetic (not geometry) and should therefore be explained on

[111]For the establishment of the Savilian professor of geometry at Oxford, see my "Scientific Studies at Oxford and Cambridge in the Seventeenth Century—A Research Review," *Historia Scientiarum*, No. 20 (1981), pp. 57–75, esp. pp. 60–66.

[112]John Wallis, *Mathesis universalis: Operum mathematicorum pars prima* (Oxford, 1657), "Dedicatio"; *Opera mathematica*, I (Oxford, 1695), p. 13.

[113]*Mathesis universalis*, p. 13; *Opera*, I, p. 18.

arithmetical rather than geometrical principles.[114] So the discipline of 'mathesis universalis' is conceived by Wallis as the first part of the mathematical sciences because it is as pure as arithmetic. It can be said that with Wallis the modern concept of real numbers began to be formed in the sense that continuous geometrical magnitudes tended to be interpreted as arithmetical numbers.[115]

Newton seems to have possessed a mathematical thought similar to that of his compatriot Wallis. He considered algebra to be constituted on the same computational basis as ordinary arithmetic, and gave the title *Arithmetica universalis* to the manuscripts which are supposed to have been drafted in the spring of 1684.[116]

Furthermore, in 1684 Newton was commencing to compose a treatise titled *Matheseos universalis specimina*. Unfortunately he stopped writing and left only the beginning part of the projected monograph. This treatise was initially planned to repond to David Gregory's newly published *Exercitatio geometrica de dimensione figurarum*, which Newton received in June 1684.[117] In writing this book, Newton seems to have intended to supplement Gregory's techniques in the quadrature of various geometrical figures which used the method of algebraic infinite series. He also tried to display some of his own analytical methods for solving similar problems in order to demonstrate the priority and superiority of his own techniques to his contemporaries, such as Gregory and perhaps Leibniz. The unfinished *Specimina* consists mainly of the method of infinite series and that of fluxions.[118] Newton does not mention the reason why he chose the title *Matheseos universalis specimina*. The treatise is not on finite algebra, with which Descartes and Van Schooten were mainly concerned, but on infinite algebra. It is clear that the Newtonian 'mathesis universalis' does not simply mean an algebraic discipline prior to arithmetic and geometry. It seems to possess a Janus face. One of its faces looked to Descartes in the sense that it was written in the manner of symbolic algebra; the other face looked to Leibniz in the sense that it was on infinitesimal mathematics transcending the traditional finite analytical discipline.

---

[114] *Mathesis universalis*, p. 73; *Opera*, I, p. 56.

[115] See my "The Acceptance of the Theory of Proportion in the Sixteenth and Seventeenth Centuries," (Ch. 5, n. 162), esp. pp. 91–95; J. F. Scott, *The Mathematical Work of John Wallis* (London, 1938; New York, 1981), pp. 66–71; Jacob Klein, *Greek Mathematical Thought and the Origin of Algebra*, Translated by Eva Brann (Cambridge, Mass., 1968), pp. 211–224. For the mathematical thought of Wallis in general, see A. Prag, "John Wallis, 1616-1703. Zur Ideengeschichte der Mathematik im 17. Jahrhundert," *Quellen und Studien zur Geschichte der Mathematik, Astronomie und Physik*, Apt. B, 1 (1931), pp. 381–412.

[116] D. T. Whiteside, ed., *The Mathematical Papers of Isaac Newton*, Vol. V (1683–1684) (Cambridge, 1972), pp. 538–621.

[117] For the situation in which Newton composed *Matheseos universalis specimina*, see D. T. Whiteside, "General Introduction," pp. xix–xx and "Introduction," pp. 414–419 of *The Mathematical Papers of Isaac Newton*, Vol. IV (1674–1684) (Cambridge, 1971).

[118] *Ibid.*, pp. 526–589.

In the eighteenth century, a small textbook of symbolic algebra was published in Leiden in 1727. It was written in Latin by the Dutch Newtonian William Jacob 's Gravesande for his teaching of mathematics at the University of Leiden, and published under the title *Matheseos universalis elementa. Quibus accedunt, specimen commentarii in Arithmeticam universalem Newtonii, ut & De determinanda Forma Seriei intiniitae adsumtae Regula nova.* This booklet explained concisely symbolic algebra like Van Shooten's *Principia matheseos universalis*, and, as the title indicates, attemted to comment on Newton's *Arithmetica universalis*, published by William Whiston in Cambridge in 1707. In Chapter 1 with the title "Generalities on Quantity and on the Aim of Analysis," we read the following significant sentence: "This art, algebra, also analysis, is called by some people 'mathesis universalis'."[119] This brusque characterization of 'mathesis universalis' shows quite beautifully in a condensed form the following historical circumstances: Proclus's commentary on Euclid began to be read in the sixteenth century, the term 'mathesis universalis' started to be used for a mathematical discipline, and it was also invoked in Descates's *Regulae*. Putting it in other words, the art of computation called algebra transmitted from Arabic mathematics was recognized as an art of analysis and called 'mathesis universalis' because of its use for the mathematical disciplines commonly and broadly. In this texbook by 's Gravesande, we find simly how to manipulate symbolic algebra with infinite series, and there are no 'deep' philosophical discussions whatsoever. This monograph aquired broad audiences: Its Dutch version was published in 1728, and its English translation in 1752.

In his collected essays on the history of Western philosophy as rigorous science with the title *Mathesis universalis* Heinrich Scholz put it: "Die Descartessche Algebra will also nicht weniger sein als eine Art von *passe-partout-Mathematik*, in der Sprache Descartes': *Mathesis universalis*."[120] Discussing various attempts to found mathematics from the formation of axiomatics in classical Greek mathematics to the formalist foundations of mathematics by David Hilbert, this book indicates how symbolic algebra was important for Descartes and how the symbolically algebraic way of thought played a central role in modern mathematics as a whole.

What becomes clear in the above examination is that the treatises by Van Schooten, Wallis, Newton, and 's Gravesande were all strictly mathematical. We see that in these books 'mathesis universalis' meant an elementary discipline of symbolic algebra on 'general' quantities and took a much more clear and substantial form as a mathematical discipline than it did in Van Roomen's

[119] G. I. 's Gravesande, *Matheseos Universalis Elementa. Quibus accedunt, Specimen commentarii in Arithmeticam Universalem Newtoni: ut & De determinanda Forma Seriei infinitae adsumtae Regula [Methodus] Nova* (Leiden, 1727), p. 4: "Ars haec, Algebra, etiam Analysis, à quibusdam Mathesis Universalis vocatur." See A. Rupert Hall, " 's Gravesande, William Jacob," *DSB*, V, p. 510.

[120] Heinrich Scholz, *Mathesis Universalis: Abhandlungen zur Philosophie als strenger Wissenschaft* (Basel/Stuttgart, $_2$1969), p. 108.

writings. Newton's rival Leibniz will proceed even beyond these mathematicians and attempt to create his very broad concept of 'mathesis universalis' and his own philosophy of mathematics to accompany it. Descartes's mathematical thought can be understood in a sufficiently deep manner only when we study Leibniz's as well. Thus, let us look at Leibniz's idea of 'mathesis universalis' and his philosophy of mathematics associated with it.

§ 2. The Leibnizian Synthesis

A. 'Mathesis Universalis' Extended

The idea of 'mathesis universalis' was considerably extended with a truly synthetic thinker of the later half of the seventeenth century, Gottfried Wilhelm Leibniz. Leibniz's attempt at broadening the concept appeared as early in his first published work, *Dissertatio de arte combinatoria* of 1666. It seems to be almost certain that he acquired the idea from Van Schooten's *Principia matheseos universalis*, contained in the two-part second edition of Descartes's *Géométrie*, which appeared in 1659 and 1661. This essay was written before applying himself to any advanced mathematics so that its results were admittedly immature. But it may be said to have been quite original and to have contained an ambitious program of logic, mathematics, and metaphysics which was to occupy his thinking for the rest of his life. Such a general statement seems to be applicable especially to the concept under discussion.

After elaborating quantity and number, Leibniz tries to widen the domain of arithmetic, which is the science of known quantity and number:

> This is the origin of the ingenious Specious Analysis (*Analysis Speciosa*) which *Descartes* was the first to work out, and which *Frans van Schooten* and *Erasmus Bartholius* later organized into precepts, the latter in what he calls the *Elements of a Universal Mathematics* (*elementa Matheseos Universalis*). *Analysis* is thus the science of ratios and proportions, or of unknown quantity, while *Arithmetic* is the science of known quantity, or numbers. But the Scholastics falsely believed that number arises only from the division of the continuum and cannot be applied to incorporeal beings. For number is a kind of incorporeal figure, as it were, which arises from the union of any beings whatever; for example, God, an angel, a man, and motion taken together are four.[121]

[121] G. W. Leibniz, *Dissertatio de arte combinatoria* (Leipzig, 1666): *Die philosophischen Schriften* (=*PS*), ed. C. I. Gerhardt, Bd. 4 (Berlin, 1880), p. 35: Leibniz, *Philosophical Papers and Letters: A Selection*, translated and edited, with an Introduction by Leroy E. Loemker (Dordrecht/Boston, $_2$1969), pp. 76–77.

"Elementa matheseos universalis" in the quoted passage must be read "Principia matheseos universalis," of course. The author is here simply explaining what Descartes and his followers accomplished with the concept of 'mathesis universalis'. That notion means, according to Leibniz's observation, a specious analysis or symbolic algebra, the science of unknown quantity.

In the latter part of the quoted paragraph, Leibniz criticizes the Aristotelian and strictly empirist concept of number, according to which number consists only of parts of the corporeal substance. It is known that the Scholastics and empiricists did not approve of algebraic abstraction without its empirico-geometrical foundation.[122] According to Leibniz's epistemological standpoint, however, any beings whatever may compose a number, for example, an angle and a man, which are different species, taken together are two.

In the 1666 monograph, Leibniz did not propose his own mathematical idea at least in terms of 'mathesis universalis' and went no further than suggesting that the Cartesian symbolic algebra might be counted as an example of his combinatorial art. The Leibnizian art of combinations belonged to the tradition stemming from the thirteenth-century scholastic encyclopedist Ramon Lull, which Paolo Rossi's *Clavis universalis* has thoroughly investigated. The idea of Lull can be further traced back to the Muslim mathematician-philosopher Naṣīr al-Dīn al-Ṭūsī of the thirteenth century. This was not unnatural, because Lull could read and write Arabic.[123]

Consequently, Leibniz's combinatorial art was not primarily derived from his early mathematical studies. The idea of that art is said to have occurred to him in his schooldays as "an alphabet of human thoughts" (*Alphabetum cogitationum humanarum*).[124] As words and sentences are combinations of the letters of the alphabet, human thoughts must be composed, Leibniz believed, of a relatively small number of simple and basic concepts. The young German philosopher's primary concern in 1666 was to reform not mathematics but logic and metaphysics, although a few fascinating results in mathematics are contained in it.[125] Beyond the domain of the Cartesian algebraic analysis, John Wallis's *Arithmetica infinitorum* (1655) was referred to, but the authorship was ascribed to Seth Ward, Savilian professor of astronomy and Wallis's (Savilian

[122]This point of view was invoked by Issac Barrow invoked in his *Lectiones mathematicae* (delivered from 1664 to 1666 and published in 1683) when he attacked Wallis's analytical standpoint in the latter's Savilian lectures the *Mathesis universalis*. See my *Op. cit.* (n. 115), esp. pp. 86–88.

[123]Roshdi Rashed, "Combinatorial Analysis, Numerical Analysis, Diophantine Analysis and Number Theory," in R. Rashed, ed., *Encyclopedia of the History of Arabic Science*, Vol. 2: *Mathematics and the Physical Sciences* (London/New York, 1996), p. 378; Further in detail in Prof. Rashed's graduate lecture in the summer semester of 1995 at the University of Tokyo.

[124]See G. H. R. Parkinson, "Introduction" to Leibniz, *Logical Papers* (Oxford, 1966), p. xii; E. J. Aiton, *Leibniz: A Biography* (Bristol/Boston, 1985), p. 18

[125]Eberhard Knobloch, "The Mathematical Studies of G. W. Leibniz on Combinatorics," *Historia Mathematica*, **1** (1974), pp. 409–430, esp. pp. 411–416.

professor of geometry) colleague at Oxford![126]

It seems to have been in a fragment under the title *Elementa nova matheseos universalis* where Leibniz first proposed his mature project of 'mathesis universalis'. The date of the composition of this tract cannot be determined precisely, but it was surely written before the rather long treatise entitled "Mathesis universalis", which is supposed to have been drafted around 1695[127] and will be examined later. Leibniz clearly intended to enlarge this fragment into book-form, as can be read in the introductory note "the idea of a book whose title will be *Elementa nova matheseos universalis*."[128] That title shows that the planned book would be a new and reformed version of Van Schooten's *Principia matheseos universalis*, which in 1666 Leibniz mistakenly wrote as *Elementa matheseos universalis.*

The author indicates in the margin of the fragment what the writing should be: "Such Elements of a Universal Mathematics should be such that they can be useful for interpreting cryptography, playing chess and the others of this sort."[129] He compares his new mathematical program with the symbolic algebra of Viète and Descartes and explains the content of his enlarged conception of 'mathesis universalis':

> These Elements of a Universal Mathematics differ much more from the hitherto known Specious [Mathematics] than the Specious [Mathematics] of Viète or of Descartes differs from the old Symbolics.
>
> Here shown is the method of extending the Geometrical Calculus to those problems as well which transcend Algebra (hitherto acquired).
>
> Propounded is Synthesis and Analysis, or Combinatorics as well as Algebra.[130]

Leibniz's extension of the traditional conception of 'mathesis universalis'

[126]Leibniz, *Op. cit.* (n. 121), p. 41; Loemker, p. 75.

[127]On the dates of Leibniz's writings, consult Kurt Müller und Gisela Krönert, hrsg., *Leben und Werk von Gottfried Wilhelm Leibniz: Eine Chronik* (Frankfurt am Main, 1969). On that of the "Mathesis universalis," see p. 136. Mittelstraß conjectures that the "*Elementa nova matheseos univeralis*" dated from about 1675. See J. Mittelstraß, "The Philosopher's Conception of *Mathesis Universalis* etc." (n. 108), p. 604.

[128]Louis Couturat, *Opuscules et fragments inédits de Leibniz: Extraits des manuscripts de la Bibliothèque royale de Hanovre* (Paris, 1903), p. 348: "⟨Idea libri cui titulus est⟩ ELEMENTA NOVA MATHESEOS UNIVERSALIS."

[129]Couturat, *Loc. cit.*: "{Elementa Matheseos ⟨Universalis⟩ talia esse debent ut prodesse possint etiam ad Cryptographemata interpretanda, ad ludum Schaccorum ludendum, et talia id genus.}"

[130]*Ibid.*: "Haec Elementa Matheseos universalis multo plus differunt a Speciosa hactenus cogita, quam ipsa Speciosa Vietæ aut Cartesii differt a Symbolica veterum. Ostendetur his Methodus Calculum Geometricum ad illa quoque problemata porrigendi quae Albgebram (hactenus receptam) transcendunt. Tradetur et Synthesis et Analysis, sive tam Combinatiria quam Algebra."

is tied up inseparably with his earlier understanding of the very general art of combinations. Algebra, he wrote to Henry Oldenburg in 1675, is only a part of his general science of symbols: "I have come to understand that everything of this kind which algebra proves in only due to a higher science, which I am accustomed to call sometimes combinatorics, sometimes characteristics."[131] With such a universal doctrine of symbols, not only analysis, or the art of invention (*ars inveniendi*), but also synthesis, or the art of judgment (*ars judicandi*) comes to constitute a newly conceived 'mathesis universalis'. In a later part of the tract *Elementa nova matheseos universalis*, he refers to the methods of synthesis and analysis:

> The method of soving a problem is either synthetic or analytic. [...] It is synthetic or combinatorial if we go through other problems and finally arrive at ours and what belongs here is the method of proceeding from simple problems to complex ones. It is analytic if, starting from our problem, we go back to arrive at those presuppositions which are sufficient for its solution.[132]

The synthetic method for Leibniz becomes combinatorial since the proceeding from simple problems to complex ones, that is, synthesis, is in principle nothing more than the rearrangement of combinations of letters. But how mathematically does he assert that his art of symbols is far beyond what Viète and Descartes have left him? He answers this question as follows:

> Universal mathematics should propound a certain Method for making exact determinations through those which fall under the imagination, or rather so that it is called Logic of imagination. Accordingly excluded from it is Metaphysics about pure intelligible things such as thinking and action. And Special mathematics about Numbers, Situation, and Motion is excluded.
>
> Imagination is generally concerned with the two, Quality and Quantity, or magnitude and form; according to which things are called similar or dissimilar, equal or inequal. It is clear that the consideration of similitude belongs to General mathematics not less

[131] "Leibniz an Oldenburg, Paris, 28. Dezember 1675," Leibniz, *Sämtliche Schriften und Briefe*, Dritte Reihe: *Mathematischer Naturwissenschaftlicher und Technischer Briefwechsel*, Bd. 1 (Berlin, 1976), p. 331: "Ego vero agnosco, quicquid in hoc genere praebat algebra, non nisi superioris scientiae beneficium esse, quam nunc combinatoriam, nunc characteristicam appellare soleo [...]." Cf. *Mathematische Schriften* (= *MS*), ed. C. I Gerhardt, Bd. 1 (Berlin, 1849), pp. 85–86; Loemker, p. 166.

[132] Couturat, *Ibid.*, pp. 350–351: "Mathodus solvendi problema est sel synthetica vel analytica. [...] Synthetica ⟨vel combinatoria⟩ est cum alia problemata percurrimus et incidimus tandem in nostrum ⟨et huc pertinet Methodus eundi à simplicibus ad composita problemata.⟩ Analytica est, cum à nostro inchoantes regredimur ut perveniamus ad conditiones quæ ad ipsum sovendum sufficiunt." For Leibniz's general account for the dual method, see his "De Synthesi et Analysi universale seu Arte inveniendi et judicandi," *PS*, Bd. 7 (Berlin, 1890), pp. 292–298; Loemker, pp. 229-234.

> than that of equality from the fact that Special mathematics such as Geometry often investigates similarities of figures.[133]

Here Leibniz defines 'mathesis universalis' as a logic of imagination. Imagination, according to him, is related with both quality and quantity, not simply with quantity, the object of usual mathematics. In the next portion of the fragment, he does not explain in detail the character of his newly projected mathematical field about quality, but it appears from what he has explained in the passage cited above that the mathematics of quality is concerned with a 'geometry of position', about which he had started to talk openly to others in 1678.[134]

Noteworthy in his idea of a geometry of position, among others, is a letter to Huygens of September 8, 1679. There Leibniz expressed his dissatisfaction with the traditional algebra. "Algebra," he wrote, "is the characteristic for undetermined numbers or magnitudes only. But it does not express position, angles, and motion directly: hence it is often difficult to analyze the properties of a figure by calculation, and still more difficult to find very convenient geometrical demonstrations and constitutions, even when the algebraic calculation is completed."[135] Leibniz's attempt to create a new geometrical characteristic was to fill the lack of a science of position, angles, and motion.

In a later tract under the title *De analysi situs*, probably written after 1693,[136] Leibniz more explicitly criticized Descartes because, in his opinion, the latter's algebra had been strictly limited to magnitudes. What was needed, according to him, was a characteristic which could treat positions of geometry. For that purpose similarity as well as equality should be subjected to an algebraic calculation. Whether his attempt was successful or not (in fact Huygens did not have a high opinion of it), the new domain of geometrical similarity became a part of his 'mathesis universalis'.

In the rest of the *Elementa nova matheseos universalis*, Leibniz shows an inclination to include his newly discovered infinitesimal calculus in his 'mathesis universalis'.[137] This intention is quite understandable as the Leibnizian

[133] *Ibid.*, p. 348: "Mathesis universalis tradere cebet Methodum aliquid exacte determinandi per ea quæ sub imaginationem cadunt, sive ut ita dicam Logicam imaginationis. Itaque hinc excluduntur Metaphysica circa res pure intelligibiles, ut cogitationem, actionem. Excluditur et Mathesis specialis circa Numeros, Situm, Motum. Imaginatio generaliter circa duo versatur, Qualitatem et Quantitatem, sive magnitudinem et formam; secumdum quæ res dicuntur similes aut dissimiles, æquales aut inæquales. Et vero similitudinis considerationem pertinere ad Mathesin generalem non minus quam æqualitatis, ex eo patet quod Mathesis specialis, qualis est Geometria, sæpe investigat figurarum similitudines."

[134] See Aiton's description in his *Leibniz: A Biography* (n. 124), pp. 96–98.

[135] "Leibniz à Christiaan Huygens. 8 septembre 1679, Appendice," *Œuvres complètes de Christiaan Huygens*, t. 8 (La Haye, 1899), pp. 219–220; *MS*, Bd. 2 (Berlin, 1849), p. 20; Loemker, p. 250.

[136] Müller u. Krönert, *Op. cit.* (n. 127), p. 126. *MS*, Bd. 5 (Berlin, 1858), pp. 178–183; Loemker, pp. 254–258.

[137] Couturat, *Ibid.*, pp. 349–350, especially the editor's note 1 on p. 350.

infinitesimal analysis is a part of his *characteristica universalis*. If that theory is mathematically sound, it becomes naturally a component of his 'mathesis universalis'.

Leibniz's full-fledged account of his idea of 'mathesis universalis' can be seen in his treatise under the title *Mathesis universalis*, which is conjectured to have been written around 1695. He drafted only the preface and the first part of it. But it may be said without much exaggeration that we can witness in it the most mature and systematic conception of his 'mathesis universalis'. He begins the monograph with the definition of 'mathesis universalis':

> Universal mathematics is the science about quantity in general, or about the method of estimating, or rather of designating limits within which something falls. And since every creature has limits, it can be hence said that as Metaphysics is the general science of created things. It has two parts: the science of the finite (which comes under the name of Algebra and will be explained first), and the science of the infinite, where the finite is determined with the intervention of the infinite.[138]

Here it is emphasized that 'mathesis universalis' is the science of created things in contrast to metaphysics, which is understood as the most general science covering all things unqualifiedly. The Leibnizian 'mathesis universalis' has two components: One is the algebra of the finite and the other is the transcendent science of the infinite.[139] The Algebra is also known by the name logistics (*Logistica*).[140] This definition of algebra supports our interpretation that Van Roomen regarded algebra as a part of 'supputatrix' = 'logistice' (pp. 351–353). The science of the infinite is also called "the superior part of *mathesis universalis*,"[141] of which his differential calculus, he claims, is the most powerful instrument.

In Leibniz's thought, moreover, the two sciences, logic and mathematics, are seen as interpreting one another. He puts it:

> And just as many have tried to explain Logic on the analogy of computation and Aristotle himself has shown in his *Analytics* in a Mathematical manner, so in turn and much more rightly especially

[138] *Mathematische Schriften*, Bd. 7 (Halle, 1863), p. 53: "Mathesis universalis est scientia de quantitate in universum, seu de ratione aestimandi, adeoque limites designandi, intra quos aliquid cadat. Et quoniam omnis creatura limites habet, hinc dici potest, ut Metaphysica est scientia rerum generalis, ita Mathesin universalem esse scientiam creaturarum generalem. Duasque habet partes: scientiam finiti (quae Algebrae nomine venit priorque exponetur), et scientiam infiniti, ubi interventu infiniti finitum determinatur." For an account of the relationship between metaphysics and universal mathematics, see Gilles-Gaston Granger, "Philosophie et mathématique leibniziennes," *Revue de métaphysique et de morale*, 86e année (1981), pp. 1–37, esp. pp. 1–5.

[139] *Ibid.*, p. 68.

[140] P. 53.

[141] P. 69.

> Universal Mathematics and even Arithmetic and Algebra can be treated in the manner of Logic, as if they were Logical Mathematics, so that in this way Universal Mathematics coincides in effect with Logistics and the Logic of Mathematics; hence our Logistics is given in some places the name, Mathematical Analysis (*Analysis mathematica*).[142]

He does not go beyond this statement on his explanation of logic, insofar as the treatise "Mathesis universalis" is concerned, but it is seen that even logic tends to constitute a part of his 'mathesis universalis'.

A clear statement that logic in Leibniz's understanding becomes a part of 'mathesis universalis' appears in his *Nouveaux essais sur l'entendement humain*, a criticism, composed around the years 1703-1704, of John Locke's empiricist *Essay Concerning Human Understanding*, first published in 1690 and translated into French by Pierre Coste in 1700: "I hold that the invention of the syllogistic form is one of the products, and indeed one of the most important, to have been made by the human mind. It is a kind of *Mathématique universelle* whose importance is too little known."[143]

Indeed the Lebnizian 'mathesis universalis' is nothing other than "mathesis speciosa."[144] And logic and the infinitesimal calculus, insofar as they manipulate symbols, become parts of 'mathesis universalis'. The symbolic algebra of Viète, Descartes, and Van Schooten is, Leibniz asserts, totally superseded by the much broader idea of symbolic mathematics, for their algebra is subordinate to his combinatorial art.[145] The ambitious (primarily non-mathematical) project of 1666, thus, crystallizes into a specifically general conception of 'mathesis universalis'. This conception is almost identical with the one which will continue to be used especially among philosophers even today.[146] As Jürgen Mittelstraß

---

[142]P. 54: "Et quemadmodum multi Logicam illustrare tentaverunt similitudine computi ipseque Aristoteles in Analyticis Mathematico more locutus est, ita vicissim et multo quidem rectius Mathesis praesertim universalis, adeoque Arithmetica et Algebra tractari possunt per modum Logicae, tanquam si essent Logica Mathematica, ut ita in effectu coincidat Mathesis universalis sive Logistica et Logica Mathematicorum; unde et Logistica nostra nomine Analyseos Mathematicae passim venit." As for the translation, we refer partly to Gottfried Martin, *Leibniz: Logic and Metaphysics*, translated by K. J. Northcott and P. G. Lucas (Manchester, 1964; The original German edition, 1960), p. 84.

[143]Leibniz, *Nouveax essais sur l'entendement humain* in *Sämtliche Schriften und Briefe*, Sechste Reihe, Bd. 6 (Berlin, 1962), Livre IV, Chap. 17, § 4, p. 478. See the English translation by Peter Remnant and Jonathan Bennett (Cambridge, 1981), which has the same pagination as the above edition.

[144]*Op. cit.* (n. 128), p. 68.

[145]P. 61.

[146]See, for example, Edmund Husserl, *Logical Investigations*, translated by J. N. Findlay, Vol. I (London/New York, 1970; The original German edition, 1900), "Prolegomena to Pure Logic," § 60, "Links with Leibniz," pp. 218–220. "Through his Combinatoria he [Leibniz] is also the intellectual father of the pure theory of manifolds (*die reine Mannigfaltigkeitslehre*), a discipline close to pure logic and in fact intimately one with it." (p. 220). On the development of Leibniz's concept of 'mathesis universalis', see Dietrich Mahnke, "Leibnizens

puts it, "the old idea of *mathesis universalis* which was still restricted to mathematical theories in their narrow sense, finds in Leibniz its modern form."[147] With Leibniz, 'mathesis universalis' came to mean 'as broadly as possible' symbolic mathematics.

B. Leibniz's 'Logicist-Formalist' Philosophy of Mathematics

How did Leibniz understand the epistemological status of mathematical truths, or especially of 'mathesis universalis', the principal part of the mathematical sciences? How did his understanding of mathematics differ from that of Descartes? Just as we did with Descartes in Section 1, so in this subsection we will consider Leibniz's philosophy of mathematics.

Interestingly, at least one of the spurs which drove the young Leibniz to formulate his own system of mathematics and philosophy of mathematics seems to have been a criticism of that science by the skeptics (for example, Francisco Suárez). He, however, responded to it in a quite different way than Descartes had. His first endeavor to refute the skeptics' attack on the certainty of mathematics appears in a manuscript "Demonstratio propositionum primarum" (A Demonstration of First Propositions), probably written between the autumn of 1671 and the beginning of the following year.[148] The very first sentence of the fragment reads: "I believe that no proposition should be accepted unless having been proved."[149] He intends to prove propositions which appear *prima facie* to be self-evidently true to the extent to which they are necessarily to be admitted "even by any skeptic whomever" (*a Sceptico etiam quantocumque*).[150] In the quoted sentence appears the word "skeptic." It is known that Leibniz became interested in skepticism before the end of 1664, i. e. at the age of eighteen.[151]

Leibniz apparently got a clue for such an attempt from Thomas Hobbes's demonstration of the Euclidean axiom: "The whole is greater than the part" in his *Elementorum philosophiae sectio prima de corpore*, published in 1655.[152]

Synthese von Universalmathematik und Individualmetaphysik," *Jahrbuch für Philosophie und phänomenologische Forschung*, Bd. 7 (1925), pp. 305–611.

[147]Mittelstraß, *Op. cit.* (n. 127), p. 610. Cf. *Idem, Neuzeit und Aufklärung: Studien zur Entstehung der Neuzeitlichen Wissenschaft und Philosophie* (Berlin/New York, 1970), § 12 "Kunstsprache und Logikkalkül," pp. 413–453; Louis Couturat, *La Logique de Leibniz d'après des documents inédits* (Paris, 1901), Chap. VII: "La mathématique universelle," pp. 283–322.

[148]Müller u. Krönert, *Op. cit.* (n. 127), p. 24. Aiton, on the other hand, ascribes this fragment to the first mathematical product during Leibniz's sojourn in Paris during the years 1672–1676. See *Op. cit.* (n. 124), p. 42.

[149]"Demonstratio proportionum primarum," *Sämtliche Schriften und Briefe*, V-2 (Berlin, 1966), p. 479: "Ego ita sentio, nullam propositionem accipiendam esse nisi probatam [...]."

[150]*Ibid.*, p. 480.

[151]E. De Olaso, "Leibniz and Scepticsm," in Richard H. Popkin, Ezequiel De Olaso, and Giorgio Tonelli, eds., *Scepticism in the Enlightenment* (Dordrecht/Bonston/London, 1997), pp. 99–130, at p. 100.

[152]*Loc. cit.*: "Totum esse maius parte, primus demonstravit Hobbius, fundamentum Scientiae de Quantitate." Cf. Thomas Hobbes, *Elementorum philosophiae sectio prima de corpore* (London, 1655), Chap. 8, § 25, p. 72. F. J. E. Woodbridge, Hobbes: Selections (New York,

Taking notice that "Gregorius St. Vincent has denied in the angle of contact and the Cardinal Pallavicino in the infinite that the whole is greater than the part,"[153] he offers his proof of the proposition: "The whole *cde* is greater than the part *de*." Leibniz's demonstration for magnitudes is as follows:

> That is greater whose part is equal to the whole of the other, according to definition of Major.
> The part of the whole *cde* (that is, *de*) is equal to the whole *de* (that is, to itself).
> Therefore the whole *cde* is greater than the part *de*.[154]

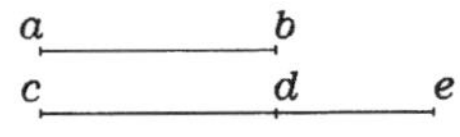

Leibniz here notes that major is a definition and that minor is an identity. He admits in general only two types of unproved presuppositions, namely definitions and identities.

The program of 'demonstrating axioms' can be said to have been Leibniz's firm conviction from the time he composed "Demonstratio propositionum primarum" until his very final statement in his correspondence with the Newtonian Samuel Clarke. In his second letter to Clarke, probably written toward the end of 1715, Leibniz writes about mathematical principles as follows:

> The great foundation of mathematics is the principle of contradiction or identity, that is, that a proposition cannot be true and false at the same time and that therefore A is A and cannot be non-A. This single principle is sufficient to demonstrate every part of arithmetic and geometry, that is, all mathematical principles.[155]

According to Leibniz, the principle of contradiction is equal to identity, and all principles of arithmetic, geometry, and undoubtedly 'mathesis universalis' can be based on the single principle of identity. Leibniz continued to nurse and to profess openly throughout his life this bold ideal of proving everything except definitions and identities.

True, Leibniz retained his program of 'demonstrating axioms' as a scientific task throughout his life. But, it would be an error to think that this intellectual project was possessed in the literal and narrow sense. To examine Leibniz's response to Descarte in detail, let us put the problem in order. First,

---

1958), pp. 92–93.

[153] "Ut Totum esse maius parte in Anglo contactus negavit Gregorius a S. Vincentio et in infinito Cardinalis Pallavicius." For a discussion on the angle of contact, see J. E. Hofmann, *Leibniz in Paris 1672–1676* (Cambridge, 1974), pp. 12–14. Note that for Leibniz the angle of contact and the infinite are not magnitudes.

[154] *Ibid.*, p. 482: "Cuius pars alteri toti aequalis est, id est maius per def. Majoris. Pars totius *cde* (nempe *de*) est aequalis toti *de* (nempe sibi ipsi). Ergo *cde* est maius quam *de*; totum parte. Quod erat demonstrandum."

[155] *PS*, Bd. 7 (Berlin, 1890), p. 355; Loemker, p. 677.

how did Leibniz understand the concept of 'clear and distinct', a characteristic through which Descartes had the veracity of God to guarantee the truth? Second, when Leibniz stated "demonstrating axioms," how should we undertand "axioms"? To these questions, two treatises written during the mid-1680s seem to extend our knowledge.

To the first question, Leibniz tried to answer in 1684 by publishing "Meditations on Knowledge, Truth, and Ideas" in the journal *Acta eruditorum*, which had just begun to appear in Leipzig. What the author made efforts to do in this short essay, the title of which reminds us Descartes's *Meditationes*, is to understand the notion 'clear and distinct' (*clair et distinct; clara et distincta*) which had been oveused by the Cartesians (such as Arnault and Malebranche) much more evidently and to distinguish it in the form of a ladder. According to Leibniz, the ladder of knowledge rises as follows:

> Knowledge (*cognitio*) is either obscure (*obscura*) or clear (*clara*), and again, clear knowledge is either confused (*confusa*) or distinct (*distincta*), and distinct knowledge either inadequate (*inadaequata*) or adequate (*adaequata*), and adequate knowledge either symbolic (*symbolica*) or intuitive (*intuitiva*), and, indeed, if knowledge were, at the same time, both adequate and intuitive, it would be absolutely perfect (*perfectissima*).[156]

This ladder of knowledge may require an explanation. First of all, the knowledge having an instrument to acknowledge things represented is "clear". Further, by distinguishing from similar things, "distinct" knowledge is obtained. Searching "distinct" knowledge thoroughly and "analysis" done to the end, the knowledge becomes "adequate". However, if "analysis" is long, things cannot be "intuitively" known, and "symbolic" procedures become necessary. The recognition only by words is this case, e. g. a regular polygon with one thousand sides. This kind of knowledge can be called "blind" (*caeca*). Most knowledge in arithmetic and algbra belongs to this category. "There is no knowledge of a distinct primitive notion except intuitive."[157]

Toward the end of the same tract, Leibniz tells that he has a predecessor in his understanding of mathematical knowledge: "Pascal, a most talented man, largely agrees with this in his excellent essay "On the Geometrical Mind" (a fragment of which appears in the admirable book of the distinguished Antoine Arnault on the art of thinking well). The geometer, he says, must define all the terms which are even a bit obscure and prove all truths which are even a bit of dubious. But I wish he had defined the limits beyond which a notion or statement is no longer even a bit obscure or dubious."[158] Pascal's treatise

[156] *PS*, Bd. 4, p. 422; Loemker, p. 291; ***Philosophical Essays***, translated by Roger Ariew and Daniel Garber (Indianapolis & Cambridge, 1989), p. 23.

[157] *PS*, Bd. 4, p. 423; Loemker, p. 292; ***Philosophical Essays***, p. 25.

[158] *PS*, Bd. 4, p. 426; Loemker, p. 294; ***Philosophical Essays***, p. 27.

in the cited sentence was "De l'esprit géométique et de l'art de persuader,"[159] originally drafted as a preface Arnault's *Nouveax élémans de géométrie*, published in 1667, but actually used for *La logique de Port-Royal* of 1662, and has been considered to be one of the greatest masterpieces on the critical method of axiomatic mathematics before David Hilbert's *Grundlagen der Geometrie* of 1899. Thus, Leibniz's understanding on mathematical knowledge, particularly on symbolic knowledge, was not 'naive' at all. We know that Kurt Gödel, who was fascinated by Leibniz, paid special attention to the depth of the tract.[160]

It is necessary to examine further a manuscript drafted around 1686 with the title *Projets et essays pour arrive à quelque certitude pour finir une bonne partie des disputes et pour avancer l'art d'inventer* in order to look for the origin and the real intention of Leibnizian program of 'demonstrating axioms'. In this essay we better comprehend what Leibniz thought by 'axioms'. There Leibniz referred to Pascal's "De l'esprit géométique," again. He states: "I looked for some marks given to us for recognizing things too dubious and too obscure. And I trust that to complete the sciences some propositions which are called axioms (*quelques propositions qu'on appelle axiomes*) must be proved, as, in fact, Apollonius tried to prove some of the propositions which Euclid had used without demonstrations. Euclid was right, but Apollonius was even more right."[161] From these sentences, we know that Leibiniz's program of 'demonstrating axioms' was, in reality, to attempt to prove not 'primitive principles' in the strict sense of the words which do not accept any kinds of demonstrations but "propositions which are called axioms."

It was by having recourse to this belief for establishing the rigorous scientific knowledege as far as possible that Leibniz criticized Descartes's later philosophy of mathematics, that is, the creation doctrine of mathematical truths. He regarded that doctrine as an intellectual absurdity. His criticism of the French philosopher's point of view can be seen frequently in his mature writings, such as an essay usually called *Discours de metaphysique* (1686), § 2,[162] *Essai de theodicée sur la bonté de Dieu, la liberté de l'homme et l'origine du mal* (1710), §§ 175–192,[163] *Monadologie* (1714), §§ 33–35 and 46.[164] However, it would be sufficient for us here to examine only his criticism of the general principles of the Cartesian philosophy titled *Animadversiones in partem generalem Principiorum Cartesianorum* (Critical Thoughts on the General Part of the Principles of Descartes), written in 1692. These notes took the form of paragraph after

---

[159] Pascal: *Selections* (n. 96), pp. 173–194; *Œuvres complètes* (n. 96), pp. 575–604.

[160] Charles Parsons, "Platonism and Mathematical Intuition in Kurt Gödel," *The Bulletin of Symbolic Logic*, **1** (1995), p. 45, n. 3.

[161] "Projets et Essays pour arriver à quelque certitude pour finir une bonne partie des disputes et pour avancer l'art d'inventer," Couturat, *Opuscules et fragments inédits de Leibniz* (n. 127), pp. 181–182.

[162] *PS*, Bd. 4, pp. 427–428; Loemker, p. 304.

[163] *PS*, Bd. 6 (Berlin, 1885), pp. 218–231.

[164] *PS*, Bd. 6, pp. 612–614; Loemker, pp. 646–647.

paragraph of critical commentaries on the first two parts of Descartes's *Principia philosophiae* and were intended to be the author's overall critique of the entire Cartesian philosophical system.

With respect to Descartes's maxim that everything must be doubted as far as possible, Leibniz is skeptical about its effect, so he presents his own strategy for establishing a firm criterion for the certainty of knowledge, in particular of mathematics. For this purpose he adopts his way of 'demonstration of mathematical principles'. He mentions his predecessors in this strategy, Apollonius and Proclus in ancient Greece, who tried to prove some propositions which Euclid assumed as self-evident, and his contemporary Gilles Personne de Roberval, who was planning the new *Elemens de geometrie*.[165] Leibniz charges Descartes with not having committed himself to such an attempt:

> I am convinced that the demonstration of the axioms is of great assistance to true analysis or the art of discovery. So if Descartes had wished to carry out what is best in his rule, he should have worked at the demonstration of scientific principles and thus achieved in philosophy what Proclus tried to do in geometry, where it is less necessary. I should not blame him for being satisfied so often with verisimilitude, if he himself had not aroused expectations with so strong a profession of exactness. I blame Euclid much less for assuming certain things without proof, for he at least established the fact that if we assume a few hypotheses, we can be sure that what follows is equal in certainty, at least, to the hypotheses themselves. If Descartes or other philosophers had done something similar to this, we should not be in difficulty.[166]

Leibniz certainly proposed an attempt at proving the axioms of mathematics, and, as has been seen already, made every effort to reduce the mathematical principles to definitions and identical propositions. A point to notice is that in the above-quoted passage he at times utilizes the terminology 'hypothesis' instead of 'axiom'. The reason for this usage seems to have been that he was familiar with the skeptics' methods of making charges against the certainty of mathematics.

Leibniz writes immediately after the above statement:

> Moreover, the skeptics, who despise the sciences on the pretext that they sometimes use undemonstrated principles, ought to regard this

[165] "Animadversiones," *PS*, Bd. 4, p. 355; Loemker, p. 383. See also *Nouveaux essais* (n. 143), p. 107 and 406–407. "Leibniz and Tschirnhaus examined Roberval's papers and decided that the *Elements of Geometry* was not worth publishing." P. Remnant and J. Bennett's translation, note "Roberval", pp. lxix–lxx. Now Roberval's treatise was published as *Éléments de géométrie de Roberval*, texte présentés par Vincent Jullien (Paris, 1996) & Kokiti Hara, "Quelques Œuvrages des Géométrie More Veterum de Roberval, (1) et (2)," *Historia Scientiarum*, Vol. 2, Nos. 1–2 (1992), pp. 13–44 & 109–117.

[166] "Animadversiones," *PS*, Bd. 4, p. 355; Loemker, p. 384.

> as said also to them. I hold, in contrast, that the geometricians should be praised because they have pinned down science with such pegs, as it were, and have discovered an art of advancing and of deriving so many things from a few. If they had tried to put off the discovery of theorems and problems until all the axioms and postulates had been proved, we should perhaps have no geometry today.[167]

Leibniz knows clearly how skeptics despise mathematical propositions. Proceeding along analytical routes from such propositions, we will necessarily attain indemonstrable starting points which are usually termed principles. On the ground of the indemonstrability of such primitive propositions, the skeptics may claim that even mathematical knowledge cannot be warranted to be absolutely certain. Leibniz, on the other hand, criticizing the opponents of mathematical truths, maintains that to assume a few hypotheses for advancing discussion is much more fruitful in constructing the sciences, such as geometry. The proponent of mathematical knowledge makes no attempt to discuss further the epistemological character of the mathematical propositions which have been obtained by starting from such a few undemonstrated hypotheses. Nor does he present his argument so as to refute thoroughly and finally the skeptics' assault on mathematics. We will later suggest what Leibniz's final answer to this problem looked like.

Leibniz censures the creation doctrine of Descartes for the same reason that the French philosopher has just rejected the skeptics' attack on mathematics. Descartes developed his creation theory of mathematical knowledge in Part I, Paragraph 5 of the *Principia philosophiae*, as we have seen in the previous section (see pp. 367–368). He there presented "the reasons for doubting even mathematical demonstrations." Leibniz, commenting on Paragraph 5, writes as follows:

> There can be no doubt in mathematical demonstrations except insofar as we need to guard against error in our arithmetical calculations. For this there is no remedy except to reexamine the calculation frequently or to have it tested by others and also to add confirmatory proofs. This weakness of the human mind arises from a lack of attention and memory and cannot be overcome completely, and Descartes's mention of it, as if he knew a remedy, is in vain. It would be enough if the state of affairs in other fields were the same as that in mathematics; indeed, all reasoning, even the Cartesian, however convincing and accurate, is subject to this doubt, whatever may be said about some powerful deceiving spirit or about the distinction between dreams and waking.[168]

---

[167] *Ibid.*

[168] *PS*, Bd. 4, p. 356; Loemker, p. 384.

In Paragraph 5, Descartes claimed that we may be able to doubt mathematical truths which are usually regarded as most certain. The doubt, according to him, can apply to two categories of mathematical conceptions: the demonstrations of mathematics and the principles which are considered to be self-evident. In his commentary on Paragraph 5, Leibniz tries to respond to the Cartesian doubt only about the first category of mathematical demonstrations. He might have thought to have resolved the second doubt about mathematical principles by means of their reduction to appropriate definitions and identities, the program of which is explained in his earlier commentary on Paragraph 1.

How has Leibniz understood and answered Descartes's suspicion that we are always deceived even in mathematical proofs? It seems clear that he understands the doubt about them primarily as errors we may make in arithmetical calculation and geometrical reasoning. The only possible remedy for such mistakes is, in his opinion, to reexamine the mathematical procedures and to have them tested by others. The human mind is so weak as to commit errors frequently. Descartes's attempt at overcoming the skeptics' assault, Leibniz concludes, is after all in vain.

It may be argued, by having examined Leibniz's critical commentaries on Descartes's creation theory, that he possibly failed to grasp Descartes's subtle formulation of skeptical doubt about mathematical demonstrations, for he interpreted his predecessor's objects of suspicion simply to be errors of individuals. But here we should be reminded that Leibniz believed throughout his life in the possibility of the final success of reducing mathematical principles to definitions and identities. Thus, we need to evaluate his overall endeavor to conquer the skeptics' criticism of mathematics by means of 'demonstrating its principles'. Only through this procedure will we be able to judge whether Leibniz's denouncing of Descartes was sound at all.

As we have noted above, one of the reasons that Leibniz formulated his proof theory of reducing mathematical principles to definitions and identities was to evade the skeptical criticism of mathematics. Judging from hindsight, Leibniz's project was a failure. Above all, he sometimes stretched his conception of definitions or identities. In the *Nouveaux essais*, for example, propositions were divided into two categories: those propositions which need proof, and those which are also redefined as "identities or nearly so "(*identiques, ou peu s'en faut*). As for the second category, he noted that "identical or immediate propositions do not admit of proof."[169]

A similar stretching of notions can be also witnessed in the logicist of the late nineteenth and early twentieth centuries Gottlob Frege, who claimed that mathematical truths "can be proved purely logically and hence are analytic,"[170] when attempting to criticize Kant's philosophy of mathematics developed in the

[169] *Nouveaux essais*, Bk I, Ch. i, § 18, p. 82

[170] Frege, *Die Grundlagen der Arithmetik* (Breslau, 1884), § 88; M. S. Mahoney's translation in P. Benacerraf and H. Putnam, eds., *Philosophy of Mathematics* (Cambridge, 1983), p. 154.

Kritik der reinen Vernunft. Frege added to the previous categorical statement that "they are in fact contained in the definitions, but like the plant in the seed, not like the rafter in the house."[171] In practice, Leibniz, when necessary, extended the concept of identities to that of "identities or nearly so." In this regard, he very nearly offended his own maxim, as Margaret Wilson has pointed out.[172]

Although his formulation, if taken rigorously, has turned out to be insufficient, Leibniz's real motive for the 'proof of mathematical axioms' was not totally nonsense. His bold program was useful for disclosing the logical structure of mathematics by trying to describe mathematical propositions in terms of the specific symbols of the *characteristica universalis*. Leibniz was never successful in achieving this project. But what he entertained in his mind when criticizing Descartes and composing his *Nouveaux essais* was clarifying the formally rigorous reformulation of mathematical principles which had been assumed simply as 'self-evident' by the young Descartes in the *Regulae* and by Locke in *An Essay Concerning Human Understanding.*[173] As we will see in our "Conclusion," the 'incompleteness' of the Leibnizian *characteristica universalis* was proved in a rigorously mathematical manner by his admirer Gödel. The Leibnizian program, although paradoxcal, can thus claim at least this one merit.

How, then, did Leibniz think of mathematical truths epistemologically and metaphysically? This question must be crucial for one who is considering his 'mathesis universalis'. We may be able to answer the question by examining again and more closely his attitude toward the skeptical attack on mathematical statements.

We have already pointed out that Leipniz was familiar with the skeptics' arguments against mathematics. His later attitude toward them, insofar as the theory of mathematics is concerned, best appears in his correspondence with the leading French skeptic Simon Foucher, part of which was published in the *Journal des sçavans.*[174] It is by analyzing the correspondence that Stuart Brown has recently characterized Leibniz's later metaphysical system as

[171] *Ibid.* Cf. W. V. Quine's criticism of logicism in his "Carnap and Logical Truth," in Benacerraf and Putnam, eds., *Op. cit.* (n. 169), p. 358. Quine puts it: "Kant's readiness to see logic as analytic and arithmetic as synthetic, in particular, is not superseded by Frege's work."

[172] Margaret D. Wilson, "On Leibniz's Explanation of 'Necessary Truth'," in Harry G. Frankfurt, ed., *Leibniz: A Collection of Critical Essays* (Notre Dame/London, 1972), pp. 401–419.

[173] I understand Margaret D. Wilson to be claiming this point in her splendid essay "Leibniz and Locke on 'First Truths'," *Journal of the History of Ideas*, **28** (1967), pp. 347–366; Also reprinted in Ch. 24 of her *Ideas and Mechanism: Essays on Early Modern Philosophy* (Princeton, 1999), pp. 353–372.

[174] On the relation between Leibniz and Foucher, see R. H. Popkin, "Leibniz and the French Sceptics," *Revue internationale de philosophie*, 20e année (1966), pp. 228–248; W. H. Barber, *Leibniz in France: From Arnauld to Voltaire* (Oxford, 1955), pp. 43–46; *Leibniz' 'New System' and Associated Contemporary Texts*, translated and edited by R. S. Woolhouse and Richard Franks (Oxford, 1997), pp. 37–60.

having a "hypothetico-deductive"[175] or "pragmatic"[176] inclination, breaking with 'Cartesian rationalism'. Brown's observation is, we judge, generally ingenious and sound. But he seems to have over-simplified his characterization of Descartes's philosophy of mathematics, as may be easily imagined from our previous discussions, and of the mathematico-philosophical standpoint which Leibniz adopted deliberately and finally.

In order to reexamine Leibniz's view of mathematical truths carefully, let us begin with his letter to Foucher of 1686. There he discussed the principle of contradiction, which is, in his view, identical with that of identity:

> It is certain, even, that some truths must be supposed, on pain of giving up hope of making demonstrations, for proofs could not go on to infinity. You cannot ask for the impossible, otherwise you would be a witness that you were not seriously searching for the truth. Hence, I will suppose boldly that two contradictions could not be true, and that what implies a contradiction could not be, and consequently that necessary propositions (that is whose contraries imply a contradiction) have not been set up by a free decree, or else we are misusing words. Nothing could be more clear for proving things. You yourself suppose them in writing and reasoning, or else at any moment you could defend the opposite of what you say.[177]

The expression: "I will suppose boldly etc.," must be a note of sarcasm directed to the excessively faint-hearted skeptic in using simple logical rules.[178] This passage indicates that Leibniz's admitting them was circumspect. In this sense, he appears to have accepted mathematical principles 'pragmatically'.

A similar argument is stated in a letter written in 1692, the year in which Leibniz drafted the *Animadversiones in partem generalem Principiorum Cartesianorum*, which we have already analyzed in detail. That letter resembles his general critical comments on Cartesian philosophy, but is more sophisticated and better shows his logic for escaping from the skeptics' assault. He writes to Foucher:

---

[175]Stuart Brown, *Leibniz* (Brighton, Sussex, 1984), pp. 11, footnote 7: "I take Rationalism to be a view not simply that a deductive metaphysics is possible but that a fully demonstrated metaphysics is possible. Leibniz's later metaphysical system is, as I shall attempt to show, hypothetico-deductive. Its principles are assumptions rather than self-evident truths."

[176]S. Brown, "Leibniz's Break with Cartesian 'Rationalism'," in A. J. Holland, ed., *Philosophy, Its History and Historiography* (Dordrecht/Boston, 1985), pp. 195–208, esp. p. 202. "One way of expressing this reform is to say that Leibniz rejected the Cartesian appeal to our ideas in favour of a pragmatic criterion of metaphysical truth. We do not have intuitive knowledge of the essence of substance, as Descartes had mistakenly supposed."

[177]"Leibniz an Foucher," *PS*, Bd. 1 (Berlin, 1875), p. 382. G. W. Leibniz, *Discourse on Metaphysics and Related Writings*, edited and translated with an Introduction, Notes and Glossary by R. Niall, D. Martin, and Stuart Brown (Manchester/New York, 1988), pp. 130–131; See also Brown, *Leibniz* (n. 175), p. 74.

[178]Brown, *Leibniz*, p. 74.

> I am of the same opinion as you that it is good to search for proofs of all the truths which can be proved. This is not because that would be absolutely necessary, nor because it would be necessary to stop oneself until one could prove all the principles by the first ones, for if the geometricians had wished to wait to seek the solutions of problems and the demonstrations of theorems until they had demonstrated their axioms and postulates, they should have been wrong and denied themselves of the advantages which geometry has brought to us. Nevertheless, it is good that there should be certain minds who endeavor to supply that which one has left behind in order to advance. And if one understands your reasonings on the art of doubt in this manner, there is nothing so reasonable. But it would be perhaps good that you would have explained distinctly (that is your intention) in order that those who do not understand sufficiently do not suppose unfortunately that the Academy is opposed to the progress of the sciences. From experience I have recognized how important the demonstration of axioms is in order to improve that which I call veritable analysis. And I am astonished that there is no one who has guarded against this sufficiently.[179]

From the above passage we recognize that Leibniz did not think the demonstration of mathematical principles to be absolutely necessary. His project of proving the axioms was primarily to improve our mathematical knowledge. So the Leibnizian program was not naive at all. If our discussion is restricted to his reasoning up to this point, we may agree in principle with Brown's judgment that Leibniz inclined toward 'pragmatism' with regard to the philosophy of mathematics.

Leibniz nevertheless goes on to accuse Descartes of not having adopted his own strategy. He argues immediately after the sentences which we have just quoted:

> M. Descartes himself, who revived the doubts of the ancients to make

[179] "Leibniz an Foucher. Janvier 1692," *PS*, Bd. 1, p. 402: "Je suis de vostre avis qu'il seroit bon de chercher les preuves de toutes les verités qui se peuvent prouver. Ce n'est pas que cela soit absolument necessaire, ny qu'il faille qu'on s'arreste jusqu'à ce qu'on puisse prouver tous les principes par les premiers, car si les Geometres avoient voulu attendre à chercher les solutions des problemes et les Demonstrations des theoremes, jusqu'à ce qu'ils eussent demonstré leur axiomes et demandes ou postulata, ils auroient mal fait et se seroient privés des avantages que la Geometrie nous a apportés. Cependant il est bon, qu'il y ait certaines esprits, qui tachent de suppleer ce qu'on a laissé en arriere pour avancer. Et si on prend ainsi vos raisonnemens sur l'art de douter, il n'y a rien de si raisonnable. Mais il seroit peut estre bon que vous expliquassiés distinctement, que c'est là vostre intention, à fin que ceux qui ne l'entendent point assés, ne s'imaginent pas mal à propos, que l'Academie s'oppose aux progrés des sciences. J'ay reconnu par experience, de quelle importance seroit la demonstration des Axiomes, pour perfectionner ce que j'appelle la veritable Analyse. Et je m'etonne que personne n'y a pris garde assés."

> us admit nothing without proof, should have pondered about the demonstrations of axioms; but instead of this he jumped recklessly on to other resonings which are much less solid but more appropriate to acquire applause and that which is called popular favor (*aura popularis*). It is not necessary to despise the difficulties which Sextus Empiricus ascribed to the dogmatists, for they serve to bring them back to principles.[180]

Certainly Leibniz devised a strategy different from Descartes's in order to escape from the skeptics' criticism of mathematics: It was by means of the demonstration of principles. Until the end he continued to be optimistic and to believe that mathematical propositions can be, in principle at least, reduced to identities. This strategy is what Leibniz accepted thoughtfully after having scrutinized the logic used by the skeptics to degrade the mathematical disciplines. On this point, Brown writes: "Leibniz, in his later writings, uses the words 'demonstration' and 'proof' in a modern way, i. e. as assumption-relative. [...] Leibniz appears to pay lip-service to the ideal of a 'full demonstration'."[181] Unfortunately, this statement is a half-truth. At least Leibniz could not appreciate Descartes's rather deep doctrine that God is capable of creating even mathematical truths. As far as this theory is concerned, the French philosopher was not necessarily so naive a rationalist as Brown has depicted. We should evaluate highly Leibniz's attempt to arrange mathematical systems properly by examining their principles and reducing them to more primitive ones. But his project of reducing them to definitions and identities does not seem to have simply been "lip-service." In this regard, the Leibnizian program was too ambitious and too optimistic to be accomplished.

We have stated above that Brown's statement is a half-truth. In fact, Leibniz's later philosophy of mathematics reached in its own way a considerable depth which is comparable with Descartes's creation doctrine and tells us something significant, although it has the certain limitation indicated above. His later view on the nature of mathematical knowledge was set forth in his *Nouveaux essais*. It can be said, we claim, that in that book Leibniz understood completely what Descartes's "Cogito, ergo sum" means and what mathematical statements signify logically and epistemologically in comparison with the *cogito*.

As to Descartes's first truth, Leibniz lets his spokesman Théophile state:

> As to primitive truths (*vérités primitives*) of fact, these are inner experiences which are immediate with the immediacy of feeling. This

[180] *Ibid.*: "M. Descartes luy même qui a renouvellé les doutes des anciens, pour nous faire rien admettre sans preuve, devoit songer aux demonstrations des Axiomes; mais au lieu de cela il se jetta à corps perdu sur d'autres raisonnemens bien moins solides, mais plus propres à acquerir des applaudissemens, et ce qu'on appelle auram popularem. Il ne faut pas mepriser les difficultés que Sextus Empiricus faisoit aux Dogmatistes, car elles servent à les ramener aux principes."

[181] Brown, *Leibniz*, p. 73.

> is where the first truth of the Cartesians and St. Augustine belongs: I think, therefore I am. That is, I am a thing which thinks. But we must realize that just as identities can be general or particular, and that they are equally evident in either case (since A is A is just as evident as *A thing is what it is*), so it is with the first truth of fact. For not only is it immediately evident to me that *I think*, but it is just as evident that *I think various thoughts*: at one time *I think about* A and at another *about* B and so on. Thus the Cartesian principle is sound, but it is not the only one of its kind. This shows that all of the primitive truths of reason and of fact have in common that we cannot prove them by anything more certain.[182] [The emphasis is in the original.]

The Cartesian first principle, "Cogito, ergo sum," according to Leibniz, is characterized as a truth which is ascertained by the immediate self-reflection. It has the same characteristic as the identity "A is A" in that we cannot prove it by anything more certain. In other words, "Cogito, ergo sum" can be substituted for by other primitive truths which have the same logical nature.

This observation is reiterated in a later chapter under the title "Des propositions qu'on nomme maximes ou axiomes" (The title itself is worthy of attention): "One can always say that the proposition *I exist* is evident in the highest degree, since it cannot be proved by anyone else—indeed, that it is an 'immediate truth'. To say *I think therefore I am* is not really to prove existence from thought, since *to think* and *to be thinking* are one and the same, and to say *I am thinking* is already to say I am."[183] (The emphases are in the original.) To paraphrase Leibniz's words, in their logical function, the primitive truths are different from other statements which are thought to be true but cannot be proved by immediate reflection or which may be reduced to more primitive ones. Almost all mathematical truths cannot be such primitive truths.

Then in the *Nouveaux essais* how did Leibniz characterize mathematical knowledge in general? He counted mathematical truths as 'eternal truths' (*vérités eternelles*) as Descartes did. He states as follows:

> As for 'eternal truths', it must be understood that fundamentally they are all conditional (*conditionnel*); they say, in effect: given so and so, such and such is the case. For instance, when I say: *Any figure which has three sides will also have three angles*, I am saying nothing more than that given that there is a figure with three sides that same figure will have three angles.[184] [The emphasis is in the original.]

Mathematical truths, a kind of eternal truth, in Leibniz's judgment, are basi-

[182] Leibniz, *Nouveaux essais* (n. 143), p. 367
[183] *Ibid.*, p. 411.
[184] *Ibid.*, p. 446

cally conditional, not unconditional or absolute.

In the *Nouveax essais*, too, Leibniz entertained the reduction mathematical propositions to definitions and identities as systematically as possible to be his task. But he frankly admitted the difficulties of this task. In the same way as he did in his letters to Foucher, he took a pragmatic stance toward mathematical knowledge. In a chapter with the title "Des moyens d'augmenter nos connaissances," he states:

> Sometimes, when you are having trouble in carrying out a complete analysis, you will have to rest content with some axioms which are less primitive, as Euclid and Archimedes did; but it is better to do this than to miss or postpone some fine discovery which you could make straight away with their help. Indeed, [...] I believe that we would have almost no geometry at all (as a demonstrative science, I mean), if the ancients had not been willing to push ahead before having demonstrated the axioms which they were compelled to use.[185]

In any case, both Descartes and Leibniz were deep thinkers in their own right with respect to the theory of mathematics, although they could not finally win in their battles with the skeptics. When 'la crise pyrrhonienne' of the seventeenth century was subsiding and the skeptical discussions about mathematics came to be restricted to a very small group of philosophers, such as David Hume, the philosophies of mathematics of the two original thinkers seem to have been simplified. For example, as Popkin reports, the dogmatic Leibnizians in Germany like Christian Wolff and Georg Bernhard Bilfinger thought that their master was obviously not serious in his argument with the skeptics.[186]

At this juncture it seems to us that Imre Lakatos's observation is applicable, among others, to the two great philosophers of mathematics in the seventeenth century and their followers. Lakatos puts it in his brilliant essay *Proofs and Refutations: The Logic of Mathematical Discovery*:

> Szabó showed that in Euclid's time the word 'axiom'—like 'postulate'—meant a proposition in the critical dialogue (*dialectic*) put forward to be tested for consequences *without* being admitted as true by the discussion-partner. It is the irony of history that its meaning was turned upside down. The peak of Euclid's authority was reached in the Age of Enlightenment.[187]

It is in order to reconstruct the original theories of mathematics of the seventeenth century and to disclose an irony of history that we have examined

[185] *Ibid.*, pp. 452–453.
[186] Popkin, "Leibniz and the French Sceptics" (n. 173), p. 247.
[187] Lakatos, *Proofs and Refutations* (Cambridge, 1976), p. 49, footnote 1.

Descartes's and Leibniz's philosophies of mathematics. This same critical observation must be applicable to their crucial concept of 'mathesis universalis'. If we are permitted to ascribe the label 'mathematicism', which is usually thought to be related to 'mathesis universalis', to Descartes and Leibniz, that word is justifiable in the sense and only in the sense that we have described it in these two sections on the two philosophers' concepts of 'mathesis universalis' and their philosophies of mathematics.

In addition Ortega y Gasset's assertion in his book on Leibniz's idea of principles may be applicable only partly to Descartes and Leibniz themselves:

> The reform which Descartes initiated, which Leibniz developed, and which has been established for nearly a century—hence the *modern* idea of deductive theory—consists formally in committing that *petitio principii.*[188]

Unfortunately, however, it is true at the same time that Descartes's and Leibniz's epigones had gradually come to forget the two original thinkers' struggle in establishing their philosophies of mathematics. We add to this that, in this process of forgetting, the manner in which philosopher-mathematicians of the seventeenth century used their concept of 'mathesis universalis' had also come almost to be forgotten.

[188] José Ortega y Gasset, *The Idea of Principle in Leibnitz and the Evolution of Deductive Theory*, translated by Mildred Adams (New York, 1971), p. 111.

CONCLUSION

Descartes and the Modern Scheme of Learning

Descartes was a product of a transitional period which has been usually characterized as an age of 'general crisis'. He was born during 'the crisis of the 1590s' and in France, where the 'crisis' was particularly deep. It is reported that in 1592, four years before he was born, a Spanish writer declared: "England without God, Germany in schism, Flanders in rebellion, France with all these together."[1] In fact, France in the latter half of the sixteenth century suffered successive brutal religious wars up to 1598, the year in which the Edict of Nantes was promulgated by Henri IV, a king newly converted to Catholicism. And it was at a Jesuit college which Henri IV had the Jesuits establish that Descartes was educated. Descartes's later career was marked by the Thirty Years' War, which continued from 1618 to 1648. The period of the Thirty Years' War is characterized as an age of the decline of the Catholic states of Spain and of the rise of the Protestant states of England and the Netherlands, which began to sweep over the seas of the world. The beginning part of the "General Introduction" by the editor of *The New Cambridge Modern History*, a collection of essays discussing this age synthetically, characterizes this period: "Undeniably the first half of the seventeenth century in Europe (or for that matter in China and to a lesser extent India) was an eventful period, full of conflicts."[2]

It was in the autumn of 1619 in the midst of his journey to observe the development of the war which had recently broken out that Descartes made up his mind to reform the traditional scheme of learning as a whole. He was then equiped with an excellent education of the Renaissance by the Jesuits who had led an educational revolution in the Counter Reformation and who attached inportance to mathematical studies for achieving missionary works

[1] Quoted in Peter Clark, ed., *The European Crisis of the 1590s: Essays in Comparative History* (London, 1985), "Introduction," p. 4. The Spaniard's name is Mario Antonio de Camos.

[2] On the historical context of Descartes's lifetime, see, among others, J. P. Cooper, ed., *The New Cambridge Modern History*, Vol. 4: *The Decline of Spain and the Thirty Years War 1609-48/59* (Cambridge, 1971).

overseas in Japan and China. More important for his intellectual career was undoubtedly the philosophical trend of skepticism, which became known and influential, along with the development of the French Counter-Reformation, especially in the 1620s. The situation in the latter period has a particular name: 'la crise pyrrhonienne'. As Henri Gouhier has contended, Descartes's thought was a specific Catholic response to the 'theological crisis' of this period.[3] Descartes's later philosophy of regarding "Cogito, ergo sum" as the first truth, thought out after having searched for indubitable certainty, was an outcome of his struggle to escape from the obsession of 'conflicts' of this age and to overcome the 'crisis'.

Descartes was living in a period moving from turmoil to relative tranquility. He fought hard to stabilize, in particular, the European learned world, which had been as unsettled as the worlds of religion and of politics. It is no accident that Descartes is one of the main *dramatis personae* of Theodore K. Rabb's excellent essay attempting to sum up the debates on the so-called 'general crisis of the seventeenth century'. The title itself is suggestive: *The Struggle for Stability in Early Modern Europe*. Rabb tries to redefine the word 'crisis' strictly as it pertains to the age of Descartes and hits upon an analogy used in pathology. In its medical meaning, according to him, 'crisis' stands for a critical point where "the outcome becomes clear: the patient either recovers or dies."[4] He claims that this use of the word is also applicable to the term as utilized in history, especially for representing the situation of the seventeenth century. Then, he states about the crisis of that period: "My claim, in a phrase, is that both the rising fever and the final resolution centered on the location of authority."[5] He observes that people of the seventeenth century eagerly searched for new authority and certainty in both the political and intellectual worlds. It seems to me that Rabb's claim is particularly applicable to the world of learning in the time of Descartes. In that period, the authority of medieval learning was being severely challenged. The efforts of Renaissance thinkers could not succeed in presenting a sufficiently stable alternative authority. Descartes asked: "Can one rely on anything?" He answered: "Mathematical knowedge at least is trustworthy" in the period during which he composed his *Regulae*, and "we rely on the first truth 'Cogito, ergo sum' and then, with the hypothesis of God's veracity, mathematical propositions," at the time of his *Discours de la méthode*. It was because he was sincerely seeking solid and stable certainty that he gave such a reply.[6]

---

[3]Henri Gouhier, "La crise de la théologie au temps de Descartes," *Revue de théologie et de philosophie*, t. IV (1964), p. 53.

[4]Theodore K. Rabb, *The Struggle for Stability in Early Modern Europe* (New York, 1975), p. 31.

[5]*Ibid.*, p. 33.

[6]For the problem of certainty in this period, see, for example, Henry G. van Leeuwen, "Certainty in Seventeenth-Century Thought," *Dictionary of the History of Ideas*, ed. by Philip P. Wiener, Vol. 1 (New York, 1973), pp. 304–311.

The criterion of certainty was not the only goal that Descartes pursued in the process of forming his mathematical thought. He sought a specific certainty, that is to say, certainty with utility or accompanied by practical utility. In general, a recent interpreter of his philosophy of science has described him as a "practising scientist who also concerned himself with methodological, theological and metaphysical questions."[7] This interpretation, if the emphasis upon practice is not exaggerated, is also applicable to his concern in the domain of mathematics. What Descartes thought important was not the 'sterile' method of synthesis but the art of discovering mathematical theorems and solving mathematical problems, i. e. the method of analysis, especially algebraic analysis. In short, it was the art of algebraic analysis that he placed in the center of his conception of mathematics and, then, of knowledge as a whole.

Such an intellectual position became possible only after the time algebra had come to be viewed as a noble science. Before the Renaissance, algebra, which originated in the Arabic-speaking Islamic world, tended to be considered as the higher part of merchant arithmetic. Thus, it was "on the shoulders of merchants,"[8] to invoke the witty title of a monograph on the history of ideas. The Renaissance humanist mathematician Regiomontanus began to claim that despite its foreign origin algebra had a Greek root as well in Diophantus's *Arithmetic*. And Ramus made efforts to bridge a gulf between practical arithmetic and philosophical speculation about mathematics, two mutually separated components of the medieval world of mathematics, by criticizing the "beautiful abstractions"[9] of scholastic philosophy of mathematics. By the endeavors of the Renaissance reformers of mathematics, algebra was elevated to the rank of an academic discipline and connected to ancient authoritative knowledge. Indeed, Descartes was a thinker who could possibly have come to the historical stage only after such reformers as Regiomontanus and Ramus. He in fact certainly read Clavius's algebra textbook and is conjectured to have read a certain book (or books) by Van Roomen, who was well versed in Viète's art of algebraic analysis. The latter three mathematicians of the late sixteenth and early seventeenth centuries were all advocates of algebra, a reform which had begun during the Renaissance. Thus, Descartes was not a philosopher who was concerned simply with establishing irrefutable metaphysical certainty. He was also a practical mathematician who made efforts to construct a reformed art of algebraic analysis, a heuristic device to solve mathematical problems.

---

[7]Desmond M. Clarke, *Descartes' Philosophy of Science* (Manchester, 1982), p. 5. The author has emphasized that Descartes was not a dogmatic rationalist but a practical scientist attaching importance to experience.

[8]Richard W. Hadden, *On the Shoulders of Merchants: Exchange and the Mathematical Conception of Nature in Early Modern Europe* (Albany, 1994).

[9]Natalie Zemon Davis, "Mathematics in the Sixteenth-Century French Academies: Some Further Evidence," *Renaissance News*, **11** (1958), p. 9; *Eadem*, "Sixteenth-Century French Arithmetics on the Business Life," *Journal of the History of Ideas*, **21** (1960), p. 36: "The teacher of 'belles abstractions' was Ramus' enemy Jacques Charpentier." (p. 36, note 58).

It was a specific form of mathematics, i. e. algebraic analysis, that Descartes singled out as a paradigm for his science-theoretical ideals: certainty and utility. Certainly what he praised was not Euclidean geometry but algebraic analysis, which took a radically new form after Viète. Notable is that the new algebra employed symbols extensively to represent algebraic operations and formulae. This characteristic was not historically a matter of course.

As we have seen in Chapter 5, § 3, E (p. 272), the German historian of mathematics of the nineteenth century Nesselmann in his study of Greek algebra distinguished three developmental stages in the form of algebra: 'rhetorical', 'syncopated', and 'symbolic'. His trichotomy may be said to be oversimplified. Granted this, it still provides us with some insight about the Cartesian mathematical thought. Descartes's mathematics marked the beginning of full-fledged modern symbolic algebra. To utilize mathematical symbols was always technically possible even before the printing press penetrated the world of mathematics. In fact, cossists of the later Middle Ages and the Renaissance expressed unknown numbers by characteristic signs. But their algebra can be called 'syncopated algebra' in Nesselmann's appropriate terminology. The exactly repeatable symbols which seem to have become realized with the advent of the printing press surely contributed to the spread of the algebra of the third stage and the standardization thereof. Remember Descartes's insistence in Rule XIV of his *Regulae* that appropriate visual signs should be used to designate mathematical concepts and formulae, and his ingenious use of symbols which were simplified to the utmost in the *Géométrie*. His algebraic mathematics with much simplified symbols appears to have the mark of the Post-Gutenberg Era, not of the age of oral culture or of the time of electric computers. The language of the mechanical world-view appeared almost in parallel with the emergence of a mathematics which in itself was fully mechanized.

Along with the mechanization of mathematical language, a significant event seems to have occurred in the general mode of thinking. Before Descartes the paradigm of mathematical thinking was the dialogue, a two-way communication on geometrical figures which were as a general rule tactile. Mathematicians discussed with others or alter egos how proposed statements could be 'analyzed' and reduced to certain propositions. In the process of analysis, they fully utilized their imagination, for example, by drawing additional lines. They frequently debated with one another about the status of those principles which seemed assured. In short, they were thinking *dialectically*, or in the manner of a critical dialogue. The final criterion of mathematical truth was, at least in an age of critical temper, *negotiation* among the community of mathematical practitioners.

With Descartes the paradigm procedure of mathematics became mechanical calculation with algebraic symbols. His method of mathematical inquiry consisted in an individual manipulating signs, i. e. a non-verbal activity. This observation is in accord with Ong's view that Descartes was a thinker of the post-Ramist age. According to Ong, Ramus brought about the decay of dialogue. But the transition "from the art of discourse to the art of reason" was not necessarily completed with Ramus. The age of the full-fledged art of reason came first with our mathematician-philosopher Descartes. For him and his followers, the new logic was "the art of thinking—that is, of individualized, isolated intellectual activity, presumably uninvolved with communication."[10] Its paradigm was—not dialectic—but pure mathematics. It was he who gave the pivotal position to pure mathematics by disregarding dialectics, the art of discussion. Recall again that in his scheme of learning in the *Regulae* the art of reason had a specific form of mathematics: algebraic analysis.

It was such a methodological device that made the world, both natural and social, rationally calculable, or in a literal sense highly 'scientific'. The following metaphysical statement by Leibniz, who particularly admired Descartes and developed creatively and critically his French predecessor's thought, can be fully understood against this intellectual background: "When God calculates [with visible signs] and converts his thinking into deed, the world arises."(*Cum Deus canculat et cogitationem exercet, fit mundus*).[11] Because of the considerable effect of that device, symbolic algebra tended to be considered as an autonomous discipline and as a foundation for the mathematical sciences. The Leibnizian *characteristica universalis* was also to function as 'calculus of reasoning' (*calculus ratiocinator*) for resolving conflicts and debates in a church and in a secular society. In an anonymous essay with the title *Initia et specimina scientiae generalis sive de instauratione et augmentis scientiarum in publicam felicitatem* (The Foundations and Exemplars of a General Science, or On the Reform and Advancement of the Sciences in Public Welfare) Leibniz proclaims: "Let us calculate in order not to advance by jeers, but to speak mutually."[12] We now understand the reason why in the seventeenth century, "an eventful period, full of conflicts," the newly formed symbolic algebraic analysis tended to be regarded as important and to be called 'mathesis universalis' or 'prima mathesis'. Namely, since the later Middle Ages it began to take the role of a rigorous way of thought, in addition to the role of practical numerical calculations and measures.

We have shown in Part I of this monograph that Descartes placed the new genuine authority of learning upon the certainty of pure mathematics, especially of 'mathesis universalis' up until the time of the *Regulae*. In Part II, we have

[10] W. J. Ong, *The Presence of the Word: Some Prolegomena for Cultural and Religious History* (New Haven, 1967), pp. 63–64. See also the same author's *Ramus, Method and the Decay of Dialogue: From the Art of Discourse to the Art of Reason* (Cambridge, Mass., 1958), pp. 307–318.

[11] Leibniz, "Dialogus" (August, 1677), *PS*, Bd. 7 (Berlin, 1890), p. 191, note; Loemker, p. 185, note 4. This is a note on the margin of the manuscript attached to the sentence: "Try, I pray, whether you can begin any arithmetical calculation without numerical signs." It is known that Kepler stated a similar sentence.

[12] *PS*, Bd. 7, p. 125: "ut non jam clamoribus rem agere necesse sit, ad alter alteri dicere possit: calculemus."

argued that after the *Regulae*, he believed he had discovered this authority in the certainty of "Cogito, ergo sum," and through the agent of this statement, in the certainty of pure mathematics. He thought that through such an intellectual struggle, the crisis of learning of his time could be overcome and that the new scheme of learning would be stabilized.

But what happened to this new authority of learning?

*

'Mathesis universalis' played an enormously important role in the formation of Descartes's early mathematical and philosophical thought and his later theory of mathematics still had a significant place in his attempt to found his novel scheme of learning. That is to say, Descartes's theory of mathematics is thought to be the root of the new authority of learning.

In concluding our discussions, let us consider here how we should evaluate Descartes's theory of mathematics especially from the viewpoint of the twentieth and twenty-first centuries. This may help us focus what we have stated in Chapter 8 much more sharply and grasp both the greatness and limitations of his mathematical thought and, then, of the modern scheme of learning which he strove to construct. Such an attempt is no doubt anachronistic. But hindsight might not be totally useless in shedding light on the past theory.

We will examine concisely Descartes's theory of mathematics together with Leibniz's and try to evaluate their theories by recourse to some metamathematical results obtained through foundational studies in the late nineteenth and early twentieth centuries. This approach has a certain significance, because some elements can be found in the philosophies of mathematics of Descartes and Leibniz corresponding to the philosophies of those who participated in the *Grundlagenstreit der Mathematik* of the early twentieth cnetury. Yvon Bélaval, for instance, seems inclined to call Descartes an intuitionist and Leibniz a formalist.[13] He has seen in the French philosopher a germ of the intuitionist philosophy of mathematics of Luitzen Egbertus Jan Brouwer and Hermann Weyl, and in the German critic of the former a similarity with the formalist David Hilbert. More precisely, it seems to me that Leibniz possessed not only a logicist element of Frege and a formalist element of Hilbert, but an insight to supersede both.

Now, we should mention first the contemporary concept of 'mathesis universalis' which was elaborated as a presupposition of the *Grundlagenstreit* of the early twentieth century. It appears to have been the founder of phenomenological philosophy Edmund Husserl who, at the beginning of the twentieth century, absorbed 'legitimately' the concept of 'mathesis universalis' of the seventeenth century mathematician-philosophers such as Descartes and Leibniz, reformu-

[13] Yvon Bélaval, *Leibniz critique de Descartes* (Paris, 1960), Chapitre 1 "Intuitionisme et formalisme," pp. 23–83.

lated that notion anew, and moreover attempted a deep epistemological consideration on it. His conception of 'mathesis universalis' was formulated by following basically Leibniz's idea, in his *Logical Investigations*, Book I *Prolegomena to Pure Logic* (1900), in *Ideas*, Book I (1913), *First Philosophy*, drafted during the years 1923–1924, successively.[14] And we find its maturest form in his *Formal and Transcendental Logic*, published in 1929. It is not too much to say that an aspect of Husserl's philosophical inquiries is, in fact, in his formulation of the concept of 'mathesis universalis' and its epistemonogical justification and critical understanding.

According to Part I "The Structure and the Sphere of Objective Formal Logic" of his *Formal and Transcendental Logic*, "The genuine discovery of the formal [mathematics] was first made, at the beginning of the modern age, by way of Viète's establishment of algebra—that is to say, by way of the reduction of the theory of numbers and quantities to a deductive technique—and then attained its pure sense through Leibniz, whose *mathesis universalis* obviously has thrust off completely every restriction to even the highest materially filled universality."[15] This was, according to Husserl, Leibniz's idea of 'mathesis universalis'. In propositional logic that idea was developed drastically by De Morgan and Boole in nineteenth century England.[16] Formal mathematics and formal logic witnessed a further remarkable development in the late nineteenth century through Riemann's and Cantor's theories of 'manifold' or 'set' (*Mannigfaltigkeit*) and Frege's symbolization or formalization of predicate logic of the first order. On this extension came Hilbert's program of formalism which formalized arithmetic and logic as a whole. This new formulation of 'mathesis universalis' in the twentieth century was in itself certainly important, but meant simply for Husserl a preliminary presupposition for a more essential epistemological consideration. His earnest task was challenged in Part II "Formal to Transcendental Logic" of *Formal and Transcendental Logic*, and *Cartesian Meditations*, lectures delivered originally in Paris in 1929.

The procedure which Husserl used was through and through Cartesian. His statement should be cited to show this: "How can evidence (*clara et distincta percepio*) claim to be more than a characteristic of consciousness within me? Aside from the (perhaps not so unimportant) exclusion of acceptance of the world as being, it is the Cartesian problem, which was supposed to be solved by divine *veracitas*."[17] As can be imagined from the cited sentences, Husserl looked

[14]E. Husserl, *Logische Untersuchungen*, Bd. 1: *Husserliana*, Bd. XVIII (Ch. 8, n. 155), §§ 60–61; *Ideen*, Bd. 1 (1913): *Husserliana*, III (1950), §§ 7, 10, & 59; *Erste Philosophie* (1923/24): *Husserliana*, VII (1956), pp. 319 & 349; *Husserliana*, VIII (1959), p. 249. In the last case referred to, Husserl significantly makes an effort to seek "Mathesis univertisalissima" rather than Leibniz's "Mathesis universalis."

[15]Husserl, *Formale und transzendentale Logik* (1929): *Husserliana*, XVII (1974), p. 84; *Formal and Transcendental Logic*, translated by Dorian Cairns (The Hague, 1969), p. 80.

[16]*Ibid.*, p. 78; The English translation, p. 74.

[17]*Cartesianische Meditationen*, *Husserliana*, Bd. I (1963), p. 116 (§ 40); *Cartesian Medi-*

for providing a novel radical "evidence" for 'mathesis universalis', dissatisfied with Descartes's attempt in the *Meditationes* which, by starting from the *cogito* and appealing to the divine veracity, had God guarantee 'clear and distict' mathematical truths.

Now, setting Husserl's reformulation of 'mathesis universalis' aside, two mathematico-philosophical consequences which were recognized among mathematicians of the early twentieth century seem to be closely relevant to our discussion here, for the notion of 'mathesis universalis' in Descartes and Leibniz certainly resembled the formulation of mathematics by mathematicians in the spirit of Hilbert.[18] Two mathematico-philosophical consequences follow: (1) One is the discussion on the consistency of formalized mathematical theories, which resulted in Kurt Gödel's mathematical proof of their incompleteness in 1930–1931. (2) The other is the recognition of competing mathematical theories, which first began to be recognized in Euclidean and non-Euclidean geometries and later in Cantorian and non-Cantorian (including intuitionistic)set theories. Present-day mathematicians will admit that in general there could be many more competitive mathematical theories added to the above two.

Let us consider the first consequence. In 1922 Hilbert proposed his program for the complete formalization and axiomatization of mathematical systems in a rigorous and mathematically refutable manner.[19] In a sense, it can be interpreted as the Hilbertian reformulation of Leibniz's and Husserl's 'mathesis universalis'. The ultimate end of Hilbert's program was to prove the consistency of elementary arithmetic and finally of mathematics as a whole. For this purpose he attempted to formalize both logic and mathematics. What Gödel demonstrated in 1930–1931 was a limitation on Hilbert's formalistic program of mathematics: For example, for a formal system ample enough to include elementary arithmetic, a consistency proof cannot be carried out without modes of inference that are not formalized in the system itself.[20] Such a metamathematical result will inevitably influence our evaluation of the ideas of 'mathesis

---

tations, translated by Dorian Cairns (The Hague, 1960), pp. 82–83.

[18]Interestingly one historian of philosophy regards Hilbert as a twentieth-century inheritor of the concept of 'mathesis universalis'. See Alexander S. Kohanski, *The Greek Mode of Thought in Modern Philosophy* (Rutherford/Madison/Teaneck, 1984), 36. "Descartes—Mathesis Universalis," pp. 124–135, esp. pp. 128–129.

[19]David Hilbert, "Neubegründung der Mathematik. Erste Mitteilung," *Gesammelte Abhandlungen*, Bd. III (Berlin/Heidelberg/New York, 1970), pp. 157–177. This paper first appeared in *Abhandlungen aus dem mathematischen Seminar der Hamburgischen Universität*, Bd. 1 (1922), pp. 157–177; "The New Grounding of Mathematics: First Report," in Paolo Mancosu, ed., *From Brouwer to Hilbert: The Debate on the Foundations of Mathematics in the 1920s* (New York/Oxford, 1998), pp. 198–214.

[20]Kurt Gödel, "Eine metamathematische Resultate uber Entscheidungsdefinitheit und Widerspruchsfreiheit," *Anzeiger der Akademie der Wissenchaft in Wien, Mathematisch-naturwissenschaftliche Klasse*, **67** (1930), p. 214; *Collected Works*, ed. by Solomon Feferman *et al.*, Vol. 1: *Publications 1929–1936* (1986), pp. 140–143. See also "Über formal unentscheidbare Sätze der *Principia mathematica* und verwandter Systeme I," *Monatshefte für Mathematik und Physik*, **38** (1931), pp. 173–193; *Collected Works*, Vol. 1, pp. 144–195.

universalis' in the seventeenth century, in particular, Leibniz's project to prove the axioms of mathematics.

The second mathematico-philosophical insight useful for reflecting upon the theories of mathematics in the seventeenth century is that doctrines of the uniqueness of mathematical reality tended to be discarded after the discovery of non-Euclidean geometries. Along with the axiomatization of various mathematical theories, of which a representative work was Hilbert's *Grundlagen der Geometrie*, published in 1899, mathematicians of the twentieth century have come to admit the multiplicity of mathematical theories, not only in geometry but also in arithmetic and analysis. Even in logic, various logical theories other than classical two-valued logic have been recognized: for example, many-valued logic, modal logic, and intuitionist logic. According to today's understanding of mathematics, mathematical principles should be taken not as self-evident truths but as mere hypotheses which might lead to fruitful theorems because of the competitiveness of mathematical theories. In Paul Bernays's insightful formulation, mathematics is understood as a science of "possible idealized structures."[21] Leibniz thought that the historical world is a 'possible world' (*monde possible; mundus possibilis*) in which we are permitted to select among various choices, in contrast with the necessary world of logical and mathematical 'eternal truths'. Mathematics, now, allows various 'possible worlds' within it.

Let us reflect upon Descartes's and Leibniz's theories of mathematics in the light of these metamathematical consequences. First of all, Descartes's creation doctrine *per se*, which seems to have been thought out with deliberation under pressure from the skeptics' assault on mathematics, must be reevaluated. The structure of such a theory may allow the existence of mutually competitive axioms, e. g. the postulate of parallel lines and its negation in various forms. It seems no accident that some modern philosophers who understand the compatible but mutually exclusive mathematical theories assess the Cartesian creation doctrine positively.

Leibniz, on the other hand, could not appreciate Descartes's methodological doubt toward mathematical principles and demonstrations, as we have seen in Section 2 of the previous chapter. Instead of adopting his predecessor's manner of philosophizing, he devised a new way: 'demonstration of mathematical principles' by splitting and reducing them finally to definitions and identities. It may be true that Leibniz anticipated to some extent the conception of formal proof that has become dominant in the twentieth century.[22] But the mathemat-

---

[21]Paul Bernays, "Concerning Rationality," in P. Schilpp, ed., *The Philosophy of Karl Popper* (1974), pp. 603–605. On competitive theories in mathematics, see Stephan Körner, "On the Relevance of Post-Gödelian Mathematics to Philosophy," in Imre Lakatos, ed., *Problems in the Philosophy of Mathematics* (Amsterdam, 1967), pp. 118–137, containing discussions with Gert H. Müller and Y. Bar-Hillel. Körner calls for constructing the philosophy of competitive mathematical theories in response to the development of foundational studies, especially since Gödel.

[22]Ian Hacking, "Proof and Eternal Truths: Descartes and Leibniz" (Ch. 5, n. 222), pp. 169–

ically ordered program of formal proof, i. e. Hilbert's metamathematics, as we have seen above, was at least partially negated by Gödel's results, which limit the Leibnizian ideal of *characteristica universalis*. Leibniz could not anticipate the labyrinth lying behind his concept of formal proof. Nevertheless here it should be promptly pointed out that his idea of proving mathematical principles itself suggests that he did not simply accept existing statements "which are called axioms" as self-evident truths, and that in later years he considered mathematical truths only to be 'conditional truths'. He was not inclined to accept them without demonstration and, therefore, in a sense he can be said to have been critical. In this and in this only he can be proud of himself as a pioneer.

Let us consider more carefully the relationship between what Leibniz and Hilbert aimed at in their conceptions of formal proof and Gödel's limitative theorems. Here we should draw attention to the fact that the impossibility of proving the unqualified consistency of mathematical systems was not pointed out by Gödel in 1930–31 for the first time. In the middle of the *Grundlagenstreit*, Brouwer, for example, criticized the formalist conception of metamathematics on the ground that the proof of the consistency of mathematics contains in principle a difficulty. In a lecture of 1927, he insisted that the formalists must recognize "the fact that the (contentual) justification of formalistic mathematics by means of the proof of its consistency contains a vicious circle (*circulus vitiosus*)."[23] His reason was that "this justification rests upon the (contentual) correctness of the proposition that from the consistency of a proposition the correctness of the proposition follows." Brouwer asserted that any procedure to justify a mathematical theory must inevitably contain at least one primitive mathematical presupposition which is considered to be contentually true. It is known that Henri Poincaré, another intuitionist mathematician, also claimed the necessity of a primitive indemonstrable mathematical rule for mathematical activities, taking as an example of mathematical induction ("raisonnement par recurrence" in Poincaré's terminology).[24]

180.

[23]L. E. J. Brouwer, "Intuitionistische Betrachtungen über den Formalismus," *Koninklijke Akademie van Wetenschappen te Amsterdam*, Proc. 31 (1928); *Collected Works*, Vol. 1: *Philosophy and Foundations of Mathematics*, ed. by A. Heyting (Amsterdam, 1975), p. 410; English translation in Jean van Heijenoort, ed., *From Frege to Gödel: A Source Book in Mathematical Logic*, 1879–1931 (Cambridge, Mass., 1967), p. 491; "Intuitionist Reflections on Formalism," in Mancosu, *Op. cit.*(n. 19), p. 41.

[24]Poincaré claims that mathematical induction cannot be translated into any merely logical procedure ; it becomes a rule proper to mathematics. Thus he states: "We shall always be brought to a full stop—we shall always come to an indemonstrable axiom, which will at bottom be but the position we had to prove translated into another language. We cannot therefore escape the conclusion that the rule of reasoning by recurrence is irreducible to the principle of contradiction." Poincaré, "On the Nature of Mathematical Reasoning," in Benacerraf and Putnam, eds., *Philosophy of Mathematics* (Ch. 8, n. 148), p. 400. Poincaré's insight was later fully supported by H. Weyl, "Diskussionsbemerkungen zu dem zweiten Hilbertschen Vortrag über die Grundlagen der Mathematik," *Abhandlungen aus dem math-*

What is the difference between Gödel's results and the intuitionists' criticism of the formalist idea of metamathematics? What is interesting in Gödel's mathematical theorems is that his demonstration of the incompleteness of formalized mathematical systems is *informally mathematical*, while relying totally upon *formalized* logic and mathematics. In other words, Gödel revealed not only a limit of the Hilbertian brand of formalism but also how the given formal systems are limited mathematically and structually. It was not simply philosophical but contentually mathematical and showed the internal limitations of formalism.[25] Leibniz's and Hilbert's project, thus, was not entirely in vain. Ironically, their program of formal proof certainly made a contribution to showing its limitedness.

This can be more clearly understood when we search for intellectual origins of Gödel's incompleteness theorems. The historical process as to how Gödel discovered these theorems has been recorded by Hao Wang, who communicated personally with Gödel in the latter's last years.[26] Gödel was suggested the first clue for his incompleteness theorems by Brouwer's lecture "Mathematik, Wissenschaft und Sprache," delivered on March 10, 1928 in Vienna.[27] Afterwards he continued solid mathematical-philosophical inquiries persistently, and proclaimed his mathematical results in 1930–31. Thus although he received an intellectual impact by the intuitionists like Brouwer, the mathematician-philosopher Gödel admired most was none other than Leibniz. His admiration of Leibniz went to the extent to which he conceived an obsession that someone might interfere with publishing the manuscripts of Leibniz and even attempt to destroy them. To Karl Menger's question, "Who had an interest in destroying Leibniz' writings?", Gödel repied, "Naurally those people who do not want man to become more intelligent."[28] Both Leibniz and Gödel appreciated the

---

ematischen Seminar der Hamburgischen Universität, **6** (1928): *Gesammelte Abhandlungen*, Bd. III (Berlin/Heidelberg/New York, 1968), p. 147; English translation in Van Heijenoort, *Op. cit.*(n. 23), p. 482.

[25]For the philosophical significance of Gödel's incompleteness theorems, see, among others, Jean Ladrière, *Les Limitations internes des formalismes: Étude sur la signification du théorème de Gödel et des théorèmes apparentés dans la théorie des fondements des mathématiques* (Louvain/Paris, 1957), Chap. X: "Suggestions philosophiques," pp. 397–444; Michael Dummett, "The Philosophical Significance of Gödel's Theorem," *Ratio*, **5** (1963), pp. 140–155, & Alain Hazan, "Kurt Gödel, ou le pur formalisme comme Odysée de l'institution mathématique et du subject de la connaissance," *Revue de metaphysique et de morale*, 82e année (1977), pp. 88–107.

[26]See Hao Wang, *Reflections on Kurt Gödel* (Cambridge, Mass./ London, 1987), pp. 50, 80–91, *A Logical Journey: From Gödel to Philosophy* (Cambridge, Mass./London, 1996), pp. 69–71. Cf. John W. Dawson, Jr., *Logical Dilemmas: The Life and Work of Kurt Gödel* (Wellesley, Mass., 1997), pp. 55–59.

[27]L. E. J. Brouwer, "Mathematik, Wissenschaft und Sprache," *Monatshefte für Mathematik und Physik*, **36** (1929), pp. 153–164: *Collected Works*, Vol. 1, pp. 417–428; "Mathematics, Science, and Language," in Mancosu, *Op. cit.* (n. 19), pp. 45–53.

[28]Karl Menger, *Reminiscences of the Vienna Circle and the Mathematical Colloquium*, edited by Louise Golland *et al.* (Dordrecht/Boston/London, 1994), "Memoires of Kurt Gödel," p. 223.

significance of formalizations sufficently, and at the same time they never forgot the importance of intuitive comprehensions. It is very probable that Gödel highly admired Leibniz's criticism of Descartes's conception 'clear and distinct', in particular the 1684 tract "Meditations on Knowledge, Truth, Ideas." Leibniz stated there, "Adequate knowledge is either symbolic or intuitive: and if knowledge were, at the same time, both adequate and intuitive, it would be absolutely perfect" (see p. 407). This thought accords well with the procedures of proof which are formal and intuitive, namely non-formal, at the same time in Gödel's incomleteness theorems on symbolized or formalized mathematical theories. Naturally we should not forget that there was a historical accumulation of mathematical thoughts for about two hundreds and fifty years between Leibniz's and Gödel's understandings of the *characteristica universalis* and the intuitive mathematical procedures.

Of course, we do not necessarily want to claim that Descartes and Leibniz, especially the former, foresaw today's consequences in the philosophy of mathematics. Descartes had an incomparable insight insofar as and only insofar as he responded seriously to the doubt commonly cast by the skeptics upon the self-evident truth of mathematical principles and demonstrations. And Leibniz conceived the importance of both symbolic thinking and its intuitive complement.

Skeptical thought does not only work to limit or qualify mathematical activities. It may sometimes temper immoderate enthusiasm, but may often suggest to mathematical practitioners the right direction of research and encourage them in attaining truly fertile and meaningful results. The latter point is what Lakatos tried to emphasize by reviving skeptical arguments in the midst of the flowering of formalist dogmatists.[29]

Mathematician-philosophers in the seventeenth century in general were well informed about the deep philosophical insights of the ancients. Descartes and Leibniz were not exceptions. Particularly, the philosopher of "Cogito, ergo sum" was more subtle and deeper than he has usually been considered to be.

Such philosophical subtleties notwithstanding, as we have suggested in Chapter 8, § 2, the course which both Descartes and Leibniz ultimately adopted was directed toward the automatization of mathematical disciplines. When all was said and done, on making his veracious God in the *Meditationes* warrant the verity of existing determinate mathematical statements which are considered to be sufficiently clear and distinct, the skeptic Descartes has become the dogmatist Descartes. At this point he has suddenly lost his brilliance. It was quite ingenious that he discovered the *cogito* by carrying out the analytical method thoroughly. But there existed a obvious logical jump in his desire to guarantee the truth of determinate propositions in mathematics and of mathematical

[29] I. Lakatos, *Proofs and Refutations* (Ch. 8, n. 186), esp. Author's Introduction, pp. 1–5*Idem*, "Infinite Regress and Foundations of Mathematics," in *Mathematics, Science and Epistemology: Philosophical Papers*, Vol. 2 (Cambridge, 1978), pp. 3–23.

physics through the synthetic method conversely from the *cogito*. Thus, in the end he restored to mathematics a privileged epistemological status. It seems to have been the philosophy of mathematics of the dogmatist Descartes that most of his successors inherited. For Leibniz, too, mathematics was after all an autonomous discipline. He put it in 1694: "It seems to me that light and certainty are more needed in metaphysics than in mathematics itself, because mathematical matters carry their own tests and verification with them, this being the strongest reason for success in mathematics."[30] For the seventeenth-century philosophers, mathematics continued to serve as the ideal paradigm discipline of certainty. There was an inclination to view their 'mathesis universalis', in turn, as the principal core of the mathematical disciplines. In this sense, 'mathesis universalis' meant for them the science of the sciences, the genuine authority of the new scheme of learning.

But here we may be permitted to ask: Can 'mathesis universalis' stand by itself? In Leibniz's words, does it carry its "own tests and verification" with it? Can it serve as the foundation of mathematics? We may be able to suggest an answer to this question by extracting a crucial and exemplary exchange between Hilbert and his able disciple Weyl in 1927, during the course of the *Grundlagenstreit.*

For the formalist Hilbert, before the discovery of Gödel's incompleteness theorems, the formalized system of elementary arithmetic and then mathematics as a whole was regarded as forming a closed and self-consistent world. Mathematical activities, according to Hilbert, are essentially purely intellectual and can be carried out in an autonomous manner. For him it is the consistency proof that corroborates this characterization of mathematics.[31]

Weyl sensed a naivite in Hilbert's assertion.[32] He criticized his teacher's notion of mathematics. How can we obtain axioms to be used in a formalization? Is consistency a sufficient condition for our meaningful mathematical activities?

Weyl asked and answered these problems as follows: If one tries to prove the consistency of a mathematical theory, that theory has to be provided with axioms which are considered to be valid at least temporarily. But where do these axioms come from? Before formalizing the axiomatic system something that is rooted in our judgment and ultimately in our history must be presupposed. Hilbert stated nothing about such a presupposition which may guarantee meaningful mathematical activities. What the proponent of formalism particularly pays attention to is the consistency of the proposed theory. In sum, according to Weyl, Hilbert's world of mathemaics is turned upside down. In

---

[30]Leibniz, "De primae philosophiae Emendatione, et de Notione Substantiae," *PS*, Bd. 4, p. 469; Loemker, p. 433.

[31]D. Hilbert, "Die Grundlagen der Mathematik," *Abhandlungen aus dem mathematischen Seminar der Hamburgischen Universität*, **6** (1928), pp. 65–85; English translation in Van Heijenoort, *Op. cit.* (n. 22), pp. 464–479.

[32]H. Weyl, *Op. cit.* (n. 24), *Gasammelte Abhandlungen*, III, pp. 147–149; English translation, pp. 480–484.

reality there must be, first of all, our life-world, and then follows contentual mathematics and finally formal mathematics. What Hilbert attempts to do is to build the world of mathematics on the 'artificially' formalized mathematics, which is absurd in Weyl's opinion.

The seventeenth century holders of 'mathesis universalis' might not have possessed so strict a dogmatic formulation of autonomous mathematical activities as Hilbert. But, as was pointed out above, it is also certain that in Descartes's and Leibniz's conceptions of 'mathesis universalis' there existed a certain resemblance to Hilbert's formulation. Hilbert sought the firm formal ground on which mathematics as a whole should be constructed. For the Descartes in the *Regulae*, 'mathesis universalis' must have been the same firm foundation for constructing general certain knowledge as Hilbert's formalized system of arithmetic. Similar assertions can be said with some qualification about the Descartes in the *Meditationes* and the Leibniz requiring 'demonstrations of axioms'. In 1927, Weyl denied such a conception of mathematics, as we have just observed.

In his paper under the provocative title "Mathematics without Foundations,"[33] Hilary Putnam argues about whether or not there is a transcendent foundation of mathematics. He suspects the 'foundational' view of mathematical knowledge from the facts that there can be competitive mathematical theories and that an absolute consistency proof for classical mathematics is impossible. He insists that there are no transcendent foundations of mathematics, strictly speaking.

If in the seventeenth century 'mathesis universalis' meant a certain foundation for mathematics, that notion in the unqualified sense would also become a mere dream. This does not necessarily imply that the mathematicians' endeavors to build the system of 'mathesis universalis' as an 'as universal as possible' *veritas rationum* (in Gueroult's words) were all in vain. It simply means that 'mathesis universalis' does not constitute a closed world and is not unqualifiedly universal.

Furthermore, the mature Ludwig Wittgenstein's speculations about certainty seem to open a significant historical perspective in the general history of ideas. One of the reasons why Descartes advocated the traditional concept of 'mathesis universalis' in his *Regulae* was that it was regarded as the ideal paradigm for certain knowledge. For him, having the same degree of certainty as pure mathematics was the goal to be attained by any knowledge that may be acquired.

In *On Certainty*, Wittgenstein provides a penetrating insight as to what should happen to the certainty of mathematics. What he suggests, reflecting upon post-Gödelian mathematics, is that there is no notion like transcendent

[33] H. Putnam, "Mathematics without Foundations," in Benacerraf and Putnam, eds, *Op. cit.* (Ch. 8, n. 170), p. 303; *Mathematics, Matter and Method: Philosophical Papers*, Vol. 1 (Cambridge, $_2$1979), p. 51.

certainty and that mathematical certainty does not necessarily have a special priviledge. He does not want to claim that mathematical statements are uncertain. What he indicates is that mathematical knowledge has a different epistemological status than Descartes thought. Mathematics is concerned with not reality itself but an aspect of it (i. e. act, or practice) and may be as fallible as any other branch of knowledge. Wittgenstein writes, for example: "If you tried to doubt everything you would not get as far as doubting anything. The game of doubting itself presupposes certainty";[34] "At the foundation of well-founded belief lies belief that is not founded."[35] In these two fragments, Wittgenstein seems to us to try to assess the greatness and limitations of the founder of modern philosophy, Descartes.

Descartes discovered, he himself claimed, the truth of the *cogito*, which even the most stubborn skeptic cannot refute. But he continued to insist after all that the certainty of mathematics is extraordinary and different from that of other knowledge. What Wittgenstein desires to affirm is that mathematical certainty is not transcendent and may differ from other kinds of certainty not in kind but in degree at most. According to him, like other fields of knowledge, mathematical theories may always be fallible and their certainty is finally ascertained in practices.

Here let us reexamine Husserl's philosophical program. Husserl, among others, in the early twentieth century, an age in which the problems on the foundations of mathematics were seriously debated, restored the 'mathesis universalis' of Descartes and Leibniz in the seventeenth century. His formulation of 'mathesis universalis' and attempt at providing an epistemological meaning for it has been followed up to our own time. *La Philosphie de l'algèbre*, published by Jules Vuillemin in 1962, is an example. In this monograph Vuillemin by reconstructing Lagrange's theory of equations, Gauß's fundamental theorems of algebra, Galois's theory of equations through the theory of groups, Klein's extension of it to geometrical theories, Lie's application of it to the theory of linear differential equations, and so on acquired the epistemological significance of the modern concept of 'mathesis universalis'.[36]

Now, by understanding the 'mathesis universalis' as specifically formalized mathematics, Husserl searched for how its 'evidence was originated'. He became to notice that the *cogito* of solipsistic monads is not clearly sufficient for the subjective bearer of its 'evidence'. Having recognized its definitive insufficiency, in the Fifth Meditation of his *Cartesian Meditations* he referred to "community

[34]Ludwig Wittgenstein, *On Certainty*, ed. by G. E. M. Anscombe and G. H. von Wright, tr. by Denis Paul and G. E. M. Anscombe (Oxford, 1969). Fragment number 115.

[35]*Ibid.*, Fragment number, 253.

[36]Jules Vuillemin, *La Philosophie de l'algèbre*, t. 1: *Recherches sur quelques concepts et méthodes de l'algèbre moderne* (Paris, 1962; 2nd ed. 1993). Unfortunately, a successive volume was not published. The title of its "Conclusion" is significant: "La mathématique universelle," pp. 465–518. Vuillemin seems to have succeeded Hussearl's attempt on the philosophy of mathematics, at least in part.

of monads" and designated it as "transcendental intersubjectivity"[37] *Cartesian Meditations* by Husserl himself has not been completed, but in Sixth Meditation written by his disciple Eugen Fink, suggested by Husserl, we read the following insightful description: "Alongside the question of *egological* wholeness, there enter in here also, as tasks for constructive phenomenology, all the problems pertaining to the form of wholeness of the *intersubjective community of monads*, and above all the totality-form of monadic *history* (*monadische* Historie)."[38] In other words, the community with others and the inquiry of historicity were required for the inquiry of 'evidence'. It was a very significant recognition for Husserl of the 1920s that the radical historicity of the object and subjet of the sciences, namely passive and active historicity, is crucial.[39] It was not accidental that philosophical apostles of Husserl after this period shifted their field of study to the history of mathematics. This shift was exemplified by Oscar Becker, a student of the history of ancient Greek mathematics, Dietrich Mahnke, a researcher of Leibniz's mathematical thought, and Jakob Klein, who wrote a monumental monograph on Greek logistic and the formation of modern algebra.

In general, the first half of the twentieth century is characterized as an age of worldwide conflicts as the seventeenth century, for it witnessed unprecedented battles of World Wars I and II.[40] It seems to me that the recognition of Husserl on the intersubjectivity and historicity of the sciences contains elements we should develop.

Once again, let us return to the age of Descartes. He demonstrated an incomparable ability and depth in devising the philosophy of "Cogito, ergo sum" and in forming the tradition of analytic mathematics and of mathematical physics. But his insightful program was not totally successful. His belief does not necessarily seem to have been well founded. What we have attempted to

[37] *Cartesianische Meditationen*: *Ibid.* (n. 17), p. 158; The Engish translation, p. 130.

[38] Eugen Fink, *VI. Cartesianische Mediationen*, Teil 1: *Husserliana, Dukumente*, Bd. II/1 (Dordrecht/Boston/London, 1988), pp. 70–71; Sixth *Cartesian Meditation: The Idea of a Transcendental Theory of Method*, with Textual Notations by Edmund Husserl, translated with an Introduction by Ronald Bruzina (Bloomington/Indianapolis, $_{19}$1995), p. 63.

[39] See, among others, David Carr, *Phenomenology and the Problem of History* (Evanston, 1974) & Elisabeth Ströker, *The Husserlian Foundations of Science* (Drodrecht/Boston/London, 1997), Ch. 7: "The Question of History and "History" in Husserl's Intentional Analysis," pp. 207–227. Ströker states as follows: As early as 1921 he [Husserl] stated—and this is in my knowledge, the first passage of this sort—that is "...every ego has its own ***history*** and exists only as a subject of history, its own history. And every communicative community of absolute egos, of absolute subjectivities—in full concretion to which belongs the constitution of the world—has its "***passive***" and "***active***" ***history*** and exists in this history only. ***History is the great fact*** (Faktum) ***of absolute being***; and the final questions...are totally the same with the questions about the absolute sense of history." (The emphases are by Husserl). The quotation from Husseral is *Erste Philosophie* (1923/24), *Husserliana*, VIII, p. 506.

[40] See Stephen Toulmin, *Cosmopolis: The Hidden Agenda of Modernity* (New York, 1990), pp. 66 & 152–153. See also, *Idem, Return to Reason* (Cambridge, Mass./London, 2001).

do is not to promote a way of thinking which is simply probable or uncertain, but rather to explain the roots of his unfounded belief in the context of an 'obsession' in an age of conflicts.

Descartes considered his program of learning to possess a solid and stable certainty, and to serve as a genuine authority. He did not attempt his intellectual revolution *in vacuo*. His overall attempt to reform the world of learning seems to have been first possible as an effort to overcome the intellectual and social crisis of his time. What he created as the result of his endeavors was conditioned into a specific form of mathematics and a specific form of philosophy. How this revolution took place can only be described and analyzed in a critical historical manner. For example, as we have suggested toward the end of Chapter 7, the Hindu-Arabic art of computaion introduced into the European world from India via the Islamic world had an impact on Descates's way of thought. And the influence of the new medium of the printing press upon the new algebra should be scrutinized in further detail.[41] We live in an age allowing this examination, for with the advent of electronic computers designed mainly by John von Neumann and Alan Turing in the middle of the twentieth century, a new form of mathematics began to take root. This form of mathematics is different from symbolic algebra, just as mathematics was new in the era after Gutenberg. It is expected to completely transform mathematics.[42] To put it in the words following Hegel's *Philosophie des Rechts*, a twilight has already occurred in which Minerva's owl of a new mathematics, different from the one Descartes contributed to, should take flight. Wittgenstein proclaimed in his posthumous *Remarks on the Foundations of Mathematics*: "Even 500 years ago a philosophy of mathematics was possible, a philosophy of what mathematics was then."[43] If Wittgenstein's observation is correct, a philosophy of mathamtics in accord with its age will necessarily be created in the twenty-first century.

Descartes created algebraically analytic mathematics and a philosophy associated with it after the Renaissance, in which ancient Greek mathematics and philosophy were almost totally reborn. The world of two hundred and fifty years before he emerged was a world 'before European hegemony' in the words of an excellent scholar of social history.[44] At that time, the European world acquired the art of algebra from the Islamic world, and later began to develop its own form, along with the trend of total measurement. Descartes placed a theoretical costume on the disciple, which had been transmitted with the outlandish name of algebra, and he called it by the elegant name of 'mathesis

[41] A discussion in Elizabeth L. Eisenstein, *The Printing Press as an Agent of Change* (Cambridge, 1979), esp. pp. 520–566, may be helpful.

[42] Bjorn Engquist and Wilfried Schmid, eds., *Mathematics Unlimited—2001 and Beyond*, 2 Vols. (Berlin/Heidelberg/New York, 2001).

[43] L. Wittgenstein, *Remarks on the Foundations of Mathematics*, ed. by G. H. von Wright, R. Rhees, and G. E. M. Anscombe (Oxford, 1967), p. 157.

[44] Janet Abu-Lughod, *Before European Hegemony: The World System A. D. 1250–1350* (New York, 1989).

universalis'. He speculated on it epistemologically. It was Descartes's task in the history of mathematics to institute algebra as a full-fledged theoretical science in the mathematical disciplines. By this intellectual revolution, modern European mathematics, centered on algebraic analysis, became the mathematics 'after European hegemony,' so to speak.

Even his contemporary thinkers and the next few generations did not all agree with Descartes's thought. It is well-known that Pascal admired insightful ideas by Descartes. However, he reportedly did not join the current of Descartes's algebraically analytic mathematics, and justified his own mathematical natural philosophy by labeling, "Cartesian thought 'the novel on nature, a bit like to the story of Don Quijote...'."[45] Nevertheless, we should not be satisfied with rejecting Descartes's thought on the basis of a witty remark from Pascal. Here we mention only one significant thinker who tried to transcend the problems of his philosophy. Giambattista Vico, who is said to have once been attracted to his thought, resisted the Cartesian metaphysics and the exclusive domination of the method of algebraic analysis over mathematics and the mathematical method over physics.

Vico cast a doubt upon the philosophical trustworthiness of Descartes's first truth "Cogito, ergo sum." Vico asked what the so-called unqualified certainty of that first truth is concerned with. He ingeniously answered that it is concerned with consciousness (*conscientia*), not with knowledge (*scientia*) at all.[46] "Cogito, ergo sum" certainly may guarantee the existence of the 'ego cogitans', but it has nothing to do with the meaningful statement of any serious mathematical and scientific knowledge.

In the section entitled "De analysi" in his inaugural oration *De notri temporis studiorum ratione*, delivered at the Royal University of Naples in 1708, Vico contended that Cartesian algebraic analysis might at times deprive students of mathematical imagination and the ingenuity needed for tackling problems themselves, which traditional Greek geometry implanted in their mind. We should not depend, according to him, totally upon algebraic analysis, since to resort to it at any time when solving mathematical problems is like invoking a 'deus ex machina' to resolve a complicated drama.[47] We should not encourage youth to take the royal road!

---

[45]Pascal, *Œuvere complètes*, présentation et notes de Louis Lafumas (Paris, 1963), p. 641: "Feu M. Pascal appelait la philosophie cartésienne *le roman de la nature, semble à peu prés à l'histoire de Dom Quichot...* (Rapporte par Menjot.)"

[46]Vico, *De Antiquissima Italorum Sapientia, Liber primus: Metaphysica* (1710), in *Opere di G. B. Vico*, Vol. 1, ed. Giovanni Gentile e Fausta Nicolini (Bari, 1914), pp. 138–140; Vico, *Selected Writings*, edited and translated by Leon Pompa (Cambridge, 1982), pp. 56–59.

[47]Vico, *De nostri temporis studiorum ratione* (1709), *Opere*, Vol. 1, pp. 87–88; *On the Study Methods of Our Time*, translated with an Introduction and Notes, by Elio Gianturco (Indianapolis/New York/Kansas City, 1965), pp. 26–30. Vico could have been influenced by his friend Paolo Mattia Doria, an eager advocate of synthetic geometry, and perhaps, by Thomas Hobbes. See David Lachterman, "Vico, Doria e la geometria sintetica," *Bolletino del Centro di Studi Vichiani*, X (1980), pp. 10–35.

Vico, in another section of the same oration, warned against taking the mathematical approach as the sole method of studying nature, i. e. physics: "We demonstrate geometrical things, because we make them; if we could demonstrate physical things, we would make them."[48] We may be able to 'invent' the mathematical. But all we can do with respect to physical things is to observe or manipulate them. We, finite, incomplete and essentially historical beings, can never create them *ex nihilo.*

Vico's insights extend beyond his philosophical writngs. He, one of the founders of modern historical science, claimed that historical knowledge is not less certain than mathematical knowledge. According to Vico, "*certum* (certain) means two things: on the one hand, what is ascertained and beyond doubt, and on the other hand, what is particular as opposed to what is common, as if what is particular were certain, but what is common were doubtful."[49] The science provided with these two sorts of certainty is historical knowledge about human beings and what they have produced. With respect to certainty, then, historical science is a better science than mathematics. We can be certain in our historical knowledge as well as mathematics, for we make both. But in the second sense of certainty, the science of history is more certain than mathematics because the former is more concrete and determinate, and accordingly closer to reality. In sum, Vico began to reexamine the intellectual authority which Descartes thought to be unquestionable.

Insight like Vico's is recognized by our contemporary advocator of 'historical philosophy of science', Thomas S. Kuhn.[50] A science which we should

[48] *Ibid.*, p. 85: "[G]eometria demonstramus, quia facimus; si physica demonstrare possemus, faceremus." See also Gianturco's translation, p. 23. This Vichian thesis may be difficult to understand. But it will become easier for us to comprehend it if we are reminded of Wittgenstein's following statements: "The mathematician is an inventor, not a discoverer." ***Remarks on the Foundations of Mathematics***, revised edition by G. H. von Wright, R. Rhees and G. E. M. Anscombe (Cambridge, Mass., 1983), p. 99; "[T]he mathematician is a man who builds new roads, or invents new ways of thinking." ***Wittgenstein's Lectures on the Foundations of Mathematics, Cambridge, 1939***, ed. by Cora Diamond (Hassocks, Sussex, 1976), p. 139. On Vico's statement, see David R. Lachterman, "Mathematics and Nominalism in Vico's *Liber Metaphysicus*," in Stephan Otto and Hermut Viechtbauer, eds., ***Sachkommentar zu Giambattista Vicos Liber Metaphysicus*** (München, 1985), pp. 47–85. Vico had predecessors on this insight. One of them was Thomas Hobbes. See his ***Six Lessons to Professors of Mathematics*** (1656): ***The English Works of Thomas Hobbes***, ed. by William Molesworth, Vol. 7 (London, 1845), pp. 183–184.

[49] Vico, ***De antiquissima Italorum sapienta*** (n. 46), p. 147; Pompa's translation, p. 63.

[50] Thomas S. Kuhn, ***The Road Since Structure*** (Ch. 5, n. 257), pp. 90, 95, & 105–120. Kuhn's historical approach to philosophy of science is contrasted with 'logical philosophy of science' of which a representative philosopher was Carnap. See Donald Gillies, ***Philosophy of Science in the Twentieth Century: Four Central Themes*** (Oxford/Cambrige, Mass., 1993), p. 68. For Kuhn's philosophy of science, a historical approach is essential. On this point, I basically agree with Wes Sharrock and Rupert Read, ***Kuhn: Philosopher of Scientific Revolution*** (Cambridge, 2002). Compare with Toulmin's call to "return to reason" in his ***Return to Reason*** (n. 40). Toulmin's "reason" seems to me to be, especially, historical reason. But, such a call was already in George Sarton, ***The History of Science and the New Humanism*** (New York, 1956: First Published in 1931). Sarton states in "The Teaching of the History of Science (Second

be able to construct may be nothing other than 'historical philosophy of mathematics'. The present lines are intended to contribute in this direction.

By his novel formulation of the method of algebraic analysis and his promotion of the idea of 'mathesis universalis', Descartes became overwhelmingly influential among modern mathematicians and philosophers. He sufficiently demonstrated his ingenuity in both mathematics and philosophy by his success in solving Pappus's problem and by challenging the skeptics' attack against human knowledge. But he made little appeal to historical knowledge. To supplement or reform his mathematical thought, we have committed ourselves to a historical study of it. If we are to alter the modern scheme of mathematical activity, we have to reconstruct how it was formed in a historical manner. Descartes's mathematical thought was in many aspects conditioned by the time in which it was born. To point this out in a critical manner is necessary for arrriving at a fresh perspective on mathematical activities. Our historical study has sought to reexamine the intellectual authority which Descartes thought to be genuine and which he tried to protect. This task seems to be suitable for a servant of Clio, a successor of Si-ma Qian, Grand Historian of ancient China, and a student of Giambattista Vico.

Article)," *Isis*, **4** (1921–22), p. 249: "The historian of science will constitute the vanguard of the New Humanism."

Bibliography

(Some items which are in the notes but are inessential have been omitted from this bibliography; on the other hand, some others which don't appear in the notes but have influenced this book have been included in it.)

I. PRIMARY SOURCES

[1] Agrippa, Heinrich Cornelius, von Nettesheim. *In artem brevem Raymundi Lullij commentaria*, in *Operum pars posterior* (Lyons, 1600; Hildesheim: Olms, 1970), pp. 319–451.

[2] Albertus Magnus. *Opera omnia*, Tomus XIV, Pars I: *Metaphysica libri quinque priores*, ed. B. Geyer (Münster: Monasterium Westfalorum in Aedibus Aschendorff, 1960); Pars II: *Metaphysica libri VI–XIII* (Münster, 1964).

[3] Alexander of Aphrodisias. *In Aristotelis Metaphysica commentaria*, ed. Michael Hayduck, *Commentaria in Aristotelem Graeca*, Vol. 1 (Berlin, 1891; Berlin/New York: Walter de Gruyter, 1956).

[4] *Idem. Commentaria in duodecim Aristotelis libros de prima philosophia*, Interprete Ioanne Genesio Sepulveda (Venice: Apud Hieronymum Scottum, 1561).

[5] Apollonius of Perga. *Apollonii Pergaei Conicorum libri quattuor*, ed. F. Commandino (Bologna: Ex Officina A. Benatii, 1566).

[6] *Idem. Les Coniques d'Apollonius de Perge*, tr. par Paul Ver Eecke, Nouveau Tirage (Paris: A. Blanchard, 1963).

[7] *Idem. Apollonii Pergaei quae Graece exstant cum commentariis antiquiis*, ed. et latine interpretatus J. L. Heiberg, Vol. 1 (Leibzig: Teubner, 1891); Vol. 2 (Leipzig: Teubner, 1893).

[8] Aquinas, Thomas. *The Division and Methods of the Sciences: Questions V and VI of His Commentary on the De Trinitate of Boethius*, tr. with Introduction and Notes by Armand Maurer (Toronto: The Pontifical Institute of Medieval Studies, $_3$1963).

[9] *Idem. In duodecim libros Metaphysicorum Aristotelis expositio* (Turin: Marietti, 1964).

[10] *Idem. Opera omnia, Vol. 4, Commentaria in Aristotelem et alios: In libro Metaphysicorum; In librum Boethii de Trinitate* (Stuttgart/Bad Cannstatt: Frommann-Holzboog, 1980).

[11] Archimedes of Syracuse. *Opera omnia cum commentariis Eutocii*, ed. J. L. Heiberg, 3 Vols. (Leibzig: Teubner, $_2$1910–1915).

[12] Aristotle. *Analytica posterior*, in *Opera*, Vol. 1, ed. Immmanuel Bekker, Academia regia Boroussica (Berlin, 1831; Berlin/New York: W. de Gruyter, 1960).

[13] *Idem. Metaphysica*, in *Opera*, Vol. 2, ed. I. Bekker (Berlin, 1831: Berlin/New York: W. de Gruyter, 1960).

[14] *Idem. Aristoteles Latinus*, XXV-2: *Metaphysica, Translatio anonyma sive 'Media'*, ed. Gudrum Vuillemin-Diem (Leiden: E. J. Brill, 1976).

[15] *Idem. The Complete Works of Aristotle: The Revised Oxford Translation*, ed. Jonathan Barnes, 2 Vols. (Princeton: Princeton University Press, 1984).

[16] Asclepius of Tralles. *In Aristolelis Metaphysicorum libros* A–Z *commentaria*, ed. Michael Hayduck, *Commentaria in Aristolelem Graeca*, Vol. 6, Pars 2 (Berlin, 1888; Berlin/New York: W. de Gruyter, 1960).

[17] *Idem.* "Commentary to Nicomachus' Introduction to Arithmetic," ed. with an Introduction and Notes by Leonardo Tarán, *Transactions of the American Philosophical Society*, New Series, Vol. 59, Part 4 (1969).

[18] Averroës (Ibn Rushd). *Aristotelis Opera cum Averrois commentariis*, Vol. I, Part. 2a: *Postriorum Resolutiorum Libri Duo* (Venice: Apud Ivnctas, 1562; Frankfurt am Main: Minerva, 1962).

[19] *Idem. Aristotelis Opera cum Averrois commentariis*, Vol. IV: *De Physico avditv libri octo* (Venice: Apud Ivnctas, 1562; Frankfurt am Main: Minerva, 1962).

[20] *Idem. Aristotelis Opera omnia*, Vol. VIII: *Metaphysicorum libri XIII cum Averrois Cordubensis in eodem commentariis* (Venice: Apud Ivnctas, 1562; Frankfurt am Main: Minerva, 1962).

[21] *Idem. Tafsir Ma Ba'd At-Tabi'at : Bibliotheca Arabica Scholasticorum*, Série Arabe, t. VI, Texte Arabe inédit établi par Maurice Bouyges, deuxième volume de texte (Beyrouth: Imprimerie Catholique, 1942).

[22] Bacon, Roger. *Opus majus*, ed. J. H. Bridges, Vol. 1 (Oxford: Clarendon Press, 1897).

[23] *Idem. Opus Majus of Roger Bacon: A Translation*, by Robert Belle Burke, Vol. 1 (Philadelphia: University of Pennsylvania Press, 1928).

[24] *Idem. Communia mathematica, in Opera hactenus inedita*, Fasc. XVI, ed. Robert Steele (Oxford: Clarendon Press/London: Apud Humpherum Milford, 1940).

[25] Barozzi, Francesco. *Opusculum, in quo una Oratio, & duae Quaestiones: altera de certitudine, et altera de medietate mathematicarum continentur* (Padua: E. G. P., 1560).

[26] Beeckman, Isaac. *Journal tenu par Isaac Beeckman*, éd. Cornelis de Waard, 4 Vols. (The Hague: Martinus Nijhoff, 1939–1953).

[27] Bombelli, Rafael. *L'Algebra* (1575; Bologna: Giouanni Rossi, 1579).

[28] Bramer, Benjamin. *Beschreibung und Unterricht, wie allerley Theylungen zu den Mathematischen Instrumenten zu verfertigen: Neben dem Gebrauch eines newen Proportional Instruments* (Marburg: Paul Egenolff, 1615).

[29] Clavius, Christoph. *Opera mathematica*, 5 Vols. (Maintz: Reinhardvs Eltz, 1611–1612): *Euclidis Elementa* (Fisrt Published in Rome in 1574; t. I); *Epitome arithmeticae practicae* (First Published in Rome in 1583), *Geometria practica* (First Published in Rome in 1604) and *Algebra* (First Published in Rome in 1608; t. II); *Refutatio Cyclometriae Iosephi Scaligeri* (First Published in Mayence in 1609; t. V).

[30] *Idem, Corrispondenza*, edizione critica a cura di U. Baldini e P.-D. Napolitani, 7 Vols. (Pisa: Università di Pisa, Dipartimento di matematica, 1992).

[31] Dasypodius, Cunradus. *Analyseis geometricae sex librorum Euclidis* (Strasburg: Josias Rihel, 1566).

[32] Descartes, René du Perron. *Discours de la Méthode Pour bien conduire sa raison, & chercher la verité dans les sciences. Plus La Dioptrique. Les Meteores. et La Geometrie. Qui sont des essais de cete Methode* (Leiden: Ian Maire, 1637); Reprinted Edition with an "Avant-propos" by Henri Gouhier (Lecce: Litostampa Conte, 1987).

[33] *Idem. Geometria à Renato Descartes, [...] Operâ atque studio Francisci à Schooten* (Leiden: Ioannis Maire, 1649).

[34] *Idem. Geometria à Renato Descartes*, ed. F. Schooten, 2 Partes (Amsterdam: Apud Ludovicum & Danielem Elzevirios, 1659–1661).

[35] *Idem. Opuscula posthuma, physica et mathematica* (Amsterdam: P. J. Blaeu, 1701).

[36] *Idem. Œuvres inédites de Descartes*, éd. par Foucher de Careil, 2 Vols. (Paris: Ladrange et Durand, 1859–1860).

[37] *Idem. Œuvres de Descartes*, éd. par Charles Adam et Paul Tannery, 11 Vols. (Paris, 1897–1913; New Edition, Paris: Vrin, 1969–1975) = *AT*.

[38] *Idem. Philosophical Works of Descartes*, 2 Vols., tr. by Elizabeth S. Haldane, and G. R. T. Ross (Cambrige: Cambridge University Press, 1911).

[39] *Idem. Correspondance*, publié par Charles Adam et Gérard Milhaud, 8 Vols. (Paris: PUF and CNRS, 1936–1963) = *AM*.

[40] *Idem. The Geometry of René Descartes*, tr. by David Eugene Smith and Marcia L. Latham (New York: Dover, ${}_2$1954).

[41] *Idem. Œuvres philosophiques*, Édition de F. Alquié, 3 Vols. (Paris: Garnier, 1963–1973).

[42] *Idem. Discourse on Method, Optics, Geometry and Meteorology*, tr. with Introduction by Paul J. Olscamp (Indianapolis: Bobbs-Merrill, 1965) =Olscamp.

[43] *Idem. Regulae ad directionem ingenii: Texte critique établi par Giovanni Crapulli avec la version hollandaise du XVIIème siècle* (The Hague: M. Nijhoff, 1966).

[44] *Idem. Philosophical Letters*, tr. and ed. by Anthony Kenny (Oxford: Basil Blackwell, 1970).

[45] *Idem. Regulae ad directionem ingenii*, kritisch revidiert, übersetzt und herausgegeben von Heinrich Springmeyer, Lüder Gäbe, und Hans Günter Zekl (Hamburg: Felix Meiner, 1973).

[46] *Idem. Descartes's Conversation with Burman*, tr. with Introduction and Commentary by John Cottingham (Oxford: Clarendon Press, 1976).

[47] *Idem. Règles utiles et claires pour la direction de l'esprit et la recherche de la vérité*, Traduction selon le lexique cartésien, et annotation conceptuelle par Jean-Luc Marion avec des notes mathématiques de Pierre Costabel (The Hague: M. Nijhoff, 1977).

[48] *Idem. The World*, Translation and Introduction by Michael S. Mahoney (New York: Abaris Books, 1979).

[49] *Idem. The Philosophical Writings of Descartes*, tr. by John Cottingham, Robert Stoothoff, Dugald Murdoch, and Anthony Kenny, 3 Vols. (Cambridge: Cambridge University Press, 1984–1991).

[50] Diogenes Lauertius. *Lives of Eminent Philosophers*, Vol. II, tr. by R. D. Hicks (London: William Heinemann/Cambridge, Mass.: Harvard University Press, 1925), Book IX, Chapter 11, "Phyrro," pp. 474–519.

[51] Diophante. *Les Arithmétiques*, t. III: *Livre IV* & t. IV: *Livres V, VI, VII*, Texte établi et traduit par Roshdi Rashed (Paris: Les Belles Lettres, 1984).

[52] Diophantus of Alexandria. *Opera omnia cum Graecis commentariis*, ed. Paul Tannery, 2 Vols. (Leipzig: Teubner, 1893–1895).

[53] Duns Scotus, John. *In XII libros Metaphysicorum Aristotelis expositio* (actually by Anthony Andreae), in *Joannis Duns Scoti Opera omnia*, ed. Wadding-Vivés, t. VI: *Metaphysica textualis libri V–XII* (Paris: Apud Ludovicum Vives, 1892).

[54] Euclid of Alexandria. *Στοικείων βιβλ*. ΙΕ̄. *ἐκ τῶν Θεῶνος συνουσίων* ed. Simon Grynaeus (Basel: Apud Ioan. Hervagivm, 1533).

[55] *Idem. Euclidis Elementorum libri XV cum scholis Antiquis et commentariis Federici Commandini Urbinatis* (Pesaro: Camillo Francischini, 1572).

[56] *Idem. Euclidis Elementorum libri XV*, ed. Ch. Clavius (Rome, 1574; Rome: Apud Sanctium & Soc., ${}_2$1589), in Clavius, I, #29.

[57] *Idem. Euclidis Elementa*, post J. L. Heiberg, ed. E. S. Stamatis, 5 Vols. (Leipzig: Teubner, 1969–1977).

[58] *Idem. The Thirteen Books of Euclid's Elements*, ed. and tr. by Thomas Little Heath, 3 Vols. (New York: Dover, ${}_2$1956).

[59] Fermat, Pierre de. *Œuvres de Fermat*, éd. Charles Henry et Paul Tannery, t. I (Paris: Gautier-Villards, 1891).

[60] *Idem. Œuvres de Fermat*, t. 1: *Théorie des nombres*, Textes traduits par Paul Tannery, introduites et commentés par R. Rashed, Ch. Houzel, G. Christol (Paris: Albert Blanchard, 1999).

[61] Fibonacci, Leonardo. *Fibonacci's Liber Abaci: A Translation into Modern English of Leonardo Pisano's Book of Calculation*, translated by Lawrence E. Sigler (New York: Springer-Verlag, 2002).

[62] Fonseca, Pedro da. *Commentaria in Metaphysicorum Aristotelis Stagiritae libros*, t. III (Cologne: Svmpievs Lazari Zetzneri Bibliopolae, 1615); t. IV (Cologne, 1629).

[63] François, Jean. *L'arithmétique et la géométrie pratique* (Paris: N. Langlois, $_4$1681).

[64] *Idem. Traité de la quantité* (Rennes: P. Hallaudays, 1655).

[65] Galileo Galilei. *Le operatione del compasso geometrico, et militare* (Padua, 1606), in *Opere di Galileo Galilei*, ed. Antonio Favaro (Florence: G. Barbèra, 1932), t. II; *Operations of the Geometrical and Military Compass*, tr. with an Introduction by Stillman Drake (Washington, D. C.: Smithonian Institution Press, 1978).

[66] Gassendi, Pierre. *Disquistio metaphysica seu Dubitationes et instantiae adversus Renati Cartesii metaphysicam et responsa*, Texte établi, traduit et annoté par Bernard Rochot (Paris: Vrin, 1962).

[67] Ghetaldi, Marino. *Opera omnia*, Redactor Žarko Dadić (Zagreb: Institut za Povijest Prirodonih, Matematickih, Medicinskih Nauka, 1968).

[68] Girard, Albert. *Invention nouvelle en l'Algebre* (Amsterdam, 1629); Reprinted Editon by D. Bierens de Haan (Leiden: Muré Frères, 1884).

[69] Gödel, Kurt. *Collected Works*, 5 Vols., ed. by Solomon Feferman, John W. Dawson, Jr., Stephen Kleene, Gregory H. Moore, Warren Goldfarb, Charles Parsons, Wilfried Sieg, Robert N. Solovay, and Jean van Heijenoort (Oxford: Clarendon Press/New York: Oxford University Press, 1986–2003).

[70] 's Gravesande, William Jacob. *Matheseos Universalis Elementa. Quibus accedunt, Specimen cemmentarii in Arithmeticam Universalem Newtoni: ut & De determinanda forma seriei infinitae adsumtae regula [methodus] nova* (Lugduni Batavorum: Apud Samulem Luchtmans, 1727).

[71] Harriot, Thomas. *Artis analyticae praxis* (London: Apud Robertvm Barker, 1631).

[72] Hobbes, Thomas. *Elementorum philosophiae sectio prima de corpore* (London: Andrea Crook, 1655).

[73] Huygens, Christiaan. *Œuvres complètes*, publiées par la Société Hollandaise des Sciences, 22 Vols. (The Hague: M. Nijhoff, 1888–1950).

[74] Iamblichus. *De communis mathematica scientia liber*, ed. Nicolaus Festa (Stuttgart: Teubner, 1975).

[75] Jordanus de Nemore. *De numeris datis*, A Critical Edition and Translation by Barnabas Bernard Hughes (Berkeley/Los Angeles/London: University of California Press, 1981).

[76] Kepler, Johannes. *Harmonices mundi* (Linz, 1619), in *Gesammelte Werke*, herausgegeben von Max Caspar, Bd. VI (München: C. H. Beck, 1940).

[77] Khaldūn, Ibn. *The Muqaddimah: An Introduction to History*, tr. Franz Rosenthal, 3 Vols. (London: Routledge & Kegan Paul, 1958).

[78] *Idem. Prolégomènes d'Ebn-Khaldoun*, texte arabe par M. Quatremère, 3 Vols. (Paris: Benjamin Duprat, 1858).

[79] Al-Khwārizmī, Muḥammd Ibn Mūsa. *Le Calcul Indien (Algorismus)*, Versions latines du XIIe siècle, par A. Allard (Paris: A. Blanchard/Namur, Société des Études Classiques, 1992).

[80] *Idem. The Algebra of Mohammed ben Musa*, ed. and tr. by Frederic Rosen (London: The Oriental Translation Fund, 1831).

[81] *Idem. Al-Kitāb mukhtaṣar fī'l-ḥisāb al-jabr wa'l-muqābalah*, ed. by ʿAlī Mustafā Masharrafa and Muḥammad Mursi Aḥmad (Cairo: The Faculty of Science, 1939).

[82] Leibniz, Gottfried Wilhelm. *Mathematische Schriften*, herausgegeben von C. I. Gerhardt, 7 Vols. (Berlin/Halle, 1849–1875; Hildesheim: Olms, 1962).

[83] *Idem. Die philosophischen Schriften*, herausgegeben von C. I. Gerhardt, 7 Vols. (Berlin, 1875–1890; Hildesheim/New York: Olms, 1978).

[84] *Idem. Opuscules et fragments inédits de Leibniz: Extraits des manuscripts de la Bibliothèque royale de Hanovre*, éd. par Louis Couturat (Paris, 1903; Hildesheim: Olms, 1966).

[85] *Idem. Sämtliche Schriften und Briefe*, herausgegeben von die Deutsche Akademie der Wissenschaften in Berlin: "Demonstratio propositionum primarum," in VI-2 (Berlin: Akademie Verlag, 1966), pp. 479–486; *Nouveaux essais sur l'entendement humain*, in VI-6 (Berlin, 1962).

[86] *Idem. Discourse on Metaphysics and Related Writings*, ed. and tr. with an introduction, notes and glossary by R. Niall D. Martin and Stuart Brown (Manchester/New York: Manchester University Press, 1988).

[87] *Idem. New Essays on Human Understanding*, tr. and ed. by Peter Remnant and Jonathan Bennett (Cambridge: Cambridge University Press, 1981).

[88] *Idem. 'New System' and Associated Contemporary Texts*, ed. and tr. by R. S. Woolhouse and Richard Francks (Oxford/New York: Oxford University Press, 1997).

[89] *Idem. Philosophical Essays*, ed. and tr. by Roger Ariew and Daniel Garber (Indianapolis/Cambridge: Hackett, 1989).

[90] *Idem. Philosophical Papers and Letters: A Selection*, tr. and ed. with an Introduction by Leroy E. Loemker (Dordrecht/Boston: D. Reidel, 1969).

[91] Lipstorp, Daniel. *Specimen philosophiae cartesianae* (Leiden: Joh. & Dan. Elzevier, 1653).

[92] Mersenne, Marin. *Les Verités des sciences contre les septiques ou Pyrrhoniens* (Paris, 1625; Stuttgart/Bad Cannstatt: Friedrich Frammann, 1969).

[93] *Idem. Correspondance du P. Marin Mersenne*, éd. par Cornelis de Waard et Armand Beaulieu, 16 Vols. (Paris: PUF and CNRS, 1945–1986) = *CM.*

[94] Montaigne, Michel de. *Œuvres complètes*, Texte établi par Albert Thibaudet et Maurice Rat, Introduction et notes par Maurice Rat (Paris: Gallimard, 1962).

[95] *Idem. An Apology for Raymond Sebond*, tr. and ed. with an Introduction and Notes by M. A. Screech (London: Penguin Books, 1987).

[96] Nifo, Agostino. *Expositiones in Aristotelis libros Metaphysices* (Venice, 1559; Frankfurt am Main: Minerva, 1967).

[97] Nuñes, Pedro. *Libro de álgebra en arithmética y geometría* (Antwerp: Arnoldo Birckman, 1567).

[98] Newton, Isaac. *The Mathematical Papers of Isaac Newton*, ed. Derek Thomas Whiteside, 8 Vols. (Cambridge: Cambridge University Press, 1967–1981).

[99] Oughtred, William. *Arithmeticae in numeris et speciebus institutio [...] quasi clavis mathematicae est* (London: Apud Thomam Harpervm, 1631).

[100] Pappus of Alexandria. *Mathematicae Collectiones a Federico Commandino Urbinate in Latinum conversae et commentariis illustratae* (Pesaro: Hieronymum Concordiam, 1588).

[101] *Idem. Collectio quae supersunt*, ed. F. Hultsch, 3 Vols. (Berlin: Apud Weidemannos, 1876–1878).

[102] *Idem. La Collection mathématique*, tr. par Paul Ver Eecke (Paris/Bruges, 1933; Paris: A. Blanchard, 1982).

[103] *Idem. Book 7 of the Collection*, ed. by Alexander Jones, Part 1: Introduction, Text, and Translation; Part 2: Commentary, Index, Figures (New York/Berlin/Heidelberg/Tokyo: Springer, 1986).

[104] Pascal, Blaise. *Œuvres complètes*, texte établi et annoté par Jacques Chevalier (Paris: Gallimard, 1954),

[105] *Idem. Œuvres complètes*, Présentation et notes de Louis Lafuma (Paris: Du Seuil, 1963).

[106] *Idem. L'Esprit de la géométrie et De l'Art de persuader*, Textes et commentaires par B. Clerté et M. Lhoste-Navarre (Paris: Éditions Pédagogie Moderne, 1979).

[107] *Idem. Pascal: Selections*, ed. by R. Popkin (New York/London: A Scribner/Macmillan Book, 1989).

[108] Pererius, Benedictus. *De communibus omnium rerum naturalium principijs et affectionibus libri quindecim* (Rome: Apud Franciscum Zanettum & Bartholomaeum Tosium Socios, 1576).

[109] Piccolomini, Alessandro. *In mechanicas quaestiones Aristotelis, paraphrasis [...] eiusdem commentarius de certitudine mathematicarum disciplinarum* (Rome: Apud Antonium Bladum Asulanum, 1547).

[110] Pitiscus, Bartholomeo (with George Joachim Rheticus). *Thesaurus mathematicus sive Canon sinuum ad radium 1.00000.00000.00000* (Frankfurt am Main: Nicolaus Hoffannus, 1613).

[111] Plato. *Opera*, t. IV, ed. Ioannes Burnet (Oxford: Clarendon Press, 1902).

[112] Proclus Diadochus. *Commentariorum Procli editio prima quae Simonis Grynaei opera addita est Euclidis Elementis Graece editis* (Basel: Apud Ioan. Hervagivm, 1533), in Euclid, I, #55.

[113] *Idem. In primum Euclidis Elementorum commentariorum libri IV a Francesco Barocio Parritio Veneto editi* (Padua: Gratiosus Perchacinus, 1560).

[114] *Idem. In primum Euclidis Elementorum librum commentarii*, ed. G. Friedlein (Leipzig: Teubner, 1873).

[115] *Idem. A Commentary on the First Book of Euclid's Elements*, tr. with Introduction and Notes by Glenn R. Morrow (Princeton: Princeton University Press, 1970; With a new foreword by Ian Mueller, 1992).

[116] Ptolemy = Claudius Ptolemaeus. *Syntaxis mathematica: Opera quae exstant omnia*, Part I, ed. J. L. Heiberg (Leipzig: Teubner, 1898).

[117] *Idem. Ptolemy's Almagest*, translated and annotated by G. J. Toomer (London: Duckworth, 1984).

[118] Ramus, Peter. *Algebra* (Paris: A. Wechel, 1560); *Algebra a Lazaro Schonero emendata* (Frankfurt am Main: A. Wechel, 1599).

[119] *Idem. Geometria*, ed. Lazarus Schöner (First Published in 1569; Frankfurt am Main: A. Wechel, 1599).

[120] *Idem. Scholae dialecticae* (First Published in 1543 as *Aristotelicae animadversiones*) and *Scholae metaphysicae* in *Scholae in liberales artes* (Basel, 1569; Reprinted Edition with Walter J. Ong's Introduction, Hildesheim/New York: Olms, 1970).

[121] *Idem. Scholae mathematicae* (1569; Frankfurt am Main: Wechel, Apud Danielem & Davidem Aubrios & Clementem Scheichium, 1627).

[122] *Idem. Via regia ad geometriam: The Way to Geometry*, tr. by William Bedwell and ed. by John Clerke (London: Thomas Cotes, 1636).

[123] Regiomontanus, Johannes. "Oratio Iohannis de Monteregio, Habita Patauij in praelectione Alfargani: Hac oratione compendiose declarantur scientiae Mathematicae, et utilitates earum," in *Opera collectanae*, ed. by Felix Schmeidler (Osnabrück: Otto Zeller, 1972).

[124] [Remmelin, Johannes.] *Numerus figuratus, sive Arithmetica analytica arte mirabili jnaudita nova constans* ([Frankfurt am Main], 1614).

[125] Ricci, Matteo. *Storia dell'introduzione del cristianesmo in Cina: Fonti Ricciane*, edite et commentate da Pasquale M. D'Elia, 3 Vols. (Rome: Libreria dello Stato, 1942–1949).

[126] *Idem. China in the Sixteenth Century: The Journals of Matthew Ricci, 1583-1610*, tr. by Louis J. Gallagher (New York: Random House, 1953).

[127] *Idem. Opere storiche del P. Matteo Ricci*, edite a cura dal P. Pietro Tacchi Ventori, 2 Vols. (Rome: Macerata, 1910–1913).

[128] Roomen, Adriaan van = Adrianus Romanus. *Ideae mathematicae pars prima, sive methodus polygonorum* (Antwerp: Apud Ioannem Keerbergium, 1593).

[129] *Idem. Problema Apolloniacum quo datis tribus circulis* (Würzburg: Typis Georgii Fleischmanni, 1596).

[130] *Idem. In Archimedis circuli dimensionem expositio et analysis. Apologia pro Archimede ad clarissimum virum Josephum Scaligerum. Exercitationes Cycliae, contra Josephum Scaligerum, Orontium Finaeum et Raynarum Urasum* (Würzburg, Actually Published in Geneva: De Candolle, 1597).

[131] *Idem. Universae mathesis idea, qua mathematicae universim sumptae natura, praestantia, usus et distributio brevissimè proponuntur* (Würzburg: Apud Georgium Fleischmann, 1602).

[132] *Idem. Mathematicae analyseos triumphus* (Louvain: Sumptibus authoris, 1609).

[133] *Idem.* "The Correspondence of Adriaan van Roomen," *LIAS: Sources and Documents Relating to the Early Modern History of Ideas*, Volumen III/ Separatum (Amsterdam: Holland University Press, 1976).

[134] Roth, Peter. *Arithmetica philosophica, oder [...] Kunstliche Rechnung der Coß oder Algebrae* (Nürnberg: Johann Gabriel Doppelmayr, 1608).

[135] Al-Samaw'al, Ibn Yaḥyā al-Maghribī. *Al-Bahir en algèbre d'As-Samaw'al*, Édition, notes et introduction par Salah Ahmad et Roshdi Rashed (Damascus: Imp. de l'Université de Damas, 1972).

[136] Schooten, Frans van. *Principia matheseos universalis*, ed. Erasmus Bartholinus (Leiden, 1651), Also in Descartes, I, #34.

[137] Sextus Empiricus. *Opera quae extant* (Paris: Petrus & Jacobus Chouët, 1621).

[138] *Idem. Outlines of Pyrrhonism*, tr. by R. G. Bury (London: Heinemann/Cambridge, Mass.: Harvard University Press, 1933).

[139] *Idem. Against the Professors*, Vol. IV, tr. by R. G. Bury (London: Heinemann/ Cambridge, Mass.: Harvard University Press, 1949).

[140] Snel, Willebrord. *Περὶ λόγου ἀποτομῆς καὶ περὶ χωπίου ἀποτομῆς resuscitata Geometria* (Leiden: Impensis Hermanni Mulleri, 1607).

[141] *Idem. Apollonius Batavus seu exsuscitata Apollonii Pergaei περὶ διωπισμένης τομῆς Geometria* (Leiden: J. a Dorp, 1608).

[142] Societas Jesu. *Monumenta Germaniae Paedagogica*, ed. K. Kehrbach, Band II: *Ratio Studiorum et Institutiones Scholasticae Soc. J.* 1. (Berlin: A. Hofmannn, 1887); Band V: *Ratio Studiorum et Institutiones Scholasticae Soc.* J. 2., t. II: *Ratio Studiorum ann. 1586. 1599. 1832*, ed. G. M. Pachtler (Berlin: A Hofmann, 1887).

[143] *Eadem. Monumenta Paedagogica Societatis Jesu*, ed. Ladislaus Lukács, 7 Vols. (Rome: Institutum Historicum Societatis Iesu, 1965–1992).

[144] *Eadem. Monumenta Paedagogica Societatis Jesu Quae primam Rationem Studiorum anno 1586 editam praecessere* (Madrid: Augustinus Avrial, 1901).

[145] *Eadem. Relação Anual das coisas que fireram os Padres da Companhia de Jesus nas suas missões, nos años 1600 á 1609*, Pelo P. Fernão Guerreiro (Coimbra: Imprensa da Universidade, 1931).

[146] Spinola, Carlo. "A Letter from Miyako to G. Alvarez in Rome on December 3, 1606," Archivum Romanum Societatis Iesu, Japonica–Sinica 36, ff. 145r, 145v, & 145-a.

[147] Spinoza, Benedictus de. *Renati Des Cartes Principiorum philosophiae pars I. et II: Opera*, t. II, hrsg von C. Gebhardt (Heidelberg: C. Winter, 1924).

[148] Vico, Giambattista. *De nostri temporis studiorum ratione* (Napoli, 1709); Also in *Opere di G. B. Vico*, Vol. 1, ed. Giovanni Gentile e Fausta Nicolini (Bari: Gius. Laterza & Figli, 1914).

[149] *Idem. De antiquissima Italorum sapientia liber primus: Metaphysica* (1710) in *Opere*, Vol. 1 (I, # 148).

[150] *Idem. On the Study Methods of Our Time*, tr. with an Introduction and Notes by Elio Gianturco (Indiannapolis: Bobbs/Merrill, 1965) = A Translation of I, # 148.

[151] *Idem. Selected Writings*, ed. and tr. by Leon Pompa (Cambridge: Cambridge University Press, 1982).

[152] Viète, François. *Universalium inspectionum ad canom mathematicum liber singularis* (Paris: Jean Mettayer, 1579).

[153] *Idem. In artem analyticem isagoge* (Tours: Jamet Mettayer, 1591).

[154] *Idem. La nouvelle algèbre de M. Viète*, traduite par J. L. Vaulézard (Paris, 1630; Tours: Fayard, 1986).

[155] *Idem. In artem anlyticem isagoge et Ad logisticem speciosam notae priores*, ed. I. de Beaugrand (Paris: Guillaume Baudry, 1631).

[156] *Idem. Opera mathematica, recognita Francisci à Schooten* (Leiden, 1646; Reprinted Edition with Joseph E. Hofmann's "Vorwort und Register," Hildesheim/New York: Olms, 1970).

[157] *Idem.* "Introduction à l'art analytique" et "Première série de notes sur la logistique spécieuse," tr. par F. Ritter, *Bulletino di bibliografia e di storia delle scienze matematiche e fisiche*, t. I (1868).

[158] *Idem. Einführung in die Neue Algebra*, Übersetzt und erläutert von Karin Reich und Hermuth Gericke (Munich: Werner Frisch, 1973).

[159] *Idem.* "Introduction to the Analytic Art," tr. by J. Winfree Smith in Jacob Klein, *Greek Mathematical Thought and the Origin of Algebra*, tr. by Eva Brann (Cambridge, Mass.: The MIT Press, 1968; New York: Dover, 1992).

[160] *Idem. The Analytic Art*, tr. by T. Richard Witmer (Kent: The Kent State University Press, 1983).

[161] Wallis, John. *Mathesis universalis: Operum mathematicorum pars prima* (Oxford, 1657); Also in *Opera mathematica*, t. I (Oxford, 1695: Hildesheim/New York: Olms, 1972).

[162] Wolff, Christian. *Mathematisches Lexicon* (1716): *Gesammelte Werke*, I. Abteilung, Deutsche Schriften, Bd. 11, herausgegeben von J. E. Hofmann (Hildesheim: Olms, 1965).

[163] *Idem. Elementa matheseos universae, 5 Vols.: Gesammelte Werke, II. Lateinische Schriften*, Bde. 29–33 (1730–41), ed. J. E. Hofmann (Hildesheim: Olms, 1968–71).

[164] Xavier, Francisco. *Epistolae S. Francisci Xaverii aliaque eius scripta*, ed. G. Schurhammer et J. Wicki, 2 Vols. (Rome: Monumanta Historica Soc. Jesu, 1944–1945).

II. SECONDARY SOURCES: BOOKS AND MONOGRAPHS

[1] Abu-Lughod, Janet L. *Before European Hegemony: The World System A. D. 1250–1350* (New York: Oxford University Press, 1989).

[2] Adam, Charles. *Vie et Œuvres de Descartes: Étude historique, Supplément à l'Édition de Descartes* (Paris: Cerf, 1910).

[3] Aiton, Eric J. *Leibniz: A Biography* (Bristol/Boston: Adm Hilger, 1985).

[4] Allard, J.-L. *Le Mathématisme de Descartes* (Ottawa: Éditions de l'Université d' Ottawa, 1963).

[5] Annas, Julia, and Jonathan Barnes. *The Modes of Scepticism: Ancient Texts and Modern Interpretations* (Cambridge: Cambridge University Press, 1985).

[6] Apostle, Hippocrates George. *Aristotle's Philosophy of Mathematics* (Chicago: The University of Chicago Press, 1952).

[7] Applebaum, Wilbur, ed. *Encyclopedia of the Scientific Revolution: From Copernicus to Newton* (New York/London: Garland, 2000).

[8] Ariew, Roger, and Majorie Grene, eds. *Descartes and His Contemporaries: Meditations, Objections, and Replies* (Chicago/London: The University of Chicago Press, 1995).

[9] Ariew, Roger, John Cottingham and Tom Sorell, eds. *Descartes'* Meditations: *Background Source Materials* (Cambridge: Cambridge University Press, 1998).

[10] Armogathe, J.-R., et J.-L. Marion. *Index des Regulae ad directionem ingenii de René Descartes avec des listes de leçons et conjectures établies par G. Crapulli* (Rome: Ateneo, 1976).

[11] Ausejo, Elena, and Mariano Hormigón, eds. *Paradigms and Mathematics* (Madrid: Siglo XXI de España Editores, 1996).

[12] Baillet, Adrien. *La Vie de Monsieur Des-Cartes*, 2 Parts (Paris, 1691; Hildesheim/New York: Olms, 1972).

[13] Baldini, Ugo. *Legem Impone Subactis: Studi su filosofia e scienza dei gesuiti in Italia, 1540–1632* (Rome: Bulzoni Editore, 1992).

[14] *Idem*, ed. *Christoph Clavius e l'Attività scientifica dei gesuiti nell'età di Galileo: Atti del Convegno Internazionale (Chieti, 28–30 aprile 1993)* (Rome: Bulzoni Editore, 1995).

[15] *Idem. Saggi sulla cultura della Compagnia di Gesù (secoli XVI–XVIII)* (Padua: Cleup Editrice, 2000).

[16] Barnes, Jonathan. *The Toils of Scepticism* (Cambridge: Cambridge University Press, 1990). (→ II, #5).

[17] Beck, L. J. *The Method of Descartes: A Study of the Regulae* (Oxford: Clarendon Press, 1952).

[18] Bélaval, Yvon. *Leibniz critique de Descartes* (Paris: Gallimard, 1960).

[19] Benacerraf, Paul, and Hilary Putnam, eds. *Philosophy of Mathematics: Selected Readings* (Cambridge: Cambridge University Press, $_2$1983).

[20] Bernard, Henri. *Matteo Ricci's Scientific Contribution to China*, tr. by Edwards Chalmers Werner (Peiping: Henri Vetch, 1935).

[21] Beyssade, Jean-Marie. *La Philosopie première de Descartes: Le temps et la cohérence de la métaphysique* (Paris: Flammarion, 1979).

[22] Biard, Joël, and Roshdi Rashed, eds. *Descartes et le moyen âge* (Paris: Vrin, 1997).

[23] Blanchet, Léon. *Les Antécédents historiques du* 《*Je pense, donc je suis*》 (Paris: Félix Alcan, 1920).

[24] Bonitz, Hermann. *Aristotelis Metaphysica: Commentarius* (Bonn, 1849; Hildesheim: Olms, 1960).

[25] Bos, Henk J. M. *Redefining Geometrical Exactness: Descartes' Transformation of the Early Modern Concept of Construction* (New York: Springer, 2001).

[26] Boutroux, Émile. *Des Vérités éternelles chez Descartes*, tr. par M. Canguilhem (Paris: Vrin, 1985).

[27] Boutroux, Pierre. *L'Imagination et les mathématiques selon Descartes* (Paris: Félix Alcan, 1900).

[28] *Idem. Les Principes de l'analyse mathématique: Exposé historique et critique*, 2 Vols. (Paris, A. Hermann, 1914–1919).

[29] *Idem. L'Idéal scientifique des mathématiques: Dans l'Antiquité et dans les Temps Modernes* (Paris: Félix Alcan: 1920).

[30] Boyer, Carl B. *History of Analytic Geometry* (New York: Scripta Mathematica at Yeshiva University Press, 1956).

[31] Breton, Stanislas. *Philosophie et mathématique chez Proclus* (Paris: Beauchesne, 1969).

[32] Brown, Stuart. *Leibniz* (Brighton, Sussex: The Harvester Press, 1984).

[33] Brunschvicq, Léon. *Descartes et Pascal lecteurs de Montaigne* (1942; Paris: Pocket, 1995).

[34] *Idem. Écrits philosophiques*, t. I (Paris: PUF, 1951).

[35] *Idem. Les Étapes de la philosophie mathématique*, Nouveau Tirage (Paris: A. Blanchard, 1972).

[36] Burkert, Walter. *Lore and Science in Ancient Pythagoreanism*, tr. by Erwin L. Minar (Cambridge, Mass.: Harvard University Press, 1972; The Original German Edition Published in 1962).

[37] Cajori, Florian. *William Oughtred: A Great Seventeenth Century Teacher of Mathematics* (Chicago/London: Open Court, 1916).

[38] Charles-Saget, Annick. *L'Architecture du divin: Mathématique et philosophie chez Plotin et Proclus* (Paris: Les Belles Lettres, 1976).

[39] Chartier, Roger, Dominique Julia, et Marie-Madeleine Compère. *L'Éducation en France du XVIe et XVIIIe siècles* (Paris: Société d'Enseignement Supérieur, 1976).

[40] Chasles, Michel. *Aperçu historique sur l'origine et développement des méthodes en géométrie* (Paris: Gauthier-Villars, $_2$1875).

[41] Cifoletti, Giovanna Cleonice. "Mathematics and Rhetoric: Jacques Peletier, Guillaume Gosselin and the Making of the French Algebraic Tradition" (Princeton University Ph. D. Dissertation, 1992).

[42] Clarke, Desmond M. *Descartes's Philosophy of Science* (Manchester: Manchester University Press, 1982).

[43] Clarke, Peter, ed. *The European Crisis of the 1590s: Essays in Comparative History* (London: George Allen & Unwin, 1985).

[44] Cleary, John J. *Aristotle and Mathematics: Aporetic Method in Cosmology & Metaphysics* (Leiden/New York/Köln: Brill, 1995).

[45] Cohen, Gustave. *Écrivains français en Hollande dans la première moitié du XVIIe siècle* (Paris: Libraire Ancienne Édouard Champion, 1920).

[46] Cohen, H. Floris. *The Scientific Revolution: A Historiographical Inquiry* (Chicago/London: The University of Chicago Press, 1994).

[47] Cooper, J. P., ed. *The Cambridge Modern History*, Vol. IV: *The Decline of Spain and the Thirty Years War, 1609-48/49* (Cambridge: Cambridge University Press, 1971).

[48] Costabel, Pierre. *Démarches originales de Descartes savant* (Paris: Vrin, 1982).

[49] *Idem.* René Descartes, *Exercices pour les éléments des solides: Essai en complément d'Euclide: Progymnasmata de solidorum elementis*, Edition critique avec introduction, traduction, notes et commentaires par Pierre Costabel (Paris: PUF, 1987).

[50] Cottingham, John. *A Descartes Dictionary* (Oxford/New York: Blackwell, 1993).

[51] Couturat, Louis. *La Logique de Leibniz d'après des documents inédits* (Paris, 1901; Reprinted Edition, Hildesheim/Zürich/New York: Olms, 1985).

[52] Crapulli, Giovanni. *Mathesis universalis: Genesi di una idea nel XVI secolo* (Rome: Ateneo, 1969).

[53] *Idem. Introduzione a Descartes* (Rome: Laterza, 1988).

[54] Crescini, Angelo. *Le Origini del metodo analitico: Il Cinquecento* (Udine: Del Bianco, 1965).

[55] Crombie, Alistair C. *Robert Grosseteste and the Origins of Experimental Science 1100–1700* (Oxford: Clarendon Press, 1953).

[56] *Idem. Styles of Scientific Thinking in the European Tradition: The History of Argument and Explanation Especially in the Mathematical and Biomedical Sciences and Arts*, 3 Vols. (London: Duckworth, 1994).

[57] Cromwell, Peter R. *Polyhedra* (Cambridge: Cambridge University Press, 1997).

[58] Cronin, T. J. *Objective Being in Descartes and in Suarez, Analecta Gregoriana*, No. 154 (Rome, 1966).

[59] Crosby, Alfred W. *The Measure of Reality: Quantification and Western Society, 1250–1600* (Cambridge/New York: Cambridge University Press, 1997).

[60] Crowley, Charles B. *Universal Mathematics in Aristotelian-Thomistic Philosophy: The Hermeneutics of Aristotelian Texts Relative to Universal Mathematics* (Washington, D. C.: University Press of America, 1980).

[61] Curley, E. M. *Descartes against the Skeptics* (Cambridge, Mass.: Harvard University Press, 1978).

[62] Dadić, Žarko, *et al.*, eds. *Actes du Symposium internationlale 《La Géométrie et l'Algèbre au début du XVIIe siècle》, À l'occasion du quatrième centenaire de la naissance de Marin Ghetaldić, Dubrovnik, 29. IX-3. X 1968* (Zagreb: Institut za Povijest Prirodnik, Matematićkih i Medicinskih Nauka, 1969).

[63] Dainville, François de. *La Géographie des humanistes* (Paris: Beauchesne, 1940).

[64] *Idem. L'Éducation des jésuites* (Paris: Les Éditions de Minuit, 1978).

[65] Daumas, Maurice. *Scientific Instruments of the Seventeenth and Eighteenth Centuries*, tr. and ed. by Mary Holbrook (London: B. T. Batsford, 1972).

[66] Day, Mary S. *Scheubel as an Algebraist* (New York: AMS Press, 1926).

[67] Dear, Peter. *Mersenne and the Learning of the Schools* (Ithaca/London: Cornell University Press, 1988).

[68] *Idem. Discipline and Experience: The Mathematical Way in the Scientific Revolution* (Chicago/London: The University of Chicago Press, 1995).

[69] *Idem. Revolutionizing the Sciences: European Knowledge and its Ambitions, 1500–1700* (Basingstoke: Palgrave, 2001).

[70] Denissoff, Élie. *Descartes, premier théoreticien de la physique mathématique: Trois essais sur la 《Discours de la méthode》* (Louvain: Publications Universitaires de Louvain/Paris: Béatrice-Nauwelaerts, 1970).

[71] Dickson, Leonard Eugene. *History of the Theory of Numbers*, Vol. 2 (Washington, D. C., 1919; New York: Chelsea, $_2$1966).

[72] Doig, James C. *Aquinas on Metaphysics: A Historico-Doctrinal Study of the Commentary on the Metaphysics* (The Hague: M. Nijhoff, 1972).

[73] Doney, Willis, ed. *Eternal Truths and the Cartesian Circle: A Collection of Studies* (New York/London: Garland, 1987).

[74] Drake, Stillman. *Galileo at Work: His Scientific Biography* (Chicago: The University of Chicago Press, 1978).

[75] Dugas, René. *Mechanics in the Seventeenth Century*, tr. by Freda Jacquot (Neuchatel: Griffon, 1958).

[76] Durkheim, Émile. *L'Évolution pédagogique de France* (1938; Paris: PUF, $_2$1969).

[77] Eisenstein, Elizabeth L. *The Printing Press as an Agent of Change* (Cambridge: Cambridge University Press, 1979).

[78] Engelfriet, Peter M. *Euclid in China: The Genesis of the First Translation of Euclid's* Elements *in 1607 & its Reception up to 1723* (Leidn/Boston/Köln: Brill, 1998).

[79] Farrell, Allan P. *The Jesuit Code of Liberal Education: The Development and Scope of the Ratio Studiorum* (Milwaulkee: Bruce Publishing, 1938).

[80] Faye, Emmanuel, éd. *Descartes et la Renaissance* (Paris: Honoré Champion, 1999).

[81] Federico, Pasquale Joseph. *Descartes on Polyhedra: A Study on the De Solidorum Elementis* (New York/Heidelberg/Berlin: Springer, 1982).

[82] Feingold, Mordechai. *The Mathematicians' Apprenticeship: Science, Universities, and Society in England, 1560–1640* (Cambridge: Cambridge University Press, 1984).

[83] *Idem*, Joseph S. Freedman, and Wolfgang Rother, eds. *The Influence of Petrus Ramus: Studies in Sixteenth and Seventeenth Century Philosophy and Sciences* (Basel: Schwabe, 2001).

[84] *Idem*, ed. *The New Science and Jesuit Science: Seventeenth Century Perspectives* (Dordrecht/Boston/London: Kluwer Academic Publishers, 2003).

[85] Ferrier, Richard Delahide. "Two Exegetical Treatises of François Viète, Translated, Annotated, and Explained" (Indiana University Ph. D. Dissertation, 1980).

[86] Fichant, Michel. *Science et métaphysique dans Descartes et Leibniz* (Paris: PUF, 1998).

[87] Fink, Eugen. *Die Idee einer transzendentalen Methodenlehre (VI. Cartesianische Meditationen)*, Teil 1: Texte aus dem Nachlass Eugen Finks (1932) mit Anmerkungen und Beilagen aus dem Nachlass Edmund Husserls (1933/34) (Dordrecht: Kluwer Academic Publishers, 1988); *Sixth Cartesian Meditation: The Idea of a Transcendental Theory of Method, With Textual Notaions by Edmund Husserl*, tr. Ronald Bruzina (Bloomington/Indianapolis, Indiana University Presss, 1995).

[88] Fitzpatrick, Edward A., ed. *St. Ignatius and the Ratio Studiorum* (New York/London: McGraw-Hill, 1933).

[89] Gäbe, Lüder. *Descartes' Selbstkritik: Untersuchungen zur Philosophie des jungen Descartes* (Hamburg: Felix Meiner, 1972).

[90] Garber, Daniel. *Descartes' Metaphysical Physics* (Chicago/London, The University of Chicago Press, 1992).

[91] *Idem. Descartes Embodied: Reading Cartesian Philosophy through Cartesian Science* (Cambridge: Cambridge University Press, 2001).

[92] *Idem*, and Michael Ayers, eds. *The Cambridge History of Seventeenth-Century Philosophy*, 2 Vols. (Cambridge: Cambridge University Press, 1998).

[93] Garin, Eugenio. *L'Éducation de l'homme moderne: La Pédagogique de la Renaissance 1400–1600*, tr. de l'Italien par J. Humbert (Paris: Fayard, 1968).

[94] Gaukroger, Stephan W., ed. *Descartes: Philosophy, Mathematics and Physics* (Brighton, Sussex: The Harvester Press/Totowa, New Jersey: Barnes & Noble, 1980).

[95] *Idem. Cartesian Logic: An Essay on Descartes's Conception of Inference* (Oxford: Clarendon Press, 1989).

[96] *Idem. Descartes: An Intellectual Biography* (Oxford: Clarendon Press, 1995).

[97] *Idem. Descartes' System of Natural Phlosophy* (Cambridge: Cambridge University Press, 2001).

[98] *Idem*, John Schuster, and John Sutton, eds. *Descartes' Natural Philosophy* (London/New York,: Routledge, 2000).

[99] Giard, Luce, éd. *Les Jésuites à la Renaissance: Système éducatif et production du savoir* (Paris: PUF, 1995).

[100] Gilbert, Neal W. *Renaissance Concepts of Method* (New York: Columbia University Press, 1960).

[101] Gillies, Donald, ed. *Revolutions in Mathematics* (Oxford: Clarendon Press, 1992).

[102] Gillispie, Charles Coulston, ed. *Dictionary of Scientific Biography*, 16 Vols. (New York: Charles Scribner's Sons, 1970–1980) = *DSB*.

[103] Gilson, Étienne. *The Unity of Philosophical Experience* (New York: Charles Scribner's Sons, 1937).

[104] *Idem. Descartes, Discours de la méthode: Texte et commentaire* (Paris: Vrin, $_5$1967).

[105] *Idem. Études sur la rôle de la pensée médiévale dand la formation du système cartésien* (Paris: Vrin, $_3$1967).

[106] *Idem. La Liberté chez Descartes et la théologie* (Paris: Vrin, 1987).

[107] Goldstein, Catherine, Jeremy Gray, and Jim Ritter, eds. *Mathematical Europe: History, Myth, Identity* (Paris: Éditions de la Maison des sciences de l'homme, 1996).

[108] Gouhier, Henri. *Les Premières Pensées de Descartes: Contribution à l'histoire de l'Anti-Renaissance* (Paris: Vrin, 1958).

[109] *Idem. La Pensée métaphysique de Descartes* (1962; Paris: Vrin, 1978).

[110] Graeser, Andreas, ed. *Mathematics and Metaphysics in Aristotle* (Bern: Haupt, 1987).

[111] Graves, Frank P. *Peter Ramus and the Educational Reform of the Sixteenth Century* (New York: Macmillan, 1912).

[112] Grimaldi, N. and J.-L. Marion, eds. *Les Discours et sa méthode* (Paris: PUF, 1987).

[113] Grootendorst, Albert W. *Jan de Witt's Elementa Curvarum Linearum, Liber Primus* (New York/Berlin/Heidelberg: Springer, 2000).

[114] Gueroult, Martial. *Descartes selon l'ordre des raisons, t. I: L'Ame et Dieu* (Paris: Aubier, 1953); t. II: *L'Ame et le corps* (Paris: Aubier, 1953); *Descartes's Philosophy Interpreted according to the Order of Reasons*, tr. by Roger Ariew, Vol. I: *The Soul and God* (Minneapolis: University of Minnesota Press, 1984), Vol. II: *The Soul and the Body* (Minneapolis, 1985).

[115] Gutas, Dimitri. *Greek Thought, Arabic Culture: The Graeco-Arabic Translation Movement in Baghdad and Early 'Abbāsid Socety (2nd–4th/8th–10th Centuries)* (London/New York: Routledge, 1998).

[116] Hadden, Richard W. *On the Shoulders of Merchants: Exchange and the Mathematical Conception of Nature in Early Modern Europe* (Albany: State University of New York, 1994).

[117] Hall, Rupert A. *Philosophers at War: The Quarrel between Newton and Leibniz* (Cambridge: Cambridge University Press, 1980).

[118] Hartmann, Nicolai. *Des Proclus Diadochus philosophische Anfangsründe der Mathematik nach den ersten zwei Büchern des Euclidkommentars* (Gießen: Alfled Töpelmann, 1909).

[119] Hawlitschek, Kurt. *Johann Faulhaber 1580–1635: Eine Blütezeit der mathematischen Wissenschaften in Ulm* (Ulm: Stadtbibliothek Ulm, 1995).

[120] Hayashi, Takao. *Indo no Sūgaku: Zero no Hatsumei* (Mathematics in India: The Invention of Zero), in Japanese (Tokyo: Chuōkōron, 1993).

[121] Haydn, Hiram. *The Counter-Renaissance* (New York: Harcourt, Brace & World, 1950).

[122] Heath, Thomas Little. *Apollonius of Perga: Treatise of Conic Sections* (Cambridge: Cambridge University Press, 1896).

[123] *Idem. Diophantus of Alexandria* (Cambridge, 1910; New York: Dover, 1964).

[124] *Idem. A History of Greek Mathematics*, 2 Vols. (Oxford: Clarendon Press, 1921).

[125] *Idem. Mathematics in Aristotle* (Oxford: Clarendon Press, 1949).

[126] Heijenoort, Jean van, ed. *From Frege to Gödel: A Source Book in Mathematical Logic, 1879–1931* (Cambridge, Mass.: Harvard University Press, 1967).

[127] Hintikka, J., and U. Remes. *The Method of Analysis* (Dordrecht: D. Reidel, 1974).

[128] Hofmann, Joseph E. *Leibniz in Paris 1672–1676: His Growth to Mathematical Maturity* (Cambridge: Cambridge University Press, 1974).

[129] *Idem. Frans van Schooten der Jüngere* (Wiesbaden: Franz Steiner, 1962).

[130] Hooykaas, Reijer. *Humanisme, science, et réforme: Pierre de la Ramée (1515–1572)* (Leiden: Brill, 1958).

[131] Houzel, Christian. *La géométrie algébrique: Recherches historiques*, Préface de Roshdi Rashed (Paris: Albert Blanchard, 2002).

[132] Howell, Wilbur Samuel. *Logic and Rhetoric in England, 1500–1700* (Princeton, 1956; New York: Russell & Russell, 1961).

[133] Høyrup. Jens. *In Measure, Number, and Weight: Studies in Mathematics and Culture* (Albany: State University of New York Press, 1994).

[134] Husserl, Edmund. *Logische Untersuchungen*, Bd. 1: *Prolegomena zur reinen Logik* (1900; Tübingen: Max Niemeyer, $_5$1968); *Logical Investigations*, Vol. 1: *Prolegomena to Pure Logic*, tr. by J. N. Findlay (London/New York: Routledge, 1970).

[135] *Idem. Ideen zu einer reinen Phänomenologie und phänomenologischen Philosophie*, Erstes Buch (1913): *Husserliana*, Bd. III, herausgegeben von Walter Biemel (Haag: Martinus Niihoff, 1950).

[136] *Idem. Cartesianische Meditationen und Pariser Vorträge: Husserliana*, Bd. I, herausgegeben von S. Strasser (Haag: Martinus Nijhoff, $_2$1963); *Cartesian Meditations: An Introduction to Phenomenology*, tr. by Dorian Cairns (The Hague: Martinus Nijhoff, 1960).

[137] *Idem. Formale und transcendental Logik: Versuch einer Kritik der logischen Vernunft* (1929): *Husserliana*, Bd. XII, herausgegeben von Paul Janssen (Den Haag: Martinus Nijhoff, 1974); *Formal and Transcendental Logic*, tr. by Dorian Dorian Cairns (The Hague: Martinus Nijhoff, 1969).

[138] *Idem. Erste Philosophie* (1923/24), Erster Teil, heraugegeben von Rudolf Boehm, *Husserliana*, Bd. VII (Haag: Martinus Niihoff, 1956), and Zweiter Teil, *Husserliana*, Bd. VIII (1959).

[139] Irwin, T. H. *Aristotle's First Principles* (Oxford: Clarendon Press, 1988).

[140] Jackson, Lambert L. *The Educational Significance of Sixteenth Century Arithmetic* (New York: Printed for the Author, 1906).

[141] Jaeger, W. *Aristotle: Fundamentals of the History of His Development*, tr. with the Author's Corrections and Additions by Richard Robinson (Oxford: Clarendon Press, $_2$1948).

[142] Jami, Catherine, Peter Engelfriet, and Gregory Blue, eds. *Statecraft & Intellectual Renewal in Late Ming China: The Cross-Cultural Synthesis of Xu Guangqi (1562–1633)* (Leiden/Boston/Köln: Brill, 2001).

[143] Jones, Howard. *Pierre Gassendi 1592–1655: An Intellectual Biography* (Nieuwkoop: B. De Graaf, 1981).

[144] Jullien, Vincent. *Descartes: La Géométrie de 1637* (Paris: PUF, 1996).

[145] Jullien, Vincent, et André Charrak. *Ce que dit Descartes touchant la chute des graves: De 1618 à 1646, étude d'un indicateur de la philosophie naturelle cartésienne* (Villeneuve d'Ascq: Presses Universitaires du Septentrion, 2002).

[146] Kästner, A. G. *Geschichte der Mathematik*, Bd. III (Göttingen, 1799; Hildesheim/New York: Olms, 1970).

[147] Kenny, Anthony. *Descartes: A Study of His Philosophy* (New York: Rondom House, 1968).

[148] Klein, Jacob. "Die griechische Logistik und die Entstehung der Algebra," I. Teil: *Quellen und Studien zur Geschichite der Mathematik*, Astronomie und Physik, Abteilung B: Studien, Bd. 3, Heft 1, pp. 18–105, und II. Teil: *Ibid.*, Bd. 3, Heft 2 (1936), pp. 122–235; An English Translation: *Greek Mathematical Thought and the Origin of Algebra*, tr. by Eva Brann (1968; New York: Dover, 1992).

[149] Knorr, Wilbur Richard. *The Ancient Tradition of Geometric Problems* (Basel/Boston/Stuttgart: Birkhauser, 1986).

[150] Kobayashi, Michio. *La Philosophie naturelle de Descartes* (Paris: Vrin, 1993).

[151] Kohanski, Alexander S. *The Greek Mode of Thought in Modern Philosophy* (Rutherford/Madison/Teaneck: Fairleigh Dickinson University Press, 1984).

[152] Koyré, Alexandre. *Essai sur l'idée de Dieu et les preuves de son existence chez Descartes* (Paris: Ernest Leroux, 1922).

[153] *Idem. Études galiléennes* (Paris: Hermann, 1939).

[154] Kretzmann, Norman, Anthony Kenny, and Jan Pinborg, eds. *The Cambridge History of Later Medieval Philosophy: From the Rediscovery of Aristotle to the Disintegration of Scholasticism, 1100–1600* (Cambridge: Cambridge University Press, 1982).

[155] Kuhn, Thomas S. *The Structure of Scientific Revolutions* (Chicago: The University of Chicago Press, 1962; $_2$1970, $_3$1996).

[156] *Idem. The Essential Tension: Selected Studies in Scientific Tradition and Change* (Chicago: The University of Chicago Press, 1977).

[157] *Idem. The Road Since Structure: Philosophical Essays, 1970–1993 with an Autobiographical Interview*, ed. by James Conant and John Haugeland (Chicago/London: The University of Chicago Press, 2000).

[158] Ladrière, Jean. *Les Limitations internes des formalismes: Étude sur la significance du théorème de Gödel et des théorèmes apparentés dans la théorie des fondements des mathématiques* (Louvain: E. Nawelaerts/Paris: Gautiers-Villards, 1957).

[159] Lakatos, Imre, ed. *Problems in the Philosophy of Mathematics* (Amsterdam: North-Holland, 1967).

[160] *Idem. Proofs and Refutations: The Logic of Mathematical Discovery*, ed. John Worrall and Elie Zahar (Cambridge: Cambridge University Press, 1976).

[161] Lattis, James. *Between Copernicus and Galileo: Christoph Clavius and the Collapse of Ptolemaic Cosmology* (Chicago/London: The University of Chicago Press, 1994).

[162] Leewen, Henry G. van. *The Problem of Certainty in English Thought, 1630–1690* (The Hague: Martinus Nijhoff, 1963).

[163] Li Yan, and Du Shiran. *Chinese Mathematics: A Concise History*, tr. by John N. Crossley and Anthony W.-C. Lun (Oxford: Clarendon Press, 1987).

[164] Liard, Louis. *Descartes* (Paris: Alcan, $_3$1911).

[165] Lindberg, David. *Roger Bacon's Philosophy of Nature*, A Critical Edition, with English Translation, Introduction, and Notes, of De *multiplicatione specierum and De speculis comburentibus* (Oxford: Clarendon Press, 1983).

[166] *Idem* and Numbers, Ronald L., eds. *God and Nature: Historical Essays on the Encounter between Christianity and Science* (Berkeley/Los Angeles/London: University of Calfornia Press, 1986).

[167] Lohr, Charles H. *Medieval Aristotle Commentators: A Bibliography of Recent Secondary Literature* (Fribourg: Éditions Universitaires Fribourg/Paris: Cerf, 1988).

[168] *Idem. Latin Aristotle Commentaries: Renaissance Authors* (Firenze: L. S. Olschki, 1988).

[169] MacDonnell, Joseph. *Jesuit Geometers: A Study of Fifty-six Prominent Jesuit Geometers During First Two Centuries of Jesuit History* (St. Louis: Institute of Jesuit Sources/Vatican City: Vatican Observatory Publications, 1989).

[170] Mahoney, Michael S. "The Royal Road: The Development of Algebraic Analysis from 1550 to 1650, With Special Reference to the Work of Pierre de Fermat" (Princeton University Ph. D. Dissertation, 1967).

[171] *Idem. The Mathematical Career of Pierre de Fermat (1601–1665)* (Princeton: Princeton University Press, 1973, $_2$1994).

[172] Mancosu, Paolo. *Philosophy of Mathematics & Mathematical Practice in the Seventeenth Century* (New York/Oxford: Oxford University Press, 1996).

[173] *Idem. From Brouwer to Hilbert: The Debate on the Foundations of Mathematics in the 1920s* (New York/Oxford: Oxford University Press, 1998).

[174] Marion, Jean-Luc. *Sur l'ontologie grise de Descartes: Science cartésien et savoir aristotélicien dans les Regulae* (Paris: Vrin, 1975). (→ II, #10)

[175] Martin, A. Lynn. *The Jesuit Mind: The Mentality of an Elite in Early Modern France* (Ithaca/London: Cornell University Press, 1988).

[176] Martin, Gottfried. *Leibniz: Logic and Metaphysics*, tr. by K. J. Northcott and P. G. Lucas (Manchester: Manchester University Press, 1964; The Original German Edition, 1960).

[177] Martzloff, Jean-Claude. *A History of Chinese Mathematics*, tr. by Stephen S. Wilson (Berlin/Heidelberg/New York: Springer, 1997).

[178] McEvoy, James. *The Philosophy of Robert Grosseteste* (Oxford: Clarendon Press, 1982).

[179] McKirahan, Richard D. Jr. *Principles and Proofs: Aristotle's Theory of Demonstrative Science* (Princeton: Princeton University Press, 1992).

[180] Méchoulan, Henry, ed. *Problématique et réception* du Discours de la Méthode *et des* Essais (Parin: Vrin, 1988)

[181] Menger, Karl. *Reminiscences of the Vienna Circle and the Mathemematical Colloquium*, ed. by Louise Golland *et al.* (Dordrecht: Kluwer Academic Publishers, 1994).

[182] Milaud, Gaston. *Descartes savant* (Paris: Alcan, 1921).

[183] Mittelstraß, Jürgen. *Neuzeit und Aufklärung: Studien zur Entstehung der neuzeitlischen Wissenschaft und Philosophie* (Berlin/New York: W. de Gruyter, 1970).

[184] Moraux, Paul. *Alexandre d'Aphrodise: Exégète de la noétique d'Aristote* (Liège: Faculté de Philosophie et Lettres/Paris: E. Droz, 1942).

[185] Morse, JoAnn S. "The Reception of Diophantus' *Arithmetic* in the Renaissance" (Princeton University Ph. D. Dissertation, 1981).

[186] Mouchot, A. *La Réforme cartésienne étendue aux diverses branches des mathématiques pures* (Paris: Gauthier-Villars, 1876).

[187] Mueller, Ian. *Philosophy of Mathematics and Deductive Structure in Euclid's Elements* (Cambridge, Mass.: The MIT Press, 1981).

[188] Müller, Kurt, und Gisela Kronert, *Leben und Werke von G. W. Leibniz: Eine Chronik* (Frankfurt am Main: Vittorio Klastermann, 1969).

[189] Murdoch, John E. *"Rationes mathematice": Un Aspect du rapport des mathématiques et de la philosophie au moyen âge* (Paris: Les Presses de l'Imprimerie Alençonnaise, 1962).

[190] Mugler, Charles. *Dictionnaire historique de la terminologie géométrique des Grecs*, 2 Vols. (Paris: Klincksieck, 1958–59).

[191] Nauert Jr., Charles G. *Agrippa and the Crisis of the Renaissance Thought, Illinois Studies in the Social Science*, No. 55 (Urbana, 1965).

[192] Needham, Joseph. *Science and Civilisation in China*, Vol. 3: *Mathematics and the Sciences of the Heavens and the Earth* (Cambridge: Cambridge University Press, 1959).

[193] Nesselmann, G. H. F. *Die Algebra der Griechen* (Berlin, 1842; Frankfurt am Main: Minerva, 1969).

[194] Nikulin, Dmitri. *Matter, Imagination and Geometry: Ontology, Natural Philosophy and Mathematics in Plotinus, Proclus and Descartes* (Hants/Berlington: Ashgate Publishing, 2002).

[195] O'Meara, Dominic J. *Pythagoras Revived: Mathematics and Philosophy in Late Antiquity* (Oxford: Clarendon Press, 1989).

[196] Ong, Walter Jackson. *Ramus, Method, and the Decay of Dialogue: From the Art of Discourse to the Art of Reason* (Cambridge, Mass.: Harvard University Press, 1958).

[197] *Idem. Ramus and Talon Inventory* (Cambridge, Mass.: Harvard University Press, 1958).

[198] *Idem. The Presence of the Word: Some Prolegomena for Cultural and Religious History* (New Heaven: Yale University Press, 1967).

[199] Ortega y Gasset, José. *The Idea of Principle in Leibnitz and the Evolution of Deductive Theory*, tr. by Mildred Adams (New York: W. W. Norton, 1971).

[200] Otte, Michael, and Marco Panza, eds. *Analysis and Synthesis in Mathematics: History and Philosophy* (Dordrecht: Kluwer Academic Publishers, 1997).

[201] Pépin, Jean, et H. D. Saffrey, eds. *Proclus: Lecteur et interprète des ancients* (Paris: Éditions du Centre Nationale de la Recherche Scientique, 1987).

[202] Peters, F. E. *Aristotle and the Arabs* (New York/London: New York University Press, 1968).

[203] Pieper, Josef. *Scholasticism: Personalities and Problems of Medieval Philosophy*, tr. from the German by Richard and Clara Winston (New York/Toronto: McGraw-Hill, 1964).

[204] Poincaré, H. *La Science et l'hypothèse* (1903; Paris: Flammarion, 1917).

[205] Pólya, George. *Mathematics and Plausible Reasoning*, Vol. 1: *Induction and Analogy in Mathematics* (Princeton: Princeton University Press, 1954).

[206] Popkin, Richard H. *The History of Scepticism from Erasmus to Spinoza* (Berkeley/Los Angeles/London: University of California Press, 1979): A Revised and Expanded Edition of *The History of Scepticism from Erasmus to Descartes*, First Published in 1960 and Revised in 1964.

[207] *Idem. The Third Force in Seventeenth-Century Thought* (Leiden: E. J. Brill, 1992).

[208] *Idem*, Olaso, De Ezequiel, and Giorgio Tonelli, eds. *Scepticism in the Enlightenment* (Dordrecht/Boston/London: Kluwer Academic Publishers, 1997).

[209] Rabb, Theodore K. *The Struggle for Stability in Early Modern Europe* (New York: Oxford University Press, 1975).

[210] Rabuel, Claude. *Commentaires sur la Geometrie de M. Descartes* (Lyon: Marcellin Duplain, 1730).

[211] Rashed, Roshdi. *Les mathématiques infinitésimales du IXe au XIe siècle*, Vol. I: *Fondateurs et commentateurs* (London: Al-Furqān, 1996), Vol. II: *Ibn al-Haytham* (1993), Vol. III: *Ibn al-Haytham, Théorie des coniques, constructions géométrique et géométrie pratique* (2000), Vol. IV: *Ibn al-Haytham, méthodes géométriques, transformations ponctuelles et philosophie des mathématiques* (2002).

[212] *Idem*, éd. *Mathématiques et philosophie de l'Antiquité à l'âge classique* (Paris: CNRS, 1991).

[213] *Idem. The Development of Arabic Mathematics: Between Arithmetic and Algebra*, tr. A. F. W. Armstrong (Dordrecht: Kluwer Academic Publishers, 1994).

[214] *Idem*, ed. in collaboration with Régis Morelon, *Encyclopedia of the History of Arabic Science*, 3 Vols. (London/New York: Routledge, 1996).

[215] *Idem*, et B. Vahabzadeh, *Al-Khayyām Mathématicien* (Paris: A. Blanchard, 1999); *Omar Khayyam, The Mathematician* (New York: Bibliotheca Persica Press, 2000).

[216] *Idem*, et Hélène Bellosta, *Ibrāhīm ibn Sinān: Logique et géométrie au Xe siècle* (Leiden/Boston/Köln: Brill, 2000). (→ Biard, II, #22)

[217] Reale, Giovanni. *The Concept of First Philosophy and the Unity of the Metaphysics of Aristotle*, ed. and tr. by John R. Catan (Albany: State University of New York Press, 1980).

[218] *Idem*. *A History of Ancient Philosophy*, II: *Plato and Aristotle, and IV: The Schools of the Imperial Age* (Albany: State University of New York Press, 1990).

[219] Robinet, André. *Aux Sources de l'esprit cartésien: L'Axe La Ramée-Descartes, De la* Dialectique *de 1555 aux* Regulae (Paris: Vrin, 1996).

[220] *Idem*. *Descartes, La Lumière naturelle: Intuition, disposition, complexion* (Paris: J. Vrin, 1999).

[221] Rochemonteix, Camille de. *Un Collège des jésuites aux XVIIe et XVIIIe siècles: Le Collège Henri IV de La Flèche*, 4 Vols. (Le Mans: Leguichex, 1889).

[222] Rodis-Lewis, Geneviève. *Descartes et le rationalisme* (Paris: PUF, 1966).

[223] *Eadem*. *L'Œuvres de Descartes*, 2 Vols. (Paris: Vrin, 1971).

[224] *Eadem*. *Descartes: Biographie* (Paris: Calmann-Lévy, 1995); *Descartes: His Life and Thought*, translated by Jane Marie Todd (Ithaca/London: Cornell University Press, 1998).

[225] *Eadem*. *Le Développement de la pensée de Descartes* (Paris: Vrin, 1997).

[226] Romano, Antonella. *La Contre-Réforme mathématique: Constitution et diffusion d'une culture mathématique jésuite à la Renaissance* (Rome: École Française de Rome, 1999).

[227] Rorty, Amélie Oksenberg, ed. *Essays on Descartes's* Meditations (Berkeley/Los Angeles/London: University of California Press, 1986).

[228] Rose, Paul Lawrence. *The Italian Renaissance of Mathematics* (Genève: Droz, 1975).

[229] Rossi, Paolo. *Clavis universalis: Arti della memoria e logica combinatoria da Lullo a Leibniz* (1960; Bologna: Mulino, $_2$1983).

[230] Ross, William David. *Aristotle's Metaphysics*, A Revised Text with Introduction and Commentary, Vol. I (Oxford: Clarendon Press, 1924).

[231] Sabra, A. I. *Theories of Light from Descartes to Newton* (London, 1967; Cambridge: Cambridge University Press, $_2$1981).

[232] Saraiva, Luis, ed. *History of Mathematical Sciences: Portugal and East Asia II, Scientific Practices and the Portuguese Expansion in Asia (1498–1759)* (Lisbon: EMAF-UL, 2001).

[233] Sarton, George. *Six Wings: Men of Science in the Renaissance* (Bloomington/London: Indiana University Press, 1957).

[234] Sasaki, Chikara. *Kagakukakumei no Rekishi Kōzō* (The Historical Structure of Scientific Revolutions), 2 Vols., in Japanese (Tokyo: Iwanami Shoten, 1985; 2nd ed. Kodansha, 1995).

[235] *Idem*. *Kindai Gakumon Rinen no Tanjō* (The Birth of Modern Scientific Ideals), in Japanese (Tokyo: Iwanai Shoten, 1992).

[236] *Idem*, Mitsuo Sugiura, and Joseph W. Dauben, eds. *The Intersection of History and Mathematics* (Basel/Boston/Berlin: Birkhäuser, 1994).

[237] *Idem*. *Nijjūseiki Sugakushisō* (Twentieth-Century Mathematical Thought), in Japanese (Tokyo: Misuzu-shobo, 2001).

[238] Schmitt, Charles B. *Gianfrancesco Pico della Mirandola (1469–1533) and His Critic of Aristotle* (The Hague: M. Nijhoff, 1967).

[239] *Idem*. *Cicero Scepticus: A Study of the Influence of the Academica in the Renaissance* (The Hague: M. Nijhoff, 1972).

[240] *Idem. Studies in Renaissance Philosophy and Science* (London: Variorum Reprints, 1981).

[241] *Idem. The Aristotelian Tradition and Renaissance Universities* (London: Variorum Reprints, 1984).

[242] *Idem*, and Quentin Skinner, eds. *The Cambridge History of Renaissance Philosophy* (Cambridge: Cambridge University Press, 1988).

[243] Schneider, Ivo. *Johannes Faulhaber, 1580–1635: Rechenmeister in einer Welt des Umbruchs* (Basel/Boson/Berlin: Birkhäuser, 1993).

[244] Schuster, John A. "Descartes and the Scientific Revolution, 1618–1634: An Interpretation," (Princeton University Ph. D. Dissertation, 1977).

[245] Schwegler, A. *Die Metaphysik des Aristoteles: Grundtext*, Übersetzung und Commentar, Bd. IV: *Des Commentars*, Zweiter Hälfe (Tübingen, 1847–1848: Frankfurt am Main: Minerva, 1960).

[246] Schwickerath, Robert. *Jesuit Education: Its History and Principles* (St. Louis: Herder, 1903).

[247] Scott, John F. *The Scientific Work of René Descartes* (London: Taylor & Francis, 1950).

[248] *Idem. The Mathematical Work of John Wallis* (London, 1938; New York: Chelsea, 1981).

[249] Sebba, Gregor. *The Dream of Descartes*, assembled from Manuscripts and edited by Richard A. Watson (Carbondale/Edwardsville: Southern Illilois University Press, 1987).

[250] Serfati, Michel, ed. *De la Méthode: Recherches en histoire et philosophie des mathématiques* (Besancon: Presses Universitaires Franc-Comtoises, 2002).

[251] Shea, William R. *The Magic of Numbers and Motion: The Scientific Career of René Descartes* (Canton: Watson Publishing International, 1991).

[252] Sirven, J. *Les Années d'apprentisage de Descartes (1596–1628)* (Albi: Impremerie coopérative du Sud-ouest, 1928).

[253] Smith, Vincent Edward. *St. Thomas on the Object of Geometry: The Aquinas Lecture, 1953* (Milwaukee: Marquette University Press, 1954).

[254] Sommervogel, Carlos. *Bibliothèque de la Compagnie de Jésus: Bibliographie*, Nouvelle édition, t. II (Bruxelles: Oscar Schepens/Paris: Alphonse Picard, 1891); t. III (Bruxelles/Paris, 1892).

[255] Sorabji, Richard, ed. *Aristotle Transformed: The Ancient Commentators and Their Influence* (Ithaca: Cornell University Press, 1990).

[256] Spinola, Fabio Ambrosio. *Vita del P. Carlo Spinola della Compagnia di Giesú morto per la Santa Fede del Giappone* (Bonogna: Benacci, 1706).

[257] Ströker, Elisabeth. *The Husserlian Foundations of Science* (Dordrecht/Boston/New York: Kluwer Academic, 1997).

[258] Struik, Dirk J., ed. *A Source Book in Mathematics, 1200–1800* (Cambridge, Mass.: Harvard University Pess, 1969).

[259] *Idem. The Land of Stevin and Huygens* (Dordrecht/Boston/London: D. Reidel, 1981).

[260] Suppes, Patrick, Julius M. Moravcsik, and Henry Menndell, eds. *Ancient & Medieval Traditions in the Exact Sciences: Essays in Memory of Wilbur Knorr* (Stanford: CSLI Publications, 2000).

[261] Szabó, Árpád. *Anfänge der griechischen Mathematik* (München/Wien: R. Oldenburg/Budapest: Akadémiai Kiadó, 1969); *The Beginning of Greek Mathematics* (Budapest: Akadémiai Kiadó/Dordrecht: D. Reidel, 1978).

[262] Taisbak, Christian Marinus. *Euclid's Data: The Importance of Being Given* (Copenhagen: Museum Tusculanum Press, 2003).

[263] Thomas-Stanford, Charles. *Early Edition of Euclid's Elements* (London: Printed for the Bibliographical Society, 1926).

[264] Toulmin, Stephen. *Cosmopolis: The Hidden Agenda of Modernity* (New York: Free Press, 1990).

[265] *Idem. Return to Reason* (Cambridge, Mass./London: Harvard University Press, 2001).

[266] Tropfke, Johannes. *Geschichite der Elementarmathematik*, 4. Auflage, Bd. 4: *Arithmetik und Algebra*, Vollständig neu bearbeitet von Kurt Vogel, Karin Reich und Hermut Gericke (Berlin/New York: De Gruyter, 1980).

[267] Turner, Gerard L'E. *Antique Scientific Instruments* (Poole/Dorset: Blandford Press, 1980).

[268] Tweyman, Stanley, ed. *René Descartes: Meditations on First Philosophy in Focus* (New York/London: Routledge, 1993).

[269] Van der Waerden, B. L. *Die gemeinsame Quelle der erkenntnistheoretischen Abhandlungen von Iamblichos und Proclos, Sitzungsberichte der Heidelberger Akademie der Wissenschaften, Philosophisch-historische Klasse* (Heidelberg: Carl Winter, 1980).

[270] *Idem. A History of Algebra: From al-Khwārizmī to Emmy Noether* (Berlin/Heidelberg/New York/Tokyo: Springer, 1985).

[271] Verdonk, J. J. *Petrus Ramus en de Wiskunde* (Assen: Van Gorcum, 1966).

[272] Victor, Stephan K. *Practical Geometry in the Middle Ages: Artis cuiuslibet consummatio and Pratike de Geometrie* (Philadelphia: The American Philosophical Society, 1979).

[273] Villoslado, Riccardo G. *Storia del Collegio Romano dal suo inizio (1551) alla soppressione della Compagnia di Gesù* (1773) (Rome: Apud Aedes Universitatis Gregorianae, 1954).

[274] Volkert, Klaus. *Geschichte der Analysis* (Mannheim/Wien/Zürich: Wissenschaftsverlag, 1988).

[275] Voss, Stephen, ed. *Essays on the Philosophy and Science of René Descartes* (New York/Oxford: Oxford University Press, 1993).

[276] Vrooman, Jack Rochard. *René Descartes: A Biography* (New York: G. P. Putnum's Sons, 1970).

[277] Vuillemin, Jules. *Mathématiques et métaphysique chez Descartes* (Paris: PUF, 1960; $_2$1987).

[278] *Idem. La Philosophie de l'algèbre*, t. I: *Recherche sur quelques concepts et méthodes de l'algèbre moderne* (Paris: PUF, 1962; $_2$1993).

[279] Wallace, William A. *Galileo's Early Notebooks: The Physical Questions* (Notre Dame: University of Notre Dame Press, 1977).

[280] *Idem. Prelude to Galileo: Essays on Medieval and Sixteenth-Century Sources of Galileo's Thought* (Dordrecht/Boston: D. Reidel, 1981).

[281] *Idem. Galileo and His Sources: The Heritage of the Collegio Romano in Galileo's Science* (Princeton: Princeton University Press, 1984).

[282] Watson, Richard A. and James E. Force, eds. *The Sceptical Mode in Modern Philosophy: Essays in Honor of Richard H. Popkin* (Dordrecht: Martinus Nijhoff, 1988).

[283] Weber, Jean-Paul. *La Constitution du texte des Regulae* (Paris: Société d'Édition d'Enseignement Supérieur, 1964).

[284] Weisheipl, James A. *The Development of Physical Theory in the Middle Ages* (1959; Ann Arbor: The University of Michigan Press, 1981).

[285] *Idem. Friar Thomas d'Aquino: His Life, Thought, and Works* (Washington, D. C.: The Catholic University of America Press, 1974).

[286] *Idem*, ed. *Albertus Magnus and the Sciences: Commemorative Essays 1980* (Toronto: Pontifical Institute of Medieval Studies, 1980).

[287] Westfall, Richard S. *Never at Rest: A Biography of Isaac Newton* (Cambridge: Cambridge University Press, 1980).

[288] Wilson, Margaret Dauler. *Descartes* (London/Henley/Boston: Routledge and Kegan Paul, 1978).

[289] *Eadem. Ideas and Mechanism: Essays on Early Modern Philosophy* (Princeton: Princeton University Press, 1999).

[290] Wittegenstein, Ludwig. *Remarks on the Foundations of Mathematics* (Oxford: Basil Blackwell, 1956).

[291] *Idem. On Certainty* (Oxford: Basil Blackwell, 1969).

[292] Wright, A. D. *The Counter-Reformation: Catholic Europe and the Non-Christian World* (London: Weidenfeld & Nicolson, 1982).

[293] Yates, Frances A. *The Art of Memory* (Chicago: University of Chicago Press, 1966).

[294] *Eadem. The Rosicrusian Enlightenment* (London: Routledge and Kegan Paul, 1972).

[295] Zeuthen, Hieronymus G. *Die Lehre von den Kegelschnitten im Altertum* (Kopenhagen, 1886; Hildesheim: Olms, 1965).

III. SECONDARY SOURCES: ARTICLES

[1] Ariew, Roger. "Christopher Clavius and the Classification of the Sciences," *Synthèse*, **83** (1990), pp. 293–300.

[2] Ashley, Benedict M. "St. Albert and the Nature of Natural Science," in Weisheipl, II, #286, pp. 73–102.

[3] Baldini, Ugo. "Christoph Clavius and the Scientific Scene in Rome," in G. V. Coyne, M. A. Hoskin, and O. Petersen, eds., *Gregorian Reform of the Calendar: Proceedings of the Vatican Conference to Commemorate its 400th Anniversary, 1582–1982* (Vatican City: Pontificia Academia Scientiarum, Specola Vaticana, 1983).

[4] Balic, Charles. "The Life and Works of John Duns Scotus," in John K. Ryan and Bernardine M. Bonansea, eds., *John Duns Scotus, 1265–1965* (Washington, D. C.: The Catholic University of America Press, 1965), pp. 1–27.

[5] Belaval, Yvon. "Intuitionnisme cartésien et psychologisme," *Revue internationale de philosophie*, 37e Année, n° 146 (1983), pp. 319–325.

[6] Bellosta, Hélène. "Ibrāhīm ibn Sinān: On Analysis and Synthesis," *Arabic Science and Philosophy*, **1** (1991), pp. 211–232.

[7] Benoit, Paul. "Calcul, algèbre et marchandise," in Serres, Michel, éd., *Éléments d'hisoire des sciences* (Paris: Bordas, 1989), pp. 197–221.

[8] Berggren, J. L. and Van Brummelen, Glen. "The Role and Development of Greek Geometrical Analysis and Synthesis in Ancient Greece and Medieval Islam," in Suppes *et al.*, II, #260, pp. 1–31.

[9] Bernays, Paul. "Concerning Rationality," in Paul Arthur Schilpp, ed., *The Philosophy of Karl Popper: The Library of Living Philosophers*, Vol. XIV, Book I (La Salle: Open Court, 1974), pp. 597–605.

[10] Bockstaele, P. "Roomen (Romanus, Romain), Adriaan (Adrianus), van," *Nationaal Biografisch Woordenboek*, **2** (Brussels, 1966), cols. 751–765.

[11] Bos, H. J. M. "Newton, Leibniz and the Leibnizian Tradition," Chapter 2 of I. Grattan-Guiness, ed., *From the Calculus to Set Theory, 1630–1910* (London: Duckworth, 1980), pp. 49–93.

[12] *Idem.* "On the Representation of Curves in Descartes' *Géométrie*," *Archive for History of Exact Sciences*, **24** (1981), pp. 295–338.

[13] *Idem.* "Arguments on Motivation in the Rise and Decline of a Mathematical Theory; the 'Construction of Equations', 1637–*ca.* 1750," *Archive for History of Exact Sciences*, **30** (1984), pp. 331–380.

[14] Bosmans, Henri. "Analyse de trois ouvrages célèbres d'Adrien Romain," *Annales de la société scientifique de Bruxelles*, 29e Année, 1904–1905 (Louvain, 1905), pp. 68–79.

[15] *Idem.* "Le fragment du Commentaire d'Adrien Romain sur l'Algèbre de Mahumed ben Musa el-Chowârezmî," *Annales de la société scientifique de Bruxelles*, 30e Année, 1905–1906 (Louvain, 1906), 2e partie, pp. 267–287.

[16] *Idem.* "Romain (Adrien)," *Biographie nationale publiée par l'Académie royale des sciences, des lettres et de beaux-arts de Belgique*, t. 19 (Bruxelles, 1907), cols. 848–889.

[17] *Idem.* "La première édition de la *Clavis mathematica (sic!)* d'Oughtred. Son influence sur la *Géométrie* de Descartes," *Annales de la société scientifique de Bruxelles*, 35e Année, 1910–1911 (Louvain, 1911), 1er partie, p. 121 et 2e partie, pp. 24–78.

[18] Bouveresse, Jacques. "La Théorie du possible chez Descartes," *Revue internationale de philosophie*, 37e Année, n° 146 (1983), pp. 293–318.

[19] Breger, Herbert. "The Mysteries of Adaequare: A Vindication of Fermat," *Archive for History of Exact Eciences*, **46** (1993/1994), pp. 193–219.

[20] Brouwer, L. E. J. "Intuitionistische Betrachtungen über den Formalismus," *Koninklijke Nederlandse Akademie van Wetenschappen te Amsterdam*, Proc. 31 (1928), pp. 374–379; Also in *Collected Works*, Vol. 1: *Philosophy and Foundations of Mathematics*, ed. by A. Heyting (Amsterdam/New York/Oxford: North-Holland, 1975), pp. 409–414; An English Translation in Mancosu, II, #173, pp. 40–44.

[21] Brown, Harcourt. "The Utilitarian Motive in the Age of Descartes," *Annals of Science*, **1** (1936), pp. 182–192.

[22] Brown, Stuart. "Leibniz's Break with Cartesian 'Rationalism'," in A. J. Holland, ed., *Philosophy, Its History and Historiography* (Dordrecht/Boston: D. Reidel, 1985), pp. 195–208.

[23] Browne, Alice. "Descartes's Dream," *Journal of the Warburg and Courtauld Institutes*, **40** (1977), pp. 256–273.

[24] Busard, H. L. L. "Clavius, Christopher," *DSB*, III, pp. 311–312.

[25] *Idem.* "Pitiscus, Bartholomeo," *DSB*, II, pp. 3–4.

[26] *Idem.* "Roomen, Adriaan van," *DSB*, XI, pp. 532–534.

[27] *Idem.* "Viète, François," *DSB*, XIV, pp. 18–25.

[28] Buson, Frédéric de. "Un Exemplaire de la *Sagesse* de Pierre Charron offert à Descartes en 1619," *Archives de philosophie*, **55** (1992), Cahier 1: "Bulletin cartésien XX," pp. 1–3.

[29] Campedelli, Luigi. "Ghetaldi (Ghettaldi), Marino," *DSB*, V, pp. 381–383.

[30] Carter, Richard B. "Descartes' Methodological Transformation of *Homo Sapiens* into *Homo Faber*," *Sudhoffs Archiv*, **68** (1984), pp. 225–229.

[31] Carugo, Adriano. and Crombie, A. C. "The Jesuits and Galileo's Ideas of Science and of Nature," *Annali dell'Istituto e Museo di storia della scienza di Firenze*, **8** (1983), Fasc. 2, pp. 3–68.

[32] Chasles, M. "Histoire de l'algèbre—Note sur la nature des opérations algèbriques (dont la connaissance a été attribuée, à tort, à Fibonacci).—Des droits de Viète méconnus," *Comptes rendus des séances de l'Académie des sciences*, t. 12 (Séance du Mercredi 5 mai 1841), pp. 741–757.

[33] Cherniss, Harold. "Plato as Mathematician," *Review of Metaphysics*, **4** (1951), pp. 395–425.

[34] Cifoletti, Giovanna C. "La *Question* de l'Algèbre: Mathématiques et rhétoriques des hommes de droit dans la France du 16e siècle," *Annales: Histoire, sciences sociales*, 50e Année, n° 6 (1995), pp. 1385–1416.

[35] Cornford, F. M. "Mathematics and Dialectic in the *Republic* VI.–VII.," *Mind*, **41** (1932), pp. 37–52 and 173–190.

[36] Cosentino, Giuseppe. "Le Matematiche nella *Ratio Studiorum* della Compagnia di Gesù," *Miscellanea storia ligure*, Anno 2 (n. s.), N. 2 (1970), pp. 171–213.

[37] *Idem.* "L'insegnamento delle matematiche nei collegi gesuitici nell'Italia settentrinale," *Physis*, **13** (1971), pp. 205–217.

[38] Costabel, Pierre. "Le théorème de Descartes-Euler," *Studia Cartesiana*, **1** (1979), pp. 25–35; Reprinted in Costabel, II, # 48, pp. 15–25.

[39] *Idem.* "L'Initiation mathématique de Descartes," *Archive de philosophie*, **46** (1983), pp. 637–646.

[40] *Idem.* "Descartes et la mathématique de l'infini," *Historia Scientiarum*, No. 29 (1985), pp. 37–49.

[41] Crombie, Alistair C. "Sources of Galileo's Early Natural Philosophy," in M. L. Righini Bonelli and W. R. Shea, eds., *Reason, Experiment, and Mysticism in the Scientific Revolution* (New York: Science History Publications, 1975), pp. 157–175.

[42] *Idem.* "Mathematics and Platonism in the Sixteenth-Century Italian Universities and Jesuit Educational Policy," in ΠΡΙΣΜΑΤΑ: *Naturwissenschaftsgeschichtliche Schriften, Festschrift für Willy Hartner*, hrsg. von Y. Maeyama und W. G. Saltzer (Wiesbaden: Franz Steiner, 1977), pp. 63–94.

[43] Cronin, T. J. "Eternal Truths in the Thought of Descartes and of His Adversary," *Journal of the History of Ideas*, **21** (1960), pp. 553–559.

[44] *Idem.* "Eternal Truths in the Thought of Suarez and of Descartes," *The Modern Schoolman*, **38** (1961), pp. 269–288 and **39** (1961), pp. 23–38.

[45] Crowe, Michael J. "Ten 'Laws' Concerning Patters of Change in the History of Mathematics," *Historia Mathematica*, **2** (1975), pp. 161–166.

[46] Dainville, François de. "L'Enseignement des mathématiques," *XVIIe siècle: Bulletin de la Société d'Étude du XVIIe siècle*, N° 30 (1956), pp. 62–68.

[47] Davis, Natalie Zemon. "Mathematicians in the Sixteenth-Century French Academies: Some Further Evidence," *Renaissance News*, **11** (1958), pp. 3–10.

[48] *Idem.* "Sixteenth-Century French Arithmetics on the Business Life," *Journal of the History of Ideas*, **21** (1960), pp. 18–48.

[49] Dear, Peter. "Marin Mersenne and the Probabilistic Roots of 'Mitigated Scepticism'," *Journal of the History of Philosophy*, **22** (1984), pp. 173–205.

[50] *Idem.* "Jesuit Mathematical Science and the Reconstitution of Experience in the Early Seventeenth Century," *Studies in History and Philosophy of Science*, **18** (1987), pp. 133–175.

[51] Diem, Gudrum. "Les traductions gréco-latines de la *Métaphysique* au moyen âge: Le problème de la *Metaphysica Vetus*," *Archiv für Geschichte der Philosophie*, **49** (1967), pp. 6–71.

[52] Donahue, William H. "The Solid Planetary Spheres in Post-Copernican Natural Philosophy," in Robert S. Westman, ed., *The Copernican Achievement* (Berkeley/Los Angeles/London: University of California Press, 1975), pp. 244–275.

[53] Dummett, Michael. "The Philosophical Significance of Gödel's Theorem," *Ratio*, **5** (1963), pp. 140–155.

[54] Egmond, Warren van. "A Catalog of François Viète's Printed and Manuscript Works," in Menso Folkerts und Uta Lindgrem, hrsg., *Mathemata: Festschrift für Hermut Gericke, Boethius*, Bd. 12 (Stuttgart: Franz Steiner, 1985), pp. 359–396.

[55] Einarson, Benedict. "On Certain Mathematical Terms in Aristotle's Logic," *American Journal of Philology*, **57** (1936), pp. 33–54 and 152–172.

[56] Etchemendy, John. "The Cartesian Circle: *Circulus ex tempore*," *Studia Cartesiana*, **2** (1981), pp. 5-42.

[57] Field, J. V. "Rediscovering the Archimedean Polyhedra: Piero della Francesca, Luca Pacioli, Leonardo da Vince, Albrecht Dürer, Daniele Barbaro, and Johannes Kepler," *Archive for History of Exact Sciences*, **50** (1996/1997), pp. 241–289.

[58] Fisher, K. A. F. "Jesuiten-Mathematiker in den deutschen Assistenz bis 1773," *Archivum Historicum Societatis Jesu*, **47** (1978), pp. 159–224.

[59] Fisher, N. W. and Sabetai Unguru. "Experimental Science and Mathematics in Roger Bacon's Thought," *Traditio*, **27** (1971), pp. 353–378.

[60] Frankfurt, Harry. "Descartes on the Creation of the Eternal Truth," *Philosophical Review*, **86** (1977), pp. 36–57.

[61] Gäbe Lüder. "La Règle XIV lien entre géométrie et algèbre," *Archives de philosophie*, **46** (1983), pp. 654–660.

[62] Galluzzi, Paolo. "Il 《Platonismo》 del tardo cinquecento e la filosofia di Galileo," in *Ricerche sulla curtura dell'Italia moderna*, a cura di Paolo Zambelli (Bari: Laterza, 1973), pp. 37–79.

[63] Garber, Daniel. "*Semel in vita*: The Scientific Background to Descartes' Meditations," in A. O. Rorty, II, # 227, pp. 81–116.

[64] Giacobbe, Guilio Cesare. "Il *Commentarium de certitudine mathematicarum disciplinarum* de Alessandro Piccolomini," *Physis*, **14** (1972), pp. 162–193.

[65] *Idem*. "Francesco Barozzi e la *Quaestio de certitudine mathematicarum*," *Physis*, **14** (1972), pp. 357–374.

[66] *Idem*. "La Rifressione metamatematica de Pietro Catena," *Physis*, **15** (1973), pp. 178–196.

[67] *Idem*. "Epigoni nel Seicento della «Questio de certitudine mathematicarum» : Giuseppe Biancani," *Physis*, **18** (1976), pp. 4–40.

[68] *Idem*. "Un gesuita progressista nella «Questio de certitudine mathematicarum» rinascimentale: Benito Pereyra," *Physis*, **19** (1977), pp. 51–86.

[69] Gibson, A. Boyce. "La «Géométrie» de Descartes au point de vue de sa méthode," *Revue de métaphysique et de morale*, **4** (1896), pp. 386–398.

[70] *Idem*. "The Regulae of Descartes," *Mind*, **7** (1898), pp. 145–158 and 332–363.

[71] *Idem*. "The Eternal Verities and the Will of God in the Philosophy of Descartes," *Proceedings of the Aristotelian Society*, N. S., **30** (1929–1930), pp. 31–54.

[72] Gilan, Christian. "Sur l'histoire du théorème fondamental de l'algèbre: Théorie des équations et calcul intégral," *Archive for History of Exact Sciences*, **42** (1991), pp. 91–136.

[73] Gilbert, Philippe. "Notice sur le mathématicien louvaniste Adrianus Romanus, Professeur à l'ancienne Université de Louvain. — (XVIe siècle)," *Revue catholique*, **17** (Louvain/Brussels, 1859), pp. 277–286, 394–409, and 522–527.

[74] *Idem*. "Les sciences exactes dans l'ancienne université de Louvain," *Revue des questions scientifiques*, **16** (1884), pp. 438–453.

[75] Gödel, Kurt. "Eine metamathematische Resultate über Entscheidungsdefinitheit und Widerspruchsfreiheit," *Anzeiger der Akademie der Wissenschaft in Wien, Mathematisch-naturwissenschaftliche Klasse*, **67** (1930), pp. 214–215; Also in I, #69, Vol. 1, pp. 140–143.

[76] *Idem*. "Über formal unentsheidbare Sätze der *Principia Mathematica* und verwandter System I," *Monatshefte für Mathematik und Physik*, **38** (1931), pp. 173–193; Also in I, #69, Vol. 1, pp. 144–195.

[77] Gomes, Joaquin Ferreira. "Perdo da Fonseca: Sixteenth Century Portuguese Philosopher," *International Philosophical Quaterly*, **66** (1966), pp. 632–644.

[78] Granger, Gilles-Gaston, "Philosophie et mathématiques leibniziennes," *Revue de métaphysique et de morale*, **86** (1981), pp. 1–37.

[79] Gulley, Norman. "Greek Geometrical Analysis," *Phronesis*, **3** (1958), pp. 1–14.

[80] Haan, D. Bierens de. "Notice sur quelques quadratures du cercle dans les Pays-Bas," *Bulletino di bibliografia e di storia delle scienze matematiche e fisiche*, **7** (1874), pp. 99–140.

[81] Hackett, Jeremiah M. G. "The Attitude of Roger Bacon to the Scientia of Albertus Magnus," in Weisheipl, II, #286, pp. 53–72.

[82] Hacking, Ian. "Proof and Eternal Truths: Descartes and Leibniz," Chapter 7 of S. Gaukroger, II, #94, pp. 169–180.

[83] Hara, Kokiti. "Comment Descartes a-t-il découvert ses ovales?" *Historia Scientiarum*, No. 29 (1985), pp. 51–82.

[84] Hartner, W. "Al-DJABR WA'L-MUKABĀLA," *Encyclopaedia of Islam*, New Edition, II (Leiden/London: Brill, 1965), pp. 360–362.

[85] Hazan, Alain. "Kurt Gödel, ou le pur formalisme comme Odysée de l'institution mathématique et du sujet de la connaissance," *Revue de métaphysique et de morale*, **82** (1977), pp. 88–107.

[86] Hellman, C. Doris. "Was Tycho Brahe As Influential As He Thought?" *The British Journal of the History of Science*, **1** (1963), pp. 295–324.

[87] Hervey, Helen. "Hobbes and Descartes in the Light of some Unpublished Letters of the Correspondence between Sir Charles Cavendish and Dr. John Pell," *Osiris*, **10** (1952), pp. 62–90.

[88] Hilbert, David. "Neubegrürundung der Mathematik. Erste Mitteilung," *Abhandlungen aus dem mathematischen Seminar der Hamburgischen Universität*, Bd. 1 (1922), pp. 157–177; *Gesammelte Abhandlungen*, Bd. III (Berlin/Heidelberg/New York: Springer, 1970), pp. 157–177; An English Translation, in Mancosu, II, #173, pp. 198–214.

[89] *Idem.* "Die Grundlagen der Mathematik," *Abhandlungen aus dem mathematischen Seminar der Hamburgischen Universität*, 6 (1928), pp. 65–85; An English Translation in Heijenoort, II, #126, pp. 464–479.

[90] Hintikka, J. "Cogito Ergo Sum: Inference or Performance," *The Philosophical Review*, **72** (1962), pp. 3–32.

[91] Hofmann, Joseph E. "Michael Stifel 1487?-1567," *Sudhoffs Archiv*, Supp., No. 9 (1968).

[92] *Idem.* "Vorwort: François Viète (1540–1603)—Leben, Wirken, Bedeutung," in Viète, I, #156.

[93] *Idem.* "Bombelli's Algebra—Eine genialische Einzelleistung und ihre Entwirkung auf Leibniz," *Studia Leibnitiana*, **5** (1972), pp. 196–252.

[94] *Idem.* "Schooten, Frans van," *DSB*, XII, pp. 205–207.

[95] Hofmann, Frederick A. "Christopher Clavius and the Renaissance of Euclidean Geometry," *Archivum Historicum Societatis Iesu*, **52** (1983), pp. 233–246.

[96] Hooykaas, Reijer. "Science and Religion in the Seventeenth Century: Isaac Beeckman (1588-1637)," *Free University Quaterly*, **1** (1951), pp. 169–183.

[97] *Idem.* "Beeckman, Isaac," *DSB*, I, pp. 566–568.

[98] Ishiguro, Hidé. "Reply to Jacques Bouveresse," *Revue internationale de philosophie*, 37e Année, n° 146 (1983), pp. 311–318.

[99] *Idem.* "The Status of Necessity and Impossibility in Descartes," in A. O. Rorty, II, #227, pp. 459–471.

[100] Jansen, Bernhard. "Die Pflege der Philosophie im Jesuitenorden wärend des 17.–18. Jahrhunderts," *Philosophisches Jahrbuch der Görres-Gesellschaft*, **51** (1938), pp. 172–215, 345–366, and 435–456.

[101] Jardine, Nicholas. "The Forging Modern of Realism: Clavius and Kepler against the Sceptics," *Studies in History and Philosophy of Science*, **10** (1979), pp. 141–171.

[102] Kenny, Anthony. "The Cartesian Spiral," *Revue internationalw de philosophie*, 37e Année, n° 146 (1983), pp. 247–256.

[103] Kirchvogel, Paul A. "Bramer, Benjamin," *DSB*, II, p. 419.

[104] *Idem.* "Faulhaber, Johann," *DSB*, IV, pp. 549–553.

[105] Knobloch, Eberhard. "Sur la vie et l'œuvres de Christophore Clavius (1538–1612)," *Revue d'histoire des sciences*, **41** (1988), pp. 331–356.

[106] *Idem.* "Christoph Clavius. Ein Namen- und Schriftenverzeichnis zu seinen *Opera mathematica*," *Bolletino delle scienze mathematiche*, **10** (1990), pp. 135–189.

[107] *Idem.* "L'Œuvre de Clavius et ses sources scientifiques," in Giard, II, #99, pp. 263–283.

[108] *Idem.* "Sur le rôle de Clavius dans l'histoire des mathématiques," in Baldini, II, #14, pp. 35–56.

[109] Körner, Stephan. "On the Relevance of Post-Gödelian Mathematics to Philosophy," in Lakatos, II, #159, pp. 118–137.

[110] Kraus, Pamela A. "From Universal Mathematics to Universal Method: Descartes's 'Turn' in Rule IV of the *Regulae*," *Journal of the History of Philosophy*, **21** (1983), pp. 159–174.

[111] Kubitz, O. A. "Scepticism and Intuition in the Philosophy of Descartes," *The Philosophical Review*, **48** (1939), pp. 472–491.

[112] Lachterman, David R. "Vico, Doria e la geometria sintetica," *Bolletino del Centro di Studi Vichiani*, **10** (1980), pp. 10–35.

[113] *Idem.* "Mathematics and Nominalism in Vico's *Liber Metaphysicus*," in Stephan Otto und Hermut Viechtbauer, hrsg., *Sachkommentar zu Giambattista Vicos Liber Metaphysicus* (München: Wilhelm Fink, 1985), pp. 47–85.

[114] *Idem.* "*Objectum Purae Matheseos*: Mathematical Constitution and the Passage from Essence to Existence," in A. O. Rorty, II, #227, pp. 435–458.

[115] Laird, W. R. "Archimedes among the Humanists," *Isis*, **82** (1991), pp. 629–638.

[116] Lakatos, Imre. "Infinite Regress and the Foundations of Mathematics," *Mathematics, Science and Epistemology: Philosophical Papers*, ed. by John Worrall and Gregory Currie, Vol. 2 (Cambridge: Cambridge University Press, 1978), pp. 3–23.

[117] Lee, H. D. P. "Geometrical Method and Aristotle's Account of First Principles," *Classical Quarterly*, **29** (1935), pp. 113–124.

[118] Leeuwen, Henry G. van. "Certainty in Seventeenth-Century Thought," Philip P. Wiener, ed., *Dictionary of the History of Ideas*, **1** (New York, 1973), pp. 304–311.

[119] Liard, Louis. "La méthode et la mathématique universelle de Descartes," *Revue philosophique de la France et l'étranger*, **10** (1880), pp. 569–600.

[120] Lindberg, David C. "On the Applicability of Mathematics to Nature: Roger Bacon and His Predecessors," *British Journal for the History of Science*, **15** (1982), pp. 3–25.

[121] Lo Bello, Anthony. "Albertus Magnus and Mathematics: A Translation with the Annotations of Those Portions of the Commentary on Euclid's *Elements* Published by Bernhard Geyer," *Historia Mathematica*, **10** (1983), pp. 3–23.

[122] MacDonnell, J. "Jesuit Mathematicians before the Suppression," *Archivum Historicum Societatis Iesu*, **45** (1976), pp. 139–147.

[123] Mahoney, Michael S. "Another Look at Greek Geometrical Analysis," *Archive for History of Exact Sciences*, **5** (1968), pp. 318–348.

[124] *Idem.* "Descartes: Mathematics and Physics," *DSB*, IV, pp. 55–61.

[125] *Idem.* "Last Years at Cambridge," A Book Review of D. T. Whiteside, ed., *The Mathematical Papers of Isaac Newton*, Vol. 7 (Cambridge, 1976), *Science*, Vol. 196, No. 4292 (20 May 1977), pp. 864–865.

[126] *Idem.* "Mathematics," Chapter 5 of David Lindberg, ed., *Science in the Middle Ages* (Chicago: University of Chicago Press, 1978), pp. 145–178.

[127] *Idem.* "Mathematics," in Joseph R. Strater, ed., *Dictionary of the Middle Ages*, **8** (New York, 1987), pp. 205–222.

[128] *Idem.* "The Beginnings of Algebraic Thought in the Seventeenth Century," Chapter 5 of Gaukroger, II, #94, pp. 141–155.

[129] *Idem.* "Changing Canons of Mathematical and Physical Intelligibility in the Later 17th Century," *Historia Mathematica*, **11** (1984), pp. 417–423.

[130] Mancosu, Paolo. "Aristotelian Logic and Euclidean Mathematics: Seventeenth-Century Developments of the *Quaestio de certitudine mathematicorum,*" *Studies in History and Philosophy of Science*, **23** (1992), pp. 241–265.

[131] Manders, Kenneth. "Descartes et Faulhaber," *Arichives de philosophie*, **58** (1995), Cahier 3, Bulletin Cartésien XXIII, pp. 1–12.

[132] Margolin, Jean-Claude. "L'Enseignement des mathématique en France (1540-1570): Charles de Bovelles, Fine, Peletier, Ramus," in Peter Scharratt, ed., *French Renaissance Studies* (Edinburgh: Edinburgh University Press, 1976), pp. 109–155.

[133] Merlan, Philip. "Alexander of Aphrodisias," *DSB*, I, pp. 117–120.

[134] Minio-Paluello, Lorenzo. "Aristotle: Tradition and Influence," *DSB*, I, pp. 267–281.

[135] *Idem.* "L' 《Aristoteles Latinus》," in *Opuscula: The Latin Aristotle* (Amsterdam: A. M. Hakkert, 1972), pp. 459–482.

[136] Mittelstraß, Jürgen. "Die Idee einer Mathesis Universalis bei Descartes," *Perspektiven der Philosophie*, **4** (1978), pp. 177–192.

[137] *Idem.* "The Philosopher's Conception of *Mathesis Universalis* from Descartes to Leibniz," *Annals of Science*, **33** (1979), pp. 593–610.

[138] Molland, A. G. "Shifting the Foundations: Descartes's Transformation of Ancient Geometry," *Historia Mathematica*, **3** (1976), pp. 21–49.

[139] *Idem.* "Mathematics in the Thought of Albertus Magnus," in Weisheipl, II, #286, pp. 463–478.

[140] Mueller, Ian. "Aristotle on Geometrical Objects," *Archiv für Geschichte der Philosophie*, **52** (1970), pp. 156–171; Also in J. Barnes *et al.*, eds., *Articles on Aristotle, 3. Metaphysics* (London: Duckworth, 1979), pp. 96–107.

[141] *Idem.* "Geometry and Scepticism," in J. Barnes, *et al.*, eds., *Science and Speculation: Studies in Hellenistic Theory and Practice* (Cambridge: Cambridge University Press/Paris: Éditions de la Maison des sciences de l'homme, 1982), pp. 69–95.

[142] *Idem.* "Iamblichus and Proclus' Euclid Commentary," *Hermes*, Bd. 115 (1987), pp. 334–348.

[143] *Idem.* "Mathematics and Philosophy in Proclus' Commentary on Book I of Euclid's *Elements*," in Pépin, II, #201, pp. 305–318.

[144] *Idem.* "Aristotle's Approach to the Problem of Principles in Metaphysics M and N," in Graeser, II, #110, pp. 241–259.

[145] *Idem.* "Aristotle's Doctrine of Abstraction in the Commentators," in Sorabji, II, #255, pp. 463–480.

[146] *Idem.* "Sur les principes des mathématiques chex Aristote et Euclide," in Rashed, II, #212, pp. 101–113.

[147] *Idem.* "Foreword to the 1992 Edition," in Proclus, tr. Morrow, I, #115, pp. ix–xxxi.

[148] Murdoch, John E. "From Medieval to the Renaissance Aristotle," in John Henry and Salah Hutton, eds., *New Perspectives on Renaissance Thought: Essays in the History of Science, Education and Philosophy in Memory of Charles B. Schmitt* (London: Duckwordth, 1990), pp. 163–176.

[149] Nathan, Henry. "Beaugrand, Jean," *DSB*, I, pp. 541–542.

[150] Naux, Charles. "Le père Christophore Clavius (1537-1612): Sa vie et son œuvre," "Première partie: L'Homme et son temps," *Revue des questions scientifique*, **154** (1983), pp. 55–67; "Deuxième partie: Cavius Astronome," *Ibid.*, pp. 181–193, and "Troisième partie: Clavius mathémaicien," *Ibid.*, pp. 325–347.

[151] Olaso, E. de. "Leibniz and Scepticism," in Popkin *et al.* II, #208, pp. 99–130.

[152] Owen, G. E. L., "The Platonism of Aristotle," in Martha Nussbaum, ed., *Logic, Science and Dialectic: Collected Papers in Greek Philosophy* (Ithaca: Cornell University Press, 1986), pp. 200–220.

[153] Philips, Edward C. "The Correspondence of Father Christopher Clavius S. I. Preserved in the Archives in the Pont. Gregorian University," *Archivum Historicum Societatis Iesu*, **8** (1939), pp. 193–222.

[154] *Idem.* "The Original Mathematical Autographs of Christopher Clavius," *Bulletin of the American Association of Jesuit Scientists, Eastern Section*, **18**, No. 1 (1941), p. 203, and "The Proposals of Father Christopher Clavius, S. J., for Improving the Teaching of Mathematics," *Ibid.*, No. 4 (May 1941), pp. 203–208.

[155] Popkin, Richard H. "Leibniz and the French Sceptics," *Revue internationale de philosophie*, **20** (1966), pp. 228–248.

[156] *Idem.* "Scepticism, Theology and the Scientific Revolution in the Seventeenth Century," in Imre Lakatos and Alan Musgrave, eds., *Problems in the Philosophy of Science* (Amsterdam: North Holland, 1968), pp. 1–28.

[157] Prag, A. "John Wallis, 1616-1703. Zur Ideengeschichite der Mathematik im 17. Jahrhundert," *Quellen und Studien zur Geschichte der Mathematik, Astronomie und Physik*, Apt. B, 1 (1931), pp. 381–412.

[158] Purnell Jr., F. "Jacopo Mazzoni and Galileo," *Physis*, **14** (1972), pp. 272–294.

[159] Putnum, Hilary. "Mathematics without Foundations," in *Mathematics, Matter and Method: Philosophical Papers*, Vol. 1 (Cambridge: Cambridge University Press, $_2$1979), pp. 43–59; Also in Benacerraf and Putnam, II, #19, pp. 295–311.

[160] Rashed, Roshdi. "L'analyse et la synthèse selon Ibn an Haytham," in Rashed, II, #212, pp. 131–162.

[161] *Idem.* "Indian Mathematics in Arabic," in Sasaki, Sugiura, and Dauben, II, #236, pp. 143–148.

[162] *Idem.* "Fermat and Algebraic Geometry," *Historia Scientiarum*, Vol. 11, No. 1 (2001), pp. 1–23.

[163] Reich, Karin, "Diophant, Cardano, Bombelli, Viète—Ein Vergleich ihrer Aufgaben," *Rechenpfennige: Aufsätze zur Wissenschaftsgeschichite* (München: Forschungs-Institut des Deuschen Museums, 1968), pp. 131–150.

[164] *Idem.* "Quelque remarque sur Marinus Ghetaldus et François Viète," in Dadić, II, #62, pp. 171–174.

[165] Rey, Abel. "De Viète à Descartes," *Travaux du IXe congrès international de philosophie: Congrès Descartes, II. Études cartésiennes*, IIme partie (1937), pp. 27–32.

[166] Ritter, Frédéric. "François Viète: Inventeur de l'algèbre moderne 1540–1603, Essai sur sa vie et son œuvre," *Revue occidentale philosophique, sociale et politique*, second série, **10** (1895), pp. 234–274 and 354–415.

[167] Robert, André. "Descartes et l'analyse des anciens," *Archives de philosophie*, **13** (1937): *Autour du "Discours de la Méthode" (1637–1937)*, pp. 221–245 [301–325].

[168] Robinson, Richard. "Analysis in Greek Geometry," *Mind*, **45** (1936), pp. 464–473.

[169] Rochot, Bernard. "Gassendi et les mathématiques," *Revue d'histoire des sciences*, **10** (1957), pp. 69–78.

[170] Rose, Paul Lawrence. "《Certitudo Mathematicarum》 from Leonardo to Galileo," *Leonardo nella Scienza e nella Tecnica: Atti del Simpositio Internationale di Storia della Scienza, Firenze-Vinci 23–26 giugno 1969* (Firenze: Giunti Barbèra, 1975), pp. 43–49.

[171] *Idem.* "Professors of Mathematics at Padua University 1521–1588," *Physis*, **17** (1975), pp. 300–304.

[172] *Idem.* "A Venetian Patron and Mathematician of the Sixteenth Century: Francesco Barozzi (1537–1604)," *Studi Venetiani*, N. S., **1** (1977), pp. 119–178.

[173] Rose, P. L. and Stillman Drake. "The Pseudo-Aristotelian *Questions of Mechanics* in Renaissance Culture," *Studies in the Renaissance*, **18** (1974), pp. 65–104.

[174] Rossi, Paolo. "The Legacy of Ramon Lull in Sixteenth-Century Thought," *Mediaeval and Renaissance Studies*, **5** (1961), pp. 182–213.

[175] Ruland, A. "Adrien Romain, premier professeur á la faculté de médicine de Wurzbourg," *Bibliophile belge: Bulletin trimestriel publié par la Société des Bibliophiles de Belgique*, **2** (1867), pp. 56–199, 161–187 and 256–269.

[176] Saito, Ken. "Book II of Euclid's *Elements* in the Light of the Theory of Conic Sections," *Historia Scientiarum*, No. 28. (1985), pp. 31–61.

[177] *Idem.* "Compounded Ratio in Euclid and Apollonius," *Historia Scientiarum*, No. 31 (1986), pp. 25–59.

[178] Saliba, George A. "The Meaning of al-jabra wa'l-muqābalah," *Centaurus*, **17** (1973), pp. 189–204.

[179] Sarton, George. "An Appeal for the Republication in Book Form of Father Bosmans' Studies on Belgian Mathematics in the Sixteenth and Seventeenth Centuries," *Isis*, **40** (1949), pp. 3–6.

[180] Sasaki, Chikara. "Ramus' Struggle against Scholastic Logic: Historical Introduction and Translation," *Journal of the Department of Liberal Arts*, College of Arts and Sciences, University of Tokyo, No. 15 (1982), pp. 119–136.

[181] *Idem.* "The Acceptance of the Theory of Proportion in the Sixteenth and Seventeenth Centuries: Barrow's Reaction to the Analytic Mathematics," *Historia Scientiarum*, No. 29 (1985), pp. 93–126.

[182] *Idem.* "The Adoption of Western Mathematics in Meiji Japan, 1853–1903," Sasaki, Sugiura, and Dauben, II, #236, pp. 165–186.

[183] *Idem.* "The French and Japanese Schools of Algebra in the Seventeenth Century: A Comparative Study," *Historia Scientiarum*, Vol. 9, No. 1 (1999), pp. 17–26.

[184] *Idem.* "L'Introduction de la science occidentale dans le Japon du *bakumatsu* et de Meiji (1840–1912)," *Daruma: Revue d'études japonaises*, N° 10/11 (Automne 2001–Printemps 2002), pp. 55–85.

[185] *Idem.* "The Emergence of the Japanese Mathematical Community in the Modern Western Style, 1855–1945," in Karen Hunger Parshall and Adrian C. Rice, eds., *Mathematics Unbound: The Evolution of an International Mathematical Research Community, 1800–1945* (Rhode Island: American Mathematical Society/London: London Mathematical Society, 2002), pp. 229–252.

[186] *Idem.* "Descartes as a Reformer of the Mathematical Disciplines," Biard and Rashed, II, #22, pp. 37–45.

[187] Schmitt, Charles B. "The Rediscovery of Ancient Skepticism in Modern Times," in Myles Burnyeat, ed., *The Skeptical Tradition* (Berkeley/Los Angeles/London: University of California Press, 1983), pp. 225–251.

[188] *Idem.* "Philosophy and Science in Sixteenth-Century Universities: Some Preliminary Comments," in J. E. Murdoch and E. D. Sylla, eds., *The Cultural Context of Medieval Learning* (Dordrecht: D. Reidel, 1975), pp. 485–537.

[189] *Idem.* "L'Introduction de la philosophie platonicienne dans d'enseignement des universités à la Renaissance," in *Platon et Aristote à la Renaissance: XVIe colloque international de Tours* (Paris: Vrin, 1976), pp. 93–104.

[190] Schneider, Ivo. "Der Proportionzirkel, ein universelles analogrechnerinstrument der Vergangenheit," *Deusches Museum Abhandlungen und Berichte*, **38** (1970), Heft 2, pp. 1–96.

[191] Schuster, John A. "Descartes' *Mathesis Universalis*: 1619–28," in Gaukroger, II, #94, pp. 41–96.

[192] Scriba, Christoph J. "Zur Lösung des 2. Debeauneschen Problems durch Descartes: Ein Abschritt aus der Frühgeschichte der inversen Tangenten-aufgaben," *Archive for History of Exact Sciences*, **1** (1961), pp. 406–419.

[193] *Idem.* "The Inverse Method of Tangents: A Dialogue between Leibniz and Newton (1675–1677)," *Archive for History of Exact Sciences*, **2** (1964), pp. 113–137.

[194] Serfati, Michel. "Les compas cartésiens," *Archives de philosophie*, **56** (1993), pp. 197–230.

[195] *Idem.* "Le développement de la pensée mathématique du jeune Descartes (l'éveil d'un mathématicien)," in Serfati, II, #250, pp. 39–104.

[196] Six, Karl. "Descartes im Jesuitkolleg von La Flèche," *Zeitschrift für katholische Theologie*, **38** (1914), pp. 498–508.

[197] Sloan, Phillip R. "Descartes, the Sceptic, and the Rejection of Vitalism in Seventeenth-Century Physiology," *Studies in History and Philosophy of Science*, **8** (1977), pp. 1–28.

[198] Smith, David Eugene. "The Place of Roger Bacon in the History of Mathematics," in A. G. Little, ed., *Roger Bacon: Essays Contributed by Various Writers on the Occasion of the Commemoration of the Seventh Centenary of His Birth* (1914; New York: Russell & Russell, 1972), pp. 153–183.

[199] Speziali, Pierre. "Mydorge, Claude," *DSB*, IX, pp. 598–599.

[200] Springmeyer, Heinrich. "Eine neue kritische Textausgabe der *Regulae ad directionem ingenii* von René Descartes," *Zeitschrift für philosophische Forschung*, **24** (1970), pp. 101–125.

[201] Stipanic, Ernest. "L'œuvre principale de Getaldic »De resolutione et compositione mathematica «," in Dadić, II, #62, pp. 91–104.

[202] Struik, Dirk J. "Snel (Snellius or Snel van Royen), Willebrord," *DSB*, XII, pp. 499–502.

[203] Sylla, Edith. "Compound Ratios: Bradwardine, Oresme, and the First Edition of Newton's Principia," in Everett Mendelsohn, ed., *Transformation and Tradition in the Sciences: Essays in Honor of I. Bernard Cohen* (Cambridge: Cambridge University Press, 1984), pp. 11–43.

[204] Szabó, Árpád. "Zum Problem der so. geometrischen Algebra in Euklids Elementen," in Maeyama und Saltzen, III, #42, pp. 373–393.

[205] Tannery, Paul. "La Correspondance de Descartes dans les Inédits du Fonds Libri pour l'histoire des mathématiques," *Mémoires scientifiques*, publiés par J.-L. Heiberg et H.-G. Zeuthen, t. IV: *Sciences modernes*, éd. par Gino Loria (Toulouse: Édouard Privat/Paris: Gautier Villars, 1926), pp. 203–228.

[206] Treutlein, P. "Das Rechen in 16. Jahrhundert," *Abhandlungen zur Geschichte der Mathematik*, Erstes Heft (Leipzig, 1877), pp. 1–100.

[207] *Idem.* "Die deutsche Coss," *Abhandlungen zur Geschichte der Mathematik*, Zweites Heft (Leipzig, 1879), pp. 1–124.

[208] Treweek, A. P. "Pappus of Alexandria: The Manuscript Tradition of the *Collectio Mathematica*," *Scriptorium*, **11** (1957), pp. 195–233.

[209] Tummers, Paul M. J. E. "The Commentary of Albert on Euclid's *Elements* of Geometry," in Weisheipl, II, #286, pp. 479–499.

[210] Van de Pitte, Frederick. "Descartes' *Mathesis Universalis*," *Archiv für Geschichite der Philosophie*, **61** (1979), pp. 154–174.

[211] Vogel, Kurt. "Stifel, Michael," *DSB*, XIII, pp. 58–62.

[212] Wallace, William A. "Aristotle in the Middle Ages," *Dictionary of the Middle Ages*, **1** (New York: Charles Scribner's Sons, 1982), pp. 456–469.

[213] Weber, Jean-Paul. "Sur la composition de la Regula IV de Descartes," *Revue philosophique de la France et de l'étranger*, **154** (1964), pp. 1–20.

[214] Weisheipl, James A. "Albert Magnus and the Oxford Platonists," *Proceedings of the American Catholic Philosophical Association*, **32** (1958), pp. 124–139.

[215] *Idem.* "The Life and Works of St. Albert the Great," in Weisheipl, II, #286, pp. 13–51.

[216] *Idem.* "Classication of the Sciences in Medieval Thought," in J. A. Weisheipl (ed. by Willam E. Carroll), *Nature and Motion in the Middle Ages* (Washington, D. C.: The Catholic University of America Press, 1985), pp. 203–237.

[217] Weyl, Hermann. "Diskussionsbemerkungen zu dem zweiten Hilbertschen Vortrag über die Grundlagen der Mathematik," *Abhandlungen aus dem mathematischen Seminar der Hamburgischen Universität*, **6** (1928), pp. 86–88; Also in *Gesammelte Abhandlungen*, hrsg. von K. Chandrasekharan, Bd. III (Berlin/Heidelberg/New York: Springer, 1968), pp. 147–149; An English Translation, in Heijenoort, II, #126, pp. 480–484.

[218] Whiteside, Derek Thomas. "Patters of Mathematical Thought in the Later Seventeenth Century," *Archive for History of Exact Sciences*, **1** (1960), pp. 179–388.

[219] Wilson, Margaret D. "Leibniz and Locke on 'First Principle'," *Journal of the History of Ideas*, **28** (1967), pp. 347–366; Also in Wilson, II, #289, pp. 353–372.

[220] *Eadem.* "On Leibniz's Explanation of 'Necessary Truth'," in Harry G. Frankfurt, ed., *Leibniz: A Collection of Critical Essays* (Notre Dame/London: University of Notre Dame Press, 1972), pp. 401–419.

[221] Yates, Frances A. "The Art of Ramon Lull: An Approach to It through Lull's Theory of Elements," *Journal of the Warburg and Courtault Institutes*, **17** (1954), pp. 115–173; Also in *Lull and Bruno: Collected Essays*, Vol. 1 (London: RKP, 1982), pp. 9–77.

Indices

Name Index

Subject Index

Treatise Index